DYNAMIC PROGRAMMING FOR OPTIMAL WATER RESOURCES SYSTEMS ANALYSIS

Prentice Hall Advanced Reference Series

Engineering

DENNO   *Power System Design and Applications for Alternative Energy Sources*
ESOGBUE, ED.   *Dynamic Programming for Optimal Water Resources Systems Analysis*
FERRY, AKERS, AND GREENEICH   *Ultra Large Scale Integrated Microelectronics*
HAYKIN, ED.   *Selected Topics in Signal Processing*
JOHNSON   *Lectures on Adaptive Parameter Estimation*
MILUTINOVIC, ED.   *Microprocessor Design for GaAs Technology*
QUACKENBUSH, BARNWELL III, AND CLEMENTS   *Objective Measures of Speech Quality*
ROFFEL, VERMEER, AND CHIN   *Simulation and Implementation of Self-Tuning Controllers*
SASTRY AND BODSON   *Adaptive Control: Stability, Convergence, and Robustness*

DYNAMIC PROGRAMMING FOR OPTIMAL WATER RESOURCES SYSTEMS ANALYSIS

Edited by

AUGUSTINE O. ESOGBUE

School of Industrial and Systems Engineering

Georgia Institute of Technology

PRENTICE HALL   Englewood Cliffs, New Jersey 07632

Library of Congress Cataloging-in-Publication Data

Dynamic Programming for optimal water resources
  systems analysis.

  (Prentice Hall advanced reference series.
Engineering)
  Selected papers presented at the Special Workshop
on Dynamic Programming and Water Resources sponsored
by the National Science Foundation, held at the Georgia
Institute of Technology, Atlanta, Ga., June 25-27,
1986.
  Includes indexes.
  1. Water resources development--Systems engineering--
Congresses. 2. Dynamic programming--Congresses.
I. Esogbue, Augustine O. II. Special Workshop on Dynamic
Programming and Water Resources (1986 : Georgia Institute
of Technology) III. National Science Foundation (U.S.)
IV. Series.
TC409.D96 1989   33.91   89-3810
ISBN 0-13-221516-0

Editorial/production supervision
  and interior design: *Gertrude Szyferblatt*
Manufacturing buyer: *Mary Ann Gloriande*

Prentice Hall Advanced Reference Series

 © 1989 by Prentice-Hall, Inc.
A Division of Simon & Schuster
Englewood Cliffs, New Jersey 07632

All rights reserved. No part of this book may be
reproduced, in any form or by any means,
without permission in writing from the publisher.

Printed in the United States of America

10 9 8 7 6 5 4 3 2 1

ISBN   0-13-221516-0

PRENTICE-HALL INTERNATIONAL (UK) LIMITED, *London*
PRENTICE-HALL OF AUSTRALIA PTY. LIMITED, *Sydney*
PRENTICE-HALL CANADA INC., *Toronto*
PRENTICE-HALL HISPANOAMERICANA, S.A., *Mexico*
PRENTICE-HALL OF INDIA PRIVATE LIMITED, *New Delhi*
PRENTICE-HALL OF JAPAN, INC., *Tokyo*
SIMON & SCHUSTER ASIA PTE. LTD., *Singapore*
EDITORA PRENTICE-HALL DO BRASIL, LTDA., *Rio de Janeiro*

DYNAMIC PROGRAMMING FOR OPTIMAL WATER RESOURCES SYSTEMS ANALYSIS

TABLE OF CONTENTS

PREFACE

I. OVERVIEW OF WATER RESOURCES PROBLEM SOLVING VIA DYNAMIC PROGRAMMING   1

On the Origin of the Name Dynamic Programming   3
Nina Bellman

Dynamic Programming and Water Resources: Origins and Interconnections   4
Augustine O. Esogbue

Water Problems and Issues of the State of Georgia: Program Goals and Priorities   11
Bernd Kahn

Dynamic Programming and Practical Water Resources Systems Engineering   16
Warren A. Hall

A Taxonomic Treatment of Dynamic Programming Models of Water Resources Systems   27
Augustine O. Esogbue

II. MULTIOBJECTIVE-MULTIPURPOSE LARGE SCALE WATER RESOURCES SYSTEMS MODELS   73

Multi-Objective Dynamic Programming in Water Resources   75
George W. Tauxe

Objective-Space Dynamic Programming Approach to Multi-dimensional Problems in Water Resources   88
John W. Labadie and Darrell G. Fontane

Knowledge Based Dynamic Programming for Water Resources Management   116
Osman Coskunoglu

Dynamic Programming and Non-Separable Water Resources Problems   128
Moshe Sniedovich

Closed-Loop Control, Balancing, and Model Reduction of Large Scale Water Resource Systems   147
José A. Ramos

III. WATER SUPPLY AND DISTRIBUTION MANAGEMENT   165

Identification of Demand Models from Noisy Observations: An Application to Water Resources   167
Luv Kumar Kher and Soroosh Sorooshian

On Optimal Pumping Policies for Groundwater   179
Toshio Odanaka

Dynamic Programming for Optimized Control of Water Supply and Distribution Systems **187**
Bryan Coulbeck

Dynamic Programming for Optimization of Pump Selection and Scheduling in Water Supply Systems **216**
Bryan Coulbeck and C.H. Orr

Optimal Design of Large Complex Water Resources Conveyance Systems Via Nonserial Dynamic Programming **234**
Augustine O. Esogbue and Chae Y. Lee

IV. WATER QUALITY, WASTE TREATMENT AND FLOOD CONTROL SYSTEMS   **259**

Improving Water Quality by Optimal Aeration Control via Dynamic Programming **261**
Hiroshi Sugiyama

A Variable State-Space Dynamic Programming Model for Optimizing Industrial Waste Treatment Sequences **276**
J. Hugh Ellis

Dynamic Programming for Flood Control Planning: The Optimal Mix of Adjustments to Floods **286**
T.L. Morin, W.L. Meier and K.S. Nagaraj

V. REAL TIME AND STOCHASTIC RESERVOIR OPERATION MODELS   **307**

A Multi-Reservoir Model With A Myopic Optimum **309**
Matthew J. Sobel

An Adaptive Control Model for Single Reservoir Operation **316**
Ramesh Sharda and M.A. El-Tayeh

Extended Linear Quadratic Gaussian Control for the Real Time Operation of Reservoir Systems **329**
Aristides Georgakakos

A Dual Approach to Stochastic Dynamic Programming for Reservoir Release Scheduling **361**
E.G. Read

Accuracy of the First-Order Approximation to the Stochastic Optimal Control of Reservoirs **373** *not in the list*
Peter K. Kitanidis and Roko Andricevic

Biases in Stochastic Reservoir Scheduling Models **386**
E.G. Read and J.F. Boshier

NOTES ABOUT AUTHORS   **399**

Author Index   **409**

Subject Index   **419**

DYNAMIC PROGRAMMING FOR OPTIMAL WATER RESOURCES SYSTEMS ANALYSIS

PREFACE

The field of water resources planning, design, and management contains an assortment of important but complex problems whose efficient treatment can be approached via various tools of operations research and systems engineering. Among the most potent and popular of such methodologies is dynamic programming. The use of this problem solving philosophy and technique has grown very rapidly in recent times. This growth is due in part to the advances in computational techniques of dynamic programming and the advent of powerful computational devices such as supercomputers, parallel computing, etc.

As part of the Bellman Continuum, a gathering of professional associates of Professor Bellman dedicated to the furtherance of his works, a Special Workshop on Dynamic Programming and Water Resources sponsored by the National Science Foundation was held at the Georgia Institute of Technology, Atlanta, Georgia from June 25 - 27, 1986. This volume is an outgrowth of that Workshop and contains selected papers invited for presentation at that Workshop. It is also a completion of a long project, under the inspiration of Professor Bellman, to write a book on the subject.

Since dynamic programming has become identified as one of the most applicable techniques in water resources systems management, the Workshop sought to facilitate its appropriate and increased usage by practitioners, model builders and technique developers. Because, as in any other field, there have been some bad and excellent papers written on the subject, we were careful to invite experts from different parts of the world. We believe that the authors have made invaluable contributions which will advance the field.

The book is divided into five sections. The first part motivates the subject of the Workshop and includes surveys of origins and uses of dynamic programming in water resources and how both fields have complemented each other. Nina Bellman gives an insider's view of the origin of the name dynamic programming, and I provide the interconnection of the subject with water resources. In particular, the paper by Warren A. Hall, Elwood Mead Professor of Engineering at Colorado State University and formerly Director, Office of Water Research and Technology, U.S. Department of the Interior, discusses the employment of the philosophy of dynamic programming in conceptualizing and solving practical water resources systems engineering problems that may arise in different lands. Just as Bellman is referred to as the father of dynamic programming so can Hall be called the father of dynamic programming and water resources. To show the relevance of the papers to practical existing problems of the sort that may be faced by different states, regions or countries, we include the paper by Bernd Kahn of the Environmental Resources Center, Georgia Institute of Technology, which highlights some of the major water resources problems of the State of Georgia. The section is appropriately concluded by me with a Taxonomy of various models occurring in the literature. It is an important and monumental task.

Most water resources problems or projects are multifaceted, multidimensional and even multistaged. Part Two deals with those aspects emphasizing multiobjective-multipurpose and large scale water resources systems, particularly the techniques for developing models of greater fidelity and intelligent methods for obtaining results from such otherwise complex models. They include papers by George Tauxe, John Labadie and Daniel Fontane, Osman Coskunoglu, Moshe Sniedovich and José Ramos, each of whom have extensive experience on their subjects.

In Part Three, problems of water demand identification, supply, distribution and efficient management are considered. New methodologies as well as successful real life applications, as for example in England and Japan, are reported. The authors include Luv Kher and Soroosh Sorooshian, Toshio Odanaka, Bryan Coulbeck and C. H. Orr, Augustine Esogbue and Chae Y. Lee.

Part Four contains papers on water quality, industrial waste treatment, and flood control systems. Included are contributions by Hiroshi Sugiyama, J. Hugh Ellis, T. L. Morin, W. L. Meier, and K. S. Nagaraj.

The volume is concluded with a section on real time operation and stochastic models of reservoirs. As can be seen from the Taxonomy paper considerable attention has been paid to the subject in the literature. It, therefore, not surprisingly attracted special interest of both model builders and users at the Workshop particularly during the round table discussion that followed. The section contains papers for on line control and successful real life applications notably in New Zealand and Egypt. The authors are Matthew Sobel, Ramesh Sharda and M.A. El Tayeh, Aristides Georgakakos, E. G. Read and J. F. Boshier, Peter Kitanidis and Roko Andricevic.

The papers in this book have, for the most part, been reviewed by me, my graduate students and colleagues. I have also written a considerable portion. They treat various problems arising in water resource systems engineering and contain approaches and data which should prove useful to managers, practitioners and model developers as well. The book summarizes and clarifies previous models as well as presents new results including some from recently completed doctoral dissertations. It brings, for the first time in one volume, information and material that hitherto had been scattered in various journals and technical libraries. Students, researchers, and all those interested in the efficient use of an important tool such as dynamic programming in the resolution of pressing and difficult problems of managing our scarce water resources, should find this volume a timely addition to the library, both reference and text of water resource systems engineering.

I acknowledge with much gratitude the support of the National Science Foundation, Nina Bellman and members of the Bellman Continuum, all the contributors to this volume as well as the conference participants, the Coca Cola Company, my mentor and friend Dick Bellman, who introduced me to the art of using dynamic programming, and Warren Hall, who excited my interest in the use of the method to address various problems occurring in water resources. The support of Dr. Thomas E. Stelson, Vice President for Research at the Georgia Institute of Technology, as well as the secretarial assistance of Ms. Joanne Lewis are also gratefully acknowledged.

Augustine O. Esogbue, Ph.D.

Atlanta, Georgia

PART ONE

OVERVIEW OF WATER RESOURCES PROBLEM SOLVING
VIA DYNAMIC PROGRAMMING

ON THE ORIGIN OF THE NAME DYNAMIC PROGRAMMING

NINA BELLMAN
Santa Monica, California

The first thing Richard Bellman always said about dynamic programming was that it was a theory of multi-stage decision processes. He said that it was the only one he knew of, but another theory might be developed sooner or later. He didn't think so, but thought it not impossible.

When asked why he called his theory of multi-stage decision processes dynamic programming, he began with a statement to the effect that it was not programming at all, and that in some ways the choice of name was unfortunate. Then he would explain the reason for the name.

The ideas which resulted in dynamic programming began to take shape after World War II. He first began to promote the theory during the first Eisenhower Administration. Charles Wilson, of General Motors, was Secretary of Defense. Wilson saw no need for most research. He had downplayed it at General Motors, and continued the same policy in government.

Dick worked for the Rand Corporation in the '50's. At that time, Rand was funded almost entirely by the Air Force, and as a think tank was particularly vulnerable to budget cuts on the ground that nothing tangible was produced there. In fact, Dick took a half-time position at UCLA in the Engineering Department in 1953, because of the shortage of money at Rand.

That is the background for the choice of name of dynamic programming. Dick wanted a name that would make it acceptable to the people who were in charge of distributing grants. Programming had a practical sound to it. Besides, the word was fast being popularized by other powerful developments such as linear programming. He liked the word dynamic programming because it is an exciting word for a powerful tool.

Years later, Dick wished he could have chosen a more descriptive name. He decided that the early '80's was too late to change a familiar name. With all the recent changes in the names of large corporations, he might have felt differently now.

DYNAMIC PROGRAMMING AND WATER RESOURCES: ORIGINS
AND INTERCONNECTIONS

AUGUSTINE O. ESOGBUE
School of Industrial and Systems Engineering
Georgia Institute of Technology
Atlanta, Georgia 30332-0205, U.S.A.

INTRODUCTION AND ORIGIN

For nearly three decades since the first reported use of dynamic programming to treat a water resources development problem appeared in the English literature [16], the method and its many variants have been employed in a variety of ways to address a gamut of problems connected with various aspects of water resources systems.

In a study for the U.S. Office of Water Resources and Technology, U.S. Department of the Interior dealing with a systems approach to integrative urban water-land management, the use of systems techniques was surveyed and documented [7]. It was found that dynamic programming was one of the most highly favored techniques in water resources systems analysis. In fact, it then ranked second to simulation but ahead of linear and nonlinear programming combined. Most important, the appeal of the method has grown rather rapidly in recent years. It is expected that the growth pattern will continue in the future. This growth is not only in number but in types and size of problems addressed.

As shown in Table 1, the problem areas ranged from reservoir operation and design to water distribution, sequencing and expansion of facilities, water resources planning, conjuctive use, water quality and irrigation management. To fully understand the variety of the techniques, models and problems addressed, we developed the matrix which is reproduced in Table 1.

In a follow up study [10] in which we critically evaluated a number of the reported studies using various systems techniques in urban water management, we were again impressed by the prominent place of dynamic programming. We were primarily interested in the transferability of those techniques and models to urban water management in a manner which will facilitate their use by the practitioners. Consequently, we focused on approaches to make published work usable by the practitioners. Figure 1 shows the suggested mechanism considered instructive for enhancing the transferability of these techniques. We note that short courses, seminars and symposia were considered important avenues for furthering the goal of developing and transferring appropriate water resources systems technology. Thus, the workshop sponsored by the National Science Foundation, is an attempt to implement one of the foregoing suggested technology transfer approaches.

Many of the foregoing sentiments were collaborated recently by Yakowitz in his review of the applications of dynamic programming in water resources [35]. In his extensive survey of the subject, both the problem areas, techniques and needs were discussed.

Ever since the development of dynamic programming and water resources systems analysis and optimization, there has been a curious

TABLE 1. TWO-WAY CLASSIFICATION OF SYSTEMS ANALYSIS TECHNIQUES APPLIED TO WATER RESOURCES MANAGEMENT
(Entries represent # of papers selected from the Literature of the period 1965 to April 1976.)

	PRIMARY TECHNIQUES				OTHER TECHNIQUES								UNCLASSIFIED			
SOLUTION TECHNIQUE / PROBLEM AREA	Linear Programming 1	Dynamic Programming 2	Simulation 3	Statistical Method 4	Nonlinear Programming 5	Integer Programming 6	Quadratic Programming 7	Geometric Programming 8	Network Analysis 9	Queueing Theory 10	Project Scheduling 11	Forecasting 12	Economic Analysis 13	Input-Output Analysis 14	Mathematical Models 15	Systems Analysis 16	Other 17	TOTALS
---	---	---	---	---	---	---	---	---	---	---	---	---	---	---	---	---	---	---
Water Supply 1	7	6	5	1	1	2	1	1	1	1			1	3	5	3	10	47
Conjunctive Use 2 (I)	2	5	1	5		1							1			2	3	15
Water Quality 3	6	6	9	1	2				1		1	1	4		10	5	19	68
Water Treatment 4	4	2	5	1									2		1	4	6	26
Waste Treatment 1		3	2					1							5	5	10	27
Water Distribution 2 (II)	2	6	6		1	2			1						2	3	2	24
Water Storage 3															1		1	2
WR Planning 4	4	4	4	1	2							1	3		1	7	9	32
WR Evaluation 1	5		1										4			1	1	12
WR Development 2 (III)		2		4							6					1		5
WR Sequencing/Expansion 3	1	6		1						1			1		1	1	1	16
WR Management 4	3	2	5	6		1	1	1		1		1	3	1	6	4	40	67
WR Data Collection 1	2																	2
WR Data Systems 2 (IV)			1															1
WR & Land Use Planning 3			4	1														1
WR Interrelationships 4																		5
Water Pricing 1		2	2	1								7	2	1	2	2	2	9
Water Demand 2 (V)	1	1	1	4											3		8	14
Flood Control 3	4		3	1	1							1			3		2	22
Urban Runoff/Watersheds 4	2	2	14	6	2	1		1							4		2	29
Streamflow 1				1											1		1	6
Sewer Design 2	3	3		1	1												1	7
Reservoir Design 3 (VI)	3	2	1	2	1												7	20
Reservoir Operation 4	7	11	3	8	2				1			1			1	1	12	47
Recreational Facilities 1	1		1	1														4
Plant Location 2 (VII)																		1
Rainfall 3			1	2														3
River Yield 4				1														1
Totals	48	63	73	38	9	6	2	4	3	3	7	12	22	5	46	39	134	153

5

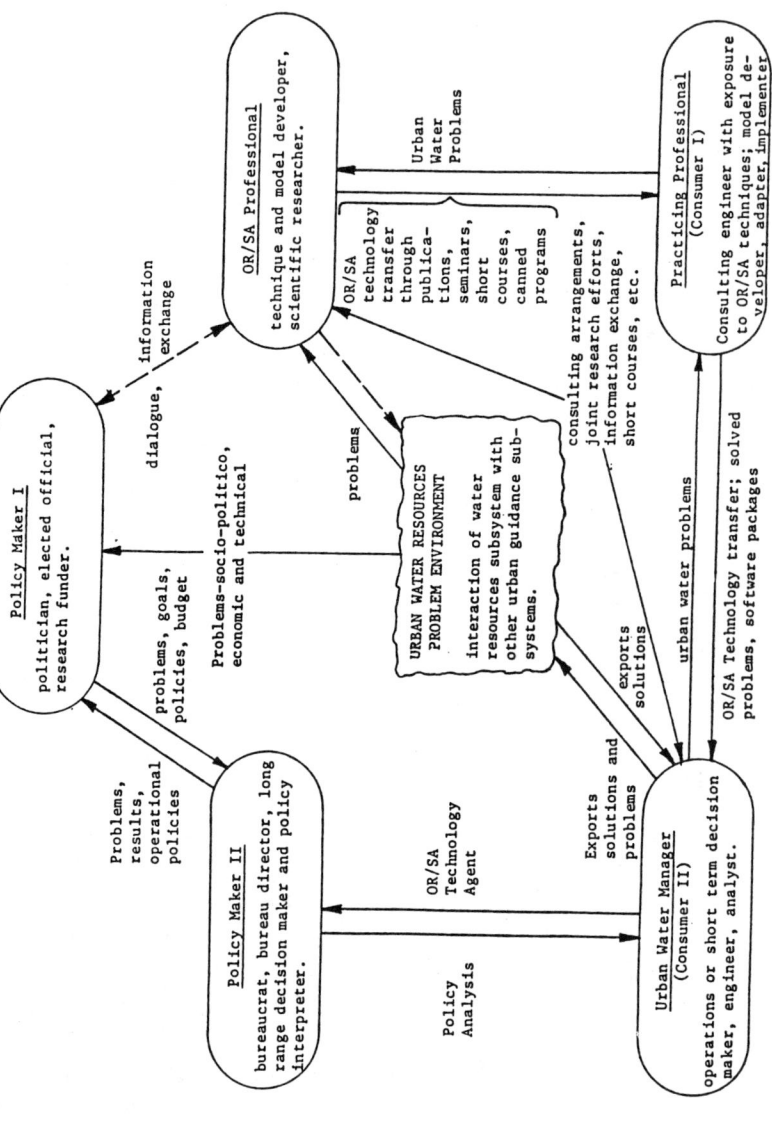

Fig 1. A Model for the Transfer of Systems Technology to Urban Water Resources (Source [10])

symbiotic relationship between the two. For example, it was correctly
pointed out by Yakowitz that water resources problems were among the
first real world laboratory for testing the applicability of dynamic
programming to non defense issues. Attempts to solve certain large
scale systems problems arising in water resources systems have led to
the development of certain computational techniques in dynamic programming. An example is the method we called "imbedded state space dynamic
programming" developed in connection to deal with the computational
difficulties of solving certain sequencing and scheduling problems in
capacity expansion of water supply projects [30]. Several doctoral
theses, both in civil and environmental engineering with emphasis on
water resources, have been written using dynamic programming as the
principal analytic tool. See, for example, Sule [33] who considers the
operations of a multiple purpose reservoir. Similarly, several doctoral
dissertations within mathematics, engineering and operations research
have been written on theoretical and computational aspects of dynamic
programming using water resources problems as their leitmotif. One
example is Morin [29]. On the practical side, the now well known Central
Valley project in California uses policies generated from research work
at the UCLA Water Resources Center. The models stem from the work of
Warren A. Hall and Ronald Shepard and their associates both at UCLA and
Berkeley. The origins of these efforts are the dynamic programming
models reported by Hall et al. [20].

Yakowitz [35] aptly summarizes this relationship in the following:

> "Water resource problems have served as a stimulus to the
> development of dynamic programming itself, and many of
> the studies to be surveyed have attracted the attention
> of workers outside hydrology. Water allocation problems
> suggested by Warren Hall are to be found in Bellman's
> [1957, p. 144] foundational book. A reservoir operations
> problem figures prominently in Larson's [1968] exposition
> of the state increment dynamic programming technique.
> Moreover, a reservoir control problem is among the few
> selected applications studied in the recent treatise on
> stochastic dynamic programming by Dynkin and Yushkevich
> [1979]. An unmistakable conclusion is that water resource
> problems serve as an excellent impetus and laboratory for
> dynamic programming developments; conversely, progress in
> making dynamic programming applications in water resources
> economically viable depends on further advances in
> theoretical and numerical aspects of dynamic programming."

A major impediment to a more widespread and universal application
of dynamic programming to water resources problems is the vastly
publicised but exaggerated problem known as the curse of dimensionality
[1]. The curse of dimensionality is due primarily to the exponential
growth in rapid access storage necessary to perform the computational
solution of the functional equation of dynamic programming, as the
number of state variables increases. Another but less inhibiting
problem is the computational time requirements of certain dynamic
programming models. Although real, there have been substantial advances
in the theory, computation and algorithms of dynamic programming all
geared to reducing dimensionability. Among these techniques are state
increment dynamic programming [Larson (27)], method of the tube [Durling
(8)], method of the tree [Wong (8)], differential dynamic programming

7

[26], discrete differential dynamic programming [23], dynamic programming and partial differential equations [Bellman & Angel (8)], terminal state dynamic programming [Collins (8)], imbedded state space dynamic programming [Morin & Esogbue (31)], nonserial dynamic programming [Esogbue (11)] and an array of methodologies for stochastic systems (24, 25). Most of all, the advances in computer technology (both hardware and software) but notably micro and super computers, have revolutionized the field of dynamic programming.

In spite of the foregoing, practitioners in the theory and applications of dynamic programming and water resources systems field have not capitalized optimally on this relationship through appropriate interaction and technology transfer. A major cause of this dilemma is that those involved in the theoretical developments are not reasonably familiar with important water resources problems and environment. Conversely, those in water resource systems are not familiar with the breakthroughs in dynamic programming. Additionally, there have been unfortunate but inevitable abuses such as erroneous applications of the methodology. We hasten to point out that other fields and methodologies have also suffered from similar excesses.

OBJECTIVES OF THE WORKSHOP

In view of the above brief analysis, a workshop bringing together practitioners, water resources model builders and developers of dynamic programming methodology, both the theory and algorithms, was considered not only timely but necessary. The objective was to bring about the much needed interaction among the key participants and to review the history, assess the state of the art, and prognosticate about future appropriate developments as well as linkages so that more breakthroughs in the application of dynamic programming to solve larger and more water resources problems can take place. An additional important objective was the identification, isolation and grouping of the major research issue needs, and problems which when successfully resolved would enhance the utility of dynamic programming in various application areas but particularly water resources systems. To aid this exploitation of the methodology by practitioners, the presentations and discussions from the workshop were compiled, reviewed and edited into this volume which will constitute a major reference work for students, researchers, practitioners, as well as funding agencies. This will be a step in the right direction for dynamic programming as suggested by Bellman and for water resources as initiated by Warren Hall.

The topics include various aspects of dynamic programming both theoretical and computational and their use to model, analyze and design a gamut of water resources systems problems from supply to irrigation as detailed in the Taxonomy presented by Esogbue [13] in this volume.

PREVIOUS MEETINGS OR DOCUMENTS ON THE SUBJECT

As far as we could determine there had never been any workshop or meeting of this kind. There had, of course, been some meetings on dynamic programming as for example the International Conference in 1977 at Vancouver, British Columbia. That led to the book, <u>Dynamic Programming and Its Applications</u> edited by M. L. Puterman and published by Academic Press in 1978 [32]. There, however, does not appear to have been any other conferences devoted totally to dynamic programming.

There have been numerous meetings on the use of systems techniques in water resources including a 1969 workshop which I directed entitled "Operations Research and Systems Engineering in Complex Water Resources Systems." This was held at Case Western Reserve University in Cleveland, Ohio and attended by professionals and educators from across the U.S. and abroad.

Similar but different short courses, both in orientation, structure and audience, include the ones on Large Scale Systems organized at UCLA and the annual short courses given by Dr. Yacov Haimes when he was at Case Western Reserve University on Multilevel Hierarchical Techniques and Water Resource Systems.

Clearly, papers dealing with the use of dynamic programming to model water resource systems abound in various journals and have been presented at various meetings of numerous professional societies. These are reviewed in the chapter on Taxonomy. Of special note is the review paper by Yakowitz [35].

REFERENCES

1. Bellman, R., Dynamic Programming, Princeton University Press, Princeton, N.J., 1957
2. Bertsekas, D., Dynamic Programming and Stochastic Control, Academic Press, New York, 1976
3. Buras, N., Scientific Allocation of Water Resources, Elsevier Press, New York, 1972
4. Butcher, W., "Stochastic Dynamic Programming for Optimum Reservoir Operation," Water Resources Bulletin, Vol. 7, No. 1, pp. 115-123, 1971
5. Dreyfus, S. and A. Law, The Art and Theory of Dynamic Programming, Academic Press, New York, 1977
6. Esogbue, A.O. and B. R. Marks, "Nonserial Dynamic Programming: A Survey", Operational Research Quarterly, Vol. 25, No. 2
7. Esogbue, A.O., Integrative Procedures for Coordinated Urban-Water Land Management: A Systems Analysis, NTIS No. PB 243-518, 1975, 219 pages
8. Esogbue, A.O. and A. J. Singh, "Reduction of Dimensionality in Dynamic Programming Revisited: A Comparative Study and Analysis of Three Algorithms", Opsearch, Vol. 12, No. 3-4, 1975
9. Esogbue, A.O., "Mathematical Programming, Computers, and Large Scale Water Resources Systems," Water Resources and Pollution, Vol. 2, 1980
10. Esogbue, A.O. and G. Willeke, Transfer of Systems Technology to Urban Water Management, NTIS No. PB 283-466, 1978, 237 pages
11. Esogbue, A.O., Dynamic Programming Algorithms and Analysis for Nonserial Networks, Technical Completion Report No. ARO DAAG 29-80-G-0010, School of Industrial and Systems Engineering, Georgia Institute of Technology, January 1983
12. Esogbue, A.O., Quantitative Models and Analyses of Urban Nonpoint Source Water Pollution Control Systems, NTIS No. PB 83-214940, 1983, 217 pages.
13. Esogbue, A.O., "A Taxonomy of Dynamic Programming and Water Resources," in this Volume.
14. Hall, W., "Aqueduct capacity under an optimum benefit policy," J. Irrig. Drain. Div. Am. Soc. Civ. Eng., 87(IR3), 1-11, 1961.

15. Hall, W.A., "Optimum design of a multiple-purpose reservoir," *J. Hydraul. Div. Am. Soc. Civ. Eng.*, 90, 141-149, 1964.
16. Hall, W. and N. Buras, "The dynamic programming approach to water resources development," *J. Geophys. Res.*, 66, 517-521, 1961.
17. Hall, W. and W. Butcher, "Optimal timing of irrigation," *J. Irrig. Drain. Div. Am. Soc. Civ. Eng.*, 94(IR-2), 267-275, 1968.
18. Hall, W. and D. Howell, "The optimization of single-purpose reservoir design with the application of dynamic programming to synthetic hydrology samples," *J. Hydrol.*, I, 355-363, 1963.
19. Hall, W. and T. Roefs, "Hydropower project output optimization," *Proceedings, J. Power Div. Am. Soc. Civ. Eng.*, PO1, 67-79, 1966.
20. Hall, W.A., W.S. Butcher, and A. Esogbue, "Optimization of the operation of a multi-purpose reservoir by dynamic programming," *Water Resour. Res.*, 4(3), 471-477, 1968.
21. Hall, W.A., R. Harboe, W. G. Yeh, and A. Askew, "Optimum firm power output from a two reservoir system by incremental dynamic programming," *Contrib.* 130, Water Resour. Center, Univ. of Calif. at Los Angeles, Los Angeles, 1969a.
22. Hall, W., G. Tauxe, and W. G. Yeh, "An alternate procedure for the optimization of operations for planning with multi-river, multiple-purpose systems," *Water Resour. Res.*, 5(6), 1367-1372, 1969b.
23. Heidari, M., V. T. Chow, P. V. Kokotovic, and D. D. Meredith, "Discrete differential dynamic programming approach to water resources systems optimization," *Water Resour. Res.*, 7(2), 273-282, 1971.
24. Hinderer, K., *Foundations of Non-Stationary Dynamic Programming With Discrete Time Parameter*, Springer-Verlag, New York, 1970.
25. Howard, R.A., *Dynamic Programming and Markov Processes*, MIT Press, Cambridge, Mass., 1960.
26. Jacobson, D. and D. Mayne, *Differential Dynamic Programming*, Academic, New York, 1970.
27. Larson, R., *State Increment Dynamic Programming*, Elsevier, New York, 1968.
28. Larson, R. and J. Casti, *Principles of Dynamic Programming*, Vol. 1, Dekker, New York, 1978.
29. Morin, T.L. "Optimization of One Shot Processes," Ph.D. Thesis, Department of Operations Research, Case Western Reserve University, Cleveland, Ohio, 1971.
30. Morin, T. and A. Esogbue, "Some efficient dynamic programming algorithms for the optimal sequencing and scheduling of water supply projects," *Water Resour. Res.*, 7(3), 479-484, 1971.
31. Morin, T.L. and A. O. Esogbue, "Imbedded state space approach to reduction of dimensionality in dynamic programming," *Journal of Mathematical Analysis and Applications*, Vol. 48, No. 3, 1974.
32. Puterman, M. (Ed.), *Dynamic Programming and Its Applications*, Academic, New York, 1978.
33. Sule, B.F., "Optimal Operation of a Multiple Purpose Reservoir," Ph.D. Dissertation, Cornell University, Ithaca, NY, January 1984.
34. Yakowitz, S., "Constrained Differential Dynamic Programming," submitted to *Math. Anal. Appl.*, 1982.
35. Yakowitz, S., "Dynamic Programming Applications in Water Resources," *Water Resources Research*, Vol. 18, No. 4, 1982, pp. 673-696.

WATER PROBLEMS AND ISSUES OF THE STATE OF GEORGIA: PROBLEM GOALS AND PRIORITIES

Bernd Kahn, Director
Environmental Resources Center
Georgia Institute of Technology
Atlanta, GA 30332, USA

Georgia has the full range of water problems commonly associated with industrial and agricultural development in its geographical setting, ameliorated by generous rainfall and the efforts of a forward-looking state water management agency. The Environmental Resources Center (ERC), as the State Water Research Institute under the Water Resources Research Act of 1984 and earlier laws dating to the Act of 1964, functions to provide support for the water management agency through a network of research at universities and colleges in Georgia.

The average rainfall of approximately 50 inches per year is sufficiently distributed throughout the year and over all parts of the state so that normally water is plentiful. The rain maintains numerous large and small streams that originate in north Georgia and flow through the state to the Atlantic Ocean or the Gulf of Mexico, and recharges an enormous aquifer underlying south Georgia. Water users generally depend on streams and shallow wells in northern Georgia and on deep wells in southern Georgia. This water supply supports industrial and agricultural development throughout the state, and population growth with concentration in the Atlanta area.

A number of water quality, quantity, and management problems have developed with growth. Small streams in north Georgia have reached capacity for supplying water and receiving waste. Some deep wells in Georgia have dropping water levels, and some in coastal Georgia have salt-water intrusion. Surface runoff and resulting sedimentation have been a longstanding problem, and herbicide/pesticide/nutrient runoff has increased with intensified agriculture. Solid and toxic waste management is being recognized as a major problem with water pollution implications. The economics and technology of building and maintaining water and wastewater treatment and distribution facilities need consideration as Federal support decreases and guidelines become more demanding.

The state agency, the Georgia Department of Natural Resources, has addressed these problems by enforcing laws that control water quality and water use. As a result, surface water quality throughout the state has improved to meet standards, and industrial development is guided to locations where water remains available at the appropriate volumes. The agency has also identified the following priority issues, prompted by the problems associated with an extensive drought in 1986.

- plan regional reservoirs in north and central Georgia;

- implement a comprehensive ground water management program for both quality and availability;

- establish a modern hazardous waste facility;

- develop and implement a toxics management strategy

These efforts should go far toward managing water in Georgia for the foreseeable future.

PROGRAM GOALS AND PRIORITIES

The Environmental Resources Center attempts to develop each annual program to meet the informational needs of managing Georgia's water resources, to the extent feasible by supporting research by university faculty members. The needs are identified by direct contact with staff of the Georgia Department of Natural Resources, by guidance of an advisory committee, by interaction with the professional community, and by discussions with institute directors from other states in the region. These needs are then formally listed as state priorities, incorporated in the regional list of priorities, circulated to researchers with the request for proposal submission, and considered in selecting projects for the annual program. The technical completion report for each project is published and made available to state water managers to provide the needed information.

The research priorities developed jointly by the directors of the institutes in the South Atlantic Gulf region, and considered applicable to all states in the region, are listed in Table 1. Several of these priorities have been carried over from previous years, and would be expected to remain for an extended period. Others were added to the list in response to recently perceived needs. Continuing review by the institute directors will evaluate the extent to which these needs are being met by the research programs in the participating states, and the relative urgency of the remaining needs.

The six projects undertaken in Georgia during FY 1986 to meet some of these needs are described briefly in the following paragraphs, and in further detail in the next section. The continuity inherent in the program can be seen in earlier annual reports and in the publication list that follows the detailed project descriptions.

Project 02 addressed the topic of water quantity with specific regard to ground water movement. It continues a series of studies that have modelled ground water flow, considering first the steady state for simple systems, then multilayer systems, and now, unsteady state induced by pumping. Finite element Galerkin matrices are used in coupled differential equations to determine piezometric head distributions. Examples pertaining to south Georgia are presented.

Project 03 was a water quantity project concerned with surface and ground water interactions. A microcomputer program that considers the extent of water withdrawal from streams due to pumping nearby wells was developed for stream flow prediction.

The program was designed for ease of use with readily available flow and pumping data. It was given a sensitivity test for conditions found in south Georgia.

Project 04 also was in the category of water quantity, considering that aspect of the water cycle related to runoff vs. infiltration. It examined the extent to which the surface impact of rain drops interferes with infiltration. Infiltration rates were determined for several soils from the Georgia Piedmont in greenhouse experiments. Decreases in infiltration rates were associated with specific soil characteristics.

Project 05 addressed water quality by considering the optimum location of monitoring wells. The project selected a region in southwestern Georgia where ground water quality monitoring is needed and utilized available flow characteristics to propose an optimum sampling network. Selecting the numbers and locations of these wells was based on a statistical approach.

Project 06 was a water quality project which tested a new analytical approach. Samples were first separated by liquid chromatography and the fractions were then analyzed with a swept-potential electrochemical detector. Considerable sensitivity was attained in detecting certain azo dyes at levels that could not be detected when the procedures were used separately.

Project 07 considered the water quality of non-point agricultural runoff from a drainage-subirrigation system used in south Georgia to farm land otherwise unsuitable for this purpose. Both surface and shallow subsurface and runoff water were samples for nitrates, phosphates, chlorides, and ammonia to determine whether the special procedures used in this approach resulted in unusual pollution detriment.

In the midst of this program cycle, the research program for FY 1987 was developed. The projects listed in Table 2 were incorporated in the FY 1987 program. Note that it includes a continuation of project 03.

Table 1: WATER RESOURCES RESEARCH PRIORITIES OF THE SOUTH ATLANTIC-GULF REGION

The eight states of the South Atlantic-Gulf Region consisting of Alabama, Florida, Georgia, Mississippi, North Carolina, South Carolina, Tennessee and Virginia generally have a large surface water supply. In addition, tremendous ground water resources are available in the extensive limestone and coastal terrains of the region. Precipitation throughout the region is abundant. Yet within the Piedmont area only limited supplies are available from wells and streams. The region constitutes a major art of the sun belt which is moving rapidly in economic growth. The attendant increasing demands on the water and other resources require resolution of the identified major water resources problems.

Critical water problems focus on preservation of water quality, increasing competition for supplies from ground water and surface water sources, and the need for improvements in the methodology and application of water resources management. The range of problems relates in part to the variations in physiography within the region that control the occurrence and variability of water.

In addressing these critical problems, research will be concentrated on the following areas:

WATER QUALITY

Research needs in the water quality area involve information, information management, and the protection of surface and ground water from degradation. It includes industrial and municipal wastewater treatment and subsurface disposal of hazardous/toxic wastes. In addition, problems from non-point sources of both municipal and agricultural sources, including soil erosion, agricultural runoff and pesticides, pertain to this area. The development and improvement of monitoring techniques and analysis are also imprtant, as well as water quality problems associated with eutrophication and weed control.

WATER MANAGEMENT

Research needs in the area of water management include legal, institutional, and financial arrangements. Specific items such as basin planning, water use control, transfers and/or diversions of water, flood control, and drought planning are all priority issues. It also includes construction of facilities, financing and pricing, and water conservation and reuse. Management includes quality protection studies, upgrading of supplies and state and/or federal and interstate interactions or compacts.

WATER QUANTITY

Research needs in the water quantity areas include studies of the basin water cycle for an understanding of prediction. They also include items of surface water flow, basin planning, low flow predictions (7Q10), flood control, water use, and water allocation. Studies of ground water availability, and the locations, movement, and volume of ground water are needed. Also of importance are use and user impacts, and surface and ground water interactions.

AQUATIC AND ENVIRONMENTAL PROTECTION

Research needs in this area include studies of wetlands, swamps and marshes, fish and other biota, and the quality of life. Studies of ecological balance, protection of endangered species, and the impacts of dredging and filling are needed.

EMERGING PROBLEMS

Studies not included in other priority areas but dedicated to solving emerging water problems which are identified as critical issues by key state water management officials in the region are included in this category.

Table 2: FY 1987 WATER RESEARCH INSTITUTE PROGRAM FOR GEORGIA

Title	Principal Investigator
Temporal and Spatial Variations in the Radon Content of Ground Water in the Vicinity of the Elberton Granite Batholith of Norheastern Georgia	Kathleen H. Cole, David D. Wenner and John E. Noakes University of Georgia
Evaluation of Methods for Measuring Stream Flow Reduction Due to Ground Water Pumping	John F. Dowd and Kathryn J. Hatcher University of Georgia
Optimal Reservoir Operation Schemes	Aristides P. Georgakakos Georgia Institute of Technology
In situ Biological Treatment of Contaminated Ground Waters	F. Michael Saunders Georgia Institute of Technology
Geophysical Methods for Ground Water Location in Crystalline Rock of the Georgia Piedmont	Dann J. Spariosu University of Georgia
International Conference on Fluid Flow in Fractured Rocks	Ram Arora Georgia State University

REFERENCES

[1] Kahn, B., Annual Report of Georgia Water Resources Research Institute Activities Under Public Law 98-242 in Fiscal Year 1986, ERC 06-87, Environmental Resources Center, Georgia Institute of Technology, pp. 1-6.

DYNAMIC PROGRAMMING AND PRACTICAL WATER RESOURCES SYSTEMS ENGINEERING

WARREN A. HALL
Department of Civil Engineering, Colorado State University
Fort Collins, Colorado 80523

1. INTRODUCTION

Some thirty years have passed since I was first introduced to Dr. Richard Bellman, and to his then new dynamic programming concepts. I had just returned to UCLA from U.C. Davis (1956) where I had initiated a program of laboratory and field research directed toward irrigation engineering and management. My new assignment, as Assistant Dean for Undergraduate Studies, required that my research program be reoriented from field work to something I could work on at my desk. Several years earlier, I had worked a bit with optimization in connection with some maritime cargo loading research, and it seemed to me that there was a wide open opportunity to apply some of these ideas to water resources planning and management. Dean Boelter suggested that I contact a "Dick Bellman" at RAND. With what in retrospect would appear to be a lot of brass, I called Dr. Bellman and asked him to come see me, and at my convenience (I explained how busy I was!).

Had I realized that his daily consulting fee at that time was greater than my monthly salary—before deductions—I doubt that I would have had the courage to call at all. Fortunately for me, ignorance was bliss, for Dick responded to my invitation immediately and showed up at my office precisely on time. We had several hours of discussion which proved to be extremely valuable to me. I can't say the value was reciprocal, but I do remember that he was highly enthusiastic the entire time and that I had to terminate the session around 7 p.m.

This was just the first of many such meetings over the following years, during which he patiently worked to remove my ignorance and misconceptions, and to help develop my understanding of optimization analysis as being something rather special and distinct from the mathematical modeling of physical behavior which dominated my own thinking.

This patience was typical of Dick's attitude toward anyone who wanted to learn and was willing to put in the effort. On the other hand, he could be very short with anyone who wanted a cookbook for optimization and even completely non-responsive to persons who wanted to argue that his theory was incorrect. I recall one paper which purported to show that dynamic programming could not possibly be valid. I could not understand the reasons, so I asked Dick about it. After he explained, I asked him why he didn't respond. His answer was also classical Bellman. "Why should I? He is the one who is wrong!" Here was a man who not only knew the mathematics of optimization but also understood the optimal use of his own time.

2. SOME DIFFICULTIES ENCOUNTERED IN THE USE OF MATHEMATICAL MODELS FOR SYSTEMS ENGINEERING

It is always pleasant to reminisce about one's personal associations with a great man, but I have been asked to prepare a paper on the use of dynamic programming for water resources planning and management. I have two options. One is to present the results of some specific systems engineering studies for the optimization (broadly defined) of the Mahanadi River Basin in east Central India. The other is to elaborate on the use of the concepts and principles of dynamic programming (as I understood them from my contacts with Dick Bellman) which have proved to be very useful for me and my colleagues in India's Central Water Commission for the process of developing appropriate formulations of the planning problems of that river basin system.

I will choose the latter, because I believe that it may be most useful to the readers of this book. During the past thirty years, in my work in India, the Middle East, Latin America and the U.S. I have been increasingly impressed with something Dick tried to tell me in the beginning. Optimization is the process of selecting a specific set of actions (decisions) implicitly or explicitly embedded in a particular problem formulation, in such a manner that no feasible change in that set will produce a resultant situation considered by the decision maker to be "better" (my paraphrase). However, as Bellman informed me, the formulation of the problem must be a correct representation of the actual planning and/or management situation and the full responsibility for that formulation and its validity rests with the user of the models, including those classified as "optimization models." He considered that I was the user (and so I visualized my role), hence it was my responsibility to provide a proper formulation in the first place. His words, almost exactly, were "If you can express your problem in mathematical form, and if that form has these essential characteristics, then dynamic programming can efficiently and effectively find the optimal decision set you seek." He also pointed out some advantages of D.P. over some of the possible alternative procedures, particularly the "invariant embedding" aspect which gave not only the specific optimal decision set for the specific magnitudes of the conditions, but also gave all other optimal decision sets for all other magnitudes of these conditions (state variables). This is clearly an important advantage for planning large scale systems where one cannot be certain in advance what those conditions might be. It is even more advantageous when complex systems must be broken down into manageable components, as will be discussed later [1].

To pursue, Bellman's admonition in a way which I hope will be taken as constructive criticism to improve analysis, the major reason for the seeming reluctance of practicing professional engineers to utilize our optimization models "off the shelf" is to be found in the inadequacies of our problem formulations. A review of the water resources systems engineering literature (including my own contributions) will show that virtually all begin with a postulated problem formulation. That formulation is only infrequently justified. Even then, it is clear that the author had his own mind's eye model of a particular planning situation. Usually that model is only implicitly presented and the reader is left to deduce the problem visualized from the mathematical treatment. Before anyone reacts negatively to this paragraph, please read the next.

The problem described above is inherent in the research process by which mathematical optimization models must be created. Real planning problems are much too variable, in both generality and specifics, to allow formulation of a Master Optimization Model (MOM) for any and all water resources problems. Most of us who have been contributing to the literature have done so in the role of scientists developing scientifically sound models for the optimization of our specific versions of the water resources problems. Regardless of how "generalized" we try to make those models, the very process of generalization rules out certain specific conditions (special cases). Thus what we have produced and presented is a smorgasbord of useful ideas and approaches which should be interpreted with Bellman's admonition: "if your problem, when you formulate it, looks like this, then this procedure should prove useful for your planning and management purposes." Again I must repeat that the general problem formulation deficiency is both inherent in the research process and necessary if any progress is to be made. If I have a criticism at all, it is that we haven't been careful to label our results with Bellman's equivalent of the Surgeon General's warning.

To back up my contention regarding the variability of both problems and problem situations I would like to cite a few examples from my own experience. In 1965 the UCLA group (Hall, Butcher, Roefs, Esogbue, Haimes, Tauxe, Shephard, Parikh, and, earlier, Buras, Burton and Howell) completed a study of the northern California river-reservoir systems and their unitized operation, in cooperation with the State of California [2]. The State officials were very pleased with our results. We sent a copy of the report to the U.S. Bureau of Reclamation (USBR). Subsequently a group of USBR engineers came to visit me, explaining that it was an interesting study but not much use to them. With less than half an hour of discussion, I had to agree that they were absolutely right! This admission, however, did not diminish the value of the study to the California Department of Water Resources one bit. Note that exactly the same physical systems and purposes were involved. Our formulation was a reasonably valid formulation of that system from the point of view of the problems faced by the State of California. It was a totally invalid formulation from the point of view of the problems faced by the U.S. Bureau of Reclamation. I personally doubt that any feasible formulation could be created which would be valid for both sets of problems. Although a lot of excellent progress has been made by the continuation UCLA group (under the direction of Dr. William Yeh), the Bureau's original problems, as they presented them to me, have not yet been adequately formulated.

Subsequently, I was asked by Harza Engineering to see what I could come up with respect to a similar multiple reservoir, multiple purpose system in Honduras. I accepted, fully confident that our California studies provided just what was needed. They did not. They did provide building blocks, but substantial modifications were necessary before the results were even close to being satisfactory.

Next I was asked to outline a systems engineering approach to a hydroelectric and irrigation project in Peru. I was sure that now I had the models in hand, but once again major revisions were required. I was also asked by the Government of Iraq to prepare a systems engineering analysis of a major multiple purpose reservoir system. Again I had to make significant revisions in what (again) I was previously willing to bet was a fully adequate problem formulation. What I did not know until later was that the Russians and a European firm had each conducted such an analysis, but both analyses had been rejected by the Government engineers as not being responsive to their problems. I suspect they may have used their unmodified models off the shelf. After the major revisions which I made to my earlier models, the results were accepted by the Government.

They then asked me to conduct another analysis on the same river but for the downstream reservoir sites and water use areas. This was done cooperatively with Motor Columbus of Baden, Switzerland. By this time I was prepared. Although the basic concepts were the same and some of the elements could be used as building blocks, once again several major revisions were necessary before either we or they were satisfied that we were resolving their planning problem.

A similar result occurred in the study by the Systems Engineering Unit, Central Water Commission, Government of India for which I served as

Senior Project Advisor, under the auspices of the United Nations. Major revisions were required to suit even the general India conditions. During these studies we tried to use, directly, most of the principle models available in the literature (including reports) but without success. This was seldom because of errors in the models. Usually, it was due to an inadequate representation of Indian conditions, problems and objectives in the original formulation of the models.

That statement includes my own models, each of which was based on formulations that were useful in California, Honduras, Peru and/or Iraq. Working as a team we were able to make modifications which permitted use of some of those models for certain problem situations encountered in the Mahanadi. Yet when we were asked to look at similar problems a bit further south in India we had still more changes to make.

These examples are presented to emphasize two points. One is that the task of problem formulation is by far the most important for practical application of systems engineering techniques and models to water resources planning and management. This phase cannot be done using mathematical models only. It can only be done satisfactorily if the formulator also understands the implications of what has not been included and adjusts his formulation so that these can be taken into account. The second is that, in my experience there is no "MOM" available now or in the forseeable future which can serve for the optimization of any specific case in water resources planning or operation as a Master Optimization Model requiring no more than programming and data input.

Even though there is no MOM for water resources, it does not follow that classical research is useless for professional practice. Recall Bellman's argument that the formulation of the optimization problem is the sole responsibility of the engineer who intends to use it to arrive at his recommendations. If he does not adequately understand his own problem (and many do not appear to have such an understanding), he probably won't come up with satisfactory recommendations with or without mathematical optimization models. If he does, then, if he can develop his own skills in problem formulation, he can use the scientific literature as building blocks for assuring maximum solvability of his formulation. As is the case in using physical building blocks, considerable cutting and fitting may still be necessary.

However, it is my contention, again based on my own experience nationally and internationally, that if any engineer tries to use anyone's optimization model (including my own) off the shelf as if it were a MOM, he is very likely to come up with something potentially disastrous for his own professional reputation. Under such circumstances it is hardly amazing that there has been little use of optimization models off the shelf from the literature. Most practicing professionals have had little opportunity to develop the skills necessary for problem formulation using mathematical optimization models. They have developed, almost by instinct, the skills for problem formulation using judgmental optimization, but these skills are constrained by the limitations on human capacity for keeping large numbers of factors and magnitudes in proper focus.

These comments, if accepted as valid, would appear to suggest more university attention on the educational aspects of water resources systems engineering, at least in the immediate future. Research is important and must be done if adequate progress is to be made. However, insofar as practical applications in professional practice are concerned, it seems to me that something is lacking in the "technology transfer" aspects. This is a university function as much as research. Although it doesn't seem to get much attention at promotion and tenure time, the long term ability of the university to attract water research funds (directly or indirectly) most assuredly will depend on how well this job is done. We don't do much bragging today.

3. THE INDISPENSABLE PHASES IN PRACTICAL APPLICATIONS OF SYSTEMS ENGINEERING

To get a better idea of what needs to be done, I should review the basic elements of the planning process as this relates to systems engineering. There are three indispensable elements. The first is the function of problem formulation, discussed at some length above. The second is the mathematical analyses of the formulation, well exemplified for water resources by two or three decades of scholarly publications plus specialized engineering reports. This is the mathematical model realm of dynamic programming (including all extensions), linear programming, quadratic programming, integer programming, various search techniques, etc. Finally, there is the function of evaluation of both the results of the analysis and, implicitly, the original formulation.

For effective professional planning, the three steps are highly interactive, with multiple feedback loops required so that the final evaluation in phase 3 has adequate practical utility for the de facto decision makers (plural) who have the final word.

At present I would argue that (1) the practicing professionals have far too little understanding or appreciation of optimization models, particularly their structure and functions--the same deficiencies Dr. Bellman and his colleague, Dr. Robert Kalaba, tried to correct in my own capabilities back in 1956. Courses in optimization theory have recently become relatively common in undergraduate and graduate curricula. However, by and large, these are courses which present the theory of the classical optimization techniques. Most get bogged down in the arithmetic processes of techniques such as linear programming. They touch inadequately or not at all on the two most important phases, problem formulation and evaluation of the analyses when made.

I must confess to a good measure of these inadequacies in my own undergraduate and graduate courses. In a large part it is because it is so much easier and rewarding to the instructor to teach the theory of analysis than the practice of problem formulation and analysis evaluation. The latter two phases can be very frustrating to a student accustomed to precision physical models. Also in large part, it is due to our role as educators and research scientists. Our days are quite well filled with research and teaching, leaving little or no time for getting practical experience with the other two phases of planning, particularly in the role of responsible professional planners. Even in our consulting work we are essentially advisors. The client has to be responsible for the ultimate evaluation of our advice and our formulation, so once again it is far easier for us to advise on the analytical phase than on either the formulation or evaluation phases.

With nothing more than my own isolated experience to offer as proof, I have come to the conclusion that the first and last phases are similar to engineering design. That is, they can be learned (mainly by experience) but they cannot be taught. As with design in our engineering curricula, we could offer a "problem formulation" course or an "analysis evaluation" course. These courses could include some pedagogical guidelines (do's and dont's) but, it would have to be up to the student to teach himself how to formulate problems and how to evaluate the analyses. The course could only provide the opportunity for such learning in an environment where mistakes and errors are teachers, not inadvertent destroyers of professional reputations.

As an alternative, such a course could be conducted (not taught) as a part of the professional in-house training programs of water resources agencies. This would have major advantages if the course conductors always included both academic types and successful practicing water resources professionals. The latter presently do not use much in the way of modern systems engineering models, but they have in fact been formulating, analyzing (with judgment, supplemented by simulation models) and evaluating the analyses. The fact that they are successful, suggests that they have mastered the three phases for themselves. The academicians can guide the formulations of the participants in the direction of incorporating the power of mathematical logic. In this way the two types of course conductors would be highly complementary in approach, outlook and ideas for getting on with the task.

4. THE PRINCIPLES OF DYNAMIC PROGRAMMING AS A GUIDE TO PROBLEM FORMULATION

In section 2 I promised I would try to show how the underlying principles and logical thought process of dynamic programming have been useful to me in the professional practice of systems engineering. The proceeding digression into the phases of professional practice was made in order to set the stage, for without those ideas, my discussion might seem somewhat void of purpose. Let me return to my basic objective.

To begin with, I am going to describe dynamic programming logic from a non-mathematical point of view, and relate it to the problems of technology transfer of any and all optimization models into practical, useful, cost effective applications. What I hope to show is that almost the same logic—in verbal descriptive form—provides most of the "do" and "don't" guiding principles for problem formulation and analysis evaluation.

Dynamic programming has been described as a "sequential decision process." I won't deny this attribute, but of far greater importance to water resources system problem formulation would be its description as a "decomposition-recomposition" or "disaggregation-reaggregation" decision process.

To help illustrate the problems faced by the engineer in the task of problem formulation, let me cite an exaggerated example. In most water development engineering reports I have read, the purpose stated is to provide the maximum acre feet or millions of cubic meters of water per year at minimum cost. Taking such a statement at face value as the properly formulated objective of water resources planning, and using it in any optimization model, the optimal decision would clearly be, don't provide any reservoir storage. If any water is stored, evaporation losses will increase, otherwise the mathematical expectation of the maximum water available is the mean annual flow. That, or any lesser amount, can be obtained at zero (minimum) cost without a reservoir. Since water is water (in the above statement) it makes no difference to the mathematical model whether the amount available at any time is above or below average by any amount.

Clearly this is an improperly formulated problem in almost any non-trivial case. What is wrong? The basic function or purpose of the water system is to take a natural occurrence of water with a highly stochastic, unreliable distribution over time and space, and convert it to a more desirable time-place distribution, with the required attributes of reliability in time and place, quality characteristics, etc., as needed for the technology of its use for other systems. These other systems in turn are intended to accomplish more basic goals and objectives.

I may have left something out, but everything in the above statement of purpose is an integral part of the performance of every water resources system. My improper, exaggerated example took certain engineering reports at face value, ignoring both time and space distributional requirements as well as the all important reliability characteristics, all of which were only implied. No resource, water or otherwise, has value unless it has the time and place reliability required by the technology of its use. For most uses, the long term investment in that technology far exceeds the corresponding share of water resources investment cost. This means that, despite a few good examples and a lot of wishful thinking, these water use technologies are not going to change at the whim of the water planner. The few examples of successful change that might be cited, all involved little or no modification of current investments. If changes are to be made, they must also be met with high, stable reliability.

To accomplish the time-place-reliability-quantity-quality transformation, certain facilities have been utilized for thousands of years. Each of these facilities has its own unique function. Reservoirs (surface or groundwater) provide storage capacity to allow the reregulation of flow with respect to reliability and temporal distribution. Pipes and open channels, natural or man-made, perform the function of spatial distribution and spatial reliability.

Now let us initially associate Bellman's activities or stages with these functions to be accomplished rather than to the individual elements and various impacts on social, economic, political and other objectives. First look at them from an overall point of view. Storage capacity is the only available, technically feasible means of modifying the temporal availability of water. Using Bellman's principle of optimality, the water resources systems cannot be optimal unless this function is performed at minimum cost in terms of other resources used, including the direct environmental, social and other important impacts of providing and operating that storage capacity.

Similarly, channels, natural or man-made, are the only technology we have for modifying the spatial distribution of water to one considered to be more desirable.

Stopping here for the moment, we have identified two separate and distinct generalized subsystems which have separate and distinct functions. From a pure mathematics point of view we could optimize each without regard to the other. However, there exists an interdependence between the two subsystems. Whatever indices are used to describe the actual quantities of resource ultimately made available over time, those particular quantities must also be reliably distributed over space. Therefore if we should optimize each system separately, as a function of this final output level, it will be a simple matter to find the optimal combinations of those two systems.

Note this statement is valid regardless of how those costs and other impacts are evaluated, and regardless of what uses might ultimately be made of the water resource provided. It is also valid for virtually every water resources development system.

I could similarly describe other water resources subsystems, each of which has its function to perform. For each, however, the central interconnecting variable is usually that same quantity of water.

Of course the phrase "quantity of water" really refers to a vector-like quantity which involves the optimal hourly, daily, monthly distribution of the "quantity" over the year. It also includes a specific set of requirements concerning the reliability characteristics (which will require a complete probability function). However, because of the "technology" constraint, these are either not decision variables or become at most potential secondary decision levels, involving decision makers normally outside the authority of water resource agencies. Thus these parameters should be included as an appropriate set of potentially alterable constraints, associated with the variable "quantity of water." Similar statements can be made for other aspects of the resource such as energy (in or out), flood peak mitigation, etc.

The above arguments were used by our (India) Systems Engineering Unit project to make a fairly general first cut formulation of the very complex, multi-objective water resource development potential (including existing works) for the Mahanadi river. This river basin system involved a potential for more than 80 reservoirs and about the same number of water service areas (but not in one to one correspondence). Flood control was an important consideration. So were avoidance of environmental impacts and achievement of acceptable levels of social equity. Different uses were involved with the high priority uses associated with substantially lower reliability requirements than the lower priority uses. These included irrigation, hydroelectric energy and urban and industrial uses [3].

This first cut was a decomposition model in the spirit of Bellman's principle of optimality. The total system could not be optimal unless these functions were accomplished optimally. Because of the small number of interacting properties, the time and reliability modified output of the river regulation subsystem is the input to the spatial distribution system. The dynamic programming concept of making an optimal combination of these two subsystems provides the model for an optimal recomposition. The many physical, environmental and social issues are easily imposed on this analysis, either as constraints or as multiple objectives.

The arguments are not quite complete. If all regulation occurred upstream of all water distribution, the decomposition-recomposition given would be complete. As for the Mahanadi in India, this is frequently not the case. It was necessary (but not difficult) to prescribe a simple iterative process, with one or two iterations, whereby individual elements of one or the other of these subsystems would be constrained by the feasibility of sending or receiving water between such elements. This is by way of explanation only. These refinements are not essential to the "guiding principles" of problem formulation being presented here, i.e., the principle of using the functions (purposes) of various parts of the complex system to disaggregate it logically into conveniently formulated sub-problems and subsystems.

If the development were limited to one reservoir site and one water use sub area we could stop here. In most cases there are a number of alternate reservoir sites and alternate areas of use. To illustrate the next steps consider that there are a number of reservoir sites, at each of which, all or part of the time and reliability regulation could be accomplished. Invoking Bellman's principles, the overall time-reliability regulation should be accomplished with maximum cost-effectiveness (however defined). Furthermore each reservoir in the system makes its own contribution to the reregulation. The nature of this contribution in fact is different for reservoirs in series or in parallel. This can be properly reflected in the decomposition-recomposition models. Suffice it to say for purposes of this presentation, each reservoir should be best capable of making whatever contribution might be expected of it in the final optimal plan and to do so at minimum cost. That is, we need the function representing the optimal contribution as a function of cost, or vice versa. The recombination of the individual reservoirs can then be accomplished using the principles, and usually the algorithms of dynamic programming. It is at each of these stages that the appropriate feasibility constraints created by the other major subsystems are to be included as suggested above. A similar statement can be made for the distribution system, the water use subsystem and all other functional subsystems.

I could describe this process in more detail, and with a broader representation of functions of subsystems, but that is not the purpose of this paper. I only want to show how the concepts of Bellman's dynamic programming were used to guide us in the proper formulation of our systems engineering problem. Obviously there were still a lot of specific characteristics that caused difficulties, not the least of which were the hydrological characteristics of the monsoon rains and consequent runoff, and the somewhat nebulous definition of "reliability" imposed on us as a judgmental optimization of that issue by higher authorities. I will only comment on the latter by admitting I felt, on the basis of my experience elsewhere, including my own irrigated farm, that the prescribed definition of reliability was far from optimal. After a lot of discussion and computational experience with the results (evaluation phase), I now consider that this judgmental optimization is going to prove to be very close to the proper conclusion for many if not most conditions in India.

5. THE DYNAMIC PROGRAMMING APPROACH AS A MAJOR SIMPLIFICATION OF MULTIPLE OBJECTIVE OPTIMIZATION

Bellman's concepts, which he expressed in dynamic programming, have had another very significant impact on our work in India. Although the stated "objectives" of water development in India are heavily flavored with references to economic development, famine prevention, and employment, there are in fact hundreds of objectives, in the sense of consequences to be enhanced or avoided. Thus any significant planning effort would or should involve multiple-objective optimization considerations.

The problem was that few if any of the expressed or implied objectives could be measured with any degree of quantitative confidence, let alone be related by mathematical functions to the specific decisions involved in water resources development. That comment includes the commonly advocated economic impact objective. One could come up with crude estimates of people affected, based on not too reliable demographic data, but for all practical purposes these basic or primary objectives were so speculative that no useful purpose could possibly be served by their inclusion in any mathematical optimization model whether a MOM or a site specific formulation.

Again Bellman to the rescue! By utilizing the above guiding principles for the formulation of our problem in terms of mathematical models, we were able to perform a satisfactory "pareto optimum" analysis to obtain the non-inferior set using only four "objective indices" none of which would be considered as a basic (underlying) objective. These were (1) quantity of water (properly regulated in time and space), (2) capacity of the reservoir system, (3) reserve capacity for flood peak mitigation and (4) hydroelectric energy, also properly regulated in time and place [3].

This particular set has some very important properties. First given the technologically required time-place distributions of water and energy and their corresponding (but different) reliability requirements, the impact of any particular non-inferior set on virtually all other "objectives" can be estimated directly from the magnitudes of the elements of that set, with the assurance that it is the "best" that can be done without degrading one or more other objectives. Second, the results demonstrate the marginal effects of those technological time, space, reliability constraints, thus allowing an evaluation of the desirability of making reasonable modifications thereof. Note that for reasons described earlier these adjustments must be long term modifications. While they can (and should) be considered as potential decision variables, they must in fact be parametrically adjusted and only in terms of long term beneficial effects. When this is done, the reoptimization of the system is a matter of a few minutes of computer analysis, plus the usual post analysis evaluation.

Finally, from the above it would seem clear that these four decision parameters constitute a relatively simple framework for multiple objective analysis, each element of which has physical significance and is readily modeled using the mathematics of cardinal numbers. All other objectives can be evaluated, either quantitatively or qualitatively, as necessary, in terms of these four objective indices.

One last remark should be made about the use of dynamic programming concepts. In addition to the assertions above, there are significant equity considerations involved in the siting and sizing of individual reservoirs, aqueducts and other facilities constituting the water resources system. In our studies we used a very efficient "hand calculation" dynamic programming procedure (SIMPDYNPRO) [4] which I developed for the Honduras study. With this procedure, after a little experience, a ten stage problem can be solved manually in a few hours. Most of the arithmetic can be mental supplemented by a hand calculator. The simple tables created in the process clearly revealed what Bellman referred to as "second best, third best," etc., options, thus allowing on the spot trade-off concessions between purely qualitative objectives and the quantitative objectives being used in the analysis. These same tables are generated by most D.P. algorithms and require only a print statement to make them available. Normally these qualitative objectives are related to specific elements of the facilities provided, hence this "multiple objective" trade-off analysis can be readily and accurately made at that point (stage) of the analysis.

Most computer programs written for D.P. do not automatically print out these intermediate computational results. However, those details have proved to be invaluable to us and, in practice, the additional computer printout cost is negligible--essentially one table for each "stage" of the combinatorial process. Our intention (not yet implemented) is to provide this capability in our D.P. subroutines at Colorado State University.

6. SUMMARY

It has not been possible, within the limitations of this paper, to touch all the things that could and perhaps should be said. Since a number of different ideas were involved, a brief summary of the points I have tried to make, not necessarily in order, should be useful. First, I tried to call attention to a serious gap existing between research and practical application of systems engineering, and argued that the heart of this gap lies in the problem formulation and analysis evaluation phases of the planning process. Second, I argued that capability (skill) in these two deficient areas can be learned but probably cannot be taught. Third, I tried to show how the ideas and concepts of dynamic programming have proved to be more useful to me in the practice of systems engineering than have the formalisms of D.P. or any other optimization technique. Fourth, I argued that those same concepts can be utilized as guidelines for effective and accurate problem formulation, in that they constitute a practical basis for simplification of the problem to be formulated, using the well defined functional (purpose) requirements for each component subsystem or facility. Finally I argued that the primary computational result need only be the non-inferior set of four basic objective indices from which all other objective impacts may be evaluated for any non-inferior set.

My overall summary is a simple statement that Bellman, through his concepts and principles of dynamic programming, has had and will continue to have far greater impacts on water resources planning and management than might be anticipated from the purely mathematical processes involved.

REFERENCES

1. W.A. Hall and N. Buras, The Dynamic Programming Approach to Water-Resources Development, Journal of Geophysical Research, 66:(2):517-520 (1961).
2. W.A. Hall and R.W. Shephard, Optimum Operations for Planning of a Complex Water Resources System, University of California Water Resources Center, Contribution No. 122 (1967).
3. W.A. Hall, Systems Engineering for Integrated Development of Water Resources in India, Systems Engineering Unit, Central Water Commission, Government of India, Final report on UNDP project IND/80/006 (1987).
4. W.A. Hall, SIMPDYNPRO, Department of Civil Engineering, Colorado State University, CE646 class notes (1978).

A TAXONOMIC TREATMENT OF DYNAMIC PROGRAMMING MODELS OF
WATER RESOURCES SYSTEMS

AUGUSTINE O. ESOGBUE
School of Industrial and Systems Engineering
Georgia Institute of Technology
Atlanta, Georgia 30332-0205, U.S.A.

ABSTRACT

This paper presents an overview of the various aspects of dynamic programming and the problem areas of water resources. The models which have appeared in the literature addressing the use of this problem solving tool in water resources systems are then reviewed and grouped using a taxonomic scheme developed for its analysis. This is summarized in tabular form and followed by a discussion of research needs and an extensive list of references.

1. INTRODUCTION: PROBLEM STATEMENT.

In discussing the origins and interconnections of dynamic programming and water resources, we pointed out that the two fields have contributed to the progress in the state of knowledge of each other. We observed, for example, that the water resources systems field has been a testing ground for some techniques and models of dynamic programming. Additionally, some of the problems encountered in solving water resources problems have led to the development of aspects of theory and computational solution of dynamic programming. For the field of water resources systems engineering and analysis, dynamic programming has enabled problems of certain structure and complexity to be handled in areas where other systems and optimization techniques had proven ineffective. This therefore has led to advances in our ability to analyze optimally the usually complex and multifaceted systems problems in water resources.

It was also stated that dynamic programming has not only become one of the most popular techniques from the standpoint of frequency of use but also from the perspective of the variety of problems addressed. The number and growth is staggering. In this paper, we wish to examine more critically and in depth our foregoing observations. We survey these applications not only with respect to their origins and genesis but considering their coverage. In other words, we seek to identify the types of water resources problems addressed, the dynamic programming techniques used - both the model and its computational solution, and the degree to which the proposed solution approach is appropriate or usable.

Finally, we hope to identify the trends and the manner in which authors have been influenced by one another. By developing a taxonomy, we hope to unify these sundry models and treat them in one framework. This could serve as an important reference source for users and future developers of dynamic programming based water resources models. Clearly, this is an ambitious task as was learned by Yakowitz [216]. Our work is a realization of an effort we reported in 1975, aided by Yakowitz's excellent review, and the continued explosive developments in the field to date.

The plan of the paper is as follows. We begin in Section 2 with a survey of the field of dynamic programming, both the theory and the computational approaches for realizing solutions to the functional equation. Clearly, it is impossible to present a complete treatment of dynamic programming in this paper. While we attempt to cover as many significant topics as possible, our focus is only on those that have impacted the field of water resources considerably. In Section 3, we identify and classify the various water resources problems discussed in the field. We next present a survey of the models of the use of dynamic programming in treating the identified problems. This is given in Section 4, the heart of the paper, as its length clearly shows. In Section 5, we present a scheme for a taxonomic classification of the issues encountered in the papers. The paper is concluded with a synthesis of our observations and findings including a summary of research needs as well as an extensive list of references. The reader is now fully prepared for the detailed subject and technique oriented models which appear in the remaining parts of this book.

2. DYNAMIC PROGRAMMING: AN OVERVIEW

An incisive analysis of a dynamic programming model is best facilitated by examining certain aspects of its structure. One way to consider the structure of a given formulation is to specify the following characteristics of the state space of the system under study. That is, a model could be classified as either discrete or continuous, deterministic or stochastic, and serial or nonserial.

A permutation of these characteristics then yields the following eight distinct structures which will be used as the leitmotif for our classification:

i) Discrete Deterministic Nonserial (DDN), ii) Discrete Deterministic Serial (DDS), iii) Continuous Deterministic Nonserial (CDN), iv) Continuous Deterministic Serial (CDS), (v) Discrete Nonserial (DSN), vi) Discrete Stochastic Serial (DSS), vii) Continuous Stochastic Nonserial (CSN), viii) Continuous Stochastic Serial (CSS).

We will now briefly describe some fundamental aspects of these structures.

2.1 Deterministic Dynamic Programming

In general, the problem addressed is as follows: Find a sequence of decisions, $\{q_1,\ldots,q_N\}$ such that an objective function G_N is optimized. Stated mathematically, we have:

$$\text{Optimize } G_N = g(p_0, p_1,\ldots,p_N; q_1,q_2,\ldots,q_N) \tag{1}$$

$$\text{subject to } p_{k-1} = T(p_k,q_k), \quad k = 1,\ldots,N \tag{2}$$

$$\text{and} \quad p_k \in P_k \tag{3}$$

$$q_k \in q(p_k,k) \tag{4}$$

Equation (2) assumes a stationary process. Recognizing that the optimum G_N depends on the initial state p and on the number of transformations over N stages, we define the following functional

$$f_N(p) = \underset{\{q_k\}}{\text{opt}} [G_N] \quad , \quad k = 1,\ldots,N \tag{5}$$

Invoking Bellman's Principle of Optimality, this optimization problem involving N stage variables can be imbedded within a family of N subproblems of varying initial states p (usually low dimensional, perhaps one) and of varying stages k, k = 1, 2,...,N. This form of imbedding is done in terms of the state variable. The resultant functional equation in terms of $f_k(p)$, a mathematical transliteration of the principle of optimality, is:

$$f_k(p) = \underset{q_k}{\text{opt}} \{g_k(p_k,q_k) \; 0 \; f_{k-1}[T(p_k,q_k)]\} \tag{6}$$

$$\text{for } k = 2,\ldots,N$$

with

$$f_1(p_1) = \underset{q_1}{\text{opt}} \{g_1(p_1,q_1)\} \tag{7}$$

subject to (2), (3) and (4) where "0" is a binary isotonic operator. For example if $G_N = \sum_{k=0}^{N} g(p_k,q_k)$, then, $0 = +$ as in allocation problems. If on the other hand, the problem is such that $G_N = \prod_{k=0}^{N} g(p_k,q_k)$ then $0 = \times$ as in reliability problems.

The formulation given in 6(a) and 6(b) implies that we can observe the system and make decisions only at a discrete set of stages (time or any epoch in a process which permits decision making), hence it is called a discrete system. Continuous systems, on the other hand, change continually over time and perhaps are defined over a continuous state space. In such cases, the transformation (transition) equation given by (2) takes the following form (substituting t for k):

$$\dot{p}(t) = h[t,p(t),q(t)] \tag{8}$$

and summation given in (7) will take the integral form, with $\dot{p}(.)$ the first derivative. Continuous systems can be transformed into discrete systems assuming that the decision variables are piecewise constant in time. This transformation is often desirable to avoid the burden of dealing with the existence and uniqueness questions usually connected with continuous processes [17]. We note that, in turn, discrete processes may be converted to continuous versions if (rarely) necessary.

So far, we have discussed systems with serial structures, in other words, processes whose inputs to any stage are outputs from only one stage, usually the preceding one. Similarly, the output of any stage is the input to only one stage, the next one. A nonserial structure, on the other hand, represents systems where at least one stage in the system either receives inputs from more than one stage or sends outputs to more than one stage [68]. More formally, nonserial dynamic programming, especially of the type considered in [23], is concerned with optimizing a function of the following type:

$$\text{Optimize: } \Phi(x) = \text{opt.} \sum_{k \in K} \xi_k(x^k), \tag{9}$$

where $x = \{x_1, x_2, \ldots, x_n\}$ is a set of discrete variables and $k = \{1, 2, \ldots, n-1\}$. If $x^k = \{x_k, x_{k+1}\}$ for all k then the problem has a serial structure. However, if the set $x_i = 1, 2, \ldots, n$ is specified as for example by $x^k \subset x$ then the problem is termed a nonserial optimization problem.

Classically, four basic nonserial systems are considered [163]. These are (a) diverging branch, (b) converging branch, (c) feedforward loop, and (d) feedback loop systems. Some more complex systems may be obtained by various combinations of these four basic nonserial systems. These are the subject of ongoing inquiry in the current literature [65]. Of particular interest are the high level computing algorithms of Esogbue and Warsi [69] and methods for processing multi branch as well as complex loop systems discussed by Lee and Esogbue [127]. An application of the resultant algorithms to complex water resources conveyance networks is given in this volume.

2.2 Stochastic Dynamic Programming

Clearly, not all processes are deterministic or can be satisfactorily approximated by deterministic models. In certain situations, the effect of the state transition equation is not known with certainty as in the deterministic case. In other words, instead of a known output state and returns, we have a distribution of states and/or returns. Decision making under such scenarios is categorized as <u>risk</u> if the probabilities of the occurrences are known, <u>uncertainty</u>, if these probabilities are unknown or cannot be determined; and if there exists an adversary, opponent, or competitor to take advantage of our decision, a <u>game</u> theoretic situation results. If the variables are imprecise or vague and the source of uncertainty not merely due to randomness of a probabilistic nature, then a <u>fuzzy</u> decision making situation arises.

The cases of risk and uncertainty including adaptive decision making are usually treated under the term stochastic dynamic programming. Here, the measure of return is usually in terms of the expected value, or in terms of the probability of achieving at least a certain level of return. Both of these measures possess certain invariant properties as far as multistage decision making is concerned with the result that the structure of the resultant functional equation is greatly simplified.

To treat stochastics induced by risk, and suppose that the uncertainty or indeterminateness is approximated by a random variable ω, then given a process represented by the following:

$$G_k = g_k(p_k, q_k, \omega_k) \quad \text{with } \omega_k \text{ a random variable and the} \tag{10}$$

state transition equation

$$p_{k-1} = \tau(p_k, q_k, \omega_k) \quad k = 1, 2, \ldots, N \tag{11}$$

and $\omega_1, \omega_2, \ldots, \omega_N$ independently distributed with probabilities $\Omega_1(\omega_1), \ldots, \Omega_N(\omega_N)$ respectively, then the total revenue

$$G_N(p_n, \ldots, p_1; q_N, \ldots, q_1; \omega_N, \ldots, \omega_1) = \sum_{k=1}^{N} g_k(p_k, q_k, \omega_k) \tag{12}$$

and

$$f_k(p_k) = \underset{q_k}{\text{opt}} \{\sum_{\omega_k} \Omega_k(\omega_k) \, \Phi_k(p_k, q_k, \omega_k)\} \tag{13}$$

where

$$\Phi_k(p_k, q_k, \omega_k) = g_k(p_k, q_k, \omega_k) + f_{k-1}(\tau_k(p_k, q_k, \omega_k)), \tag{14}$$
$$2 \leq k \leq N$$

$$\Phi_1(p_1, q_k, \omega_1) = g_1(p_1, q_1, \omega_1) \tag{15}$$

An important type of stochastic dynamic programming is the class known as stationary dynamic programs in which the transitions and returns are governed by finite Markov chains. We assume that the state variable p and decision variable q are defined as follows:

$$\vec{P} = \{1, 2, \ldots, i, \ldots, M\}$$

$$\vec{Q} = \{1, 2, \ldots, k, \ldots, m\} \text{ with } \Delta i \subset Q, \, \forall \, i \, \varepsilon \, P$$

Note that $Q = \underset{p \varepsilon \vec{P}}{\cup} \Delta p$ of the possible decisions. In other words, p is an integer between 1 and M and q is one between 1 and m belonging to Δi. When at the beginning of a period the system is in state i and decision $q = k_i$ is taken, the distribution of probability of state p' at the end of the period is given by

$$dH(j/i,k) = p_{ij}^{(k)}, \, j \, \varepsilon \, \vec{P} \tag{16}$$

and $\quad dH(p'/i,k) = 0, \, p' \notin \vec{P} \tag{17}$

Suppose that we assume that the return $g_k(.)$ can be expressed as $r_{ij}^{(k)} = v(i,k,j)$, then by the principle of optimality, we have

For n = 1

$$f_1(i) = \underset{k \varepsilon \Delta i}{\text{opt}} \, \Phi(i,k) \tag{18}$$

For n > i

$$f_n(i) = \underset{k \varepsilon \Delta i}{\text{opt}} [\Phi(i,k) + \sum_{j=1}^{M} f_{n-1}(j) \, p_{ij}^{(k)}] \tag{19}$$

with

$$0 \le p_{ij}^{(k)} \le 1, \quad \forall\; i,j,k \quad \text{and} \quad \sum_{j=1}^{M} p_{ij}^{(k)} = 1, \quad \forall\; i,k \qquad (20)$$

where

$$\Phi(i,k) = \sum_{j=1}^{M} p_{ij}^{(k)} r_{ij}^{(k)} \qquad (21)$$

This is expressed by saying that $\forall\; i \in \vec{P}$ the vectors

$$[p_{ij}^{(k)}] = [p_{i,1}^{(k)}, p_{i,2}^{(k)}, \ldots, p_{i,M}^{(k)}] \qquad (22)$$

are stochastic vectors. The decision functions $\vec{q}(p)$ are then represented as decision vectors $\{k\} = \{k_1, k_2, \ldots, k_m\}^T$ in which the decision k_i to be taken when the state at the beginning of the period is i. A strategy is therefore a sequence of decision vectors. For a given strategy, the system is ruled by a finite Markovian chain. Because the sets $\vec{P}$ and $\vec{Q}$ have a finite number of elements, the immediate return $\Phi(p,q)$ is bounded and at the same time possesses a maximum. A well known method for solving problems of this genre, including discounted and non-discounted models, is the so called Howard's algorithm [107]. Further discussions of these models and algorithms are found in Bertsekas [24] and Heyman and Sobel [99].

Another class of stochastic dynamic programs of interest is that termed decision making under complete ignorance or uncertainty. Various methods for its treatment abound depending on the accepted criterion of performance. It is known that different criteria can lead to different policies. Among the criteria usually considered are: maximizing the minimum possible return, Hurwitz α which weights the degrees of optimism and pessimism, Savage's minimization of maximum regret, etc. Stochastic dynamic programs of the game theoretic or conflict formalism while interesting in their own right have not been reported widely in the water resources literature and are therefore omitted here.

A yet another group of stochastic dynamic programs is discussed under the title of adaptive control processes. This is a probabilistic decision process where the uncertainty is usually about the value of one or more constants defining the problem. Here, a decision making problem under uncertainty is transformed or adapted into one of risk by estimating the probabilities from the information derived from the observation of the performance of the process. Since this is done sequentially and learning is usually involved, dynamic programming becomes a natural optimization approach. An excellent introduction to the subject is provided by Bellman and Kalaba [19] while a good report of its application to water resources can be found in Stedinger et al. [193].

To represent another form of uncertainty in stochastic systems due not to randomness but imprecision or vagueness in the decision maker's state of knowledge, the theory of fuzzy sets [220,221] may be invoked. A central question in the modeling and control of any system is that of representativeness and fidelity. For example, we traditionally make the tacit assumption that a system and its elements, attributes and interconnections can be observed, measured, described and controlled with precision. Clearly, this is not true in reality.

Fuzzy dynamic programming, therefore, is an attempt to fuzzify the elements of a control process such as the alternatives and decisions, the constraints, the performance function as well as the transition function to reflect the different degrees in our abilities to represent them. These fuzzy variables and the functional are represented in terms of their membership functions which give the degrees of belongingness to their fuzzy sets.

The literature of fuzzy dynamic programming begins with the seminal paper by Bellman and Zadeh [21]. A comprehensive review of the subject and its extensions is given in Esogbue and Bellman [66] which contains various references of its applications to water resources and other systems.

2.3. Computational Procedures for Dynamic Programming

Numerical solution of the functional equation, usually via the predictor-corrector algorithm, is normally routine for problems involving one or two state variables. For high dimensional problems, however, the computational requirements (space and time complexity) are taxing. The principal problem, usually referred to as the curse of dimensionality, has been responsible for limited use of dynamic programming in water resources systems. To circumvent or mitigate these problems, numerous computational procedures or algorithms have emerged while others are being developed. Although time complexity is an important consideration, the critical constraint revolves around the space complexity or the usually exponential growth in high speed memory requirement with the number of state variables. The thrust of various algorithms is therefore to minimize this requirement.

A discussion of the comparative effectiveness of these methods in solving problems of different structures and sizes is beyond the scope of this paper. However, a general discussion of the computational aspects is provided by Morin in a volume edited by Puterman [168]. In general, however, the solution procedures available can be grouped into several categories, based on their efficiency in handling the problems in the realm of each of the following types: (i) deterministic, (2) stochastic, (3) non-serial. Note that because of the extra difficulties (as mentioned earlier) caused by continuous systems, they are most often approximated by discrete systems using piecewise linearization. Hence, we concentrate on methods for discrete systems.

Some of the classical approximation techniques reported in the literature for resolving aspects of the storage problems are: approximation in policy space, approximation in state space and approximation in function space. All of these are based on the method of successive approximation. Other approaches include Durling's Method of the Tube, Larson's State Increment Dynamic Programming, Differential Dynamic Programming, Morin and Esogbue's Imbedded State Space Approach, and Collins' Diagonal Decomposition Method. The concepts behind these approaches will next be briefly reviewed.

2.3.1. Computational Procedures for Deterministic Problems

(a) <u>Conventional DP (CDP [19])</u>:

At any stage, $k \in N$, given the quantized values of $f_{k-1}(\cdot), p_k, q_k$ at a finite number of grid points, it is required to compute $f_k(p)$ from Equation 6. This can be done in two steps using the predictor-corrector algorithm. First, the optimum $f_k(p)$ is determined by backward solution of the functional equation and by straightforward comparison over discretized points of q_k and stored on each of the predetermined grid points. This is the predictor move. Then, in the second step, the corrector move or reconstruction phase, the state transition equation is solved and any other values of $f_{k-1}(\cdot)$ not precomputed at previously identified grid points are obtained by an appropriate interpolation or extrapolation scheme, to improve $f_k(p)$. Realization of this algorithm via the digital computer is usually accomplished via a computer program consisting of three nested DO loops.

(b) <u>State Increment DP (SIDP [124])</u>:

This method is sometimes more aptly referred to as dynamic programming with controlled time increments. SIDP, like the CDP, is based on iterative application of Bellman's principle of optimality. However, instead of using a fixed time interval (stage) as in CDP, SIDP determines the time interval as the minimum time required for at least one of the state variables to change by one increment. This choice of interval ensures that the next state for any given control lies within a small neighborhood of the point at which control is applied. Another important aspect of this approach relates to the transfer of large amounts of data from low-speed to high speed memory (10^{-3} seconds).

The computations are carried out in blocks. Blocks are processed in an order which exploits knowledge of the preferred direction of motion. In general, the characteristics of a given problem situation will give a clear indication of this direction. Lack of this knowledge usually results in increased computational effort. In general, however, it can be demonstrated that the computational burden and memory requirement of this algorithm does not depend on the state dimension, thus making it a competitive method for reducing the so called curse of dimensionality.

(c) <u>Discrete Differential DP (DDDP [109])</u>:

This is basically a successive approximation method with a computational orientation to nonlinear dynamical systems, hence its appeal to water resource systems. The method starts with an initial guess of the optimal values of the state variables. Then a small band of values on either side of this initial guess is used as the state space of the problem. Applying the recursive equation over this reduced state space, if locally improved solution is obtained, the procedure is repeated with the new solution replacing the initial solution. This iterative method stops when the new solution does not improve the objective function, hence a local optimum is usually reached. The method therefore guarantees a global solution only when a problem's structure is such that any local solution is also a global solution. The utility is thus dependent on our ability to determine this structure.

(d) <u>Diagonal Decomposition Method (DDM [41])</u>:

This is another successive approximation technique in which the problem of solving a functional equation with a domain space of dimension N is converted to the iterative solution of N-functional equations with a domain space of dimension one. The underlying idea behind this method is the fact that a matrix A can be decomposed into its diagonal and off-diagonal parts. This decoupling procedure, applicable to state and decision equations, transforms the N dimensional optimization problem into N one-dimensional optimization problems.

(e) <u>Functional Decomposition Method [176]</u>:

The computative product of a contraction mapping leads to a nested sequence of multi-stage processes. This process can be considered as dynamic programming processes within dynamic programming processes. This forms the basis for decomposing the functional equation into a series of one dimensional functional equations which can then be solved routinely. The objective of this approach is therefore akin to that of the diagonal decomposition approach described earlier.

(f) <u>Durling's Concept of a Tube (COT) [56])</u>: The basic concept of storing the optimal value function in fast access memory is replaced by representing a function of several variables by a series of polynomials. The function at grid points is reconstructed from the stored values of the coefficients of the approximating series orthogonal function. The trajectory approximation procedure is initiated by constructing an approximate feasible trajectory, over coarse grid points.

A tube is constructed enclosing the neighborhood region of this trajectory, and using CPD, the new trajectory is evaluated over the region with finer grid points bounded by the tube. If calculations show that an optimal point lies outside the tube, that point within the objective function is chosen instead as the best approximation to the optimal path. The procedure stops when an approximate trajectory is found which does not violate boundaries.

(g) <u>Imbedded State Space Dynamic Programming (ISSDP [159])</u>:

By exploiting the discontinuity properties of the maximum convolution, the usual M dimensional search over the entire state space can be reduced to a one-dimensional search over an imbedded state space. This approach suggested by Morin and Esogbue successfully mitigates the curse of dimensionality in many cases. In particular, when the return functions are discontinuous or can be transformed into step functions, then the return $g(p,q)$ obtained by starting in p and using policy q as well as the optimal return $f_k(p)$ can also be shown to be step functions of their arguments. This implies that calculation of the recurrence relation of the resultant functional equation reduces to a process of simply updating a list of efficient points and a list merging operation.

This algorithm developed in connection with a capacity expansion problem involving water resources projects has been applied to a wide variety of problems such as the separable nonlinear multidimensional knapsack problem and generalized combinatorial optimization problems.

The method not only performs very efficiently with regards to core
storage and computational time requirements but also obtains correct
solutions to problems where brute force application of "conventional"
dynamic programming will yield incorrect results.

(h) <u>Other Techniques with Reduced Computational Requirements [18]</u>:

There are a number of other solution procedures for dynamic
programming formulations, but because of space constraints we will not
review them here. These include quasilinearization [20], Lagrange
multipliers [18], approximation in function space and in policy space
[18] terminal state dynamic programming [41,42] and the continuous
version of differential dynamic programming [109]. Although the
differential dynamic programming approach is very efficient possessing
quadratic convergence and has the capability to circumvent the
limitations due to the curse of dimensionality, it is in general
difficult to implement. Further, it requires certain differentiability
properties which may not be guaranteed in certain situations.

2.3.2. Computational Procedures for Stochastic Systems

One of the many attractions of dynamic programming is that the
functional equation formulations for both deterministic and stochastic
processes are very similar in structure. Bellman's "approximation in
function space" and "approximation in policy space techniques" can be
used for solving most stochastic dynamic programming problems. To
discuss solution of stochastic dynamic programs it is fruitful to
consider specific structures.

A method developed particularly for stochastic dynamic programming
with finite Markov chains structure and which has been used in analyzing
water resources systems is Howard's algorithm. A popular version is
known as the policy iteration algorithm or the iteration in policy space
procedure (IPS) [107]. We may consider the discounted and non-discounted
cases, but for illustrative purposes let us consider the problem of
maximizing the discounted total returns over a finite time or infinite
time horizon. This procedure, which is related to linear programming, is
composed of two parts. The first is a "value determination operation,"
where for a given policy k, the value of the "gain" is determined. Using
this "gain" a new optimal control function is found in the second step,
called the "policy improvement routine." This new policy is sent to the
first step and so on. The procedure stops when there is no strict
improvement in the value ("gain"). The algorithm is quite simple and
elegant with convergence to the optimal strategy being fairly rapid. A
schematic representation of this algorithm is given in Figure 1. As can
be seen, the solution reduces to solving sets of linear simultaneous
equations as well as subsequent comparisons with each succeeding policy
possessing a higher gain than in the previous one.

In summary, from the above, we see that a) the value determination
operation yields the g and v_i corresponding to a given choice of q_i and
p_{ij}. In other words, it yields values as a function of policy; b) the
policy improvement routine yields the p_{ij} and q_i that increase the gain

for a given set of v_1. In other words, it yields the policy as a function of the values. We note that in general one may enter the iteration in any box. If, for example, one chooses (A), one needs the initial policy while if one chooses (B), then one needs the set of returns.

Fig. 1: Schematic Representation of Howard's Algorithm

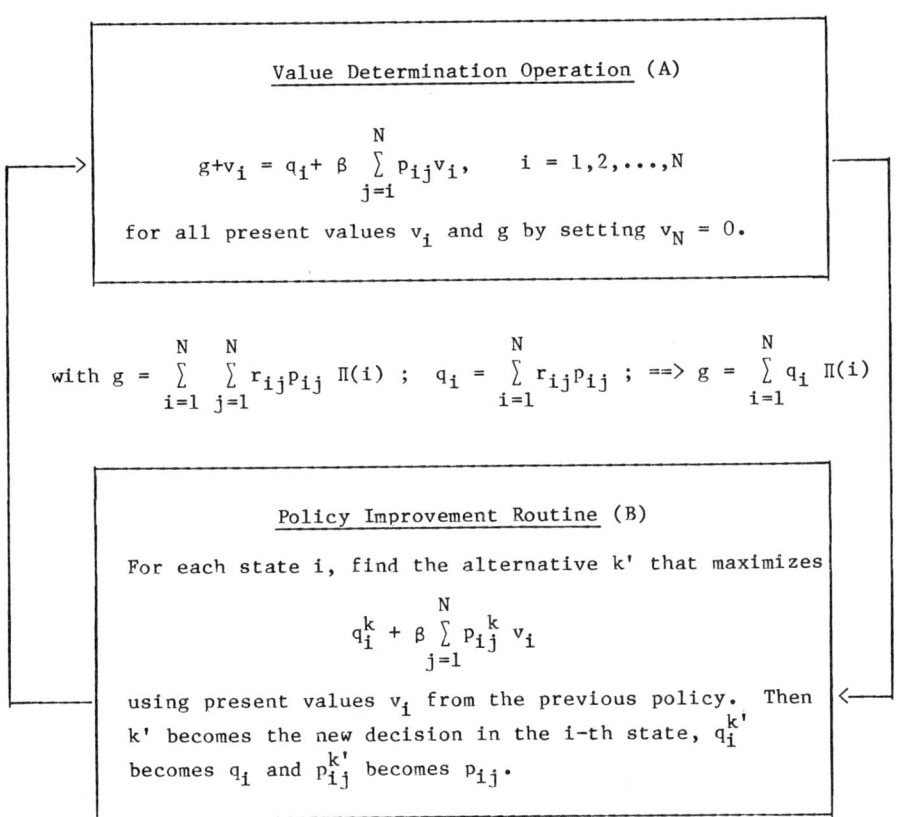

2.3.3 Computational Procedures for Nonserial Systems

Although the conventional dynamic programming algorithm can be used to solve nonserial systems from a computational feasibility standpoint, only four of the various schemes which were discussed for serial deterministic systems in Section 2.3.1 seem to be applicable. The most promising for large systems are: Collins' diagonal decomposition method (DDM), Rose's functional decomposition method (FDM), discrete differential dynamic programming (DDDP) and the imbedded state space approach (ISS).

The decomposition method (DM) is developed specifically for the nonserial dynamic programming problems, especially the type referred to as secondary optimization problem [23]. Basically, dynamic programming tries to eliminate variables in a nonserial dynamic programming problem in such a manner as to minimize the amount of storage and the number of computations required. This idea is accomplished by choosing an order of elimination of variables, by a secondary optimization problem, for which the largest degree of the eliminated variables is minimal.

The ISSDP method can be used to reduce a multidimensional state variable problem to a sequence of one dimensional problems. To minimize the storage problems which would be normally involved if the conventional algorithm were employed to process data, Esogbue and Warsi [69] report success with a special high level algorithm developed specifically for the four basic nonserial structures. Of particular interest are the computational complexity analyses provided for the algorithms developed.

3. CATEGORIZATION AND TAXONOMY OF WATER RESOURCES PROBLEMS

3.1. Generation of Problem Groups of Areas

The state of an economy is one of the principal determinants of the water related problems treated in any area or region at a given point in time. For example, in the United States during the nineteenth century, improvement of the navigable streams was the main water problem [60]. In the twentieth century, however, many other problems emerged, necessitating a reclassification. A sample of such problems include: irrigation of land in the dry areas of the West, control of floods, generation of hydroelectric power, supply of water for domestic and industrial uses, provision of recreational facilities, reduction of stream pollution from point and non-point sources, desalination, and so on. Another important determinant of the water problems considered by any community is the climate of the area under study. The role of social and political concerns such as those propagated by environmentalists cannot be overemphasized.

3.2. Categorization Scheme

In the late 1960s up to the early 1980's, the Office of Water Research and Technology of the U.S. Department of the Interior was charged with the responsibility for planning water resources development in the United States. It regularly developed a list of the main water problems in each region of the U.S. using primarily the two factors stated above. We hasten to point out that such a list or basis is not the only one possible. For example, by associating water problems with water services available or demanded in a region, the following list may be generated [12]: (a) water supply; (b) power generation; (c) low flow regulation; (d) recreation; fish and wildlife, and (e) flood control. For our purpose, however, this grouping is not complete primarily because problems pertinent to methodology and planning as well as technical problems (such as water quality) are missing. A more general and complete scheme given by the COWRT [165] is described in the following:

(i) <u>Improving Water Resources System Planning and Management Processes:</u> This will include studies delving into relationships of water resource systems with other regional and functional planning guidance subsystems, methods of analysis for predicting the effects of water programs, institutional matters, and other processes revolving around planning methods, evaluation of costs and benefits, means of cost distribution, demands for water, and laws governing water rights, etc.

(ii) <u>Control of Heated Water Discharge</u>: This category involves identifying the various consequences of thermal loading in rivers, lakes and reservoirs and the control of damaging amounts of heat in these waters as well as the ways of keeping heated water discharges within safe bounds, or limits.

(iii) <u>Control of Sediment</u>: Sediment is a major pollutant of the nation's water resources, both surface and ground water. This category includes studies of the physiologic effects of sediment on animals and plants, their relationships to waterborne pollutants, and physical problems of downstream deposition of sediment.

iv) <u>Water Quality</u>: A major component of any water resources is the quality aspect. All efforts which aid in the process of identifying pollutants, sources and fate of pollution, water quality control, effects of pollution, waste and water treatment processes, etc. are considered.

v) <u>Meeting Increased Water Supply Requirements</u>: This addresses aspects of water quantity such as demand for water and includes conservation of fresh water, increasing amounts of fresh water available, and improving the distribution of fresh water to reduce the discrepancies between gains and losses, demands and availability. Reliability and availability issues are also part of this category.

vi) <u>Mitigation of Water-Caused Damages</u>: Water is a resource which has both beneficial and destructive dimensions. Strategies for the minimization of the negative effects including structural and non-structural prevention and remedies of the effects of floods are considered.

vii) <u>Conserving Ecologic Values in Water Resources Planning</u>: This category includes studies whose purpose is the determination and recognition of danger signals, responses to emergencies, and prediction as well as safeguards for the future.

viii) <u>Metropolitan Area Water Systems</u>: This group involves studies of techniques for the coordinated development of water and sewer systems in metropolitan areas, so as to meet the quantity and quality demands created by urbanization. They possess sharper focus as well as geographical and political boundaries.

(ix) <u>Conservation of the Estuarial Water Resource</u>: The section includes problems relative to a better understanding of the complex problems and solutions centered on and created by estuaries.

(x) <u>Dissemination and Application of Knowledge</u>: Disclosure of the conclusions reached after experimentation, analysis, and interpretation. It relates mostly to the problem of knowledge transfer to users. This category is not germane to our subject matter and is thus omitted in the analysis that follows.

(xi) <u>Hydrologic Cycle</u>: This group involves the gamut of problems centered around an improved understanding of the basis of water quantity, quality, movement, and distribution so central to the harnessing and management of water resources.

(xii) <u>Engineering Structures</u>: The development of new designs for construction and operation of water resources engineering systems is the focus of this group.

(xiii) <u>Other Problems</u>: In the complex world of water resources, many important research endeavors do not comfortably fit into the first twelve of OWRT's problem areas. Although many of the topics covered in this miscellaneous area impinge upon studies covered in the preceding compartments, these topics do not conform to the taxonomy.

4. SURVEY OF THE APPLICATIONS

4.1 Reservoir Operation Studies

An overview of the dynamic programming approaches employed in deriving optimal operating policies for reservoir systems is presented in this section. The term "reservoir" will be used throughout the section, for the sake of consistency, without differentiating whether it is a man-made dam or a lake, and so forth. Reservoir operation policy, in the same general sense, will mean a time schedule of releases from reservoirs, of pumpages from aquifers and/or reservoirs, and of aquifer recharge operations [33].

The two properties of reservoir operation problems which make the dynamic programming approach particularly attractive are: its stage-wise structure and the nonlinearity of the system. The central problem may then be summarized as "whether resources available at present (water, in our context) should be used in the current period or saved for later employment" [50]. This is clearly a multistage problem with stages being the sequence of time periods. Nonlinearities, on the other hand, are of two types: economic and physical [174]. Economic nonlinearities can sometimes be handled by piece-wise linearization, as they are usually concave and univariate. The same thing can not be said, however, about physical constraints. They abound in groundwater management and hydropower output functions optimization. In the surface water case, there are nonlinearities associated with both flood control operation and hydroelectric energy production [90,174].

The foregoing characteristics are the properties which make the use of dynamic programming so attractive in addressing this problem efficiently. Our intention is not to show or claim that dynamic programming is the most efficient mathematical programming approach for this problem. There are, for example, recent claims of competitive nonlinear programming based approaches which should be explored. Discussions which compare the efficiencies of several different techniques can be found elsewhere [33,126,135,177]. We assert, however, that dynamic programming is a potent weapon for solving these problems.

Applications of dynamic programming to reservoir operation problems will be reviewed under three categories: (1) deterministic; (2) stochastic; (3) nonserial. Our classification is dynamic programming formulation rather than the nature of the problem oriented. In some studies, although the stochastic nature of the problem is realized, it is not incorporated in the dynamic programming formulation but is taken into consideration by some other means, such as using critical period hydrology input data [93,202]. Such formulations will be considered under the category of deterministic reservoir operation.

Finally, a complete review of all studies encountered in the literature is not intended here. It is beyond the scope of our current effort. Since we are also interested in the genesis of these models, we pay special attention to most of the important initial efforts. For the later ones, we review only those which are radically different from or better than the early studies.

4.1.1. Deterministic Reservoir Operation

For problems in this category, it is only natural to consider water storage in the reservoir as the state variable in the resultant dynamic programming formulations. This choice of state variable often yields very efficient solution procedures for the formulated problem if there is only one reservoir under consideration. When considering a set of reservoirs as a single system, the optimization problem becomes more complex. Increase in the number of reservoirs and hence increase in the state variables leads to a situation where the well known "curse of dimensionality" problem becomes prohibitive. To alleviate this problem, some decomposition methods are usually used. One such decomposition, a variation of the Dantzig-Wolfe approach to large scale systems optimization, leads to the adroit use of linear programming in conjunction with dynamic programming. This can be done in one of two ways: a dynamic programming-linear programming technique or a linear-programming-dynamic programming technique. In the dynamic programming-linear programming approach, dynamic programming is used to optimize the subproblems. These suboptima are then combined to form a master problem which, in turn, is solved by linear programming. The alternate approach, linear programming-dynamic programming, does just the opposite: the subproblems are solved by linear programming while the master problem is optimized by dynamic programming.

Hall, et al. [93] exploiting this decomposition method, employed the dynamic programming-linear programming technique to find the optimum water releases for firm water, dump water, peak power and off-peak power. The objective function sought to maximize the total monetary return from the sale of water supply released by the reservoirs and the hydroelectric power generated through the plants. Releases of water for fish and wildlife enhancement, recreation, navigation, previously existing water rights, salinity control, minimum and maximum releases suggested by the physical characteristics of the reservoirs, and similar downstream uses were considered as constraints. The dynamic programming-linear programming technique employed operates as follows: First, each individual reservoir system is optimized separately by dynamic programming, using a set of prices supplied by the master; then using the optimal responses (releases from each reservoir), linear programming is used to optimize the overall system operation. The solution of the dual of the latter gives a new set of shadow prices which provide a second set of input prices to be fed subsequently to the reservoirs. If these prices are the same as the ones already used by dynamic programming, the process terminates at the optimum, otherwise dynamic programming is employed again and the iteration goes on. Hydrologic risks are addressed by making use of the so called "critical period hydrology" concept. This study generated an optimal monthly reservoir operation plan for the system of reservoirs in a region. The dynamic programming model and routine is reported by Hall et al. [93].

In a follow up study [14] a linear programming-dynamic programming technique is employed ala Hall et al. to operate a multipurpose multi reservoir system so as to minimize the potential energy loss resulting from the releases of stored water for the satisfaction of the various uses. Storage reservations for flood control, minimum storages for recreational use and mandatory releases for irrigation, navigation, fish and wildlife survival and contracted hydroelectric energy generation are the constraints imposed on the system. A linear programming formulation is used to determine the optimal reservoir releases and ending storage vectors for a period that corresponds to all alternative paths to the feasible cumulative on-peak energy generated at the beginning of the next period. Then dynamic programming is used to select from the several alternative paths to any of the feasible energy generation points of the "next" period that path which produces the ending storage vector characterized by the greatest sum of its component storages. The state variable chosen in this dynamic programming formulation is the cumulative energy, while the decision variable is the feasible energy constraint for the period. This procedure is repeated for each period of the planning horizon. The use of the linear programming routine for each period effectively permits a period-by-period linearization of the system equations with only very minimal error. Although a large number of linear programming solutions is necessitated by this procedure, the total computation time is relatively low because of the efficiency of the linear programming codes employed and the sparseness of the coefficient matrix at each period.

Decomposition of multistate dynamic programming can be done by some approximation methods also. One such method is the successive approximation method of Bellman [17]. This method is used to determine the release policy of a multi reservoir (six) system in California [202]. The sole purpose of the system is the development of the firm water contract. Demand is considered to be concentrated at a single demand point below the last reservoir. Flood control, evaporation losses, mandatory releases are imposed as constraints. The original multiple state variable dynamic programming problem is decomposed into a series of subproblems of one state variable in such a manner that the sequence of optimizations over the subproblems converges to the solution of the original problem. Each subproblem is then solved by incremental dynamic programming. In particular, Trott and Yeh's formulation is of the forward dynamic programming variety. The main disadvantage of this method, however, is that convergence to the global optimum is, in general, not guaranteed.

The dynamic programming recursive equation is also used as a means of decomposing an N-season time operation problem consisting of a system of M reservoirs in both series and parallel connection. The problem is first formulated to consider spatial and temporal optimizations. Using storage and diversion as the decision, the complexity of the resultant

problem is shown to be 2MN. However, by means of the recursive
functional equations, this problem is decomposed into N sequentially
optimized two M-decision problems. The first optimization problem
involves the within stage operations which affect only that particular
time period. Since the river system is, in itself, a multiple stage
serial system with sequential structure, the problem is further
decomposed. The approach utilized is akin to that discussed in Sections
2.3.1.e.

One of the earliest reports of attacking a water resources problem
as nonserial branched systems are due to Meir and Beightler [150,151],
who use the branch compression principle to combine branches into the
main serial chain. This then allows the complex system to be optimized
by a one-state, one decision dynamic programming technique. For every
combination of initial and final storage levels the diversion is
determined so as to optimize the return. The second problem involves
spatial and temporal optimizations and requires simultaneous selection of
the decision variables. The Hooke and Jeeves pattern search technique of
nonlinear programming [126] is used to solve this problem.

A multi variable polynomial function is used to approximate the
functional equation, to avoid the burden of carrying over the information
from one stage to another in tabular form. This approach is applied to a
system of five lake activity points and three on-stream activity points
(reservoirs). The system serves to meet water demands for irrigation,
municipal and industrial water supply, hydropower generation, navigation,
low flow augmentation, and recreation as well as being operated to provide
for flood control. Operating rules that would best utilize the available
water to satisfy the various target diversion demands and would maximize
hydropower generation as well as recreation activities, are required to
be obtained from the model.

We note that one of the earliest applications of dynamic programming
to water resources and, in particular, to a single reservoir system [87]
involved a direct assault on the problem without resort to the use of
decomposition techniques as in the foregoing models. So in a sense this
represents an example of how our ability to solve interesting water
resources problems via dynamic programming is enhanced by developments in
the computational aspects of the method. In fact, Fults and Hancock [75]
also circumvent the use of any decomposition technique in their
formulation for a five reservoir system. They only consider the joint
operation of two reservoirs and solve the problem with two state
variables using state increment dynamic programming. This again is
progress.

Another reasonably powerful method which is used efficiently to
solve multi reservoir systems without resorting to decomposition
techniques is discrete differential dynamic programming [98]. The method
and its limitations have been discussed earlier. In this application,
the authors illustrate the power of the method on a prototypical four
reservoir nonserial system but then the problem is solved serially. This
same problem originally introduced by Larson [24] to demonstrate the
computational efficiency of his state increment dynamic programming, is
also used by Nopmongcol and Askew [164] to test their multi level
incremental dynamic programming, and by Murray and Yakowitz [16] for their
constrained differential dynamic programming model.

Although most of the formulations so far have been of the serial dynamic programming variety, multi reservoir systems in reality form networks which possess a nonserial structure. Even though some of the studies [131] had recognized this fact, as was mentioned previously, one of the earliest reported works that explicitly considers the nonserial structure of the multi reservoir operation problem was given by Meier and Beightler [150]. A river basin which is composed of a main stream and one tributary is considered. Two of the five reservoirs proposed are located on the tributary stream. To decide whether to keep all the water at each reservoir or allow part of it to flow downstream, the nonserial river basin system is decomposed into equivalent serial systems amenable to analysis by dynamic programming. There is, however, almost a complete absence of the treatment of the multi reservoir operations planning or design problem via nonserial dynamic programming algorithms.

4.1.2 Stochastic Reservoir Operation

In reality, the inflows of water to reservoirs are, of course, not known deterministically. These inputs to the system can, however, be expressed as a stochastic function based on the available historical data and the problem then becomes a stochastic optimization problem. Once the problem is recognized as being stochastic, dynamic programming formulation can either be stochastic or reasonably approximated as deterministic. In the first case, the stochastic nature of the problem is handled explicitly. This approach is called explicit stochastic optimization (ESO) while the later one is referred to as implicit stochastic optimization (ISO) [43].

ESO models use probability distributions of streamflow directly, to optimize the expected value of the criterion function, rather than samples drawn in streamflow synthesis procedures. ISO approaches start with determining several equally probable time series realizations of system input, using streamflow synthesis. The optimum operating rule for each realization is, then, determined by dynamic programming. Finally, decision functions, to be used to estimate the optimum decision, are defined through regression analysis.

There are mainly two difficulties in stochastic reservoir operation models: (i) Streamflow dependencies: In a single reservoir case, inflows are not usually dependent but serially correlated. If this correlation is included in the dynamic programming formulation, inflows in the preceding month become part of the state space and hence the dimensionality of the problem increases. In a multi reservoir case this problem is aggravated by the fact that inflows to each reservoir exhibit some degree of cross-correlation between each other. Neglecting the dependencies between inflows (serial, cross or both) results in estimates of reservoir failure probability that are usually lower than those encountered in the more realistic simulation model subjected to the same operating policy [143]. (ii) Reliability: It is usually required that water demands must be satisfied with a given level of reliability. This leads to the concept of reservoir operation optimization based on the concept of a permissible probability of failure. The economic and social costs of failure of the system are, however, rarely known. This uncertainty hinders a proper representation of reservoir reliability constraints in the dynamic programming model.

One of the earliest attempts reported in the English language to solve a stochastic one reservoir operation problem via dynamic programming is the ESO model formulated by Buras [32]. Both correlation and reliability issues are omitted in the model, which maximizes expected return from the conjunctive operation of a surface reservoir and an aquifer supplying water for irrigation. Three state variables, the amount of water available in the surface reservoir, aquifer and recharge facility, are used to describe the system.

The dynamic programming model developed by Mawer and Thorn [143] for a single reservoir, is essentially similar to Buras' model. The major difference is that Buras used probabilities associated with each inflow whereas Mawer and Thorn, recognizing the Markovian nature of the process, used transition probabilities dependent on the month considered to optimize the operation problem subject to reliability constraints. A simulation model is used in conjunction with a one state variable dynamic programming model to circumvent the problem of non-explicit penalties associated with failure of the system and omission of serial correlations.

One of the earliest models based on the ESO approach to a multi reservoir system, with explicit consideration given to the dependencies for streamflows, is due to Schweig and Cole [183]. Expected values of the net benefits from two linked reservoirs are maximized. Serial correlation of inflows as well as cross correlation between the inflows to each reservoir is incorporated into the model with the assumption of very simple streamflow dependencies. A penalty for failure of the system is included in the objective function as a deficit cost. They describe a discrete stochastic dynamic programming model and apply it to a two reservoir problem around Lake Vyrnwy, Wales. Severe computational difficulties with the approach are noted in the paper.

An attempt to overcome the lack of control over system failure associated with optimum operating policies is made by Askew [7,8]. It is first shown that stochastic dynamic programming exercises no direct control over the probability of system failure, if the only economic incentive for avoiding failure in stochastic dynamic programming is the reduction in net benefits that failure causes. Then, an extra penalty is imposed that is incurred if the system fails, and optimal policies are determined for different values of this penalty. A simulation model tests the policies to provide estimates of their associated probabilities of failure and reduction in expected net benefits [7].

The elimination of the time consuming simulation component in this iterative procedure is the motivation behind the second procedure developed by Askew [8] in the so called chance constrained dynamic programming model. This model has the ability to restrict the probability of a given variable taking on values outside a fixed range. This is achieved by imposing constraints that are functions of various pertinent probabilities.

A second modification is the inclusion of a penalty function in the objective function, which somewhat represents the "socio-political" cost of failure, instead of economic cost of not being able to provide water. This approach places a bound on the expected number of failures within the demand at any time. Both procedures were applied to a single purpose single reservoir system.

The first ISO model noted in the literature is the Monte Carlo dynamic programming proposed by Young [218,219]. Several equally probable sequences of streamflows which could occur in the future are generated via streamflow synthesis. A forward-looking dynamic programming algorithm is then used to map these data on possible future streamflows into data on future storage levels and release schedules. To determine the optimal operating rule, a regression analysis model in which release is the dependent variable and storage level and previous inflow are independent variables, is utilized. This takes care of the streamflow dependencies. Reliability of the system, however, is omitted.

A comprehensive critique of this method as well as a comparison with some ESO models is given by Croley [43], who then proposes an alternate stochastic optimization (ASO) approach as a combination of ISO and ESO. This approach as stated by Croley "involves the empirical transformations of the probability distribution for inputs into that for optimum decision at each stage, similar to ESO analysis. However, instead of maximizing the expected total benefit by the selection of the decision probabilities at each stage, the total benefit is maximized for each of several possibilities for future inputs, as in ISO, to determine the empirical distribution of the optimum decision at each stage."

More recent studies are reviewed by Yakowitz [216] and Stedinger et al. [193]. In particular, they include the works of Gal [78] and Turgeon [204], who discuss the use of parametric stochastic dynamic programming, aggregation – decomposition, and aggregation stochastic dynamic programming algorithms to derive optimal operating policies for multiple reservoir systems. Gal chooses parameter vectors which approximate a quadratic return function over a substantial portion of the state space and chooses the set of controls that minimize the returns. One common thread in these models is the use of previous period's inflow as a hydrologic state variable in deriving stationary policies.

It appears that the thrust of the recent models is on how to improve the models of the reservoir inflows. Bras et al. [30], for example, considering the distribution of inflows over the next T periods and beyond to be conditioned upon all information available at the beginning of the current period, use a non-stationary adaptive control stochastic dynamic programming algorithm to optimize the operation of the reservoir for the planning horizon. They apply their model to the High Aswan Dam exploiting the large times of flow within the Nile Basin in deriving a much more accurate inflow forecast than those based on the preceding period's inflow as was done by previous authors.

The problem with the foregoing approach, however, is that because the state variable is all the available flow information vector, the dimension of the problem can be quite considerable. Stedinger et al. [193] remedy this problem by introducing an appropriate surrogate such as the minimum variance forecast of the (hydrologic state variable) reservoir's inflow during all future periods using all available information in all periods up to the present. Their model consists of two conditional expectations. The modifications introduced result in two models, one a non-stationary adaptive control model and the other a "nearly stationary 'perfect forecast' operating model." They apply their models to the High Aswan Dam and compare their performance with various other stochastic dynamic programming algorithms. The point to their work is that the use of all available inflow information in generating the hydrologic state variable leads to better reservoir operating policies than the traditional use of the preceding period's information only. However, it is noted that the the cost of implementing the algorithms, both stationary and non-stationary, may be quite prohibitive.

An interesting departure from the foregoing approaches is presented by Sobel [190], who considers a four reservoir nonserial (converging branch) system located in a river basin. It is, however, assumed as in previous models that the inflows are auto-correlated. A discrete time model with one period delay for a discharge to flow downstream to the next reservoir is presented. The model is simplified by ignoring evaporation and leakage at the reservoirs. Exploiting the affine structure of the model, Sobel presents a solution to the resultant eight state variables and four decision problem by optimizing four scalar one period problems. The solution is shown to possess a myopic optimum. Although the assumptions are somewhat restrictive, it shows that certain large scale problems can be solved even in a closed form manner by resorting to some adroit manipulation of the models.

4.1.3 Reservoir Sequencing and Sizing

An important set of decisions in water resources development is to determine the size and sequence of each project undertaken. There is a close interaction between these two categories of decisions. Not much attention, however, had been given to this interaction. Sizing decisions are often determined in a static context, while sequencing decisions are made for an already fixed size of projects.

In the first dynamic programming approach [35] to the problem of finding the minimum cost sequencing of simple independent projects with fixed scale to meet a projected demand, it is assumed that projects may be described only by their investment cost and fixed capacity. The number of projects to meet the demand for a certain time horizon is also assumed to be given. The optimum (minimum present cost) construction sequence of these given projects is determined via dynamic programming.

This formulation is modified in [159] to solve a more general problem such that the number of projects to meet the demand for a given time horizon is not necessarily known in advance. The number of independent projects is assumed to be given. A subset of these projects is to be chosen to meet the demand and then their completion times are sequenced. After showing that this can be done by a modification of the dynamic programming formulation given in [35], a more efficient dynamic programming formulation is proposed which uses the fact that an optimal schedule is a permutation schedule. Morin and Esogbue call their approach an imbedded state space dynamic programming which is developed further and generalized in [160].

The simple sequencing problem discussed in the previous paragraphs is extended to incorporate the interdependencies between the hydroelectric projects [62,63]. Two types of interdependencies are considered: technological and market. The former one results from the fact that the downstream projects will be the beneficiaries of technological external economies from upstream storage outputs. Also, since each project supplies the same commodity, electricity, to a limited market, the market interdependence among the projects is also included in the model. It is demonstrated that accommodating these more realistic interdependencies does not increase the difficulty of analysis significantly.

A more comprehensive and generalized version of the foregoing models is proposed by Becker and Yeh [13,14]. The distinction between reservoir capacity and incremental firm water output is clearly made in [13] and multiple purposes of reservoirs are included in the subsequent model [14]. Furthermore, the sequencing model is expanded to include project sizing considerations.

In another attempt to integrate the sizing and sequencing decisions [110] linear programming, dynamic programming and simulation techniques are used in conjunction with each other. Projects are first assigned sizes by linear programming and then sequenced by dynamic programming. Simulation is used to evaluate the results and adjust the sizes if any improvement can be obtained. This last phase is particularly necessary to be able to check any errors that might have been introduced by the approximations done in linear programming and dynamic programming models.

4.2 Sewer Systems

Optimization of wastewater collection network systems generally involves the solution of two problems: the (i) determination of the optimal layout; and the (ii) determination of the optimal design (diameters, elevations, slopes) for a fixed layout.

In the studies reported to date both of these problems are treated deterministically. In the first reported work which recognized sewer design as a sequential decision process, Haith [84] proposes a dynamic programming formulation to the problem. The total headloss along the sewer network is considered as a resource to be allocated to each link. The problem, therefore, is formulated as a resource allocation problem. The cost associated with each link is assumed to be a function of the amount of head used in the links, for given discharge, topography, and subsurface conditions.

This simplifying assumption is criticized by Walsh and Brown [207], who propose another formulation based on the same concept discussed above. Here, however, the arrays of unit costs for different excavation conditions, pipes, manholes, and pavement replacements are used instead of a single cost function.

On a problem formulated similar to the ones discussed above, the computational efficiency is tested via a computer program that incorporates dynamic programming optimization [153]. Merritt and Bogan report the results obtained as satisfactory. These approaches, however, are inadequate for large systems because of inefficiencies in the computational procedures used. Vast improvements in the algorithmic methods for finding the optimal design of large storm sewer systems are advanced by Mays and Yen [148]. A naturally occurring nonserial system is decomposed into serial systems and solved via two methods, dynamic programming and DDDP, for comparative purposes of their efficiencies. It is shown that DDDP is vastly superior, at least from a computational time requirements standpoint. In the paper to be found in this volume, Esogbue and Lee [67] propose the use of an efficient multi converging branch nonserial algorithm based, on dynamic programming, for the sewer problem and similar nonserial systems in water resources. The nonserial dynamic programming algorithm is shown to be much more efficient both from computational time and space complexity standpoints than any other method reported to date.

The models discussed so far assume a given preselected topography and layout. Thus, the thrust is on the optimal size, slope, and depth of soil coverage of sewers. The general assumption has been that the layout for particular network can be satisfactorily found by the judgment and experience of the engineer. The only reported work which incorporates the layout into the problem attempts to find the optimal layout, pipe diameters, elevations and slopes in a single computational procedure. This work is due to Argaman et al. [4]. They conclude, however, that the dynamic programming formulation presented for this purpose can be applied to only small sewer systems. We note, additionally, that Mays et al. [149], considering the complexity of the joint problem in one simultaneous optimization effort, propose a heuristic optimization model to be used for screening purposes. Their model consists of two conjunctive parts each of which employs DDDP as the optimization technique.

In a related subject dealing with optimal management of urban runoff particularly that due to storm water, one of the best management practices developed for mitigating the effects of the attendant pollution is the use of detention basins or ponds to drain tracts, especially in areas whose channel capacities have been taxed by urban runoff. It has been shown also that integration of detention storage with channel rectification in a central design can result in an improved cost effective approach to storm water pollution abatement over single and separate approaches.

Mays and Bediment [145] consider the problem of optimal design of detention basins by using DDDP to determine the basin size and location as well as the minimum cost. This model is improved on by Bennett [22], who extends the approach to the problem of optimal basin design including the outlet structures and receiving channels. The improvement over Mays and Bediment stems from the fact that the cost of interaction of various locations and sizes of detention basins as well as the associated outlet structures and the channels connecting the basins are considered. Additionally, the model specifies the sizes and locations of the basins, the number of outlet pipes as well as the size and characteristics of the connecting channels. It is thus a more flexible and suitable tool for design purposes. An application to the Brays Bayou Watershed in Houston, Texas is reported.

4.3 Irrigation Systems

An irrigation system has basically two subsystems of interest in modelling: irrigation water and irrigated area. The decision problems associated with an irrigation water subsystem include: (i) amount (rate) and timing of irrigation, (ii) capacity of irrigation system (canals, pumps, etc.), (iii) selection of irrigation technology and projects, and (iv) allocation of water among several agricultural areas.

To solve any of these problems, the crop pattern should have already been specified. The type of crops to be grown and the acreage of land to be allotted to each crop, in short crop pattern, then become decision problems associated with the irrigated area subsystem. The determination of the optimum pattern necessitates the identification of the crop yield - water use relations. Because these topics are quite beyond the water management subject, the studies pertinent to them are excluded in this study.

Considering the problem of determining the optimal timing of irrigation over a season in an uncertain environment as a short-run problem, Dudley et al. [55] propose a dynamic programming model for its treatment. In the absence of suitable experimental results, a plant growth-soil moisture simulation model is incorporated into a two state variable stochastic dynamic programming model to determine the optimal intraseasonal allocation pattern for irrigation water in a variable environment. The decision variable in the dynamic programming model is the level to which the available soil water content is allowed to fall before irrigating. The two state variables are soil water content and irrigation reservoir level per acre of irrigated crop.

DeLucia [45] also takes into account both stochastic crop demand (based on stochastic rainfall but deterministic evapotranspiration) and stochastic reservoir inflows when he uses dynamic programming to allocate water optimally from a hypothetical surface reservoir and underground aquifer over a season, given a specific area of a single crop. The result is a three state variable and two decision variable model. His model consists of a number of simplifying assumptions such as that only one irrigated crop is considered.

An interseasonal as well as intraseasonal irrigation system is handled in an integrative formulation via a dynamic programming model by Dudley and Burt [53]. The objective of their model is to determine an optimal decision rule with respect to the following three classes of crop irrigation decisions: (1) intertemporal water application rates, (2) whether or not some acreage should be abandoned from further irrigation for the remainder of the season, and (3) the optimal acreage to plant for potential irrigation at the beginning of the season. The solution of the problem is said to be a basis for optimizing the levels of three design variables, namely, developed irrigation acreage, reservoir capacity, and distribution system capacity.

In a related study, irrigation water quality is incorporated into any economic analysis of irrigation with saline water. A Markovian dynamic programming model is developed which can generally be applied to the optimal analysis of alternative irrigation decisions in regard to quantity and quality of irrigation water at any time of the year. The decision variables in the analysis presented are restricted to quantity of water used for leaching, with the quality of water considered an exogenous variable.

There are a number of studies which formulate a reservoir operation problem considering irrigation in a dynamic programming model. Such models have already been discussed in Section 4.1 dealing with "Reservoir Operation" and will therefore not be repeated here.

4.4 Hydropower Systems

The problem of development of the hydroelectric potential of a river basin, with the objective of meeting power requirements projected over time at minimum discounted cost has been analyzed by means of dynamic programming. For a given set of hydropower plants proposed, with the annual cost and the capacity of each of them, a dynamic programming formulation is suggested by Kuiper and Ortolano [120] to determine the optimum sequence of installation. The states used in the formulation correspond to alternative possible combinations in each stage (time). If all the possible levels of capacities for each plant had been considered, the size of the problem would have been astronomical. To avoid this, it is assumed that hydro plants are either built at their full levels of development or not built at all. The model formulated is then solved by a conventional dynamic programming algorithm. A simulation model is used to find the optimum operation of the hydro power system scheduled.

Hydroelectric projects within the same river basin tend to have two types of interdependencies: technological and market. Technological interdependence is the result of technological external economies created by an upstream storage capacity for downstream sites. Market interdependence arises since each project supplies the same commodity to a limited market. A dynamic programming formulation, very similar to the one discussed in the foregoing, is presented in such a way as to accommodate these interdependencies [63].

The problem of sequencing and operation of multi purpose reservoirs, with one purpose being hydroelectric energy production, has already been discussed under the "Reservoir Operation" and "Reservoir Sequencing and Sizing" sections, and hence will not be repeated here.

4.5 Water Quality Problems

Aerators can be used to raise the level of dissolved oxygen in a polluted stream through in-stream artificial aeration augmentation. A discretized dynamic programming model is formulated for finding the optimal allocation of artificial aeration along a polluted stream [36].

Recently, Sugiyama [195] has proposed an efficient and readily implementable dynamic programming model of a version of the foregoing problem. The particular problem considered is the control of the aeration rate at a very short local portion of a polluted river of a slow stream inside the city of Osaka in Japan. It is formulated as an optimal control problem governed by a set of differential equations with a view to decreasing the biochemical oxygen demand (BOD) and increasing the dissolved oxygen (DO) simultaneously. The author reduces dimensionality by solving the resultant functional equation via a Box Jenkin's hill climbing method. The implementation of his algorithm on real life sized problems using a relatively inexpensive device such as a hand held programmable calculator is reported.

4.6 Water Treatment:

A problem in a municipal water treatment system is to determine the sizes of new treatment plants and the times at which these new plants are added to the system. This is a form of capacity expansion planning

problem. Hinomoto [101] has formulated a dynamic programming model for
this problem. The objective is to determine the optimum sizes and
installation times of the new plants.

4.7 Urban Water Resources:

Water quality and quantity are closely related to each other. Modern
methods for managing urban water resources call for an adroit integration
of both. Such an integrated approach based on dynamic programming is
proposed by Shih and Meier [148]. The major decision variables are the
degree of waste treatment required at each waste treatment plant and the
release required at each reservoir to maintain acceptable quality levels.

4.8 Transbasin Diversion:

When water supplies of a river basin are drawn on for importation
into other regions, the value of services provided by the river to the
economy of the exporting region may, in general, be considered to
diminish. Moncur [153] proposes a method using dynamic programming in
conjunction with linear programming to evaluate the opportunity costs of
such a diversion. The method is based on a procedure developed
previously, to find the optimum operation of a set of reservoirs.

5. THE TAXONOMY SCHEME FOR THE PROBLEMS

The problem areas given by COWRT are rather too general for our
purpose. Here, we shall be concerned with more specific problems
addressed in the literature. We wish to develop a taxonomy scheme for
classifying them. Later, the taxonomy scheme will be merged by the COWRT
problem areas.

Because of the vast literature in water resources systems analysis
and the burgeoning growth in dynamic programming models, it is
instructive to possess a scheme for their classification. The underlying
philosophy of such a taxonomy scheme is:

a) To identify the specific problem areas in water resources as well as
the important decision factors which have been or should be
considered in available models for the solution of the specific
problems.

b) To propose a framework for classifying existing models as sub-cases
of the general models envisioned.

We begin by establishing what we mean by specific water resources
problem areas and decision factors.

5.1.a. Specific Water Resource Problem Areas:

Water resource problems can be grouped such that each group contains
problems pertinent to a water resource development project with a specific
function. Each group may then be identified by the function of the
project. A classification based on this definition is as follows:

1. Reservoir (R)
2. Irrigation (G)
3. Sewer (W)
4. Hydropower (H)
5. Quality (Q)
6. Recreation (E)
7. Flood (F)

5.1.b. <u>Decision Factors</u>:

A way of understanding this concept is to consider what type of decisions are involved in a water resource development project for either one or more functions stated above as problem areas. Decision factors are the answer to this question. Each problem area may have some decision factors peculiar to only that particular problem. However, the following factors are common to all of them:

1. Design (location, size, structural properties, etc.) (D)
2. Operations (O)
3. Investment (scheduling, sequencing, etc.) (I)
4. Stochastic (if the model includes acts of nature besides courses of actions). (S)

Additional factors peculiar to only one of the problem areas are given below:

1. Reservoir:
$$\text{Reliability } (R_1)$$

2. Irrigation:
$$\left. \begin{array}{l} \text{Water Quality } (G_1) \\ \text{Distribution System Capacity } (G_2) \end{array} \right\} \quad (G_3)$$

3. Sewer:
$$\text{Layout } (W_1)$$

The most general model which can be conceived will cover all problem areas from reservoirs to flood, and include all these decision factors involving both the general and sepcific ones. It may thus be symbolized by the acronym:

$$R_1 \; G_3 \; W_1 \; H \; Q \; E \; F \; D \; O \; I \; S$$

Less complicated models can be obtained by deleting a symbol which is not incorporated in the model. Thus, the simplest possible model will be, for example, WD, i.e. for a given layout, design of a sewer system. We note that $W_1 D \supset WD$, since $W_1 D$ includes layout also, whereas WD does not necessarily have that component in it.

5.2 <u>Taxonomy Scheme for the DP Methodology</u>

Dynamic programming formulations can basically be considered under two categories: serial (S_e) and nonserial (N_s). Each of these categories can further be divided into the following subcategories:

(A) Deterministic (D): (i) Finite (F) and ii) Infinite (I)

(B) Stochastic: (a) Markov Transition Type (M) - (i) Finite (ii) Infinite (I) and

(b) Probabilistic Systems (P) - (i) Finite (F) and (ii) Infinite (I)

Any particular model will be described by the collection of the letters given in the parenthesis. For example, the acronym, $S_e SMI$, stand for as Serial Markovian Stochastic Infinite model.

5.3 Entries of the Cells

For a given problem and dynamic programming formulation, there exists one cell. The entries of each cell describe the pertinent characteristics of the studies which fall into that cell. These characteristics and the way they are symbolized are as follows:

1. Frequency: number of the papers that fall into that cell, (n_1: a numeral)

2. Problem dimensionality: the number of state variables in the problem as it is stated originally in the paper are determined and then the maximum of these numbers is put in the cell, (n_2: a numeral)

3. Dynamic programming dimensionality: the number of state variables in the dynamic programming formulation that is solved. Again, only the maximum of them is indicated in the cell, (n_3: a numeral)

4. Computational procedure: the algorithms used to solve the dynamic programming formulations in that cell, (ℓ_1: collection of letters)

5. Other Techniques: If there is any other operations research technique used in conjunction with the dynamic programming then it is indicated (ℓ_2: collection of letters). If no such technique is used then it is indicated by "0".

6. Implementation status: It is also indicated if the studies falling into a particular cell are implemented or not (ℓ_3: letters) as reported or known from other sources consulted.

Hence, each cell could have six entries.

5.4. The Classification Matrix Notation for the Studies

The notation used for the six entries of each cell is given below (First Entry/Second Entry/Third Entry/Fourth Entry/Fifth Entry/Sixth Entry/)

First entry: a numeral, the number of papers observed in the survey.

Second entry: a numeral, the maximum dimensionality observed in the problem's natural environment.

Third entry: a numeral, the maximum dynamic programming formulation dimensionality observed in the model.

Fourth entry: a symbol for the computational procedure used. Thus, a numeral in front of the symbol is the number of times that computational procedure is used. The description of the symbols are as follows:

    CDP: Conventional Dynamic Programming
    APS: Approximation in the policy space
    AFS: Approximation in the functional space
    DDDP: Discrete Differential Dynamic Programming
    ISS: Imbedded State Space
    LM :   Lagrangian Multiplier Method
    IPS: Iteration in Policy Space
    U :   Unclear - not easily determined
    _ :   Not solved, merely formulated

Fifth entry: a symbol for the techniques used in conjunction with any or some combination of the following:

    LP: Linear Programming
    PS: Hooke's pattern search
    S :   Simulation
    O: none

Sixth entry: The paper's implementation status as far can be determined from the contents of the paper or from first hand knowledge.

    I: Implemented
    NI: Not implemented

5.5 An Overview of the Classification Scheme

A classification scheme is developed in this section. Because this paper is interested in two of the dimensions of the studies under concern namely (1) problems encountered in the planning, design and management of, water resources systems; and (2) the dynamic programming models developed to solve these problems, our classification matrix will have two dimensions. In the following, the list of the problems is first given and then the list of the dynamic programming formulation types follows. A description of the problems, based on the Office of Water Research and Technology (OWRT) report and the dynamic programming formulation types have already been given in previous sections. They are used as the row and column respectively of the matrix.

The entries of the classification matrix are derived from the literature review treated in Section 4 of this paper. Finally, the classification matrix is presented in a tabular form. (See Table 1) Some of the problems and the dynamic programming formulations discussed in Sections 2 and 3 respectively may not have been considered in the literature review of Section 4. Consequently, they are not included in the tabular representation of the classification matrix of Table 1.

TABLE 1: CLASSIFICATION MATRIX

DP FORMULATION	SERIAL						NONSERIAL		
	DETERMINISTIC		STOCHASTIC				DETERMINISTIC		STOCHASTIC
			Markov Transition Type		Probabilistic System				
WR PROBLEM AREAS	Finite	Infinite	Finite	Infinite	Finite	Infinite	Finite	Infinite	
OWRT PROBLEM AREA 1:									
(A) RESERVOIR SYSTEMS									
(1) Operation									
(i) Deterministic	7/N_s/6/4/ 3CDP,1APS,1AFS 1DDDP/LP,PS/ 11,6NI 1/S_e/1/1/CDP/ S/NI						2/N_s/U/U/ U/O/NI		
(ii) Stochastic			1/S_e/1/1/ IPS/O/NI		2/S_e/1/1/ U/S/NI				
(2) Sizing									
(i) Deterministic	1/N_s/1/1/CDP/ O/NI								
(ii) Stochastic									
(3) Location									
(i) Deterministic	1/N_s/1/1/CDP/ O/NI								
(ii) Stochastic									
(4) Sequencing & Scheduling									
(i) Deterministic	4/S_e/M/1/2ISS, 1CDP/O/3NI								
(B) HYDRO POWER SYSTEM									
(1) Sequencing & Scheduling									
(i) Deterministic	2/S_e5/1/CDP/ S/NI,I								
(ii) Stochastic									
(2) Operation									
(i) Deterministic	1/S_e1/1/CDP/ O/NI								

TABLE 1: cont'd

WR PROBLEM AREAS	SERIAL							NONSERIAL		
	DETERMINISTIC		STOCHASTIC					DETERMINISTIC		STOCHASTIC
			Markov Transition Type		Probabilistic System					
DP FORMULATION	Finite	Infinite	Finite	Infinite	Finite	Infinite		Finite	Infinite	
(C) IRRIGATION										
(1) Water Quality										
(i) Deterministic										
(ii) Stochastic			$1/S_e/1/1/U/O/NI$							
(2) Irrigated Acreage										
(i) Deterministic										
(ii) Stochastic					$1/S_e/5/4/APS/O/NI$					
(3) Technology Selection										
(i) Deterministic	$1/S_e/1/1/U/O/U$									
(ii) Stochastic										
(4) Timing										
(i) Deterministic	$1/S_e/2/2/-/O/NI$									
(ii) Stochastic										
OWRT PROBLEM AREA 4:										
(A) STREAM AERATION										
(i) Deterministic	$1/S_e/1/1/LM/O/NI$									
(ii) Stochastic										
(B) TREATMENT PLANT										
(1) Capacity Expansion										
(i) Deterministic	$1/S_e/1/1/CDP/O/NI$									
(ii) Stochastic										

TABLE 1: cont'd

DP FORMULATION	SERIAL						NONSERIAL		
	DETERMINISTIC		STOCHASTIC				DETERMINISTIC		STOCHASTIC
			Markov Transition Type		Probabilistic System				
WR PROBLEM AREAS	Finite	Infinite	Finite	Infinite	Finite	Infinite	Finite	Infinite	
OWRT PROBLEM AREA 5:									
(A) DESALINATION (i) Deterministic	$1/S_e/2/2/U/S/NI$								
(ii) Stochastic									
OWRT PROBLEM AREA 8:									
(A) SEWER SYSTEMS (1) Design (i) Deterministic	$3/N_s/1/1/DDDP/O/NI$								
(ii) Stochastic									
(2) Design & Layout (i) Deterministic	$1/S_s/N/1/U/O/NI$						$1/N_s/1/U/O$ NSDP		
(ii) Stochastic									
(B) INTEGRATED QUALITY & QUANTITY (i) Deterministic	$1/S_e/2/1/U/O/NI$								
(ii) Stochastic									
OWRT PROBLEM AREA 13:									
(A) TRANSBASIN DIVERSION (1) Economic Effect (i) Deterministic	$N/S_e/1/1/U/LP/NI$								

6. SUMMARY AND RESEARCH NEEDS

From the foregoing survey, it is obvious that water resources problems and the techniques of dynamic programming have been popular among researchers. Clearly, reservoir operation and design problems predominate. This is also reflected in the distribution of papers in the sections of this volume. In reservoir studies, the emphasis had been on operation but the design aspects and efforts to combine both have also appeared. Multi purpose and multi reservoir systems have been of interest. This is as it should be. However, the models have often been simplified considerably almost ad absordum. Large reservoir systems located on several different streams and converging or diverging at a delta in a basin, for example, have been decomposed and analyzed separately because of computational tractability. The result is that often times errors and inaccuracies are introduced in the process. This difficulty can now be circumvented or minimized by utilizing efficient nonserial dynamic programming algorithms available in current literature. As more efficient algorithms are developed, our ability to analyze more complex realistic systems will inevitably be enhanced.

Another issue is that for a long time, the most popular dynamic programming computational algorithm has been DDDP. Yakowitz has shown that differential dynamic programming, although difficult to implement, is by far more efficient. We have also shown that in the case of nonserial systems, our nonserial dynamic programming algorithms developed for various types of nonserial systems and complexities, are vastly superior. The hope is that optimal exploitation of these algorithms will be pursued to generate models of better fidelity, solve larger scale as well as more realistic problems and provide more accurate and less expensive solutions. Our capability to explore various policy issues should also expand.

We also note that, especially in reservoir systems, stochastically occurring problems had initially been approximated by their deterministic analogues primarily to ease computational burden. However, with developments in and better understanding of stochastic dynamic programming, algorithms for stochastic models have begun to crop up. Here again, adroit modeling and familiarity with the literature of stochastic dynamic programming can lead to more realistic and easily implementable models. In this regard, the papers by Sobel, Sharda and El Tayeh, Georgakakos, Read and Boshier, Kitanides and Andricevic - all in this volume, represent important contributions to the literature.

The other popular water resources problem addressed is the area of sewer systems. Initially, the emphasis was on design with some layout profile given. As our ability to deal with more complex dynamic programming systems increased, the layout and design problem is being handled simultaneously as it should be. Again the principal technique used is DDDP - a tool which is familiar to the model developers. Further, the problem is treated by decomposition. We believe that as stated before, more efficient dynamic programming algorithms will be invoked so that large networks can be handled efficiently. Also both problems can be treated simultaneously without resorting to decomposition.

Our comments relate to three primary aspects. The first deals with the water resources problem areas that have not been modelled correctly or satisfactorily. Here, we are concerned with incorrect description of the problem as it occurs in real life settings. We are also concerned about erroneous dynamic programming formulation. Sometimes, the models are simplistic and oversimplified. Other times, they are correct and realistic but very little regard is paid to the issue of computational efficiency. This results in difficulties when a practitioner, for example, wants to implement the model. It is simply too expensive. We have noticed also that the sensitivity analysis aspect of dynamic programming, one of its sterling properties, tends to be sublimated in these models. We expect to see some remedies which address these concerns.

A related problem is that of implementation and computational efficiency. Some of this concern has been touched on albeit tangentially in the previous paragraph. The temptation to employ algorithms familiar to an author, whether they are the most efficient available or not, should be resisted. An effort to keep abreast of the technology of problem solving via dynamic programming should be a profitable investment. Efforts to take advantage of the presence of various types of computing devices including the super computer and micro computers should be made. For example, Sugiyama's paper in this book shows how, by employing the proper solution algorithm, an otherwise expensive problem can be solved via an inexpensive device such as a programmable hand-held calculator.

Additionally, certain water resources problems such as reservoir operation, have proved to be a fertile ground for model developers, while various other areas have been shunned or treated as sacrosanct. Examples are irrigation, stream aeration, transbasin diversion, water quality especially nonpoint source pollution problems, etc. While these areas may be suitably and appropriately modeled via other techniques and sometimes very difficult to model, we believe that the principal reason for the paucity of attention given to them is due to our unfamiliarity with the potency of dynamic programming. For example, that water resources is typified by multiobjective, multicriteria decision making cannot be gainsaid, yet very few explicit treatments of this problem exist in the literature. The surrogate worth method described by Hall and Haimes is an attempt to deal with an aspect of this concern. Tauxe's paper in this volume considers this via multi objective dynamic programming. Labadie and Fontane also address it in their paper.

Finally, the developments in the area of computer hardware such as the arrival of the super computers, advances in computer technology and information processing such as via parallel computing, efficient high level programs and data manipulation/transfer systems, all augur well for the eventual eradication of the curse of dimensionality in dynamic programming. This translates into the analyst's increased ability to model, solve and design water resources systems as they occur in their natural habitat.

REFERENCES

1. Alarcon, L., and D. Marks, "A Stochastic Dynamic Programming Model for the Operation of the High Aswan Dam," Tech. Rep. 246, Ralph M. Parsons Lab., Dept. of Civ. Eng., Mass. Inst. of Technol., Cambridge, 1979.
2. Amandes, C. B., and Bedient, P. B., "Storm Water Detention in Developing Watersheds," Journal of the Environmental Engineering Division, ASCE, 106(EE2), Proc. Paper 15335, 403-419, 1980.
3. Anis, A., and E. Lloyd, "Reservoirs with Mixed Markovian-Independent Inflows," SIAM J. Appl. Math., 25(1), 68-76, 1972.
4. Argaman, Y., U. Shamir, and E. Spivak, "Design of Optimal Sewerage Systems," ASCE, Journal of the Environmental Engineering Division, 99(EE5), 703-716, October 1973.
5. Arunkumar, S., "Optimal Regulation Policies for a Multipurpose Reservoir with Seasonal Input and Return Function," J. Optim. Theory Appl., 21, 319-328, 1977.
6. Arunkumar, S., and K. Chon, "On Optimal Regulation for Certain Multi-Reservoir Systems," Oper. Res., 26(4), 551-562, 1978.
7. Askew, A. J., "Optimum Reservoir Operating Policies and the Imposition of a Reliability Constraint," Water Resour. Res., 10(1), 51-56, 1974a.
8. Askew, A. J., "Chance-Constrained Dynamic Programming and the Optimization of Water Resources System," Water Resour. Res., 10(6), 1099-1106, 1974b.
9. Askew, A. J., "Use of Risk Premium in Chance-Constrained Dynamic Programming," Water Resour. Res., 11(6), 862-866, 1975.
10. Askew, A., "Comments on 'Reliability-Constrained Dynamic Programming and Randomized Release Rules in Reservoir Management' by Lewis A. Rossman," Water Resour. Res., 14(1), 159-160, 1978.
11. Bather, J. A., "A Diffusion Model for the Control of a Dam," J. Appl. Probl., 5, 55-71, 1968.
12. Beard, L. R., "Status of Water Resource Systems Analysis," Journal of the Hydraulics Division, ASCE, 99(HY4), 559-565, April 1973.
13. Becker, L., and W. W. G. Yeh, "Optimal Timing, Sequencing and Sizing of Multiple-Reservoir Surface Water Supply Facilities," Water Resour. Res., 10(1), 57-62, 1974a.
14. Becker, L., and W. W. G. Yeh, "Optimization of Real Time Operations of Multiple-Reservoir System," Water Resour. Res., 10(6), 1107-1112, 1974b.
15. Becker, L., W. W. G. Yeh, D. Fults and D. Sparks, "Operations Models for Central Valley Project," J. Water Resour. Planning Management Div., ASCE, 102(WR1), 101-115, 1976.
16. Becker, L., W. W. G. Yeh, D. Fults and D. Sparks, "Timing and Sizing of Complex Water Resource Systems," Water Resour. Res., 100(HY10), 1457-1470, October 1974.
17. Bellman, R., Dynamic Programming, Princeton University Press, Princeton, N.J., 1957.
18. Bellman, R., and S. Dreyfus, Applied Dynamic Programming, Princeton University Press, Princeton, N.J., 1962.
19. Bellman, R. and R. Kalaba, Dynamic Programming and Modern Control Theory, Academic Press, New York, N.Y., 1965.
20. Bellman, R. and R. E. Kalaba, Quasilinearization and Nonlinear Boundary-Value Problems, American Elsevier, New York, N.Y., 1965.
21. Bellman, R. E. and L. A. Zadeh, "Decision Making in a Fuzzy Environment," Management Science, 17, B141-164, 1970.

22. Bennett, M. S., "Dynamic Programming Model for Determining Optimal Sizes and Locations of Detention Storage Facilities," <u>Proceedings of the International Symposium on Urban Hydrology, Hydraulics and Sediment Control</u>, Lexington, KY, 461-468, July 1983.
23. Bertele, D. and F. Brioschi, <u>Nonserial Dynamic Programming</u>, Academic Press, New York, 1973.
24. Bertsekas, D., <u>Dynamic Programming and Stochastic Control</u>, Academic Press, New York, 1976.
25. Bertsekas, D., and S. Shreve, <u>Stochastic Optimal Control</u>, Academic Press, New York, 1978.
26. Bhat, U., <u>Elements of Applied Stochastic Processes</u>, John Wiley, New York, 1972.
27. Blackburn, J., "Critical Appraisal of the Texas Law of Drainage," unpublished paper, 1979.
28. Bogle, M., and M. O'Sullivan, "Stochastic Optimization of a Water Supply System," <u>Water Resour. Res.</u>, 15(4), 778-786, 1979a.
29. Bogle, M. and M. O'Sullivan, "Stochastic Optimization of a Water Supply Expansion," <u>Water Resour. Res.</u>, 15(5), 1229-1237, 1979b.
30. Bras, R. L., R. B. Buchanan, and K. C. Curry, "Real Time Adaptive Closed Loop Control of Reservoirs with the High Aswan Dam as a Case Study," <u>Water Resour. Res.</u>, 19(1), 33-52, 1983.
31. Buchanan, R. B., and R. L. Bras, "Study of a Real-Time Adaptive Closed Loop Control Algorithm for Reservoir Operation," <u>Tech. Rep. 265</u>, Ralph M. Parsons Lab. Dept. of Civ. Eng., Mass. Inst. of Technol., Cambridge, July 1981.
32. Buras, N., "Conjunctive Operation of Dams and Aquifers," <u>Proc. Am. Soc. Civ. Eng.</u>, 89(HY6), 111-131, 1963.
33. Buras, N., <u>Scientific Allocation of Water Resources</u>, Elsevier, New York, 1972.
34. Butcher, W., "Stochastic Dynamic Programming for Optimum Reservoir Operation," <u>Water Resour. Bulletin</u>, 7(1), 115-123, 1971.
35. Butcher, W., Y. Haimes, and W. Hall, "Dynamic Programming for the Optimal Sequencing of Water Supply Projects," <u>Water Resour. Res.</u>, 5(4), 1196-1204, 1969.
36. Chang, S., and W. Yeh, "Optimal Allocation of Artificial Aeration Along a Polluted Stream Using Dynamic Programming," <u>Water Resour. Bulletin</u>, 9(5), 985-997, 1973.
37. Chernoff, H., and A. Petkau, "Optimal Control of a Brownian Motion," <u>SIAM J. Appl. Math.</u>, 34(4), 717-731, 1978.
38. Chow, V. T., and Cortes-Rivera, G., "Applications of DDDP in Water Resources Planning," Illinois Water Resources Center, Urbana, Research Report No. 78, January 1974.
39. Chow, V. T., D. R. Maidment, and G. W. Tauxe, "Computer Time and Memory Requirements for DP and DDDP in Water-Resources Systems Analysis," <u>Water Resour. Res.</u>, 11(5), 621-628, 1975.
40. Cohen, J., and M. Rubinovitch, "On Level Crossings and Cycles in Dam Processes," <u>Math. Oper. Res.</u>, 2(4), 297-310, 1977.
41. Collins, D. C., "Reduction of Dimensionality in Dynamic Programming via the Method of Diagonal Decomposition," <u>Journal of Mathematical Analysis and Applications</u>, 30, 223-234, 1970.
42. Collins, D. C., and A. Lew, "Dimensional Approximation in Dynamic Programming by Structural Decomposition," <u>Journal of Mathematical Analysis and Applications</u>, 30, 375-386, 1970.
43. Croley, T. E., II, "Sequential Stochastic Optimization in Reservoir Operation," <u>J. Hydraul. Div. Am. Soc. Civ. Eng.</u>, 100(HY3), 443-459, 1974.

44. Curry, K. C. and R. L. Bras, "Multivariate Seasonal Time Series Forecast with Application to Adaptive Control," Tech. Rep. 253, Ralph M. Parsons Lab., Dep. Civ. Eng., Mass. Inst. of Technol., Cambridge, 1980.
45. DeLucia, R. J., "Operating Policies for Irrigation Systems Under Stochastic Regimes," Ph.D. Thesis, Division of Engineering and Applied Physics, Harvard University, Cambridge, MA, 1969.
46. Denny, J., and S. Yakowitz, "Admissible Run-contingency Type Tests for Independence and Markov Dependence," J. Am. Stat. Assoc., 73, 177-181, 1978.
47. Denny, J., C. Kisiel, and S. Yakowitz, "Procedures for Determining the Order of Markov Dependence in Streamflow Records," Water Resour. Res., 10(5), 947-954, 1974.
48. Derman, C., Finite State Markovian Decision Processes, Academic Press, New York, 1970.
49. Doran, D. G., "An Efficient Transition Definition for Discrete State Reservoir Analysis: The Divided Interval Technique," Water Resour. Res., 11(6), 867-873, 1975.
50. Dorfman, R., "Basic Economic and Technological Concepts: A General Statement," in Design of Water-Resource Systems (A. Maas et al., principal authors), Harvard University Press, Cambridge, Mass., 154, 1962.
51. Dracup, J., and T. Fogarty, "Optimal Planning for a Thermal Discharge Treatment System," Water Resour. Res., 10(1), 67-71, 1974.
52. Dreyfus, S., and A. Law, The Art and Theory of Dynamic Programming, Academic Press, New York, 1977.
53. Dudley, N. J., and O. R. Burt, "Stochastic Reservoir Management and System Design for Irrigation," Water Resour. Res., 9(3), 507-522, 1973.
54. Dudley, N. J., D. T. Howell, and W. F. Musgrave, "Optimal Intra-seasonal Irrigation Water Allocation," Water Resour. Res., 7(4), 770-788, 1971.
55. Dudley, N. J., D. Reklis, and O. R. Burt, "Reliability, Trade-offs, and Water Resources Development Modeling with Multiple Crops," Water Resour. Res., 12(6), 1101-1108, 1976.
56. Durling, A. E., "Computational Aspects of Dynamic Programming in Higher Dimensions," Technical Report TR-64-3, Electrical Engineering Department, Syracuse University, Syracuse, N.Y., May 1964.
57. Duru, J. O., "On Site Detention: A Storm Water Management or Mismanagement Technique?" Proceedings: 1981 International Symposium on Urban Hydrology, Hydraulics and Sediment Control, U. of Kentucky, WRRI. 297-302, 1981.
58. Dyer, P. and S. McReynolds, The Computational Theory of Optimal Control, Academic Press, New York, 1970.
59. Dynkin, E., and A. Yushkevich, Controlled Markov Processes, Springer-Verlag, New York, 1979.
60. Eckstein, Otto, Water-Resource Development, Harvard University Press, Cambridge, MA, 1958.
61. Eisel, L. M., "Chance-Constrained Reservoir Model," Water Resour. Res., 8(2), 339-347, 1972.
62. Erlenkotter, D., "Sequencing Expansion Projects," Oper. Res., 21(2), 542-553, 1973a.
63. Erlenkotter, D., "Sequencing of Interdependent Hydroelectric Projects," Water Resour. Res., 9(1), 21-27, 1973b.

64. Esmaeil-Beik, S., and Y. S. Yu, "Optimal Operation of Multipurpose Pool of Elk City Lake," J. Water Resour. Plann. Manage. Div. Am. Soc. Civ. Eng., 110(WR1), 1-14, 1984.
65. Esogbue, A. O., "Dynamic Programming Algorithms and Analyses for Nonserial Networks," Technical Report J-83-3, School of Industrial and Systems Engineering, Georgia Institute of Technology, Atlanta, GA, 1983.
66. Esogbue, A. O., and R. E. Bellman, "Fuzzy Dynamic Programming and Its Extensions," in Fuzzy Sets and Decision Analysis, Zimmermann, H. J., L. A. Zadeh and B. R. Gaines (eds.), TIMS Studies in Management Sciences, 20, 147-167, 1984.
67. Esogbue, A. O. and C. Y. Lee, "Optimal Design of Large Complex Water Resources Conveyance Systems via Nonserial Dynamic Programming," Proc. NSF Special Workshop on Dynamic Programming and Water Resources, Georgia Institute of Technology, Atlanta, GA, June 1986.
68. Esogbue, A. O., and B. R. Marks, "Non-Serial Dynamic Programming - A Survey," Operational Research Quarterly, 25, 253-265, 1974.
69. Esogbue, A. O., and N. Warsi, "A High Level Computing Algorithm for Diverging and Converging Branch Nonserial Dynamic Programming Systems," J. of Computers and Mathematics With Applications, 12A(6), 719-732, 1986.
70. Faddy, M. J., "Optimal Control of Finite Dams: Continuous Output Procedure," Adv. Appl. Probl., 6, 689-710, 1974.
71. Feller, W., An Introduction to Probability Theory and Its Applications, Vol. I, John Wiley, New York, 1968.
72. Flinn, J., and W. Musgrave, "Development and Analysis of Input-Output Relations for Irrigation Water," Aust. J. Agric. Econ., 11(1), 1-19, 1967.
73. Flores, A. C., "Dynamic Optimization of Detention Storage in Urbanizing Areas," Masters Thesis, Rice University, Houston, TX, 1981.
74. Fogel, M., L. Duckstein, and C. Kisiel, "Optimum Control of Irrigation with a Water Application," J. Hydrol., 28, 343-358, 1976.
75. Fults, D. M., and L. F. Hancock, "Optimum Operations Model for Shasta-Trinity System," J. Hydraulics Div. Am. Soc. Civ. Eng., 98, 1497-1514, 1972.
76. Gablinger, M., and D. P. Loucks, "Markov Models for Flow Regulation," J. Hydraulics Div. Am. Soc. Civ. Eng., 96(HY1), 165-181, 1970.
77. Gagnon, C. R., R. H. Hicks, S. L. S. Jacoby, and J. S. Kowalik, "A Non-linear Programming Approach to a Very Large Hydroelectric System Optimization," Math. Program., 6, 28-41, 1974.
78. Gal, S., "Optimal Management of Stochastic Multireservoir Water Supply System," Water Resour. Res., 15(4), 737-749, 1979.
79. Gani, J., "Problems with Probability Theory of Storage Systems," J. R. Stat. Soc. Ser. B, 19, 181-206, 1957.
80. Georgakakos, A. P., "Extended Linear Quadratic Gaussian Control for the Real Time Operation of Reservoir System." Proc. NSF Special Workshop on Dynamic Programming and Water Resources, Georgia Institute of Technology, Atlanta, Georgia, June 1986.
81. Gessford, J. and S. Karlin, "Optimal Policy for Hydroelectric Operations," in Studies in the Mathematical Theory of Inventory and Production, K. J. Arrow, S. Karlin, and H. Scarf (eds.), 179-200, Stanford University Press, Stanford, CA, 1958.

82. Gill, S., "A Process for the Step-by-Step Integration of Differential Equations in an Automatic Computing Machine," Proceedings of the Cambridge Phil. Soc., 47, 96-108, 1951.
83. Hadley, G., and T. Whitin, Analysis of Inventory Systems, Prentice-Hall, Englewood Cliffs, N.J., 1963.
84. Haith, D., "Vertical Alignment of Sewers and Drainage Systems by Dynamic Programming," M.S. Thesis, Massachusetts Institute of Technology, Cambridge, MA, 1966.
85. Halkin, H., "A Maximum Principle of Pontryagin-type for Systems Described by Nonlinear Difference Equations," SIAM J. Control, 4(1), 90-111, 1966.
86. Hall, W., "Aqueduct Capacity Under an Optimum Benefit Policy," J. Irrig. Drain. Div. Am. Soc. Civ. Eng., 87(IR3), 1-11, 1961.
87. Hall, W. A., "Optimum Design of a Multiple-Purpose Reservoir," J. Hydraul. Div. Am. Soc. Civ. Eng., 90, 141-149, 1964.
88. Hall, W. A., and N. Buras, "The Dynamic Programming Approach to Water Resources Development," J. Geophys. Res., 66, 517-521, 1961.
89. Hall, W., and W. Butcher, "Optimal Timing of Irrigation," J. Irrig. Drain. Div. Am. Soc. Civ. Eng., 94(IR2), 267-275, 1968.
90. Hall, W. A., and J. A. Dracup, Water Resources Systems Engineering, McGraw-Hill, New York, 166, 1970.
91. Hall, W., and D. Howell, "The Optimization of Single-Purpose Reservoir Design with the Application of Dynamic Programming to Synthetic Hydrology Samples," J. Hydrol., I, 355-363, 1963.
92. Hall, W., and T. Roefs, "Hydropower Project Output Optimization," Proceedings, J. Power Div. Am. Soc. Civ. Eng., PO1, 67-79, 1966.
93. Hall, W. A., W. S. Butcher, and A. Esogbue, "Optimization of the Operation of a Multiple-purpose Reservoir by Dynamic Programming," Water Resour. Res., 4(3), 471-477, 1968.
94. Hall, W. A., R. Harboe, W. W. G. Yeh, and A. Askew, "Optimum Firm Power Output From a Two Reservoir System by Incremental Dynamic Programming," Contrib. 130, Water Resour. Center, Univ. of Calif. at Los Angeles, Los Angeles, 1969a.
95. Hall, W., G. Tauxe, and W. W. G. Yeh, "An Alternate Procedure for the Optimization of Operations for Planning with Multiple-River, Multiple-Purpose Systems," Water Resour. Res., 5(6), 1367-1372, 1969b.
96. Hashimoto, T., J. R. Stedinger, and D. P. Loucks, "Reliability, Resiliency, and Vulnerability Criteria for Water Resource System Evaluation," Water Resour. Res., 18(1), 14-20, 1982.
97. Haussmann, U. G., "Some Examples of Optimal Stochastic Controls or: The Stochastic Maximum Principle at Work," SIAM Rev., 23(3), 292-307, 1981.
98. Heidari, M., V. T. Chow, P. V. Kokotovic, and D. D. Meredith, "Discrete Differential Dynamic Programming Approach to Water Resources Systems Optimization," Water Resour. Res., 7(2), 273-282, 1971.
99. Heyman, D. P. and M. J. Sobel, Stochastic Models in Operations Research, Vol. II, McGraw Hill, New York, 1984.
100. Hinderer, K., Foundations of Non-Stationary Dynamic Programming with Discrete Time Parameter, Springer-Verlag, New York, 1970.
101. Hinomoto, H., "Dynamic Programming of Capacity Expansion of Municipal Water Treatment System," Water Resources Research, 8(5), 1178-1187, 1972.

102. Hipel, K. W., A. I. McLeod, and E. A. McBean, "Comment on 'Hydrologic Estimation and Economic Regret' by R. V. Jettmar and G. K. Young," Water Resour. Res., 13(3), 687-688, 1977.
103. Hipel, K. W., E. A. McBean, and A. I. McLeod, "Hydrologic Generating Model Selection," J. Water Resour. Plann. Manage. Div. Am. Soc. Civ. Eng., 105(WR2), 223-242, 1979.
104. Hirsch, R. M., "Synthetic Hydrology and Water Supply Reliability," Water Resour. Res., 15(6), 1603-1615, 1979.
105. Hoshi, K., and S. J. Burges, "Incorporation of Forecasted Total Seasonal Runoff Volumes Into Reservoir Management Strategies," in Reliability in Water Resources Management, E. A. McBean, K. W. Hipel and T. E. Unny (eds.), 137-166, Water Resources Publication, Fort Collins, Colo., 1979.
106. Hoshi, K., and S. J. Burges, "Seasonal Runoff Volumes Conditioned on Forecasted Total Runoff Volume," Water Resour. Res., 16(6), 1079-1084, 1980.
107. Howard, R. A., Dynamic Programming and Markov Processes, MIT Press, Cambridge, MA, 1960.
108. Howard, R., Comments on the 'Origin and Application of Markov Decision Processes,' in Dynamic Programming and Its Applications, M. Puterman (ed.), Academic Press, New York, 1978.
109. Jacobson, D., and D. Mayne, Differential Dynamic Programming, Academic Press, New York, 1970.
110. Jacoby, H. D., and D. P. Loucks, "Combined Use of Optimization and Simulation Models in River Basin Planning," Water Resour. Res., (6), 1401-1414, 1972.
111. Jacoby, S., and J. Kowalik, Mathematical Modelling with Computers, Prentice-Hall, Englewood Cliffs, N.J., 1980.
112. Joeres, E. F., J. C. Leibman, and C. S. Revelle, "Operating Rules of Joint Operations of Water Sources," Water Resour. Res., 7(2), 225-235, 1971.
113. Kaplan, M., and Y. Haimes, "Dynamic Programming for Optimal Capacity Expansion of Wastewater Treatment Plants," Water Resour. Bulletin, 11(2), 278-293, 1975.
114. Kendall, D., "Some Problems in the Theory of Dams," J. R. Stat. Soc. Ser. B, 19, 207-212, 1957.
115. Kirk, D., Optimal Control Theory, Prentice-Hall, Englewood Cliffs, N. J., 1970.
116. Kitanidis, P. K., and R. Andricevic, "Accuracy of the First-Order Approximation to the Stochastic Optimal Control of Reservoir." Proc. NSF Special Workshop on Dynamic Programming and Water Resources, Georgia Institute of Technology, Atlanta, GA, June 1986.
117. Klemes, V., "Discrete Representation of Storage for Stochastic Reservoir Optimization," Water Resour. Res., 13(1), 149-158, 1977.
118. Klemes, V., Comments on 'Optimum Reservoir Operating Policies and the Imposition of a Reliability Constraint,' by A. J. Askew, Water Resour. Res., 11(2), 365-368, 1975.
119. Korsak, A. and R. Larson, "A Dynamic Programming Successive Approximations Technique with Convergence Proofs," Automatica, 6, 245-252, 1970.
120. Kuiper, J., and L. Ortolano, "A Dynamic Programming-Simulation Strategy for the Capacity Expansion of Hydroelectric Power Systems," Water Resour. Res., 9(6), 1497-1510, December 1973.
121. Labadie, J., N. Grigg, and P. Trotta, "Minimization of Combined Sewer Overflows by Large-Scale Mathematical Programming," Comput. Oper. Res., 1, 421-435, 1974.

122. Labadie, J., N. Grigg, and B. Bradford, "Automatic Control of Large Scale Combined Sewer Systems," J. Environ. Eng. Div. Am. Soc. Civ. Eng., 101, 27-39, 1975.
123. Larson, R. E., "Computational Aspects of Dynamic Programming," in: 1967 IEEE International Convention Record, Part 3, March 1967.
124. Larson, R., State Increment Dynamic Programming, Elsevier, New York, 1968.
125. Larson, R., and J. Casti, Principles of Dynamic Programming, Vol. 1, Dekker, New York, 1978.
126. Lasdon, L., and A. Waren, "Survey of Nonlinear Programming Applications," Oper. Res., 28, 1029-1073, 1980.
127. Lee, Y., and A. O. Esogbue, "Optimal Procedures for Dynamic Programs with Complex Loop Structures," J. Math. Analysis and Applications, 119(1/2), 300-339, 1986.
128. Leibman, J., and W. Lynn, "The Optimal Allocation of Stream Dissolved Oxygen," Water Resour. Res., 2(3), 581-591, 1966.
129. Lindgren, B., Elements of Decision Theory, MacMillan, New York, 1971.
130. Little, J. D. C., "The Use of Storage Water in a Hydroelectric System," J. Oper. Res., 3, 187-197, 1955.
131. Liu, C. S., and A. C. Tedrow, "Multilake River System Operation Rules," J. of the Hydraulics Division, A.S.C.E., 99, 1369-1381, 1973.
132. Lloyd, E., "Stochastic Reservoir Theory," Adv. Hydrosci., 4, 281-339, 1967.
133. Lloyd, E., "Some Aspects of Stochastic Reservoir Theory - A Further Comment," J. Hydrol., 36, 187-191, 1978.
134. Loucks, D. P., "Computer Models for Reservoir Regulation," J. Sanit. Eng. Div. Am. Soc. Civ. Eng., 94(SA4), 657-669, 1968.
135. Loucks, D. P., and L. M. Falkson, "A Comparison of Some Dynamic, Linear and Policy Iteration Methods for Reservoir Operation," Water Resour. Bulletin., 6(3), 384-400, 1970.
136. Loucks, D. P., J. R. Stedinger, and H. A. Haith, Water Resource Systems Planning and Analysis, Prentice-Hall Englewood Cliffs, N. J., 1981.
137. McCuen, R. H., "Downstream Effects of Stormwater Management Basins," Journal of the Hydraulic Division, ASCE, 105(HY11), Proc. Paper 14977, 1343-1356, 1979.
138. McCuen, R. H., "Stormwater Management Policy and Design," Journal of Civil Engineering Design, Marcek-Kekkar, Inc., 1(1), 21-42, 1979.
139. Maidment, D. R., and V. T. Chow, "Stochastic State Variable Dynamic Programming for Reservoir Systems Analysis," Water Resour. Res., 17(6), 1578-1584, 1981.
140. Malcolm, H. R., "A Study of Detention in Urban Stormwater Management," Water Resources Research Institute of the University of North Carolina, Report No. 156, 1980.
141. Masse, P. B. D., Les Reserves et la Régulation de l'Avenir dans la Vie Economique, Hermann and Cie, Paris, 1946.
142. Matanga, G., and M. Marino, "Irrigation Planning, 2, Water Allocation for Teaching and Irrigation Purposes," Water Resour. Res., 15(3), 679-683, 1979.
143. Mawer, P., and D. Thorn, "Improved Dynamic Programming Procedures and Their Practical Application," Water Resour. Res., 10(2), 183-190, 1974.

144. Mays, L. W., "Sewer Network Scheme for Digital Computations," *Journal of the Environmental Engineering Division, ASCE*, 104(EE3), 1978.
145. Mays, L. W., and P. B. Bedient, "Model for Optimal Size and Location of Detention," *Journal of the Water Resources Planning and Management Division, ASCE*, 108(NOWR3), 270-284, 1982.
146. Mays, L. W., and H. G. Wenzel, "Optimal Design of Multi-Level Branching Sewer Systems," *Water Resour. Res.*, 12(5), 1976.
147. Mays, L. W. and Y. K. Tung, "State Variable Model for Sewer Network Flow Routing," *Journal of the Environmental Engineering Division, ASCE*, 104(EE1), 15-30, 1978.
148. Mays, L. W., and B. C. Yen, "Optimal Cost Design of Branched Sewer Systems," *Water Resour. Res.*, 11(1), 37-47, February 1975.
149. Mays, L. W., H. G. Wenzel, and J. C. Liebman, "Model for Layout and Design of Sewer Systems," *Journal of the Water Resources Planning and Management Division, ASCE*, 102(WR2), 385-405, 1976.
150. Meier, W. L., Jr., and C. S. Beightler, "An Optimization Method for Branching Multistage Water Resource Systems," *Water Resour. Res.*, 3(3), 645-652, 1967.
151. Meier, W. L., Jr., and C. J. Beightler, "Branch Compression and Absorption in Nonserial Multistage Systems," *Journal of Mathematical Analysis and Applications*, 21, 426-430, 1968.
152. Merritt, L. B., and R. H. Bogan, "Computer-Based Optimal Design of Sewer Systems," *ASCE, Journal of the Environmental Engineering Division*, 99(EE1), 35-53, February 1973.
153. Moncur, J. E. T., "Opportunity Costs of a Transbasin Diversion of Water I. Methodology," *Water Resour. Res.*, 8(6), 1415-1422, December 1972.
154. Moore, N., and W. Yeh, "Economic Model for Reservoir Planning," *J. Water Resour. Plann. Manage. Div. Am. Soc. Civ. Eng.*, 106(WR2), 383-400, 1980.
155. Moran, P., "A Probability Theory of Dams and Storage Systems," *Aust. J. Appl. Sci.*, 5, 116-124, 1954.
156. Moran, P., *The Theory of Storage*, John Wiley, New York, 1959.
157. Morin, T. L., "Pathology of a Dynamic Programming Sequencing Algorithm," *Water Resour. Res.*, 9(5), 1178-1185, 1973.
158. Morin, T., "Optimality of a Heuristic Sequencing Technique," *ASCE J. Hydraul. Div. Am. Soc. Civ. Eng.*, 100(HY8), 1195-1202, 1974.
159. Morin, T. L., and A. O. Esogbue, "Some Efficient Dynamic Programming Algorithms for the Optimal Sequencing and Scheduling of Water Supply Projects," *Water Resour. Res.*, 7(3), 479-484, June 1971.
160. Morin, T. L., and A. O. Esogbue, "The Imbedded State Space Approach to Reducing Dimensionality in Dynamic Programs of Higher Dimensions," *J. of Mathematical Analysis and Applications*, 48(3), 801-810, December 1974.
161. Murray, D., and S. Yakowitz, "Constrained Differential Dynamic Programming and Its Application to Multireservoir Control," *Water Resour. Res.*, 15(4), 1017-1027, 1979.
162. Murray, D., and S. Yakowitz, "The Application of Optimal Control Theory to Nonlinear Programming," *Math. Program.*, 21, 331-347, 1981.
163. Nemhauser, G. L., *Introduction to Dynamic Programming*, Wiley, New York, N.Y., 1966.
164. Nopmongcol, P., and A. Askew, "Multilevel Incremental Dynamic Programming," *Water Resour. Res.*, 12(6), 1291-1297, 1976.

165. Office of Water Resources Research, *1971 Annual Report*, U.S. Department of the Interior, 5-52.
166. O'Laoghaire, D., and D. Himmelblau, *Optimal Expansion of a Water Resources System*, Academic Press, New York, 1974.
167. Oven-Thompson, K., L. Alarcon, and D. H. Marks, "Agricultural versus Hydropower Trade-offs in the Operation of the High Aswan Dam," *Water Resour. Res.*, 18(6), 1605-1614, 1982.
168. Puterman, M. (ed.), *Dynamic Programming and Its Applications*, Academic Press, New York, 1978.
169. Read, E. G., and J. F. Boshier, "Biases in Stochastic Reservoir Scheduling Method," *Proc. NSF Special Workshop on Dynamic Programming and Water Resources*, Georgia Institute of Technology, Atlanta, GA, June 1986.
170. Revelle, C., E. Joeres, and W. Kirby, "The Linear Decision Rule in Reservoir Management and Design, I, Development of the Stochastic Model," *Water Resour. Res.* 5(4), 767-777, 1969.
171. Rhenals, A., and R. L. Bras, "The Irrigation Scheduling Problem and Evapotranspiration Uncertainty," Manuscript, Parsons Lab., Mass. Inst. of Technol., Cambridge, 1981.
172. Riordan, C., "A General Multistage Marginal Cost Dynamic Programming Model for the Optimization of a Class of Investment-Pricing Decisions," *Water Resour. Res.*, 7(2), 245-253, 1971.
173. Rockafeller, R. T., and R. Wets, "The Optimal Recourse Problem in Discrete Time: L^1-multipliers for Inequality Constraint," *SIAM J. Control Optim.*, 16, 16-36, 1978.
174. Roefs, T. G., and L. D. Bodin, "Multireservoir Operation Studies," *Water Resour. Res.*, 6(2), 410-420, 1970.
175. Roefs, T. G. and R. A. Guitron, "Stochastic Reservoir Models: Relative Computational Effort," *Water Resour. Res.*, 11(6), 801-804, 1975.
176. Rose, C. J., "Dynamic Programming Processes Within Dynamic Programming Processes," *Jrnl. of Mathematical Analysis and Applications* 26, 669-683, 1969.
177. Rosenthal, R., "The Status of Optimization Models for the Operation of Multireservoir Systems with Stochastic Inflows and Nonseparable Benefits," *Rep. 75*, Tenn. Water Resour. Res. Center, Knoxville, May 1980.
178. Rossman, L. A., "Reliability-Constrained Dynamic Programming and Randomized Release Rules in Reservoir Management," *Water Resour. Res.*, 13(2), 247-255, 1977.
179. Rossmith, R. L., "A Philosophy for Urban Storm Water Management," *Proceedings: 1981 International Symposium on Urban Hydrology, Hydraulics and Sediment Control*, U. of Kentucky WRRI, 273-280, 1981.
180. Russell, C. B., "An Optimal Policy for Operating a Multipurpose Reservoir," *Oper. Res.*, 20(6), 1181-1189, 1972.
181. Sathaye, J., and W. Hall, "Aqueduct Optimization with Intermediate Storage," *J. Irrig. Drain. Div. Am. Soc. Civ. Eng.*, 102(IR2), 249-264, 1976a.
182. Sathaye, J., and W. A. Hall, "Optimization of Design Capacity of an Aqueduct," *J. Irrig. Drain. Div. Am. Soc. Civ. Eng.*, 102(IR3), 295-305, 1976b.
183. Schweig, Z. and A. Cole, "Optimal Control of Linked Reservoirs," *Water Resour. Res.*, 4(3), 479-498, 1968.

184. Shih, C. S., and W. L. Meier, Jr., "Integrated Management of Quantity and Quality of Urban Water Resources," Water Resour. Bulletin, 8(5), 1006-1017, 1972.
185. Smith, David P., Jr., and P. B. Bedient, "Detention Storage for Urban Flood Control," Journal of the Water Resources Planning and Management Division, ASCE, 106(WR2), Proc. Paper 15555, 413-425, 1980.
186. Sniedovich, M., "Reliability-Constrained Reservoir Control Problems, I, Methodological Issues," Water Resour. Res., 15(6), 1574-1582, 1979.
187. Sniedovich, M., "A Variance-Constrained Reservoir Control Problem," Water Resour. Res., 16(2), 271-274, 1980.
188. Sniedovich, M., "Some Comments on Preference Order Dynamic Programming Models," Jrnl. of Mathematical Analysis and Applications, 79(2), 489-501, 1981.
189. Sniedovich, M., and D. R. Davis, "Comments on 'Chance-constrained Dynamic Programming and Optimization of Water Resource Systems,' by Arthur J. Askew," Water Resour. Res., 11(6), 1037-1038, 1975.
190. Sobel, M. J., "Multi Reservoir Optimization Problems with Myopic Optima," in this volume.
191. Srinivasan, S., "Analytic Solution of a Finite Dam Governed by a General Input," J. Appl. Probl., 11, 134-144, 1974.
192. Stedinger, J. R., "Comment on 'Real Time Adaptive Closed Loop Control of Reservoirs with the High Aswan Dam as a Case Study,' by R. L. Bras, R. Buchanan, and K. C. Curry," Water Resour. Res., 20(11), 1985.
193. Stedinger, J. R., B. F. Sule, and D. P. Loucks, "Stochastic Dynamic Programming Models for Reservoir Operation Optimization," Water Resour. Res., 20(11), 1499-1505, 1985.
194. Su, S. Y., and R. A. Deininger, "Modeling the Regulation of Lake Superior Under Uncertainty of Future Water Supplies," Water Resour. Res., 10, 11-25, 1974.
195. Sugiyama, Hiroshi, "Some Approaches to Optimal Aeration Control for Improving Water Quality via Dynamic Programming," Proc. NSF Special Workshop on Dynamic Programming and Water Resources, Georgia Institute of Technology, Atlanta, GA, June 1986.
196. Sule, B. F., "Optimal Operation of a Multiple Purpose Reservoir," Ph.D. Dissertation, Cornell University, Ithaca, N. Y., 1984.
197. Takeuchi, K., and D. H. Moreau, "Optimal Control of Multiunit Interbasin Water Resources Systems," Water Resour. Res., 10(3), 407-414, 1974.
198. Tamura, H., "Multistage Decomposition Algorithms for Optimizing Discrete Dynamic Systems with Applications," Handbook of Large Scale Systems Engineering Applications, M. Singh and A. Titli (eds.), North-Holland, Amsterdam, 1979.
199. Tauxe, G., R. Inman, and D. Mades, "Multiobjective Dynamic Programming with Applications to a Reservoir," Water Resour. Res., 15(6), 1403-1408, 1979a.
200. Tauxe, G., R. Inman, and D. Mades, "Multiobjective Dynamic Programming, A Classic Problem Redressed," Water Resour. Res., 15(6), 1398-1402, 1979b.
201. Torabi, M., and F. Mobasheri, "A Stochastic Dynamic Programming Model for the Optimum Operation of a Multi-Purpose Reservoir," Water Resour. Bulletin, 9(6), 1089-1099, 1973.

202. Trott, W., and W. Yeh, "Optimization of Multiple Reservoir System," J. Hydraul. Div. Am. Soc. Civ. Eng., 99(HY10), 1865-1884, 1973.
203. Turgeon, A., "Optimal Operation of Multireservoir Power Systems with Stochastic Inflows," Water Resour. Res., 16(2), 275-283, 1980.
204. Turgeon, A., "A Decomposition Method for the Long-Term Scheduling of Reservoirs in Series," Water Resour. Res., 17(6), 1565-1570, 1981.
205. Verhaeghe, R., "On the Determination of Stochastic Reservoir Operating Strategies Incorporating Short- and Long-term Information in Real-Time," Ph.D. Dissertation, Mass. Inst. of Technol., Cambridge, MA, 1977.
206. Wakamori, F., S. Masui, F. Funabashi, and M. Ohnari, "Optimal Operation of a Multireservoir During Flood Periods," Proceedings of IFAC Control Science and Technology 8th Triennial World Congress, Kyoto, Japan, 3835-3840, 1981.
207. Walsh, S., and L. C. Brown, "Least Cost Method for Sewer Design," ASCE, Journal of the Environmental Engineering Division, 99(EE3), 333-345, June 1973.
208. Walski, T. M., and A. Pelliccia, "Preliminary Design and Cost Estimating for Reservoir Projects," Water Resour. Bulletin, AWRA, 17(1), 49-56, 1981.
209. White, D. J., "Dynamic Programming and Probabilistic Constraints," Oper. Res., 22(3), 654-664, 1974.
210. Yakowitz, S., Mathematics of Adaptive Control Processes, Elsevier, New York, 1969.
211. Yakowitz, S., "A Stochastic Model for Daily River Flows in an Arid Region," Water Resour. Res., 9(5), 1271-1285, 1973.
212. Yakowitz, S., "Small Sample Hypothesis, Tests of Markov Order, with Application to Simulated Hydrologic Chains," J. Am. Stat. Assoc., 7(353), 132-136, 1976.
213. Yakowitz, S., "A Nonparametric Markov Model for Daily River Flow," Water Resour. Res., 15(5), 1035-1043, 1979.
214. Yakowitz, S. and D. Murray, "Constrained Differential Dynamic Programming and Its Application to Multireservoir Control," Water Resources Research, 15(5), 1017-1027, 1979.
215. Yakowitz, S., "Convergence Rates of the State Increment for the Dynamic Programming Method," Automatica, Vol. 19, No. 1, 53-60, 1983.
216. Yakowitz, S., "Dynamic Programming Applications in Water Resources, Water Resources Research, 18(4), 673-696, 1982.
217. Yeh, W., and L. Becker, "Real-time Hourly Reservoir Operation," J. Water Resour. Plann. Manage. Div. Am. Soc. Civ. Eng., 105(WR2), 187-303, 1979.
218. Young, G. K., "Techniques for Finding Reservoir Operating Rules," Ph.D. Thesis, Harvard University, Cambridge, MA, 1966.
219. Young, G. K., "Finding Reservoir Operating Rules," J. Hydraul. Div. Am. Soc. Civ. Eng., 93(HY6), 297-321, 1967.
220. Zadeh, L. A., "Fuzzy Sets," Information and Control, 8, 338-353, 1965.
221. Zimmerman, H. J., L. A. Zadeh, and B. R. Gaines, (eds.) Fuzzy Sets and Decision Analysis, TIMS Studies in Management Sciences, North Holland, Vol. 20, 1984.

PART TWO

MULTIOBJECTIVE-MULTIPURPOSE LARGE SCALE WATER RESOURCE SYSTEMS MODELS

MULTI-OBJECTIVE DYNAMIC PROGRAMMING IN WATER RESOURCES

Dr. George W. Tauxe
Associate Professor, School of Civil Engineering and Environmental Science,
The University of Oklahoma, Norman, OK 73019

ABSTRACT

While Dynamic Programming is generally construed to be a single objective process, the introduction of additional state variables transforms it into a powerful multi-objective optimization technique. The beauty of this method is its capability to develop the entire pareto optimal set with just one pass. In this paper two applications will be presented. The first deals with continuous functions and the second deals with a discretized reservoir problem.

INTRODUCTION

Dynamic Programming is classically thought to be a single objective optimization technique, and as a result has only seen limited application to multiobjective problems. This paper describes Multiobjective Dynamic Programming (MODP) and then presents two applications: the first to a three objective problem with continuous functions after Reid and Vemuri [Reid and Vemuri, 1971; Tauxe, Inman, and Mades, 1979] and the second a two objective reservoir problem [Tauxe, Mades and Inman, 1980] which must be discretized to solve.

One of the attractive features of Multiobjective Dynamic Programming is its capability of generating the non-inferior solution set as well as trade-off ratios between all objectives. Furthermore, once one of the non-inferior solutions has been selected by a decision maker, the entire policy that provided those levels of the objectives can easily be found using a traceback.

Let the multiobjective problem be defined to be

$$\min_{X} \{f_1(X), f_2(X), \ldots, f_i(X), \ldots, f_n(X)\} \quad (1)$$

subject to

$$g_k(X) \leq \hat{G}_k \quad k = 1, 2, \ldots, m$$

where X is an N-dimensional vector of decision variables; $f_i(X)$, $i=1, 2, \ldots, n$, are n objective functions and $g_k(X)$, $k=1, 2, \ldots, m$, are m constraint functions, where G_k is the limiting resource, and all functions may be non-linear in X. Problems of this form are abundant in the literature [Goicoechea et al., 1982, Haimes, 1977, Cohon, 1973] with many approaches to solutions presented.

The MODP approach can best be characterized as being of the constraint approach where all but one of the objectives are treated as constraints. Thus the problem becomes

$$\min_X f_1(X) \qquad (2)$$

subject to

$$f_i(X) \le \epsilon_i \qquad i = 2, 3, \ldots, n$$

$$g_k(X) \le \hat{G}_k \qquad k = 1, 2, \ldots, m$$

where ϵ_i are acceptable target levels of the n-1 objectives. These target levels must then be parametrically varied, each change necessitating a complete solution to (2), in order to find the non-inferior solution set. Insight into the problem as well as interaction with a decision maker can reduce the number of times that a complete solution to (2) need be obtained. One such method is the Surrogate Worth Trade-Off Method (SWT) [Haimes et al. 1974, Haimes et al. 1975].

The SWT method basically entails finding several solutions to (2), usually with the aid of a digital computer, and from these solutions the trade-off ratios, (essentially LaGrange Multipliers between objectives) are determined. Next, a decision maker is presented with this information in order to articulate preferences. Next, more computer solutions are obtained to (2) with different ϵ_i reflecting the decisionmaker's performances and the process is repeated with the decision maker until he is no longer able to find a more preferred solution.

MODP has two distinct advantages over most techniques that are used to solve problems formulated using the constraint approach. First, the entire non-inferior solution set is obtained in one computer run, and second, the trade-off ratios are a by product of the MODP solution. Thus, a method, such as the SWT can be applied without having to return repeatedly to a computer for more updated information.

Multiobjective Dynamic Programming has the same properties as conventional Dynamic Programming, [Bellman, 1957] regarding separability, convexity and continuity. It is, however, slightly more restrictive on the properties of the objective functions that are transformed into the constraint form. It is more difficult to treat min-max objectives as constraints and thus if one of the objectives of of this form, it should be selected as f_1. Functions must be monotonic and generally additive. These restrictions are not considered a serious limitation in the water resources field as objectives of such form are common.

DEVELOPMENT OF MODP

Equation (3) shows an expansion of both the objectives and the constraints in terms of the individual x_j.

$$f_i(X) = f_{i1}(x_1) + f_{i2}(x_2) + \ldots + f_{iN}(x_N) \qquad i=1, 2, \ldots, n$$

and (3)

$$g_k(X) = g_{k1}(x_1) + g_{k2}(x_2) + \ldots + g_{kN}(x_N) \qquad k=1, 2, \ldots, m$$

The individual functions $f_{ij}(x_j)$ and $g_{kj}(x_j)$ are shown combined by addition although this is not necessarily a requirement. The min-max form has been used [Orlovski, et al., 1984 and Tauxe, 1973] in water resources applications and the product form is used in the first example. Substituting (3) into (2) produces

$$\min_X f_{11}(x_1) + f_{12}(x_2) + \ldots + f_{1j}(x_j) + \ldots f_{1N}(x_N)$$

subject to

$$F_i = f_{i1}(x_1) + f_{i2}(x_2) + \ldots + f_{ij}(x_j) + \ldots + f_{iN}(x_N) \qquad (4)$$

with $0 \leq F_i \leq \varepsilon_i \quad i = 2, 3, \ldots, n$

$$G_k = g_{k1}(x_1) + g_{k2}(x_2) + \ldots + g_{kj}(x_j) + \ldots + g_{kN}(x_N)$$

with $0 \leq G_k \leq \hat{G}_k \quad k = 1, 2, \ldots, m$

By minimizing the objective in (4) for all $0 \leq F_i \leq \varepsilon_i$, i=2, ..., N and $0 \leq G_k \leq \hat{G}_k$, k=1, ..., m, the desired non-inferior solution set can be obtained, provided it is within the domain specified by $0 \leq F_i \leq \varepsilon_i$, i=2, ... N. In essence, MODP treats all objectives except the first as state variables. Although the problem does suffer from Bellman's "curse of dimensionality" it has been shown not to be a hindrance to solving water resources problems because while the number of state variables increases with each additional objective, the number of decision variables remains the same. The ramifications of this are described in [Chow, et al., 1975].

Transforming (4) into the recursive form yields

$$F_{1j}(F_{2j}, F_{3j}, \ldots, F_{nj}, G_{1j}, G_{2j}, \ldots, G_{mj}) \qquad (5)$$

$$= \min_{x_j} \{f_{1j}(x_j) + F_{1,j+1}(F_{2j} - f_{2j}(x_j), F_{3j} - f_{3j}(x_j),$$

$$\ldots, F_{nj} - f_{nj}(x_j), G_{1j} - g_{1j}(x_j), G_{2j} - g_{2j}(x_j), \ldots,$$

$$G_{mj} - g_{mj}(x_j))\}$$

subject to

$$0 \leq f_{ij}(x_j) \leq F_{ij}, \quad i = 2, 3, \ldots, n$$

$$0 \leq F_{ij} \leq \varepsilon_i$$

$$0 \leq g_{kj}(x_j) \leq G_{kj}$$

$$0 \leq G_{kj} \leq \hat{G}_k \quad k = 1, 2, \ldots, m$$

for $j = N, N-1, \ldots, 2, 1$

with $F_{iN+1} = 0$ for $i = 1, 2, \ldots, n$

in which $F_{1j}(\ )$ is the primary objective expressed as a function of objectives F_{ij}, i=2, 3, ..., N and state variables G_{kj}, k=1, 2, ..., m. Note that there is only one decision variable shown in (5) which is easily handled by a myriad of search techniques, [Hillier and Liberman, 1986, for example], however, problems with more than one decision variable can be handled by MODP.

Having solved (5) the first result of interest is $F_{11}(F_{21}, F_{31}, \ldots, F_{N1})$. This defines the non-inferior solution space. To determine the trade-off ratios all that is necessary is to evaluate the first partial derivatives of F_{11} with respect to all other F_{i1}. This can be accomplished analytically if the problem is continuous as in the first example or numerically if it is discretized as in the second.

APPLICATIONS: Example 1

To illustrate the application of MODP on a continuous basis the Reid-Vemuri example [Reid-Vemuri, 1971, Tauxe, et al., 1979] was chosen. Their statement of the problem is:

> A dam of finite height impounds water in the reservoir and that water is required to be released for various purposes such as flood control, irrigation, industrial and urban use, and power generation. The reservoir may also be used for fish and wildlife enhancement, recreation, salinity and pollution control, mandatory release to satisfy riparian rights of downstream users, and so forth. The problem is essentially one of determining the storage capacity of the reservoir so as to maximize the net benefits accrued...

There are two decision variables: x_1, the total man-hours devoted to building the dam, and x_2, the mean radius of the lake impounded. There are three objectives: $F_1(x_1, x_2)$, the capital cost of the project; $F_2(x_2)$, the water loss (volume/year) due to evaporation; and $\hat{F}_3(x_1, x_2)$, the total volume capacity of the reservoir. In order to change the volume objective to a minimization problem a reciprocal function $F_3(x_1, x_2)$ was formed.

$$F_3(x_1, x_2) = 1/\hat{F}_3(x_1, x_2) \tag{6}$$

The three objectives expressed in terms of the decision variables x_1 and x_2 are

$$F_1(x_1, x_2) = e^{0.01x_1}(x_1)^{0.02}(x_2)^2 \tag{7}$$

$$F_2(x_2) = 0.5(x_2)^2$$

$$F_3(x_1, x_2) = e^{-0.005x_1}(x_1)^{-0.01}(x_2)^{-2}$$

All decisions and objectives are constrained to be non-negative, with no other constraints upon the system. Treating $F_1(x_1, x_2)$ as the objective to be optimized and the others as constraints (7) becomes

$$\min F_1 = e^{.01x_1} x_1^{.02} \cdot x_2^2 \tag{8}$$

$$x_1, x_2 \geq 0$$

$$F_2 = 1 \cdot 0.5\, x_2^2$$

with $0 \leq F_2 \leq \varepsilon_2$

and $\qquad F_3 = e^{-.005x_1} x_1^{-.01} \cdot x_2^{-2}$

with $0 \leq F_3 \leq \varepsilon_3$

Because of the nature of the problem no values for ε_2 and ε_3 are specified. Rewriting (8) in recursive form gives

$$F_{1j}(F_{2j},F_{3j}) = \min_{x_j} \{f_{1j}(x_j) \cdot F_{1j+1}(F_{2j+1},F_{3j+1})\} \quad j=2,1 \qquad (9)$$

where

$$F_{2j+1} = \frac{F_{2j}}{f_{2j}(x_j)} \qquad F_{3j+1} = \frac{F_{3j}}{f_{3j}(x_j)}$$

$$f_{11}(x_1) = e^{.01x_1} x_1^{.02} \qquad f_{12}(x_2) = s_2^2 \qquad F_{13}(F_{23},F_{33}) = 1.0$$

$$f_{21}(x_1) = 1.0 \qquad f_{22}(x_2) = 0.5x_2^2 \qquad F_{23} = 1.0$$

$$f_{31}(x_1) = e^{-.005x_1} x_1^{-.01} \qquad f_{32}(x_2) = x_2^{-2} \qquad F_{33} = 1.0$$

The solution to this problem is outlined as follows: Starting in stage 2, solve for the optimal x_2 to find F_{12} as a function of F_{22} and F_{32} we obtain

$$F_{12}(F_{22},F_{32}) = \min_{x_2 \geq 0} \{x_2^2 \cdot 1\} \qquad (10a)$$

where

$$F_{23} = \frac{F_{22}}{f_{22}(x_2)} = \frac{F_{22}}{.5x_2^2} = 1 \qquad (10b)$$

and

$$F_{33} = \frac{F_{32}}{f_{32}(x_2)} = \frac{F_{32}}{x_2^{-2}} = 1 \qquad (10c)$$

The only solution for this last stage that satisfies both (10b) and (10c) is

$$x_2^* = \sqrt{2F_{22}} = \sqrt{\frac{1}{F_{32}}} \qquad (11)$$

where x_2^* is the optimal decision as a function of F_{22} and F_{32}. The optimal value of the objective is therefore

$$F_{12}(F_{22},F_{32}) = \begin{cases} 2F_{22} = 1/F_{32} & \text{if } F_{22} = \dfrac{1}{2F_{32}} \\[2ex] \infty & \text{if } F_{22} \neq \dfrac{1}{2F_{32}} \end{cases} \qquad (12)$$

Having found the optimum $F_{12}(F_{22},F_{32})$ for all values of F_{22} and F_{32}, the stage is decremented. The recursive equation for stage 1 is

$$F_{11}(F_{21},F_{31}) = \min_{x_1 \geq 0} \{e^{.01x_1} x_1^{.02} \cdot F_{12}(F_{22},F_{32})\} \qquad (13a)$$

where

$$F_{22} = \frac{F_{21}}{f_{21}(x_1)} = \frac{F_{21}}{1.0} \tag{13b}$$

$$F_{32} = \frac{F_{31}}{f_{31}(x_1)} = \frac{F_{31}}{e^{-.005x_1} x_1^{-.01}} \tag{13c}$$

and from the previous stage

$$F_{12}(F_{22}, F_{32}) = \begin{cases} 2 F_{22} = 1/F_{32} & \text{if } F_{22} = \dfrac{1}{2 F_{32}} \\ \infty & \text{if } F_{22} \neq \dfrac{1}{2 F_{32}} \end{cases} \tag{13d}$$

Substituting (13b) and (13c) into (13d) and then the result into (13a)

$$F_{11}(F_{21}, F_{31}) = \tag{14}$$
$$\begin{cases} \min_{x_1 > 0} (e^{.01x_1} \cdot x_1^{.02} \cdot 2 F_{21}) & \text{if } F_{21} = \dfrac{1}{2 F_{31} e^{.005x_1} x_1^{.01}} \\ \infty & \text{otherwise} \end{cases}$$

The solution to (14) can be easily obtained by forming the Lagrangian and applying the Kuhn-Tucker necessary conditions for a minimum.

$$L(x_1, \lambda) = e^{.01x_1} x_1^{.02} \cdot 2F_{21} + \lambda(2F_{21} \cdot F_{31} e^{.005x_1} x_1^{.01} - 1) \tag{15}$$

$$\frac{\partial L}{\partial x_1} = 0, \quad \frac{\partial L}{\partial \lambda} = 0$$

The solution to (15) is

$$e^{.005x_1^*} x_1^{*.01} = \frac{1}{2F_{21} F_{31}} \tag{16}$$

which when substituted into (14) gives the non-inferior solution set for all three objectives:

$$F_{11}(F_{21}, F_{31}) = \frac{1}{2F_{21}(F_{31})^2} \tag{17}$$

Since both F_2 and F_3 are positive, only one condition of (14) appears in (17), and, since stage 1 has been reached, the solution to (8) has now been obtained. It should be observed that the optimal value of the primary objective, F_1, is expressed solely in terms of the secondary objectives, F_2 and F_3, rather than in terms of the decision variables. This expression (17) defines the non-inferior solution set, as is apparent from its form, in that no single objective may be reduced without an increase in at least one of the other objectives.

One last step is required before the preferred solution is found, namely, finding the trade-off ratios. For this problem they may be found by computing the partial derivatives of (17). Doing so yields

$$\lambda_{12} = \frac{\partial F_1}{\partial F_2} = -\frac{1}{2(F_2 F_3)^2} \qquad (18)$$

$$\lambda_{13} = \frac{\partial F_1}{\partial F_2} = -\frac{1}{F_2(F_3)^3}$$

$$\lambda_{23} = \frac{\partial F_2}{\partial F_3} = -\frac{2F_2}{F_3}$$

Given any set of values of F_2 and F_3 the first objective is determined from (17) and the trade-off ratios from (18). Thus, the non-inferior solution set can now be described completely as $(F_1, F_2, F_3, \lambda_{12}, \lambda_{13}, \lambda_{23})$.

Discrete points from this non-inferior set and corresponding trade-off ratios may be calculated by assuming various combinations of constraining levels for the secondary objectives, F_2 and F_3, and solving for F_1^*, λ_{12} λ_{13}, λ_{13}. Several such solutions are listed below in Table 1.

TABLE I. Sample solutions to the Reid-Vemuri example

F_2	F_3	F_1	λ_{12}	λ_{13}	λ_{23}
250	.00100	2000.0	-8.000	-4.00×10^6	-5.0×10^5
500	.00057	3062.5	-6.125	-1.07×10^7	-1.74×10^6
750	.00040	4166.7	-5.555	-2.08×10^7	-1.16×10^6
1060	.00021	10554	-9.956	-9.98×10^7	-1.0×10^6
500	.00040	6250.0	-12.500	-3.13×10^7	-3.9×10^5

These results are identical to those presented by Reid and Vemuri. The trade-off function λ_{12}, λ_{13} and λ_{23} are all relative because at a Pareto optimum, one objective's level cannot be increased without a decrease in any of the other objectives' levels.

This ends the task as far as the MODP solution is concerned with the exception of the traceback. In order to perform the traceback a preferred solution needs to be determined from the non-inferior solution set. Assume, therefore, that the Surrogate Worth Trade-Off Method [Haimes and Hall, 1974] has been utilized and the result of that interaction with the decision maker has yielded the following as the preferred solution:

$$F_1 = 4166.7 \qquad F_2 = 750 \qquad F_3 = .0004$$

for which

$$\lambda_{12} = -5.55 \qquad \lambda_{13} = -2.08 \times 10^7 \qquad \lambda_{23} = -1.16 \times 10^6.$$

The final task is now to identify the values of the two decision variables x_1 and x_2 that produce this preferred solution. First x_1^* is determined from (16) which yields $x_1^* = 93.10$ using an iterative solution technique.

From x_1^*, it is necessary to determine F_{22} and F_{32} so that x_2^* can be determined. These are determined from (13b) and (13c) and then substituted into (11) which gives $x_2 = 38.73$. Thus, in the decision makers eyes, the preferred solution is to devote 93.1 man hours building a lake of mean radius of 38.73 units which costs 4166.7 units, losses 750 volume units per year from an impoundment of 2500 (1/.0004) volume units.

Example 2

In this example a continuous problem was examined. Most water resource problems do not lend themselves to solutions as above, but, must be discretized in order to find a solution. Thus a reservoir example will be presented that has two objectives.

This second illustration of MODP considers the operation of Shasta Reservoir in California with the two objectives: 1) to maximize the cumulative dump energy generated above a prescribed level of firm energy, and 2) to minimize the cumulative evaporation loss. The primary objective, that of cumulative dump energy, was maximized in the recursive equation; and the secondary objective, the minimization of cumulative evaporation losses, was represented by a state variable. Other objectives, for example firm energy maximization, would be formulated as a physical constraint on the two objective MODP problem and then parametrically varied outside the MODP problem. MODP was used to generate the non-inferior set of solutions (which includes all feasible levels of cumulative dump energy and cumulative evaporation) for a fixed level of firm energy. For this reservoir problem the MODP solution consisted of solving a two-state variable, one decision variable problem with the systems' only structural state variable being the amount of water in reservoir storage. The decision variable was the monthly volume of reservoir release. In recursive form the problem is as follows:

$$F_j(S_j, V_j) = \max_{q_j} \{E_j(q_j, S_j, FE_j) + F_{j-1}(S_{j-1}, V_{j-1})\} \quad (19a)$$

$$V_{j-1} = V_j - v_j(q_j, S_j) \quad (19b)$$

$$S_{j-1} = S_j - Y_j + q_j + v_j(q_j, S_j) \quad (19c)$$

$$q_j \geq q_{min_j} \quad (19d)$$

$$S_{min} \leq S_j \leq S_{max}$$

for $j = 1, 2, \ldots, 96$

with $F_0(S_0, V_0) = 0$

$S_0 = S_{96} = 2976$

$V_0 = 0$

where $F_j(\cdot)$ dump energy accumulated through month j, MWH, (million watt hours)

$E_j(\cdot)$ dump energy produced in month j, MWH

FE_j firm energy required in time period j, MWH

S_j volume of storage at end of month, KAF, (thousand acre feet)

V_j cumulative evaporation through month j, KAF

q_j volume of release during month j, KAF

Y_j volume of inflow during month j, KAF

v_j volume of evaporation during month j, KAF

The function E_j represents excess energy produced above the firm level that results from a release of q_j at the average monthly storage.

The evaporation accumulation is handled by (19b) while continuity, reflecting the systems only state variable is handled by (19c).

Historical monthly inflows were used that contained the eight year critical period. A dead storage of 600 KAF and a maximum storage of 4600 KAF were assumed. Mandatory releases were included for salinity control at the Sacramento-San Joaquin delta and for downstream river navigation, (19d). The monthly evaporation was computed as the product of the monthly mean lake area and the monthly evaporation rate as shown in (20).

$$v_j(q_j, \bar{S}_j) = \alpha_m \cdot A(\bar{S}_j) \qquad (20)$$

where

$$A(\bar{S}_j) = -5.36 \times 10^{-7} (\bar{S}_j^2) + 8.4 \times 10^{-3} \bar{S}_j + 2.79$$

$$\bar{S}_j = \frac{S_j + S_{j-1}}{2}$$

α_m monthly evaporation rate coefficient (feet per month), where m=j mod 12

A reservoir surface area, KA

$\bar{S}_j$ mean monthly reservoir volume, KAF

The monthly energy production function (21) is limited by the power plant capacity of 419 MW to a maximum of 3.0168×10^5 megawatt-hours per month.

$$TE(\bar{S}_j) = \min \begin{cases} q_j \cdot ER(\bar{S}_j) \\ 3.0168 \times 10^5 \end{cases} \qquad (21)$$

where

$$ER(\bar{S}_j) = 66.5 + (21.3\bar{S}_j - 3365.0)^{\frac{1}{2}}$$

with ER the rate of energy production, MWH/KAF

and TE total energy produced during month j, MWH

Dump energy, $E_j(\cdot)$ in Eq. (19a), was calculated from (22) as follows

$$E_j(q_j, S_j, FE_j) = TE(\bar{S}_j) - FE_j \qquad (22)$$

where

$$FE_j = \beta_m \cdot AFE$$

with β_m monthly firm energy distribution coefficient m = j mod 12 and AFE annual firm energy requirement, MWH.

Initial and final storages of 2976 (KAF), the actual October mean storage of Shasta Reservoir, were specified at the beginning of the analysis.

A FORTRAN program was written to solve the 96 stage, 2 state variable 1 decision variable MODP problem. The results shown in Figure 1 presented the non-inferior solution for a run with an annual Firm Energy requirement of 10.5×10^5 MWH. The optimal release policies resulting in cumulative evaporations of 712.9 KAF, 683.6 KAF, and 642.5 KAF corresponding to points A, B, and C in Figure 1 are shown in Figure 2.

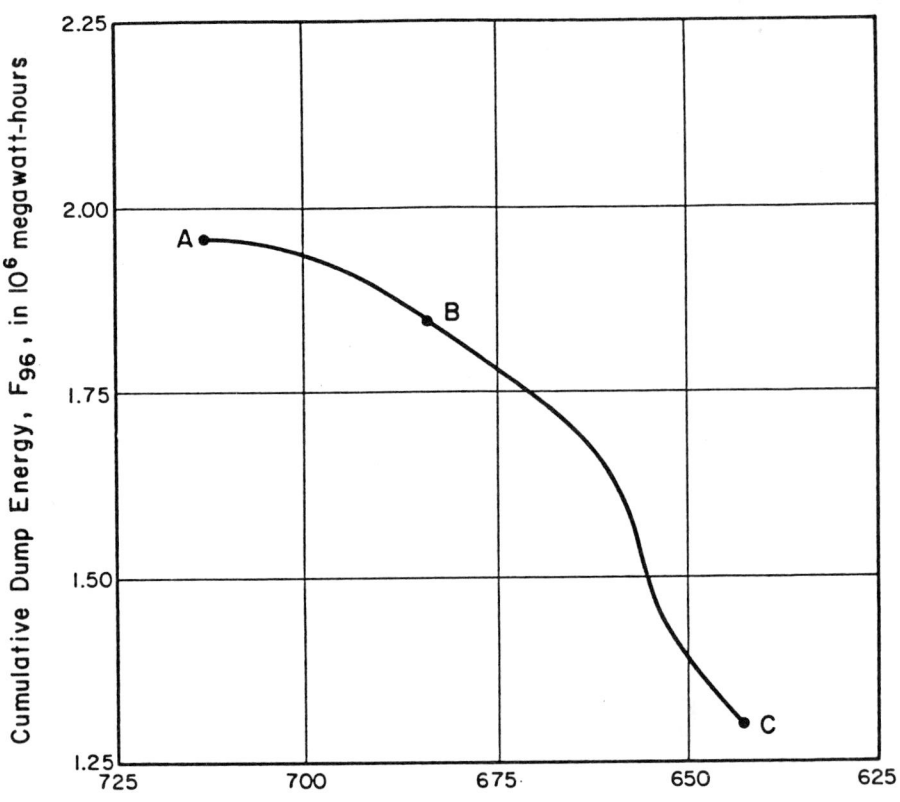

FIG. 1. Noninferior solution for firm energy of 10.5×10^5 megawatt-hours
(from Mades and Tauxe, 1980)

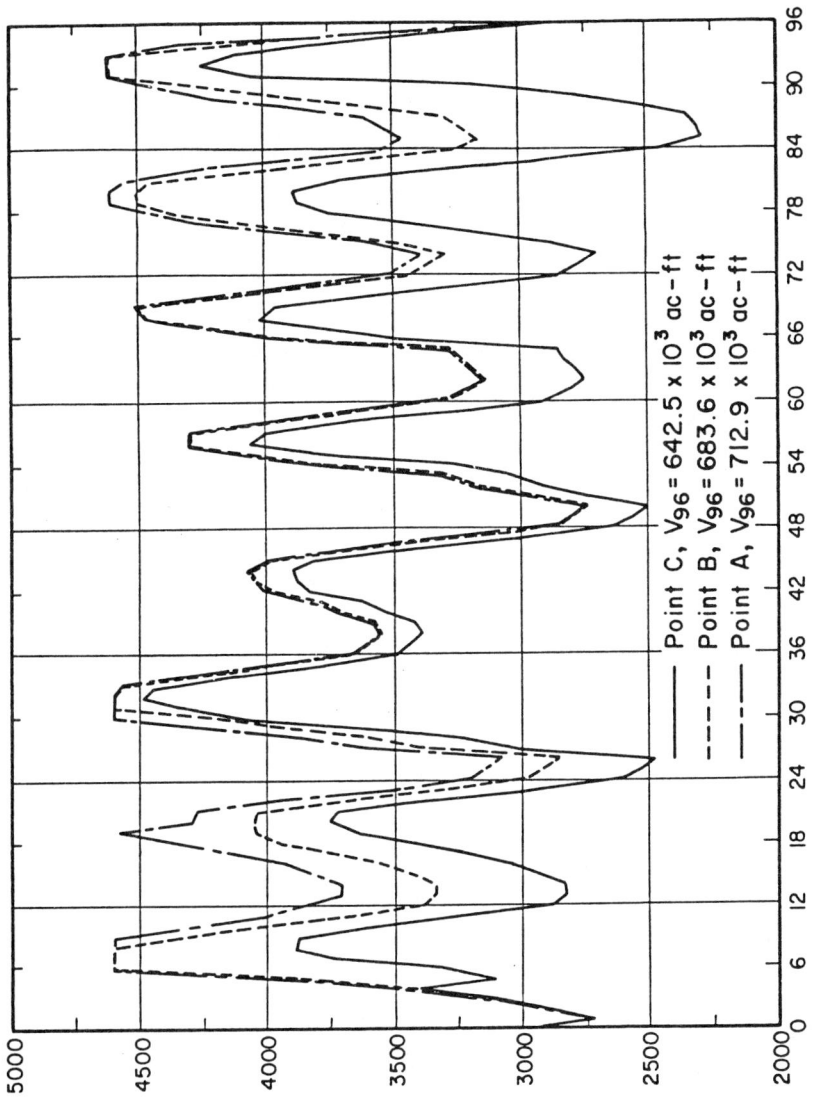

Fig. 2 Operation Policies for Select Points on Figure 1 (from Mades and Tauxe, 1980)

Discussion

The trade-off is that of dump energy versus evaporation. Since more energy and less evaporation are both desirable, the scale of the abscissa of Figure 2 is reversed to better show these trade-offs. Trade-offs along this negatively sloped curve may be interpreted as increasing "evaporation saved" during the period of analysis at the expense of dump energy. Beginning at the maximum dump energy, one observes from Figure 1 that relatively little energy is sacrificed for the first increase in evaporation saving but later on, when less evaporation requires limiting the operational policy to keep the storages down, more energy must be sacrificed. This process may be continued to the point where the minimum evaporation (maximum savings) dictates that the storage level be raised only to obtain sufficient head to obtain the firm energy as through-put energy. At this point, the dump energy results from run-of-the-river type flows through the power plant whose storage changes are dictated by firm energy production requirements.

Finally the decision maker must be concerned with the operational policies that generate the desired levels of each objective. Figure 2 was developed from information generated by a single execution of the MODP model followed by three traceback runs. The policy corresponding to point A in Figure 1 seeks to keep the reservoir as full as possible. This is expected since point A corresponds to little evaporation "saved", i.e. this policy results in the greatest evaporation. Conversely, point C's policy keeps the reservoir low so that evaporation is minimized. From Figure 2, the decision maker readily observes that increasing evaporation "savings" is at the expense of decreasing the total dump energy produced. Points A and B share a common policy during months 32 and 70 because dump energy production is severely limited by low inflows. There is no water to trade-off between evaporation "savings" and dump energy production during this period of an annual firm energy commitment of 10.5×10^5 MWH. Such information can help the decision maker select alternative combinations of objectives from the non-inferior solution set of Figure 1.

This illustrative example has shown how MODP readily provides the non-inferior set of solutions to multi-objective problems and the trade-offs between the objectives. In addition, the decision maker can easily determine the operational policies associated with these solutions.

CONCLUSIONS

Multi-Objective Dynamic Programming has been presented as a technique for quantitatively analyzing a variety of water resources problems involving non-commensurable objectives. The technique make possible the analysis of objectives that may not be handled as easily or accurately with other optimization methods. In application, the MODP problem formulation is straightforward and the problem solution is computationally feasible. Although MODP adds one additional dimension to the state space for each additional secondary objective, it generally adds fewer decision variables than required by other techniques.

REFERENCES

1. Bellman, R., Dynamic Programming, Princeton Univ. Press, New Jersey, 1957.
2. Chow, V.T., D.R. Maidment, and G.W. Tauxe, Computer Time and Memory Requirements for DP and DDDP in Water Resources Systems Analysis, Water Resour. Res., 11(5), 621-628, 1975.

3. Cohon, J.L., Ann Assessment of Multiobjective Solution Techniques for River Basin Planning Problems, PhD Thesis, M.I.T., Cambridge, Mass., 1973.
4. Goicoechea, A., D. Hansen and L. Duckstein, <u>Multiobjective Analysis with Engineering and Business Applications</u>, 519 pp., John Wiley, New York, 1982.
5. Haimes, Y.Y., <u>Heirarchical Analysis of Water Resources Systems</u>, McGraw-Hill Book Co., New York, NY, 1977.
6. Haimes, Y.Y. and W.A. Hall, Multiobjectives in Water Resources Systems Analysis: The Surrogate Worth Trade-Off Method, <u>Water Resources Res.</u>, 10(4), 1974, pp. 615-624.
7. Haimes, Y.Y., W.A. Hall and H.T. Freedman, <u>Multiobjective Optimization in Water Resources Systems: The Surrogate Worth Trade-Off Method</u>, Elsevier Scientific Publishing Company, The Netherlands, 1975.
8. Hillier, F.S. and G.J. Lieberman, <u>Operations Research</u>, Holden Day, San Francisco, 1986.
9. Mades, Dean M. and G.W. Tauxe, <u>Models and Methodologies in Multi-Objective Water Resource Planning</u>, Research Report No. 150, University of Illinois Water Resources Center, 1980.
10. Orlovski, S., S. Rinaldi, and R. Soncini-Sessa, A Min-Max Approach to Reservoir Management, <u>Water Resource Research</u>, 20(11), 1506-1514, 1984.
11. Reid, R.W. and V.V. Vemuri, On the Non-inferior Index Approach to Large-Scale Multi-Criteria Systems, <u>J. Franklin Inst.</u>, 291(4), 241-254, 1971.
12. Tauxe, G.W., Joint Operation of a Multiple Purpose Linked Reservoir System with Multi-State Incremental Dynamic Programming, PhD Thesis, University of California, Los Angeles, 1973.
13. Tauxe, G.W., R.R. Inman and D.M. Mades, Multi-Objective Dynamic Programming with Application to a Reservoir, <u>Water Resour. Res.</u>, 15(6), 1979, pp. 1403-1408.
14. Tauxe, G.W., R.R. Inman and D.M. Mades, Multi-Objective Dynamic Programming: A Classic PRoblem Redressed, <u>Water Resour. Res.</u>, 15(6).

OBJECTIVE-SPACE DYNAMIC PROGRAMMING APPROACH TO
MULTIDIMENSIONAL PROBLEMS IN WATER RESOURCES

JOHN W. LABADIE* AND DARRELL G. FONTANE*
*Department of Civil Engineering, Colorado State University,
Fort Collins, Colorado 80523

ABSTRACT

The "curse of dimensionality" continues to represent the greatest obstacle to full application of dynamic programming to complex sequential decision problems in water resources planning and management. A new approach to solving high dimensional problems is presented which conditions solutions on the one-dimensional objective-space rather than the high dimensional state-space. Sufficient conditions for global optimality are presented which are based on certain uniqueness requirements in the optimization. Aside from specification of the countability of the finite subset of decision variables, no other assumptions on problem structure or functional characteristics are necessary, including differentiability, convexity, or even continuity. Case studies in optimal reservoir operations and irrigation scheduling are presented to demonstrate successful application of objective-space dynamic programming to problems involving up to a 30-dimensional state-space.

INTRODUCTION

Dynamic programming is a powerful and versatile tool for solving a wide range of sequential decision problems in water resources. The technical literature abounds in an enormous variety of applications of discrete dynamic programming to water resource systems planning, design and operations. Labadie [24] has compiled several applications with emphasis on water resources management and Yakowitz [42] has summarized a large number of studies which demonstrate the flexibility and robustness of dynamic programming in attacking highly diverse problems.

The popularity of dynamic programming arises from a number of important advantages that it holds over many other mathematical programming techniques. These include: (i) efficient enumeration for sequential decision problems defined over discrete or integral decision sets; (ii) modest (approximately linear) increase in computational effort as a function of the number of stages in the problem, while most other methods display a geometric increase; (iii) attainment of globally optimal solutions (in a discrete sense) in the presence of functional nonlinearity, nonconvexity, and even discontinuity; (iv) exploitation of state-space and policy or decision-space constraints as a means of actually alleviating the computational burden rather than aggravating it as in other optimization methods; (v) provision of flexible feedback or closed-loop decision policies as a byproduct of the recursive calculations, whereas most other methods produce open loop policies only; (vi) particular facility with stochastic optimization problems and direct inclusion of conditional risk constraints in the optimization problem (Sniedovich [39]). The reader is referred to Dreyfus and Law [11] and Cooper and Cooper [9] for more complete discussion of the characteristics and advantages of discrete dynamic programming.

In spite of these advantages, the "curse of dimensionality" (Bellman [4]) continues to be the primary obstacle to full application of dynamic programming to realistic problem formulations in water resources. The computer storage and processing requirements of dynamic programming increase dramatically with the state-space dimension, where problems in excess of three state variables are normally considered computationally infeasible. These difficulties have been encountered by a number of researchers involved in solving large dimensional dynamic programming problems in water resources such as Cohen [10], Pereira and Pinto [35], and Grygier and Stedinger [20].

A number of techniques have been proposed for ameliorating the dimensionality problems associated with dynamic programming. A compendium of these methodologies provided by Morin [31] is still an excellent reference source, and includes techniques such as:

1. **fathoming**, which eliminates unnecessary system states by establishing bounds on the dynamic programming optimal value or return function [see Akileswaran, et al. [1] for an application to project capacity expansion].

2. **reaching procedures**, whereby only feasible or "reachable" states are retained in computer storage [see Pleban, et at. [36] for application of reaching dynamic programming to a 10 dimensional problem in real-time multi-field irrigation scheduling]

3. **approximation techniques** involving (a) functional approximation of the optimal return function over a coarse state-space grid [see Labadie, et. al. [23] for an application of orthogonal polynomials in an approximation scheme applied to linked flood detention reservoirs]; (b) sequential quadratic approximation of the optimal return function as employed in differential dynamic programming [see Murray and Yakowitz [33] for application to optimization of multireservoir systems]; (c) successive approximation approaches which optimize over the state-space one coordinate direction at a time [Giles and Wunderlich [17] apply this approach to linked reservoir systems]; and (d) coarse grid methods which start with a large state discretization interval spanning the original state-space and then successively "splice" the interval while defining increasingly more restrictive corridors around successive solutions.

4. **nearest neighbor techniques** such as state increment dynamic programming and discrete differential dynamic programming [see Larson [26], Hall, et al. [21] and Heidari, et al. [22] for applications in water resources].

A variety of other methods have also been employed for circumventing the dimensionality problem, such as use of Lagrange multipliers [see Rossman [38] for joint application of generalized duality theory and dynamic programming in water resources], state-space decomposition methods, minimum state representations, and efficient data management structures. Morin and Esogbue [32] developed a method for solving a particular class of large dimensional dynamic programming problems where the objective function is composed of step functions. They prove that in this case, the dynamic programming optimal value or return function will also be a discontinuous step function. Identification of the points of discontinuity enables a drastic reduction in the number of points that need to be evaluated in the state-space. This approach appears to be particularly well-suited to large-scale capacity expansion problems in water resources.

The major difficulty with popular methods employed thus far is that they are either still quite sensitive to the dimensionality problem, although not to as great an extent, or rely heavily on heuristics and exploitation of special problem structures. The danger in the various approximation methods is that the original problem is not really being solved, but rather one which is approximate to it in varying degrees of accuracy.

Labadie, et al. [25] developed a new category of techniques for solving high dimensional dynamic programming problems that condition optimal solutions on the one dimensional objective-space rather than the multidimensional state-space. In this approach, a one dimensional dynamic programming formulation in objective-space replaces a high dimensional dynamic programming problem involving the usual discretization of the state-space. Cooper [8] developed a technique which uses objective-space concepts which is applicable to a certain class of multidimensional resource allocation problems. The method is referred to as a "hypersurface search technique" which, for maximization problems, involves stage-wise selection of discrete, initially infeasible objective values which converge non-monotonically from above to the greatest feasible lower bound in a finite number of steps. Tauxe, et al. [40] have also employed objective-space concepts in dynamic programming for solving multiobjective problems in water resources, but have conditioned solutions on the state-space as well. This, unfortunately, results in further intensification of the dimensionality problem. Becker and Yeh [3] employed a technique similar in concept to objective-space dynamic programming, but further comparative studies by Grygier and Stedinger [20] have shown that this method results in suboptimal solutions.

Sufficient conditions for global optimality of objective-space dynamic programming solutions are presented. These conditions are based only on certain uniqueness requirements which are believed to be applicable to a large class of multidimensional problems in water resources. No other assumptions about problem structure are necessary. However, without the uniqueness requirements, optimality cannot be guaranteed. The applicability of objective-space dynamic programming is demonstrated through two case studies involving complex problems in reservoir operations for water quality management and irrigation scheduling. In spite of these applications being characterized by high dimensional state-spaces involving up to 30 state variables, they have been successfully solved with this approach.

OBJECTIVE-SPACE DYNAMIC PROGRAMMING

Problem Definition

Consider the following finite stage sequential decision problem:

$$\min \quad \phi[f_1(x_1,u_1),\ldots,f_N(x_N,u_N)] \tag{1}$$

subject to:

$$x_{i+1} = g_i(x_i,u_i) \in X_{i+1} \tag{2}$$

$$x_1 = c \text{ (given)} \tag{3}$$

$$u_i \in U_i \tag{4}$$

$$h_i(x_i,u_i) \leq 0 \tag{5}$$

$$(i = 1,\ldots,N)$$

where the operator ϕ satisfies conditions of separability, monotonicity, and decomposition (Nemhauser [34]); $x_1 \in E^n$, $x_{i+1} \in E^n$, $u_i \in E^m$, $g_i : E^{m+n} \to E^p$, $h_i : E^{m+n} \to E^q$, and $f_i : E^{m+n} \to (E^1)^+$; and X_{i+1} and U_i are compact with subset U_i further assumed to be finite and countable for all stages $i = 1, \ldots, N$. No restrictions on the structure of the mappings f_i or g_i are assumed, except that they are real-valued and a finite minimum exists. In addition, it is assumed that finite bounds on the objective values are available for each individual stage:

$$0 \le f_{min,i} \le f_i(x_i, u_i) \le f_{max,i} \qquad (6)$$

$$(i = 1, \ldots, N)$$

The most common operator ϕ for the objective function is the additive structure

$$\phi[f_1(x_1, u_1), \ldots, f_N(x_N, u_N)] = \sum_{i=1}^{N} f_i(x_i, u_i) \qquad (7)$$

but other operators such as min-max can be used with the objective-space approach. The additive operator will be used here to facilitate explanation of the algorithm, but the use of other operators will be highlighted in one of the applications presented in a subsequent section. Also, many sequential decision problems have terms in the objective function related to the final state of the system of the form $\bar{f}_{N+1}(x_{N+1})$. Here we assume that this term is incorporated into $f_N(x_N, u_N)$ through use of the state transformation equation in (2) for stage $i = N$.

In this formulation, we have assumed that the feasible decisions u_i belong to a finite countable set a priori; i.e., the decision variables are already discrete or integral and no arbitrary discretization is applied. In order to solve this problem by discrete dynamic programming, a discretization scheme for the state-space must be defined where the original sets X_i are replaced with countable subsets $\bar{X}_i$. Bertsekas [6] has proved the convergence of discrete dynamic programming to the global optimal solution in the limit as the state-space discretization approaches zero under the assumption that the functions f_i and g_i satisfy the Lipschitz conditions.

The backwards dynamic programming formulation of this problem defines the real-valued dynamic programming optimal cost or return functional $F_i(.)$ which is calculated recursively starting with $F_{N+1}(.) \triangleq 0$:

$$F_i(x_i) = \min_{u_i \in U_i} \{\phi[f_i(x_i, u_i), F_{i+1}(x_{i+1})]\} \qquad (8)$$

subject to:

equations (2), (4), (5) and (6)

The dimensionality difficulties are evident in the need to compute and store $F_i(\mathbf{x}_i)$ for all discrete combinations of $\mathbf{x}_i$. For an average of M discretizations of each state coordinate per stage, allocations for M^n values of $F_i(.)$ must be made available in the computer main memory, which is generally infeasible for a reasonable number of discrete values (say 50 to 100) and state dimension exceeding three. Although high speed disk storage can be used, the access time is several orders of magnitude slower than retrieval of information from main memory, so excessive computer time becomes the primary reason for the infeasibility of solving high dimensional problems. Again, it is important to notice that claiming achievement of the global optimal solution to equations (1) to (6) is only valid in the limit as the discretization interval approaches zero and therefore $M \rightarrow \infty$. Practically speaking, however, a reasonable discretization interval is generally established in dynamic programming from which a conditional (discrete) global optimum is claimed.

Concept of g-Uniqueness

Assuming an additive operator is utilized for the objective function, suppose we knew beforehand what the optimal objective values f_i^* were for each stage such that their summation resulted in the minimum total objective value, but we were given no other information on the optimal solution. Could we reconstruct the optimal decisions and states from this information?

Let:

$$F_{i+1}^* = F_i^* + f_i^* \qquad (9)$$

$$(i = 1, \ldots, N)$$

where we initialize

$$F_1^* \triangleq 0 \qquad (10)$$

Assuming

$$f_i^* \geq 0 \qquad (11)$$

then

$$F_{i+1}^* \geq F_i^* \quad \forall i \qquad (12)$$

Since the optimal policy produced the F_i^* values, then any other policy generating accumulated objective values F_i must result in

$$F_{N+1} \geq F_{N+1}^* \qquad (13)$$

since F_{N+1}^* is the minimum objective value or greatest lower bound L_{N+1}^*. This is illustrated in Figure 1. Note that there is no loss in generality in this analysis by using a minimizing operation. For maximization problems, we would simply be defining the least upper bound instead.

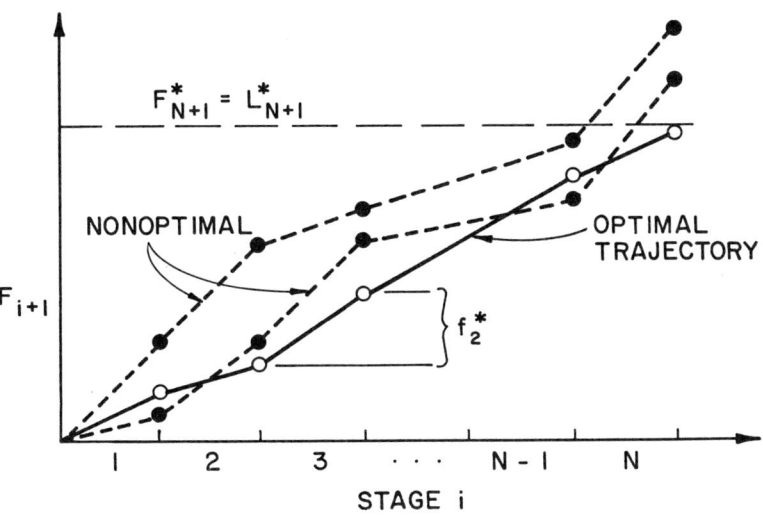

FIG. 1. Illustration of optimal trajectory in objective-space.

Suppose we start at $x_1 = c$ (given) and solve:

$$\min_{u_1 \epsilon U_1} \{f_1(x_1, u_1) + F_1^*\}$$

subject to:

$$f_1(x_1, u_1) + F_1^* \geq f_1^* = F_2^* = L_2^*$$

$$x_2 = g_1(x_1, u_1) \epsilon X_2$$

$$h_1(x_1, u_1) \leq 0$$

where F_1^* is initialized at 0. There may be more than one optimal solution u_1^* to this problem, but they must belong to a "g-unique" set.

<u>Definition</u>: A nonempty set $A \subset E^n$ is called "g-unique" if a map $g(.)$ is constant over A.

The term g-uniqueness was defined by Banerjee [2] and has been useful in analyzing differentiability requirements for the generalized dual function in mathematical programming. Notice that a unique optimum (i.e., set A is composed of a single element) is a special case of g-uniqueness.

For **stage 1**, we assume the mapping $g_1(x_1, u_1^*) = x_2^*$ remains constant over the set A_1 of all nonunique u_1^*. Clearly, without the condition of g-uniqueness, there might be a large number of candidate states x_2^* which meet the target objective bound and would therefore need to be carried forward to the next stage. Combinatorially, it would not take very many of these "ties" before available computer memory is exhausted or computer time becomes prohibitive.

Under the assumption of g-uniqueness, we must have

$$f_1(x_1, u_1^*) = f_1^*$$

For **stage 2**, let:

$$F_2^* = F_1^* + f_1^*$$

and solve

$$\min_{u_2 \in U_2} \{f_2(x_2^*, u_2) + F_2^*\}$$

subject to:

$$f_2(x_2^*, u_2) + F_2^* \geq F_3^* = L_3^*$$

$$x_3 = g_2(x_2^*, u_2) \in X_3$$

$$h_2(x_2^*, u_2) \leq 0$$

where x_2^* is the unique state computed from stage 1. Again, assuming g-uniqueness, we find

$$x_3^* = g_2(x_2^*, u_2^*)$$

Continuing in this manner, we finally reach **stage N** and solve:

$$\min_{u_N \in U_N} \{f_N(x_N^*, u_N) + F_N^*\}$$

subject to:

$$f_N(x_N^*, u_N) + F_N^* \geq F_{N+1}^* = L_{N+1}^*$$

$$x_{N+1} = g_N(x_N^*, u_N) \in X_{N+1}$$

$$h_N(x_N^*, u_N) \leq 0$$

and find optimal u_N^*.

The key point in this analysis is g-uniqueness of the system state transformation relations. Without it, we cannot find optimal $u_1^*, \ldots, u_N^*$ and $x_2^*, \ldots, x_{N+1}^*$ with just knowing $f_1^*, \ldots, f_N^*$.

Objective-Space Formulation

Suppose we now relax that assumption that we know $f_1^*, \ldots, f_N^*$. Let us then discretize the objective-space into intervals Δf and define a corridor bounded according to given $f_{min,i}$ and $f_{max,i}$ ($i = 1, \ldots, N$) such that we are sure the optimal solution lies within it (e.g., dictate smoothness conditions), as shown in Figure 2.

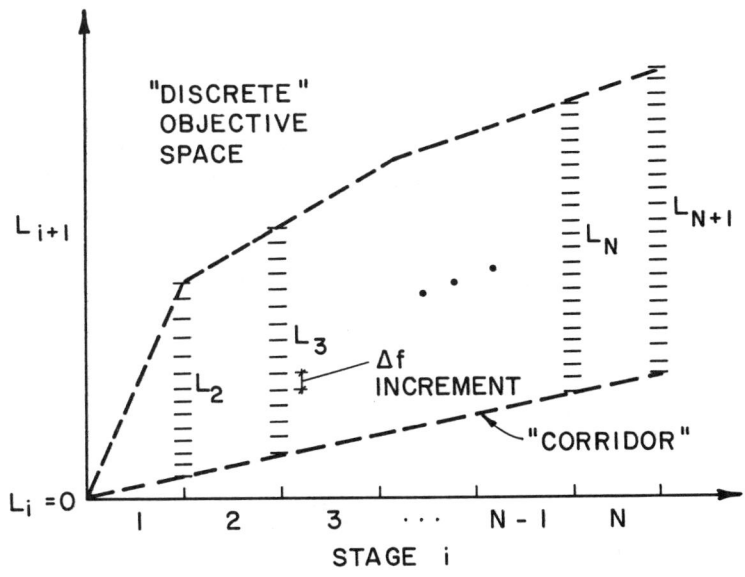

FIG. 2. Discretization of objective-space within prescribed "corridor".

Starting with $L_1 \triangleq 0$, we have:

$$\sum_{j=1}^{i} f_{min,j} \leq L_{i+1} \leq \sum_{j=1}^{i} f_{max,j} \qquad (14)$$

$$(i = 1,\ldots,N)$$

Stage 1: for all discrete lower bounds L_2 in intervals Δf satisfying (14) find

$$F_2(L_2) = \min_{u_1 \in U_1} \{f_1(x_1,u_1) + F_1\}$$

subject to:

$$f_1(x_1,u_1) + F_1 \geq L_2$$

$$x_2 = g_1(x_1,u_1) \in X_2$$

$$h_1(x_1,u_1) \leq 0$$

where F_1 is initialized at 0. We store the actual system states $x_2^*(L_2)$ as a function of discrete bounds L_2, which will be unique for each bound if g-uniqueness applies to the state transformation relations.

Because of the objective-space discretization Δf, we will have:

$$F_2(L_2) \geq L_2$$

Stage 2: for all discrete L_3, find

$$F_3(L_3) = \min_{L_2,\ u_2 \in U_2} \{f_2(x_2^*(L_2),u_2) + F_2(L_2)\}$$

subject to:

$$f_2(x_2^*(L_2),u_2) + F_2(L_2) \geq L_3$$

$$x_3 = g_2(x_2^*(L_2),u_2) \in X_3$$

$$h_2(x_2^*(L_2),u_2) \leq 0$$

For this stage, and all subsequent stages, the concept of g-uniqueness is extended to the assumption that for given discrete bound L_3 the mapping g_2 is assumed to be constant over all (possibly) nonunique optimal solutions L_2^* and u_2^*. For each discrete bound L_3, therefore, we obtain optimal (and possibly nonunique) $L_2^*(L_3)$, $u_2^*(L_3)$ and unique optimal $x_3^*(L_3)$ and store in computer memory. The latter are particularly important since we will need to know the "true" state of the system for the stage 3 calculations for each discrete bound L_3. Again, the uniqueness of $x_3^*(L_3)$ is a direct consequence of the g-uniqueness assumption.

Continuing in this manner through the remaining stages, the problem for **stage N** is solved for all discrete bounds L_{N+1}:

$$F_{N+1}(L_{N+1}) = \min_{L_N,\ u_N \in U_N} \{f_N(x_N^*(L_N), u_N) + F_N(L_N)\}$$

subject to:

$$f_N(x_N^*(L_N), u_N) + F_N(L_N) \geq L_{N+1}$$

$$x_N = g_N(x_N^*(L_N), u_N) \in X_{N+1}$$

$$h_N(x_N^*(L_N), u_N) \leq 0$$

A typical function $F_{N+1}(L_{N+1})$ is illustrated in Figure 3.

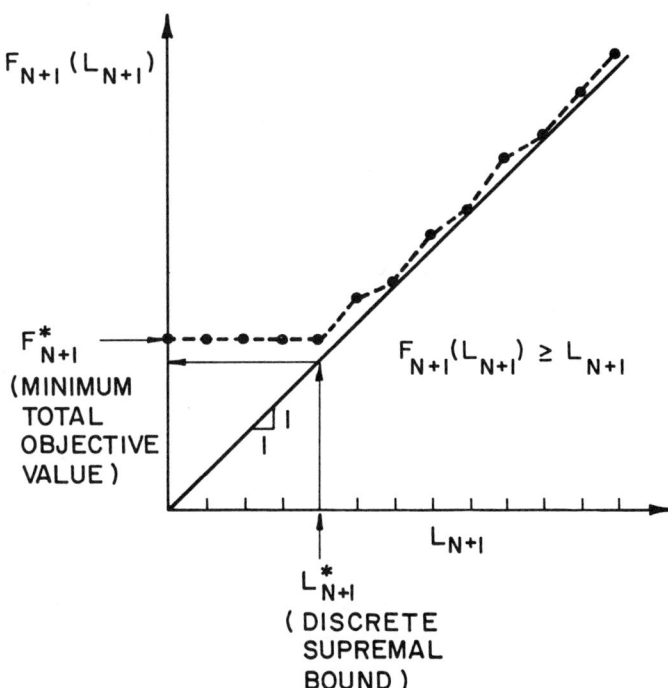

FIG. 3. Illustration of final stage optimal cost functional conditioned on bound L_{N+1}.

We will now attempt to find the supremal bound L_{N+1}^* on the minimum total objective value by solving:

$$\sup_{L_{N+1}} F_{N+1}(L_{N+1}) \quad (15)$$

where $F_{N+1}(L_{N+1}^*) \geq L_{N+1}^*$ is the minimum total objective value.

We can now initiate a traceback procedure to find the optimal decision policy. Having found ... $u_N^*(L_{N+1}^*)$ and $L_N^*(L_{N+1}^*)$, then

find $u_{N-1}^*(L_N^*)$ and $L_{N-1}^*(L_N^*)$, then

$\vdots$

find $u_2^*(L_3^*)$ and $L_2^*(L_3^*)$, then

find $u_1^*(L_2^*)$

A limited set of feedback decision policies can be computed for each stage from stored arrays $u_i^*(L_{i+1})$, $x_{i+1}^*(L_{i+1})$, and $L_i^*(L_{i+1})$. That is, we can define:

$$u_i^*(x_i) = u_i^*(x_i^*(L_i^*(L_{i+1}))) \text{ for all discrete } L_{i+1}. \quad (16)$$

$$(i = 1, \ldots, N)$$

Proposition

Consider a sequential decision problem of the form of equations (1) to (6) which is solved by the following recursive algorithm in forward direction over stages $i = 1, 2, \ldots, N$ over all discrete bounds L_{i+1} satisfying equation (14) in discrete intervals Δf:

$$F_{i+1}(L_{i+1}) = \min_{L_i, u_i \in U_i} \{\phi[f_i(x_i^*(L_i), u_i), F_i(L_i)]\} \quad (17)$$

subject to:

$$\phi[f_i(x_i^*(L_i), u_i), F_i(L_i)] \geq L_{i+1} \quad (18)$$

$$x_{i+1} = g_i(x_i^*(L_i), u_i) \in X_{i+1} \quad (19)$$

$$h_i(x_i^*(L_i), u_i) \leq 0 \quad (20)$$

(for $i = 1, \ldots, N$)

where x_1 is given, $L_1 \triangleq 0$, $F_1(.) \triangleq 0$, and optimal $x_{i+1}^*(L_{i+1})$ are stored for each stage and each discrete bound L_{i+1}. A global optimal solution u_i^* ($i = 1,\ldots,N$) will result from application of these recursive calculations in the limit $\Delta f \to 0$ **if** g-uniqueness applies at all stages of the calculations.

proof: It is sufficient to show that if an optimal feasible solution exists for the problem defined by (1) to (6) which produces optimal objective values at each stage of $f_1^*,\ldots,f_N^*$, then under the stipulated conditions of g-uniqueness, the recursive objective-space algorithm defined by (17) to (20) will be able to find this solution in the limit $\Delta f \to 0$. We will begin by defining the sets:

$$S_i = \{F_i \mid F_i = F_i(L_i) \text{ for all } L_i\} \quad (21)$$

$$(i = 2,\ldots,N+1)$$

and

$$S_i^*(L_{i+1}) = \{F_i \mid F_i = F_i(L_i^*) \text{ for all } L_i^* \text{ solutions to (17) to (20) for given } L_{i+1}\} \quad (22)$$

$$(i = 1,\ldots,N)$$

Beginning at stage 1, it is clear that

$$f_1^* = F_2^* \in S_2 \quad \text{in the limit } \Delta f \to 0 \quad (23)$$

If this were not true, it would contradict the assumption that f_1^* is an attainable optimum for stage 1. By the assumption of g-uniqueness, there must exist a unique state $x_2^*(F_2^*)$. Moving now to stage 2, it follows by the same argument used at stage 1 that

$$\phi(f_1^*, f_2^*) = F_3^* \in S_3 \quad \text{in the limit } \Delta f \to 0 \quad (24)$$

Can we now show that $f_1^* = F_2^* \in S_2^*(F_3^*)$ in the limit? Because of the g-uniqueness assumption, we know there exists a unique state $x_3^*(F_3^*)$. This implies that there exists u_2^* such that:

$$\phi[f_2(x_2^*(F_2^*), u_2^*), F_2(F_2^*)] = F_3^* \quad \text{in the limit } \Delta f \to 0 \quad (25)$$

and

$$x_3 = g_2(x_2^*(F_2^*), u_2^*) \in X_3 \quad (26)$$

which implies that $f_1^* = F_2^* \in S_2^*(F_3^*)$ in the limit $\Delta f \longrightarrow 0$. By induction, we can continue this argument through the remaining stages and show that indeed the optimal values $f_1^*, \ldots, f_N^*$ will be selected by the objective-space algorithm in the limit. ||

Practical Considerations

In order for the objective-space dynamic programming procedure to have any practical applicability in solving multidimensional problems in water resources, the following conditions must be met:

1. The optimization problem defined by equations (17) to (20) must be easily solvable for all discrete bounds L_i. The countability of the U_i subsets requires that some type of efficient enumerative procedure be used. Direct enumerative or "brute-force" techniques may perform reasonably well if the constraints (14) and (18) to (20) are suitably restrictive so as to greatly limit the number of feasible combinations. This is actually the case in many realistic water resource systems where there may exist large numbers of hydraulic, administrative, legal and other restrictions on operation and management of the system. In addition, imposition of realistic constraints will also provide greater assurance of satisfying the g-uniqueness conditions. In effect, it seems that the more complex the system, the better the objective-space dynamic programming approach is likely to perform.

2. Although countability of the finite subsets U_i is not strictly necessary for solution by objective-space dynamic programming, it is unlikely that solution of the problem defined by equations (17) to (20) will result in g-unique policies without this requirement. That is, there may be an infinite variety of ways to exactly meet the specified discrete objective bounds L_i, in which case the g-uniqueness assumption no longer applies. As an example, Becker and Yeh [3] presented an algorithm for determining the maximum firm energy produced from a large interconnected hydropower system. As mentioned previously, an approach very much like the objective-space dynamic programming procedure was utilized, but without stipulation of a countable set U_i or g-uniqueness requirements. This would normally result in an infinite set of optimal reservoir release policies that could achieve any particular objective target or bound specification. In order to circumvent this problem, Becker and Yeh [3] solve an imbedded linear programming subproblem which essentially introduces an additional objective not present in the original formulation of the problem. The purpose is to find a unique optimal solution for the specified objective target bounds. A potential energy criterion is maximized subject to satisfying specified energy targets for the current stage of the decision process. This criterion appears reasonable enough, but unfortunately it cannot guarantee that fully dynamic optimal solutions are being found over the entire decision horizon. This has been documented by Grygier and Stedinger [20] through comparisons of this procedure with other optimization methods. Again, the key difficulty is the lack of g-uniqueness of the original problem formulation, which brings into question whether objective-space dynamic programming can be used to solve this class of multidimensional problems.

3. Since the objective-space dynamic programming formulation involves discretizations over an "accumulated" objective-space (assuming the additive operator is used), it is important that the algorithm is relatively insensitive to the level of discretization Δf. That is, the practicality of the algorithm is diminished if the interval must be reduced to an extremely

small value before reasonably good solutions are obtained. It should be noted, however, that sensitivity to discretization interval and grid size can also be a serious difficulty in state-space dynamic programming and is certainly not a unique problem with the objective-space approach. Ideally, it would be beneficial to be able to establish error bounds as a function of Δf, but this would likely require certain smoothness conditions to be imposed on the problem structure. If sensitivity to the objective discretization is a problem, it may still be beneficial to use the objective-space approach for obtaining initial solutions which can be further refined by other methods.

4. Problems should be solved under various discretization schemes and the resulting solutions compared. As in any optimization procedure requiring discretization, a good deal of experience may be required to define appropriate levels.

APPLICATIONS

Reservoir Operations for Water Quality Control

Problem definition: The construction of a reservoir in a river system can result in dramatic changes in the natural physical, biological, chemical and thermal characteristics of the river both upstream and downstream of the project. In order to better meet downstream water quality goals, several dams in the U.S. have been fitted with special selective withdrawal structures whereby water can be withdrawn at various levels in the project, mixed and then released downstream. Since water quality goals are considered secondary in the operation of most projects, the daily quantity of water to be released is generally predetermined on the basis of the primary project purposes such as flood control, hydroelectric generation, and water supply. With the daily release quantity specified, the problem is to find the best way of operating the ports in the selective withdrawal structure so as to meet downstream temperature and water quality targets as closely as possible.

Selective withdrawal facilities are generally operated in a myopic fashion whereby attempts are made to satisfy thermal and water quality requirements in reservoir releases for the current day. The difficulty is that this strategy may result in problems later in the season. As an example, Figure 4 shows an average seasonal temperature profile in the Allegheny River in northwest Pennsylvania before and after construction of Kinzua Dam immediately upstream. The selective withdrawal capability is effective in releasing warmer upper level water that matches the natural stream temperatures during June and July. By mid-August, however, cold water in the reservoir becomes depleted and large deviations from the natural stream temperature occur. Although the extent of these deviations is greatly influenced by hydrologic and meteorological conditions in the basin, ecologists and fishery biologists maintain that the rapid change in temperature conditions late in the season can be more deleterious to the stream ecology than the absolute deviation of the temperature profile from the target profile. This means that attempts to "smooth out" the deviations over time would be more beneficial from an ecological viewpoint. Therefore, fully dynamic optimal control strategies for selective withdrawal structure operations are needed for minimizing the detrimental impacts of river regulation projects on the stream ecology. A solution procedure utilizing objective-space dynamic programming has been presented by Fontane, et al. [15] and is summarized as follows.

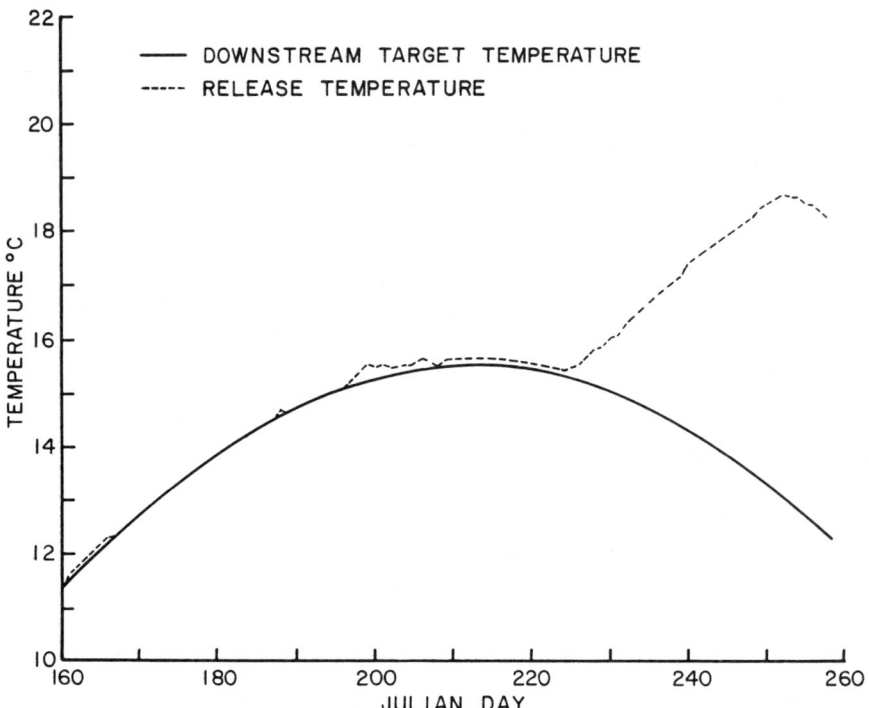

FIG. 4. Comparison of release temperature and target temperature for the case study reservoir. (Fontane, et al. [15])

<u>Objective Function</u>: The problem as stated is to find optimal strategies for regulating the selective withdrawal ports in a reservoir on a daily basis such that seasonal deviations from target temperatures are minimized. In this case, the most appropriate objective formulation is of the min-max type:

$$\min \{\max [|T_1(x_1,u_1) - \hat{T}_1|, \ldots, |T_N(x_N,x_N) - \hat{T}_N|]\} \qquad (27)$$

where $\hat{T}_i$ is the target release temperature for day i and $T_i(x_i,u_i)$ is the average release temperature during day i as a function of the current thermal state in the reservoir x_i at the beginning of day i and the port selection decisions u_i during day i.

<u>State Transformation Relation</u>: In order to solve this problem, a suitable means of simulating the thermal characteristics of the reservoir must be found. The simulation model used in this study called WESTEX was developed at the Waterways Experiment Station, U.S. Army Corps of Engineers, Vicksburg, Mississippi, and is fully documented by Loftis [27]. The model essentially provides a finite difference solution to the basic partial differential equation describing thermal dynamics in an impoundment. The model has also been used for predicting the concentration of conservative and nonconservative water quality constituents in reservoir releases.

Solution is accomplished by dividing the lake into discrete horizontal layers of homogeneous temperature, as depicted in Figure 5. Loftis [27] has shown that one-dimensional models of this type are generally acceptable if the primary use of the model is to predict release temperature. The model simulates heat transfer at the air-water interface, advective heat transfer of inflow and outflow, and internal dispersion of thermal energy. It also incorporates hydraulic models for predicting the withdrawal zone of each port and any mutual interference that might occur. This is essential for obtaining an accurate prediction of release temperature.

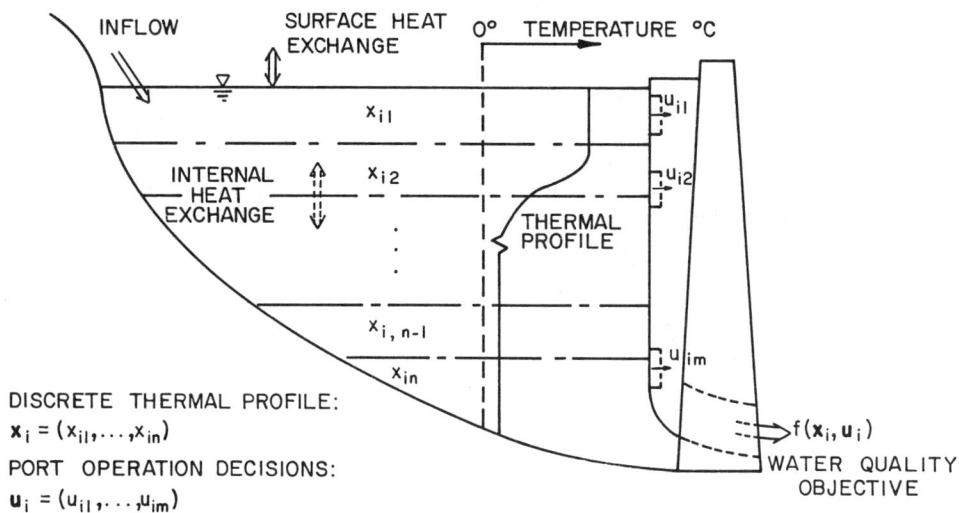

FIG. 5. Discretization of impoundment into discrete horizontal layers for thermal simulation. (Fontane, et al. [15])

The basic partial differential equation numerically solved by WESTEX is:

$$\frac{\partial T}{\partial t} = \frac{T_I Q_I}{A \Delta Z} - \frac{T_O Q_O}{A \Delta Z} + \frac{1}{A} \frac{\partial}{\partial Z} KA \frac{\partial T}{\partial Z} - \frac{1}{A} \frac{\partial (Q_v T)}{\partial Z} + \frac{1}{\rho C_p A} \frac{\partial H}{\partial Z} \quad (28)$$

with appropriate initial and boundary conditions, where T = temperature of a layer, °F; t = time, days; T_I = inflow temperature, °F; Q_I = inflow rate into layer, ft^3/day; A = horizontal cross sectional area, ft^2; ΔZ = layer thickness, ft; T_O = outflow temperature, °F; Q_O = withdrawal rate from layer, ft^2/day; Z = elevation, ft; K = vertical diffusion coefficient, ft^2/day; Q_v = net vertical flow into or out of a layer, ft^3/day; ρ = density of water, lbs/ft^3; C_p = specific heat of water, btu/lb/°F; $\partial H/\partial Z$ = external heat source, btu/ft/day.

With use of the WESTEX model, the state vector x_i represents the average temperature in each layer at the beginning of day i. The model was modified by Farber [13] so that several target temperatures and port selection strategies could be efficiently examined before the model moves to the next day in the simulation. In essence, then, the modified WESTEX model represents the state transformation relation (19). Since the finite difference solution may require a large number of horizontal layers for suitable accuracy, it follows that the vector x_i may be of extremely high dimension which will negate solution of this problem by standard dynamic programming and its extensions. In addition, the state transformation relation (19) represented by the WESTEX model is internally complex with conditional simulation logic that mitigates against analytically-based solution procedures such as optimal control theory.

Constraints: WESTEX includes an internal optimization capability in a subroutine called DECIDE which performs an enumerative procedure to determine a unique strategy for opening ports in order to meet a given release temperature target for the current day. Uniqueness is assured by the existence of a number of constraints on port operation which are necessary in order to avoid a condition called "thermal blockage" which has been observed by Fontane and Loftis [14]. These constraints represent the decision set U_i. This set is countable since once a port is selected, it is generally fully opened. Because of these constraints, it may not be possible to exactly meet a specified release temperature target. All other constraints as represented by the sets X_i and equation (20) are embodied within the simulation model.

Solution by Objective-Space Dynamic Programming: The WESTEX model is essentially driven by the specified target temperature for the current day. The key to solution of this problem is to supply targets other than the given ideal target $\hat{T}_i$ such that the seasonal maximum deviation is minimized. Within the context of the objective-space dynamic programming formulation of (17) to (20), optimization is performed over objective bound L_i as conditioned on bound L_{i+1}. For this problem, equation (18) can be written as:

$$\max\ [|T_i(x_i,u_i) - \hat{T}_i|, F_i(L_i))] \geq L_{i+1} \qquad (29)$$

It is reasonable to assume that

$$T_i(x_i,u_i) \geq \hat{T}_i \qquad (30)$$

since optimization will result in releasing slightly warmer water than the ideal target level over the season in order to "smooth out" the deviations.

If we further assume that providing any target release temperature $\bar{T}_i$ to WESTEX other than the actual ideal target level $\hat{T}_i$ will result in

$$T_i(x_i,u_i) \geq \bar{T}_i \geq \hat{T}_i \qquad (31)$$

then the target release temperature to be supplied to WESTEX can be found from

$$\bar{T}_i(L_i, L_{i+1}) = \max \{F_i(L_i), L_{i+1}\} + \hat{T}_i \qquad (32)$$

where target temperature $\bar{T}_i$ is actually a function of L_i and L_{i+1}. WESTEX then finds the optimal port selection:

$$u_i^*(\bar{T}_i) = u_i^*(L_i, L_{i+1}) \qquad (33)$$

and computes the resulting release temperature: $T_i(x_i, u_i^*(\bar{T}_i))$. Therefore, the optimization in (17) is performed over discrete L_i bounds, with an imbedded optimization on the port strategies u_i carried out internally in WESTEX.

Results: The objective-space dynamic programming algorithm was applied to finding daily optimal port selection strategies for Allegheny Reservoir during a typical 14 week period from June to September. Data for this case study were provided by the Waterways Experiment Station. The reservoir was divided into 30 discrete horizontal layers, representing a 30 dimensional dynamic programming problem in state-space. Details on data requirements and calibration of the WESTEX model can be found in Fontane, et al. [16]. The ideal target temperatures $\hat{T}_i$ were obtained from an analysis by Drummond and Robey [12] and are considered to be "an appropriate temperature regulation guideline for releases from Kinzua Dam and Seneca Powerplant if the objective is to maintain the pre-project temperature regime of the Allegheny River." Figure 6 compares the release temperatures resulting from the objective-space dynamic programming solution with the actual historical temperatures for that period. The primary difference in port operation is to release more water in the upper levels such that some deviation occurs early. This results in savings of cold water in the reservoir such that less deviation occurs later in the season.

Tests were conducted to insure that the algorithm was finding g-unique policies. Tie-breaking procedures were incorporated which confirmed that unique solutions were being computed. Computational requirements for solving this problem by the objective-space approach were modest both in main memory and execution time. Although a large main frame computer system was used for this study, it is believed that microcomputer implementation of the algorithm is feasible. It was possible to define relatively tight bounds on the feasible objective-space (equation (6)) primarily because historical data on actual operation of the reservoir were available. Extremely small objective-space discretizations were not necessary for this problem. It should also be noted that the optimization of equations (17) to (20) was actually performed in this study using a least-squares penalty term as a measure of proximity to the target temperature bound. Under this criterion, it is possible to obtain a solution slightly under the specified bound. It is believed, however, that enforcement of a strict bound would have had little effect on the solution.

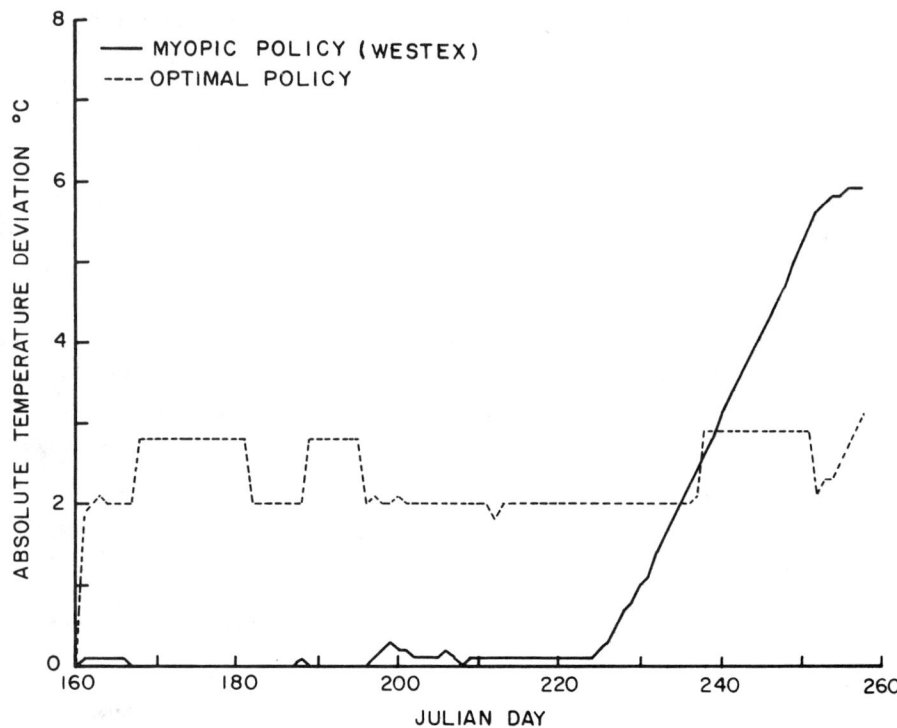

FIG. 6. Deviations of release temperature from target temperature. (Fontane, et al. [15])

The solution obtained in this study is based on perfect foreknowledge of future hydrologic and meteorological conditions. It is envisioned that this algorithm could be applied to actual real-time operations where port control strategies are updated using the objective-space algorithm on a weekly or even daily basis as new forecasts become available. Second-order statistics on objective values could be included in the total objective-space if statistical information was available on forecast errors. It would also be possible to directly include additional water quality constituents, such as dissolved oxygen and suspended solids, in a multidimensional objective-space where the bounds are treated as vectors L_i. Pareto optimal solutions could then be examined and tradeoffs between the objectives analyzed.

Optimal Irrigation Scheduling over a Season

Problem definition: As competition for water and the costs of applying it increase, more attention is being directed towards optimization of on-farm irrigation practices. It is critical that the maximum possible net return per unit of water applied is realized. Martin, et al. [29] have presented a study which employs objective-space dynamic programming for finding optimal scheduling policies for center pivot irrigation over an entire season. The uniqueness of this study is characterized by an attempt to model the complex plant-water-irrigation system as realistically as possible and to include restrictions on total water volume available over the season. Most previous studies have avoided these aspects because of the computational complexity of the resulting optimization formulation. Volume restrictions are becoming necessary as available water resources are being depleted by heavy usage. This is particularly true of deep groundwater resources such as the Ogallala formation in the High Plains region of the U.S.

Model formulation: The various components of the model used in this study include:

1. daily **soil moisture balance** for the fields which accounts for the effects on soil moisture state x of evaporation E, transpiration T, net irrigation application I, effective rainfall R, and drainage lost from the crop root zone D. The soil profile is represented by several horizontal layers in order to effectively model nonhomogeneous soil characteristics and properly account for soil moisture depletion and its effect on crop growth at various stages. "Piston flow" drainage concepts are employed in the model where upper layers are always filled to capacity first. The soil moisture balance for each layer is accounted for daily, but actual decision stages are in groups of τ days. Within a particular decision stage i, we have:

$$x_{ijk,t+1} = x_{ijkt} + \alpha(I_{ikt} + R_{ikt}) - \beta E_{ijkt} - T_{ijkt} - D_{ijkt} + (1-\alpha)D_{i,j-1,kt} \qquad (34)$$

with

$$x_{i+1,jkl} \stackrel{\Delta}{=} x_{ijk,\tau+1} \qquad (35)$$

$(i = 1,\ldots,N; \; t = 1,\ldots,\tau; \; j = 1,\ldots,J; \; k = 1,\ldots,K)$

where subscript t represents the discrete time period used in the simulation (in this case, daily), j is the discrete soil layer, and k is a specific irrigated area under the pivot. It is assumed that initial soil moisture levels are given. This equation corresponds to the state transformation relation of (19). Note that $\alpha = 1$ for the top surface layer $j = 1$, and 0 for all other layers. For evaporation, $\beta = 1$ for the top two layers (4 cm and 11 cm deep, respectively) and 0 for all other layers.

Potential evapotranspiration is calculated using the modified Penman method. Evaporation and transpiration are then separated into individual terms using a method based on the work of Childs and Hanks [7]. Actual evaporation is estimated using the concept of "energy-limiting" and "soil-limiting" phases as proposed by Ritchie [37]. Uptake of soil moisture by the crops is calculated for each soil layer using the actual transpiration demand and a weighting factor taking into account extractable amount of water in that layer and the crop root density distribution. The latter was based on a function proposed by Grimes, et al. [19]. Drainage D_{iJkt} leaving the bottom layer is assumed to be lost.

2. a **crop yield** model for predicting actual crop yield (in this case, corn) by the end of the season based upon transpiration deficits occurring at various stages of crop growth. The growth stages for corn are generally labeled vegetative, reproductive, and grain-fill. Transpiration is assumed to be controlled by plant growth status and the amount of soil moisture in each layer. Several yield models were examined, but a simple linear production function was found to predict yields as well as the more complicated models. For each area k:

$$Y_k = Y_{max,k} [1 - \gamma(\sum_{i=1}^{N} \sum_{t=1}^{\tau} \Delta T_{ikt} / T_{max,k})] \qquad (36)$$

where Y_k is actual crop yield, $Y_{max,k}$ is maximum potential yield, ΔT_{ikt} is the actual transpiration deficit incurred during decision stage i and day t, where the deficit is defined as the difference between the total transpiration from all soil layers and the maximum potential transpiration for the crop at the current stage of growth; $T_{max,k}$ is the total potential transpiration accumulated over the season, and γ is an estimated yield reduction coefficient (> 1). Once the cropping pattern is fixed and meteorological conditions are forecasted, it is assumed that $T_{max,k}$ and $Y_{max,k}$ can be calculated a priori. The transpiration deficit incurred at each stage is actually a function of the current soil moisture state of the system and the irrigation decisions, although the decision and state vectors are also composed of other terms as will be explained subsequently.

3. a **center pivot model** for predicting water and pressure requirements for irrigating various areas under a center pivot. An irrigation pattern which is typical of high-pressure impact-sprinkler pivots in western Nebraska is given in Figure 7. The assumed characteristics of the pivot are given in Table I. Allocation of water to various areas under the pivot throughout the irrigation season is the basic decision requirement in this model. The three areas are: (1) the first nine inner spans, (2) the outer 10th span, and (3) corner areas irrigated by an end gun. At any given decision stage, there are four basic discrete decision policies:

- Policy #1: no irrigation
- Policy #2: irrigation of the inner nine spans only
- Policy #3: irrigation of the inner nine spans and the 10th span
- Policy #4: irrigation of all spans plus the end gun

The current mechanics of center pivots prevents further combinations of these areas to be irrigated independently. It is assumed that a variable speed irrigation pump is available to maintain constant gross capacity of the pivot per unit area no matter which of the above areas are being irrigated. This means that when the end gun operates, certain areas under the inner spans denoted by the shaded region in Figure 7 will be operating at higher pressure in order to supply the needed pressure at the end gun. There is also an additional "throw loss" associated with the end gun which reduces total system efficiency when the end gun is operating.

Again, decision stages are assumed to occur in intervals of τ days, as opposed to system simulation stages which are daily. Once the basic irrigation policy is selected, one of several discrete irrigation application depths is selected and is assumed to be constant over the current decision stage. The pivot revolution time needed to apply this amount of water is calculated and no other irrigation policy can be initiated until the current one is finished. In this decision model, it is assumed that irrigations are actually triggered by the soil moisture depletion, which is measured as the percentage of total extractable water in the soil layers in relation to the maximum total extractable water. In effect, this means that instead of independent selection of optimal irrigation decision u_i, an optimal control law $u_i(x_i)$ is assumed which is predicated on soil moisture depletions reaching various a priori defined levels.

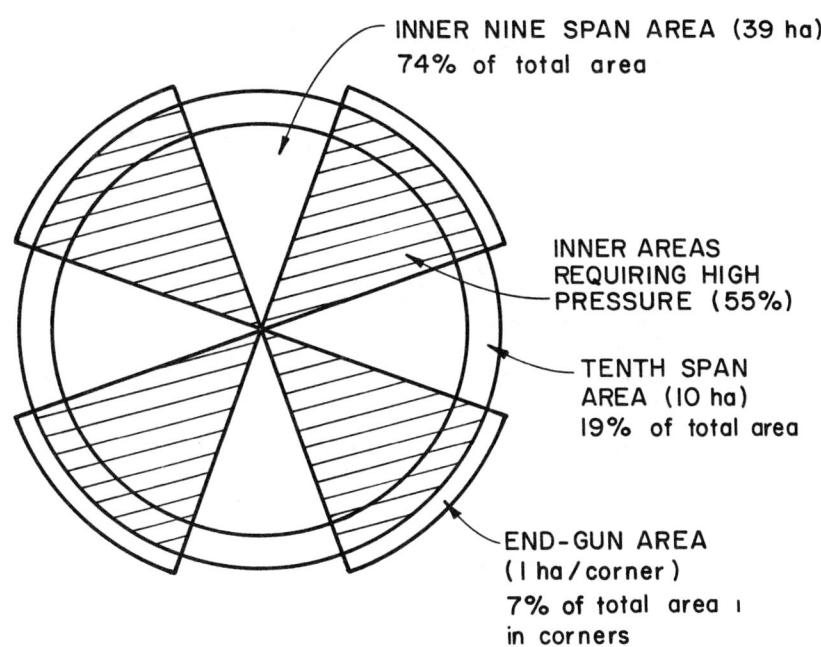

FIG. 7. Irrigation pattern for typical high-pressure impact-sprinkler center pivot. (Martin, et al. [28])

TABLE I. Center pivot characteristics.

Gross Capacity	0.94 cm/day
End Pressure [a]	345 kp_a
Pumping Lift	30 m
Application Efficiency	85%

[a] assuming the end gun is inoperative (von Bernuth [5])

In general notation, we have

$$u_i \in U_i(x_i) \quad \text{if} \quad x_i \in \bar{X}_i \tag{37}$$

where $\bar{X}_i$ is a finite and countable subset of X_i. This approach is much more realistic for center pivot operation than specifying independent decision u_i. Note that this in no way violates any assumptions associated with successful application of objective-space dynamic programming.

Under this discrete decision structure, it is possible that more than one irrigation can occur during a particular decision stage. For example, if an irrigation is initiated and completed during a decision stage and the soil moisture deficit returns to a level that triggers an irrigation prior to the end of the decision period, then another irrigation can be started. Note that the state-space must be increased because of this to include any leftover irrigations from previous periods that must be completed in the current period. This further aggravates the state-space dimensionality of the problem.

4. an **economic model** for estimating the total income from crop production, less the costs of applying irrigation water, fertilizers, and insecticides. The objective function is:

$$\max \quad p \sum_{k=1}^{K} Y_k - \sum_{i=1}^{N} \{ \sum_{t=1}^{\tau} \sum_{k=1}^{K} [c_{Ik} A_k I_{ikt} + c_D A_k D_{iJkt}] \} \tag{38}$$

where price p is the market value of the grain, \$/kg; c_{Ik} is the cost of irrigating area k, \$/ha-cm; A_k is surface area for irrigated area k, ha; I_{ikt} is the gross irrigation application depth during day t in decision stage i, cm; c_D is the cost of leaching fertilizers and insecticides which is adjusted according to whether the area is irrigated or not, \$/ha-cm; D_{iJkt} is the drainage lost from the bottom soil layer, cm. The irrigation cost c_{Ik} is represented as a function of area k since if the end gun area is irrigated, an additional cost is incurred due to throw loss and required over-irrigation in the inner spans because of increased pressure requirements.

In order to indirectly account for limitations in total seasonal water supply, a generalized Lagrange multiplier technique was employed whereby the irrigation cost is simply increased in various increments as a means of discouraging water use. A range of solutions can then be obtained according to expected seasonal limitations in water availability.

<u>Application of Objective-Space Dynamic Programming:</u> This problem as formulated is extremely difficult to solve by state-space methods because of the potentially large state dimension $J \times K$ corresponding to the soil moisture levels in each irrigated area and in each soil layer. In addition, calculation of crop status, evaporation, transpiration, soil moisture levels, and center pivot status can be time consuming computationally.

In order to demonstrate the applicability of objective-space dynamic programming to this problem, data for western Nebraska were compiled to represent average year conditions for potential evapotranspiration and total rainfall. The typical sandy soil conditions of this region were used for drainage and loss calculations. The coefficient γ for the yield model and the maximum potential grain yield are based on data provided by Maurer [30] for corn. The yield loss coefficient was calibrated at 1.29 with maximum grain yield estimated at 10,200 kg/ha. Cost data for fertilizer and insecticides are taken from Watts and Martin [41]. A basic irrigation cost of $2.92/ha-cm was assumed using the calculation technique of Gilley and Watts [18]. This assumes a diesel fuel price of $0.33/liter. Market value for corn was assumed to be $0.12/kg at the time of this study.

This particular formulation is expressed as a maximization problem (38), which requires that the L_i be treated as upper bounds in the objective-space algorithm of (17) to (20). Martin, et al. [29] actually compute a maximum total benefit under maximum yield assumptions at the beginning of stage 1 and then successively reduce the benefits over subsequent stages as various costs and reduced yields are introduced. This gives a nonincreasing objective-space trajectory which begins at stage 1 at a positive value rather than zero.

A total of 12 decision stages in 10 day increments were defined for the period June through September. The decision set in the model specified two possible net application depths (2 cm or 3 cm) and five levels of soil moisture depletion for triggering irrigations (40, 50, 60, 70 and 80 percent). These were of course predicated on the four basic decision policies described previously. This represents a total of $2 \times 5 \times 4 = 40$ discrete decision combinations at each stage. Objective-space increments selected for this problem ranged from $1000 to $100. Reduction to the smaller increment resulted in approximately a 4% increase in total net return. Further reductions were not considered necessary. A "splicing" approach was used whereby an initially coarse interval is selected and then successively reduced along with a narrowing feasible corridor in objective-space. It should be mentioned that a penalty term was included in this study for defining proximity to target bounds and these were actually carried forward through the decision stages. It is now believed that inclusion of these penalties was not necessary and future work should provide computational experience with the penalties removed.

Results: A Lagrange multiplier was added to irrigation cost in increments of 0, 4, 10, and 20 $/ha-cm in order to reflect what were defined, respectively, as unlimited, slight, moderate, and severe limitations on total seasonal water availability. Figure 8 presents the final results of the objective-space optimization to show the tradeoff between total amount of water used and the total net return resulting from the various Lagrange multiplier values. In addition, one run was made with the objective of maximizing crop grain yield only with no economic considerations or water limitations. It can be seen that the maximum yield case required 45% more water than the maximum net return case under no water limitation, with the latter providing slightly more net return. These results imply that the typical farm objectives of maximizing crop yield may actually be giving the farmer less net return while placing heavy pressure on existing water supplies.

Table II gives the optimal open-loop decision policies for the two extreme water supply scenarios. As alluded to previously, it is possible to obtain limited feedback decision policies conditioned on the system state using the objective-space approach, but such policies were not derived for this study. It is difficult to ascertain a consistent pattern in the optimal policies. This suggests that it might be best to implement the objective-space dynamic programming model on a microcomputer for daily or

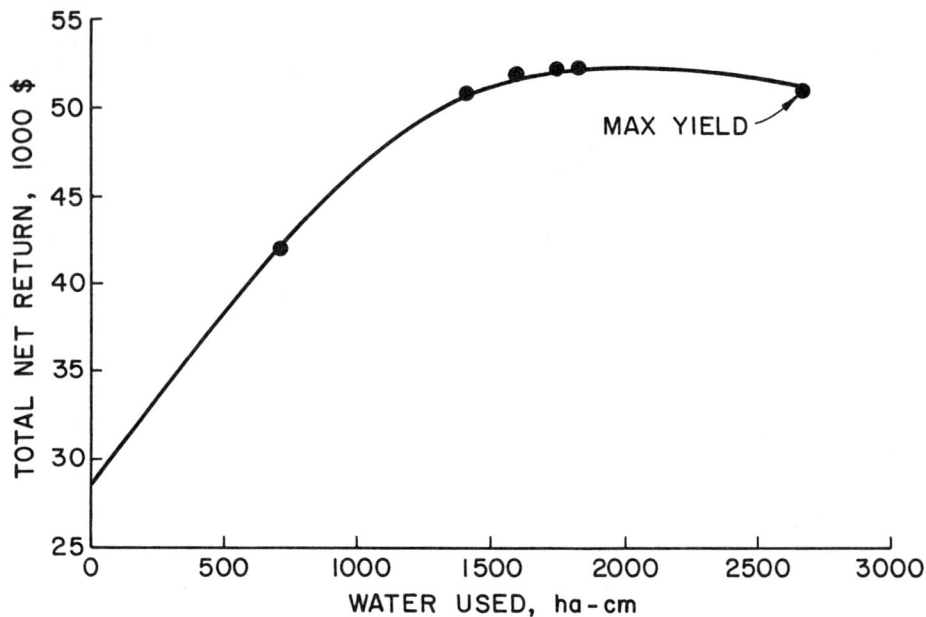

FIG. 8. Final tradeoff of total net return to seasonal water use based on Lagrange multiplier analysis. (Martin [28])

TABLE II. Optimal management decisions for the two extreme seasonal water supply scenarios, 10 day stages starting June 1. (Martin, et al. [29])

	Depth per Application, cm		Allowable Depletion, %		Irrigation Policy	
Stage	No Limit	Severe Limit	No Limit	Severe Limit	No Limit	Severe Limit
1						
2	3	2	40	50	1	1
3		3		50		2
4						
5	2	2	60	60	3	2
6	2	2	40	70	3	2
7	3		60		2	
8	3		50		1	
9	3		60		3	
10						

weekly runs in the field in order to respond to sporadic rainfall events and highly variable evapotranspiration conditions.

Figure 9 provides an indication of how the model would allocate water between the various irrigated areas under the center pivot. It can be seen that the model turns off the end gun completely during conditions of severe water limitation.

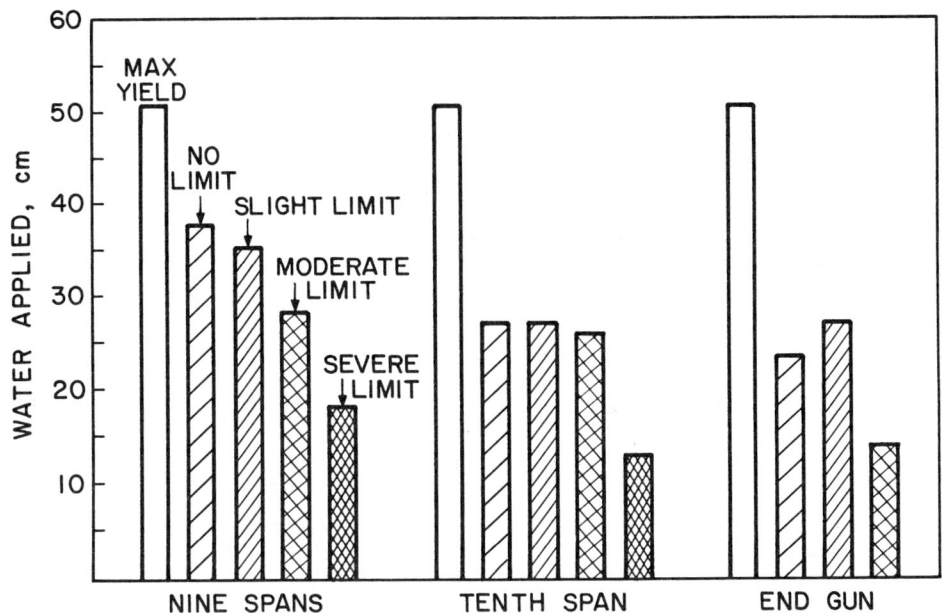

FIG. 9. Variation of total seasonal water application on each area for the two extreme water limitation scenarios. (Martin, et al. [29])

REFERENCES

1. V. Akilesaran, T.L. Morin, and W.L. Meier (presented at International Conference on Systems Modelling in Developing Countries, Asian Inst. of Tech., Bangkok, Thailand 1978).
2. K. Banerjee, Generalized Lagrange Multipliers in Dynamic Programming (Rep. ORC 71-12, Oper. Res. Cent., Univ. of Calif., Berkeley 1971).
3. L. Becker and W. Yeh, Water Resour. Res. $\underline{10}$, 1107-1112 (1974).
4. R.E. Bellman, Dynamic Programming (Princeton University Press, Princeton, N.J. 1957).
5. R.D. von Bernuth, Trans. Am. Soc. Agric. Eng. $\underline{21}$, 419-422 (1983).
6. D.P. Bertsekas, IEEE Trans. on Auto. Control $\underline{20}$, 415-419 (1975).
7. S.W. Childs and R.J. Hanks, Soil Sci. Soc. Am. Proc. $\underline{39}$, 617-622 (1975).
8. M.W. Cooper, An Improved Algorithm for Nonlinear Integer Programming (Rep. IEOR 77005, Dept. of Indus. Eng. and Oper. Res., Southern Methodist Univ., Dallas, Tex. 1977).
9. L. Cooper and M.W. Cooper, Introduction to Dynamic Programming (Pergamon Press, Oxford 1981).
10. G. Cohen in: Optimization and Control of Dynamic Operational Research Models, S.G. Tzafestas, ed. (North-Holland, Amsterdam 1982).

11. S.E. Dreyfus and A.M. Law, The Art and Theory of Dynamic Programming (Academic Press, New York 1977).
12. G.R. Drummond and D.L. Robey, Natural Stream Temperature Analysis for Allegheny River Near Kinzua, Penn. (Spec. Rep. No. 8, U.S. Army Div., Ohio Riv., Res. Cont. Cent., Cincinnati, Ohio 1975).
13. M.A. Farber, Use of Reservoir Selective Withdrawal for Water Quality Management (M.S. thesis, Dept. of Civ. Eng., Colo. State Univ., Ft. Collins, Colo. 1978).
14. D.G. Fontane and B. Loftis, Selective Withdrawal Port Location (Rep. prepared for Ford, Thornton, Norton and Assoc., Waterways Exp. Sta., Vicksburg, Miss. 1982).
15. D.G. Fontane, J.W. Labadie, and B. Loftis, Wat. Resour. Res. $\underline{17}$, 1594-1604 (1981).
16. D.G. Fontane, J.W. Labadie, and B. Loftis, Optimal Control of Reservoir Selective Discharge Quality through Selective Withdrawal (Tech. Rep E-82-1, U.S. Army Eng. Waterways Exp. Sta., Vicksburg, Miss. 1982).
17. J.E. Giles and W.O. Wunderlich, J. Water Resour. Plann. Manage. Div. Am. Soc. Civ. Eng. $\underline{107}$, 495-511 (1981).
18. J.R. Gilley and D.L. Watts, J. Irrig. and Drain. Div. , Am. Soc. Civ. Eng. $\underline{103}$, 445-457 (1977).
19. D.W. Grimes, R.J. Miller, and P.L. Wiley, Agron. Jour. $\underline{67}$, 519-523 (1975).
20. J.C. Grygier and J.R. Stedinger, Water Resour. Res. $\underline{21}$, 1-10 (1985).
21. W.A. Hall, R.C. Harboe, W. Yeh, and A.J. Askew, Optimum Firm Power Output from a Two Reservoir System by Incremental Dynamic Programming (Contrib. 130, Water Resour. Cent., Univ. of Calif., Los Angeles 1969).
22. M. Heidari, V.T. Chow, P.V. Kokotovic, and D.D. Meredith, Water Resour. Res. $\underline{7}$, 273-283 (1971).
23. J.W. Labadie, N.S. Grigg, and P.D. Trotta, J. Comput. and Oper. Res. $\underline{1}$, 421-435 (1974).
24. J.W. Labadie, Application of Dynamic Programming to Water Resources Management (Short Course Notes, Dep. of Civ. Eng., Colo. State. Univ., Ft. Collins, Colo. 1980).
25. J.W. Labadie, D.G. Fontane, and B. Loftis, presented at Symp. on Surface-Water Impound., Am. Soc. Civ. Eng., Minneapolis, Minn. June 2-5, 1980).
26. R.E. Larson, State Incremental Dynamic Programming (Elsevier, New York 1968).
27. B. Loftis, WESTEX-A Reservoir Heat Budget Model: User's Manual (Rep., U.S. Army Eng. Waterways Exp. Sta., Vicksburg, Miss. 1980).
28. D. Martin, Using Crop Yield Models in Optimal Irrigation Scheduling (Ph.D dissertation, Dept. of Agric. and Chem. Eng., Colo. State Univ. 1984).
29. D.L. Martin, D.F. Heermann, J.R. Gilley, and J.W. Labadie, Optimal Seasonal Center Pivot Management (presented at Summer Meeting, Am. Soc. Agric. Eng., Montana State Univ., Bozeman 1983).
30. R.E. Maurer, Effect of Timing and Amount of Irrigation on Drought Stress Conditioning in Corn (Ph.D dissertation, Univ. of Neb., Lincoln 1981).
31. T.L. Morin, in: Dynamic Programming and Its Applications, M.L. Puterman, ed. (Academic Press, New York 1978) pp. 53-90.
32. T.L. Morin and A.M.O. Esogbue, J. Math. Anal. and Applic. $\underline{48}$, 801-810 (1974).
33. D.M. Murray and S.J. Yakowitz, Water Resour. Res. $\underline{15}$, 1017-1027 (1979).
34. G.L. Nemhauser, Introduction to Dynamic Programming (John Wiley, New York 1966).
35. M.V.F. Pereira and L.M.V.G. Pinto, IEEE Trans. on Power Appar. and Systems $\underline{102}$, 3611-3618 (1983).
36. S. Pleban, D.F. Heermann, J.W. Labadie, and H.R. Duke, Water Resour. Res. $\underline{20}$, 887-895 (1984).
37. J.T. Ritchie, Water Resour. Res. $\underline{8}$, 1204-1213 (1972).
38. L. Rossman, Water Resour. Res. $\underline{13}$, 247-255 (1977).

39. M. Sniedovich in: Reliability in Water Resources Management, E.A. McBean, K.W. Hipel, and T.E. Unny, eds. (Water Resources Publications, Ft. Collins, Colo. 1979).
40. G.W. Tauxe, R.R. Inman, and D.M. Mades, Water Resourc. Res. 10, 1403-1408 (1979).
41. D.G. Watts and D.L. Martin, Trans. Am. Soc. Agric. Eng. 24, 911-916 (1981).
42. S. Yakowitz, Wat. Resour. Res. 18, 673-696 (1982).

KNOWLEDGE BASED DYNAMIC PROGRAMMING FOR WATER RESOURCES MANAGEMENT

Osman Coşkunoğlu
Department of General Engineering, University of Illinois,
Urbana, IL   61801

ABSTRACT

Conventional optimization models become inapplicable to those problems that lack structure. A problem may become unstructured either because of the ambiguity in the goal structure, or because of the incomplete (or analytically untractable) knowledge on cause/effect relationships. However, the human problem solving process, albeit limited, can still accomplish results in unstructured problem situations. This observation can partially be ascribed to the richness of the individual's procedural as well as substantive knowledge within the problem domain. Integrating such a knowledge base into optimization models also improves their effectiveness. This assertion is demonstrated for dynamic programming in the context of water resources management. A declarative form of knowledge representation using first order predicate logic is proposed as an effective way of encapsulating technical knowledge and qualitative constraints. The programming language PROLOG permits such a representation to be executed directly. The resulting knowledge-base can then be integrated with a dynamic programming procedure, whereas the latter might have been coded in a computationally efficient code like Pascal.

INTRODUCTION

One of the earliest application areas for the dynamic programming (DP) approach is the water resources management field [1]. For over a quarter century, the field has remained a popular application area for mathematical optimization techniques, including DP. New algorithms, formulations, and computer implementations have flourished. The impetus of these efforts has been threefold; it has served to: improve the efficiency of solution algorithms, enhance the relevance of the models through refined formulations, and boost the appeal to users by developing friendly and interactive computer interface systems.

The purpose of this paper is along a somewhat different line. The emphasis is on increasing the <u>effectiveness</u> of DP as an optimization approach; that is, increasing the range of decision problems in the water resources field for which DP can successfully be implemented.

When and why do mathematical decision models fail to live up to their purpose of aiding human decision-making processes? The Operations Research/Management Science (OR/MS) literature [2,3], as well as water resources literature [4], provide answers for the "when" part of the question more conclusively than the "why" portion. Mathematical models are successfully utilized when the decision problem is a structured one; that is, when the objective, the data, and the constraints can be prespecified unambiguously. The utility of such models diminishes when the problem becomes unstructured. In latter cases, however, human beings can still achieve results without using formal models. In understanding the reasons behind this simple, yet crucial, observation, the OR/MS or water resources literature provides little guidance other than some ill-defined human traits, such as experience, intuition, gut feeling, hunch, and so forth.

It is the thesis of this paper that <u>knowledge</u> is the major power source of human problem-solving and decision-making activities. Conventional mathematical optimization techniques reflect the structured and orderly domain for which they were developed. For a given input there is a single and fixed computational path which produces an output. In contrast, human beings are equipped with an armory of overlapping techniques, hence computational paths, for handling a problem. If an individual forgets one technique or finds a technique unsatisfactory, s/he can still strive towards a satisficing solution, albeit non-optimal. The richness, pertinence, and redundancy of the knowledge possessed by humans seems to be one important reason behind their effectiveness (not necessarily efficiency or optimality) in achieving results when analytical methods fail.

The foregoing conclusion was recognized in the Artificial Intelligence (AI) arena during the late 60's. Earlier, the focus of AI researchers was exclusively on search techniques in their quest for developing intelligent programs that can be used for general purpose as well as being powerful. Later, these general problem-solvers were recognized as being too weak for use as the basis for building high-performance systems. Instead, programs with rich domain-specific knowledge, even if poor in method, proved to be more powerful in problem solving [5].

In a similar vein, the main thrust of this paper is to investigate the role of knowledge in optimization models. Specifically, the following question is investigated: Is it possible to increase the effectiveness, hence, the range of application of dynamic programming (DP) in water resources management by integrating this approach with a domain-specific knowledge base?

DECISION MAKERS, PROBLEMS, AND MODELS

The aim of this section is to posit the limitations as well as potentials of mathematical decision models utilized in any optimization approach, including DP. Therefore, the intention is to set the stage for the proposed new modeling approach to be discussed in the next section.

The appropriate features of a decision model are primarily dictated by the type of decision problem (including the nature of the decision environment), and the cognitive aspects of the decision maker.

<u>Decision-Maker</u>

Cognitive aspects of the human decision-making process have recently been extensively studied [6,7,8,9]. One general finding of this research is that individuals exhibit serious limitations when they face a non-trivial problem situation. These limitations result in almost consistent biases in their decisions; a succinct list of 29 such biases is given in [10, pp. 166-170].

Implications of these findings to decision modeling are still controversial [11]. However, the comparative power of individuals and models has been somewhat clarified in certain areas. For example, it appears that people are much better at selecting and coding information (i.e., what variables and factors to put in the model) than they are at integrating the information through a sequence of deductive reasoning processes [12]. Even though one should avoid making premature generalizations from such findings, there is enough evidence to make a strong case against the human vs. model dichotomy and in favor of a symbiotic approach. However, much research is needed to understand the requisites of such a symbiosis.

Problem Structure

Depending upon the degree of analytical tractability, decision problems are often classified as structured, semi-structured, and unstructured [2]. However, for the purposes of this paper, an alternative classification procedure is proposed. The motivation behind the proposed classification scheme is to provide an operational way to identify the sources of complexity in a problem.

A decision problem is structured; hence, its solution can be automated to the extent that it has a clearly defined goal and to the extent that a definite procedure exists to achieve that goal. A decision problem is unstructured; hence, complex to the extent that the goal(s) are ambiguous and/or procedures are not well known. The foregoing distinction, inspired by Simon's programmable versus unprogrammable problem descriptions [13, p. 46], introduces two dimensions along which the complexity of a problem can be measured: goals and cause/effect relationships. Goals can be crisp; hence, unambiguous and uncontroversial, or they can be multiple, conflicting, and unquantifiable. Cause/effect relationships define the procedures through which consequences (effects) of a causal action can be determined. Effects of a causal action can be well known and deterministic, or they may be uncertain due to the probabilistic environment or due to the lack of knowledge.

By combining these two dimensions and working only with their extreme values, four possible decision situations emerge (Figure 1). The following examples illustrate two extreme cases:

		Cause/Effect Knowledge	
		Complete	Incomplete
Goals	Clear	I	II
	Ambiguous	III	IV

FIG. 1. A classification scheme for decision problems

Region I: How much (X) water to release in order to generate Y amount of hydro-electric energy. The goal (Y) is well defined. The cause/effect relationship determines Y in terms of X; it is given by the knowledge of potential energy and other characteristics of the reservoir and power turbines.

Region IV: How many reservoirs of what capacity should be built to utilize the water resources in a river basin for multiple purposes? Neither goals (minimum cost, maximum energy, minimum adverse environmental impacts, etc.) can be clearly

set, nor are the cause/effect relationships known
(incomplete knowledge on environmental impacts,
probabilistic nature of hydrologic and economic parameters,
etc.)

The purpose of the foregoing discussion is to highlight the structural makeup of the complexity of a problem and its implications for decision modeling.

It is important to distinguish the complexity caused by incomplete knowledge (eastern half of Fig. 1) from that caused by the ambiguity of the goals (southern half of Fig. 1). The former is of technical and scientific nature and can be alleviated, even eliminated, through accumulation of data and knowledge. The latter, however, is more subjective and requires a socio-political process of establishing a value system (e.g., is it worthwhile to upset ecology in order to produce so much hydro-electric energy?).

To summarize, the structural makeup of the decision problem complexity has three implications for mathematical optimization techniques (refer to Fig. 1):

(i) Conventional techniques can be successfully implemented for the problems within region I.

(ii) Research in multiple criteria decision-making can increase the range of application of the techniques in the southern region.

(iii) In order to push the range of application towards the eastern half, the conventional techniques should be integrated with domain specific knowledge-base systems.

KNOWLEDGE

In general terms, factual knowledge consists of descriptions, relationships, and procedures in some domain of interest [14, p. 12]. Descriptions identify and differentiate objects in the knowledge-base; e.g., "Reservoir A's purpose is irrigation." Relationships, a particular type of description, express dependencies and associations between items in the knowledge-base; e.g., "Rainfall of X amount will raise the level of reservoir by Y amount." Procedures specify operations to perform when attempting to reason or solve a problem; e.g., "If water is released from a reservoir, then the content of the reservoir decreases by that amount."

Intelligent behavior of an individual is generally ascribed to his or her knowledge [15, p. 143]. During a decision-making process, in addition to the factual knowledge described above, human beings utilize two other types of knowledge [14, p. 19]: beliefs and heuristics. Beliefs represent the information an individual has about an object [16, p. 12]. For example, the belief "When reservoir A is 80% full, it provides an adequate recreational facility" could have been achieved after a user poll about the object of "recreation." The differences between beliefs and facts [17; 18, pp. 65-66] do not allow them to be treated identically in a knowledge-base. In general, it is more complicated to make reasoning from beliefs. Heuristic knowledge, on the other hand, simplifies the reasoning process. There is no agreement on the definition of heuristics in Artificial Intelligence circles [15, pp. 28-30]. Its meaning appears to have varied both among researchers and over time. Loosely speaking, heuristic knowledge "constitutes the rules of expertise, the rules of good practice, the judgmental rules of the field, the rules of plausible reasoning. These rules collectively constitute what the mathematician George Polya has called the 'art of good guessing.' In contrast to the facts of a field its rules of good guessing are rarely written down." [19, p. 37]

Knowledge versus Data

Increases in data as well as in knowledge serve the purpose of expanding the range of applicability of mathematical optimization techniques. However, the effect of the latter is much more profound than the effect of the former. To take advantage of this, it is important to distinguish between data and knowledge. Furthermore, for computer implementation of model-data, model-knowledge, and data-knowledge links, knowledge and data-base systems must be carefully separated. Different data structures, control strategies, even different languages apply to each subsystem.

A "litmus test" for distinguishing knowledge versus data: "If we can trust an automatic process or clerk to collect the material then we are talking about data .... If we look for an expert to provide the material then we are talking about knowledge" [20, p. 79]. In general, data, when processed, provides information about specific instances. Whereas knowledge identifies information about general concepts and deals with generalizations; hence, it refers to entity types rather than to entity instances. At the time of decision-making an individual and/or an optimization model, armed with knowledge, considers the data that has been selected as relevant to the problem at hand, and makes the decision. Therefore, knowledge and data complement each other.

Another instance where the roles of knowledge and data significantly differ from each other is that when the latter is missing, the former can be activated to generate the necessary information using a different data set. For example, evaporation from a reservoir is a significant data that may not be available. One can still generate a plausible evaporation data by applying a combination of factual as well as heuristic knowledge to other climatic data.

Knowledge and Optimization Models

An optimization model makes use of two types of knowledge: descriptive (or substantive) and procedural. Descriptive knowledge is comprised of the expressions that encompass the analytically tractable components of a problem domain. The mathematical expression relating the volume of water and head to the hydroelectric energy is an example of descriptive knowledge. Procedural knowledge is encapsulated in the implicit inference mechanism and solution algorithm. Current optimization models provide a limited latitude for modifying descriptive knowledge in a model by allowing the user to specify the parameters of the model, while procedural knowledge remains fixed.

Providing the values for a predetermined type and number of parameters in a model is an inadequate way of injecting domain-specific knowledge into a model. Often these values are not completely known. The NUDGE system, for example, is a knowledge-based system for scheduling events that can take incomplete requests for events to be scheduled and transforms them into more complete requests by filling in information normally taken for granted [21, p. 40]. The developers of NUDGE contrast their approach with traditional techniques as follows:

> Traditionally, scheduling programs apply simple but powerful decision analysis techniques to finding the optimal schedule under a well-defined set of constraints .... For well-defined, formal situations the traditional power-base approach is appropriate. But for the problem of defining these formal situations when given only informal specifications, a knowledge-based approach is necessary. [23] (as cited in [21, p. 40]).

Parameters may also be unknown due to inherent uncertainties in the problem domain. In such cases decision makers often employ an analyst to reduce the uncertainty, not to acknowledge and quantify it [22] (as cited in [7, p. 27]). Reduction of uncertainty requires additional information and substantive knowledge about the problem domain. Requisite facilities for encoding such knowledge and information generation capabilities (e.g., inference from existing data) do not exit in current optimization approaches.

Increasing the facilities of optimization models to incorporate more descriptive knowledge does not only enhance the effectiveness of the model but may also improve the efficiency. Current optimization models' fixed procedural knowledge essentially contains rules and operators to search for a solution to the problem. The addition of knowledge tends to reduce the search space, and hence, the computation time. "This relationship between search and knowledge is a basic principle of Artificial Intelligence" [24, p. 4].

DYNAMIC PROGRAMMING AND KNOWLEDGE

In order to appreciate (DP) as a problem solving approach, not merely as an optimization technique, it is important to differentiate its formulation from the solution phase. DP formulation has two important features. First, a given problem is embedded into a state space. Second, the resulting equations are recursively defined. These two features also form the basis of virtually all of the problem solving approaches in Artificial Intelligence [25]. Once the problem is embedded into a state space, the next phase involves employing a search technique to obtain a solution. At this phase the problem-solving techniques of Artificial Intelligence differ from the optimization techniques employed to solve a DP formulation. The major difference is that those techniques employed in intelligent problem-solving approaches explicitly utilize the knowledge about the problem at hand, which is not the case in optimization techniques employed for a DP formulation. Hence, these latter type of approaches are referred to as "uninformed" [26, p. 95].

The purpose of this section is to identify general types of water resources-related knowledge that can be incorporated into a DP in order to render the latter as "informed."

DP Formulation

Let x_k be an m-dimensional vector denoting the state of a given system at the end of stage k. Also, let y_k be an n-dimensional vector denoting the decisions that can be made during stage k. Then, a fairly general forward DP formulation for a K stage problem can be written as:

$$f_k(x_k) = \underset{y_k}{\text{maximize}} \ [r_k(x_{k-1},x_k,y_k)+f_{k-1}(x_{k-1})], \ k=1,2,\ldots,K \quad (1.1)$$

$$x_o = \bar{x}_o \quad (1.2)$$

$$f_o(\bar{x}_o) = 0 \quad (1.3)$$

$$x_{k-1} = t_k(x_k, y_k), \quad k = 1,2,\ldots,K \quad (1.4)$$

$$x_k \in X_k, \quad k = 1,2,\ldots,K \quad (1.5)$$

$$y_k \in Y_k(x_{k-1}, x_k), \quad k = 1,2,\ldots,K \quad (1.6)$$

Where,

$r_k(x_{k-1}, x_k, y_k)$: return from stage k starting with x_{k-1}, ending with x_k, and during which the decision y_k is implemented.

$f_k(x_k)$: optimal return from stages 1 through k, if the state at the end of stage k is x_k.

$\bar{x}_o$: given initial conditions.

$t_k(x_k, y_k)$: transformation function which yields the beginning state at stage k as a function of the terminal state and decision implemented in stage k.

X_k: set of admissible values of x_k.

$Y_k(x_{k-1}, x_k)$: admissible values of y_k, for a given x_{k-1} and x_k pair.

The above formulation can be related to the water resources field as follows (see [27] for details):

x_k: levels of water at the surface and ground water reservoirs.

y_k: surface water releases for hydroelectric generation, irrigation, aquifer recharge and downstream requirements purposes; and ground water releases for irrigation purposes.

r_k: return from hydro-electricity and irrigation.

t_k: change in the reservoir levels due to release policies, evaporation, rainfall, and seepage.

X_k: upper and lower bounds on the reservoir levels

Y_k: capacities of reservoirs, power plant, and canals; minimum and maximum release requirements; and definitional constraints (such as, relating energy generation to water released through turbines).

The formulation (1) can be rearranged as follows:

$$f_k(x_k) = \underset{x_{k-1} \in X_{k-1}}{\text{maximize}} \ \underset{y_k \in Y_k(\cdot,\cdot)}{\text{maximize}} \ [r_k(x_{k-1}, x_k, y_k) + f_{k-1}(x_{k-1})]$$

$$= \underset{x_{k-1} \in X_{k-1}}{\text{maximize}} \quad \underset{y_k \in Y_k(\cdot,\cdot)}{[\text{maximize}} \; r_k(x_{k-1}, x_k, y_k) + f_{k-1}(x_{k-1})]$$

or

$$f_k(x_k) = \underset{x_{k-1} \in X_{k-1}}{\text{maximize}} \; [r_k^*(x_{k-1}, x_k) + f_{k-1}(x_{k-1})] \quad (2)$$

in which

$$r_k^*(x_{k-1}, x_k) = \max r_k(x_{k-1}, x_k, y_k) \quad (3.1)$$

subject to:

$$x_{k-1} = t_k(x_k, y_k) \quad (3.2)$$

$$y_k \in Y_k(x_{k-1}, x_k) \quad (3.3)$$

Formulations (2) and (3), together, are equivalent to (1). This decomposition allows one to choose the most appropriate computational methods as well as knowledge requirements for (2) and (3) independent of each other. Other computational advantages of the above decomposition were explained in [27, pp. 523-524].

Without going into algorithmic details and precision, a general set of steps for solving the problem (2) can be informally outlined as follows:

Step 1: Fix $x_k = \bar{x}_k \in X_k$, where $\bar{x}_k$ is not considered before.

Step 2: Solve (3) to determine $r_k^*(x_{k-1}, \bar{x}_k)$ for each $x_{k-1} \in X_{k-1}$.

Step 3: Determine that $x_{k-1} \in X_{k-1}$ which maximizes $r_k^*(x_{k-1}, \bar{x}_k) + f_{k-1}(x_{k-1})$.

Step 4: If all the elements of X_k are considered, go to stage k+1; otherwise go to Step 1.

<u>Relevance of Knowledge</u>

In Step 1 of the above solution outline, the elements of set X_k are often predetermined through state discretization. The discretization process is usually governed by the computational considerations. However, identifying interesting and potentially significant state levels from the initial conditions also requires qualitative reasoning; hence, judgement and experience are necessary. For example, if X_k is the set of possible levels of a group of reservoirs operated jointly, then a balance may be required between the storage levels and surface fluctuations in each reservoir. Such a relationship can perhaps be analytically represented. However, over time if these relationships have to be changed, one faces the familiar difficulty of modifying an algorithmic program.

Step 2 involves solving the mathematical programming problem given by (3). Such an endeavor requires a large amount of data as well as explicit, complete, and precise analytical representation of all the expressions. Often the data have been collected or generated for purposes other than use in such models. Therefore, a mismatch between the existing data base and the model is a common occurrence. This mismatch may occur either because of having data in a different format than the one required by the model or because an essential piece of data may simply be missing. Such cases call for heuristics, beliefs, and factual knowledge to generate missing data from existing ones.

The state transformation function given by Eq. (3.2) often requires hydrologic dynamics of a river basin, hence extensive factual knowledge. Capturing this knowledge through analytical expressions may cause mathematical difficulties (e.g., incorporating differential equations). It is essential to utilize judgemental heuristic knowledge in such cases. Likewise, constraints given by Eq. (3.3) may not be analytically expressed either because of the mathematical complexities or because of the qualitative constraints. For example, institutional constraints related to water rights, project ownership, and contractual agreements would not lend themselves for a mathematical formulation. Knowledge about these constraints, combined with the priorities in water rights render the situation complex. This complexity, in a way, is akin to the complexity of the chess game; i.e., there are no sophisticated knowledge requirements but the difficulty is in developing good rules based on hundreds or thousands of interacting factors. Research in Artificial Intelligence in developing computer programs that can play chess demonstrated the power of common sense and heuristic knowledge in obtaining "good" rules.

The constraints (3.2) may include those that require technical knowledge as well as managerial knowledge. For example, water released for hydropower generation may cause erosion of the reservoir shoreline and downstream channel banks [29]. Incorporating a constraint which imposes an upper bound on such an erosion requires technical knowledge as well as managerial judgment (i.e., heuristic knowledge).

In Steps 3 and 4 of the above solution outline, a search over X_{k-1} and X_k, respectively, is conducted. Mathematical structure, such as convexity, of the formulation may allow one to reduce the size of this search space. However, such a reduction often is either not possible because the formulation lacks the desirable structure or not desirable because the solution algorithm becomes too inflexible and opaque. Many problems, however, can benefit from the experience and common sense of a problem domain expert in reducing the search space. Indeed, such heuristic knowledge is likely to yield better results than the mathematical heuristics developed by the person in charge of the algorithm.

Knowledge in Action

While a person may be good at solving problems in some domain, s/he may not be able to describe a step-by-step procedure in sufficient detail for a computer programmer to be able to develop a useful flow chart. A major step towards a solution to the difficulty of structuring the tasks in human problem solving process for the computer came from the MYCIN [30] experience of developing a medical diagnosis program. It was proposed that the knowledge that a person uses to solve a problem or to make a decision can be written down as a collection of assertions and "If-then" rules. Each assertion or rule can have a confidence factor, indicating the individual's belief in the validity of the assertion.

Indeed, a large part of decisions to be made in organizations are based on rules of the following type: "If condition A and condition B and ... are met, consequence X and consequence Y and ... apply." For example, in a reservoir management, the operating rules can be expressed as:

a. If a water right has called for water or the soil moisture is depleted below the critical value, supply the amount required to return the soil moisture to field capacity or the maximum diversion rate, whichever is less.

b. If no irrigation is required or if one is not called for, compute daily consumptive use and decrease the soil moisture. [28]

In addition to a knowledge-base, which contains facts like, "Reservoir A is upstream to reservoir B," and rules like described above, a mechanism is needed to manipulate these facts and rules to form inferences. This mechanism is referred to as the "inference engine." Early in expert system development studies it was recognized that the knowledge-base and inference engine should be developed independently. The former may change over tasks and over time, while the latter remains relatively invariant of the problem domain. This architecture provides the much needed flexibility in modifying the knowledge-base.

One form of knowledge representation is a restricted form of first-order predicate logic. The programming language PROLOG [31] has a built-in inference engine which permits such representations to be executed directly. A PROLOG program consists of a collection of logic statements; therefore, all knowledge is reduced to a collection of Horn clauses of the form:

$$R \leftarrow A_1, A_2, \ldots, A_x$$

The semantics of such a Horn clause can be thought of as being either declarative

"if A_1, A_2, ..., A_x all hold so does R"

or procedural

"if you want R then do A_1, then A_2, ..., then A_x"

For the purposes of this paper, the most significant advantage of PROLOG is that it allows us to concentrate representing the domain specific knowledge in a logically coherent way, without being concerned about the inference mechanism to manipulate the knowledge. Furthermore, the recent PROLOG compilers allow interfacing a PROLOG program with those programs written in procedural languages which are more appropriate for numerical computations (e.g., Pascal).

CONCLUDING REMARKS

This paper dealt in general with a number of issues evolving around the decision-making process and the role of decision models. More specifically, the main focus was on improving the effectiveness of dynamic programming, as a problem solving approach, in the domain of water resources management problems.

One issue that has gained much attention recently is the role of human cognitive aspects in decision-making, and its implications to decision modeling. On the one hand, it has been shown that individuals are seriously flawed in their decision-making activities, on the other hand it is well known that individuals can handle problems that are prohibitively complex for computerized models. This human vs. model dichotomy, however, is superficial. This paper's position is that a model's effectiveness can be enhanced through incorporating the knowledge (whether heuristic or factual) of the managers in the problem domain. Such an endeavor can also ease the tension between the models and users. It is often stated that the latter's objections to models include two claims: "models are too complex to be useful" and "models are not useful because they exclude many important factors." These ostensibly contradictory arguments are actually very revealing. The hypothesis of this paper is that, the analyst developing models and algorithms may utilize his or her procedural knowledge to the limits in excruciating details while leaving out the substantive (descriptive) knowledge of the manager operating within the problem domain.

In addition to developing a symbiosis between the human problem solver and the model, incorporating knowledge into the latter renders the model more effective. That is, many problems which lack the analytical structure that is requisite for the mathematical models, can effectively be handled by a knowledge augmented model. Furthermore, coding the knowledge in a declarative form, independent from the procedural aspects of the model, can alleviate inflexibility and opaqueness of the models. Relatively flexible structure of a dynamic programming formulation appears to more readily allow knowledge-model integration. These arguments are further demonstrated in a forthcoming paper which also includes a specific application of knowledge-base (in PROLOG) augmented dynamic programming (in Pascal) to a replacement problem.

In the water resources area, the past construction era has been replaced, at least in the U.S.A., by the current management era. Consequently, an increased focus on optimizing the operation of existing projects to meet increasing demands can be expected. Significance of developing knowledge-based systems to this end has already been recognized [32]. This paper attempts to put the knowledge-based optimization models to the research agenda.

REFERENCES

1. W. A. Hall and N. Buras, J. Geophys. Res. 66, 517-520 (1961).
2. J. D. C. Little, Management Science 16, B466-B485 (1971).
3. P. G. W. Keen and M. S. Scott Morton, Decision Support Systems: An Organizational Perspective (Addison-Wesley, Reading, Massachusetts 1978).
4. A. K. Biswas, Water Supply and Management 3, 1-7 (1979).
5. E. A. Feigenbaum, B. Buchanan, and J. Lederberg, in: Machine Intelligence 6, B. Meltzer and D. Michie, Eds. (American Elsevier, New York 1971), pp. 165-190.
6. D. Kahneman, P. Slovic, and A. Tversky, eds., Judgement Under Uncertainty: Heuristics and Biases (Cambridge University Press, Cambridge, England 1982).
7. P. Slovic, B. Fischoff, and S. Lichtenstein, Ann. Rev. Psychol. 28, 1-39 (1977).
8. H. J. Einhorn and R. M. Hogarth, Ann. Rev. Psychol. 32, 53-88 (1981).
9. G. F. Pitz and N. J. Sachs, Ann. Rev. Psychol. 35, 139-163 (1984).
10. R. M. Hogen, Judgement and Choice (Wiley, New York 1980).
11. G. P. Huber, Management Science 29, 567-579 (1983).
12. R. M. Dawes, American Psychologist 34, 571-574 (1979).

13. H. A. Simon, The New Science of Management Decision (Prentice-Hall, Englewood Cliffs, New Jersey 1977).
14. F. Hayes-Roth, D. A. Waterman and D. B. Lenat, eds., Building Expert Systems (Addison-Wesley, Reading, Massachusetts 1983).
15. A. Barr and E. A. Feigenbaum, The Handbook of Artificial Intelligence, $\underline{I}$ (William Kaufmann, Los Altos, California 1981).
16. M. Fishbein and I. Ajzen, Belief, Attitude, Intention and Behavior (Addison-Wesley, Reading, Massachusetts 1975).
17. R. P. Abelson, Cognitive Science $\underline{3}$, 355-366 (1979).
18. A. Barr and E. A. Feigenbaum, The Handbook of Artificial Intelligence $\underline{III}$ (William Kaufmann, Los Altos, California 1981).
19. E. A. Feigenbaum, Knowledge Engineering: The Applied Side, in: Intelligent Systems, J. E. Hayes and D. Michie, eds. (Wiley, New York 1984), pp. 37-55.
20. G. Wiederhold, Knowledge Versus Data, in: On Knowledge Base Management Systems, M. L. Brodie and J. Mylopoulos, eds. (Springer-Verlag, New York 1986), pp. 77-82.
21. J. R. Quinlan, Fundamentals of the Knowledge Engineering Problem, in: Introductory Readings in Expert Systems, D. Michie, ed. (Gordon and Breach, New York 1982), pp. 33-46.
22. R. V. Brown, A. S. Kahr, and C. Peterson, Decision Analysis for the Managers (Holt, Rinehart and Winston, New York 1974).
23. I. P. Goldstein and B. Roberts, Using Frames in Scheduling, in: Artificial Intelligence: An MIT Perspective, $\underline{1}$, P. H. Winston and R. H. Brown, eds. (MIT Press, Cambridge, Massachusetts 1979).
24. M. S. Fox, Knowledge Representation for Decision Support, in: Knowledge Representation for Decision Support Systems, L. B. Methlie and R. H. Sprague, Jr., eds. (North Holland, Amsterdam 1985), pp. 3-26.
25. A. Newell and H. A. Simon, Human Problem Solving (Prentice-Hall, Englewood Cliffs, New Jersey 1972).
26. N. J. Nilsson, Principles of Artificial Intelligence (Tioga, Palo Alto, California 1980).
27. O. Coşkunoğlu and C. M. Shetty, J. Water Res. Plan. and Manag., $\underline{107}$, 513-531 (1981).
28. R. W. Koch and R. L. Allen, J. Water Res. Plan. and Manag., $\underline{112}$, 527-541 (1986).
29. R. A. Wurbs, J. Water Res. Plan. and Mang. $\underline{113}$, 130-148 (1987).
30. E. H. Shortliffe, ed., Computer-Based Medical Consultation: MYCIN (Elsevier, New York 1976).
31. W. F. Clocksin and C. S. Mellish, Programming in PROLOG (Springer-Verlag, New York 1984).
32. J. G. Gaschnig, R. Reboh, and J. Reiter, Development of a Knowledge-Based Expert System for Water Resource Problems, SRI Project 1619 (SRI International, Artifical Intelligence Center, Menlo Park, California 1981).

DYNAMIC PROGRAMMING AND NON-SEPARABLE WATER RESOURCES PROBLEMS

MOSHE SNIEDOVICH
National Research Institute for Mathematical Sciences
CSIR, P.O.Box 395, Pretoria, 0001, South Africa

ABSTRACT

A solution strategy designed for certain types of non-separable dynamic programming problems is proposed. The strategy is demonstrated through its treatment of a water resources problem which in terms of a dynamic programming formulation is rendered non-separable due to economies of scale factors. The merit of the proposed strategy is in its ability to deal effectively with problem formulations that faithfully depict those real-world features that are behind the non-separability of the objective function.

INTRODUCTION

In an extensive survey of dynamic programming applications in water resources, Yakowitz [1, p. 673] pointed out the following:

> "...An unmistakable conclusion is that water resources problems serve as an excellent impetus and laboratory for dynamic programming developments; conversely, progress in making dynamic programming applications in water resources economically viable depends on further advances in theoretical and numerical aspects of dynamic programming. At the present time the influence of dynamic programming on water resource practice is modest. Attempts will be made in this survey to point out where further mathematical modelling efforts are needed...".

Not surprisingly, one of the issues singled out by him in this connection, obviously, on account of its far-reaching implications for the application of dynamic programming techniques in water resources management, was the Curse of Dimensionality. Yet, for all the advances in computational mehtods since the survey's publication, not a single breakthrough has been achieved on this front to unfetter dynamic programming applications from the Curse of Dimensionality.

Indeed, the Curse of Dimensionality remains the irksome sore point it has always been in dynamic programming, thus continuing to present the single most serious impediment to the use of dynamic programming in the solution of real-world problems, water resources problems included.

Therefore, it is our objective in this paper to examine, in detail, one of the manifestations of the Curse of Dimensionality's hampering a conventional use of dynamic programming, and to propose an alternative solution strategy for problems thus affected. And to be precise, we want to show how problems whose non-separable objective functions make them potential prey to the Curse of Dimensionality, can be handled effectively by a strategy fusing dynamic programming and c-programming techniques.

Our discussion therefore proceeds as follows. We begin with a definition of a prototype dynamic programming problem and show that in cases where the pertinent objective function is rendered non-separable, the problem in question threatens to fall victim to the Curse of Dimensionality. We then go on to describe in very broad terms the sort of tactics that can be deployed in such situations to counter-act the difficulties brought on by the non-separability of the objective function. This leads to an outlining of the c-programming method ([2]-[6]) where we sketch its essential ingredients and explain the position that it takes vis-a-vis difficult optimization problems such as the above. Following that, we consider a prototype separable c-programming problem and we identify the stock elements required for the design of algorithms for problems of this type. We then take up again the non-separable dynamic programming problem defined at the outset and we discuss some of the issues bearing on the use of c-programming techniques in the solution of problems of this format. We end with a numeric example involving a problem that economies of scale factors render non-separable and we demonstrate how a c-programming alogrithm assists in its solution.

NON-SEPARABILITY

Consider the following deterministic multistage decision problem:

Problem $Q(s)$: $\quad f(s) := \underset{(x_1,\ldots,x_N)}{\text{opt}} g(s, x_1, x_2, \ldots, x_N)$, $s \in S_1 \subset S$ (1)

subject to

$$x_n \in D(n, s_n) \quad , \quad n = 1, 2, 3, \ldots, N, \quad (2)$$

where

$$s_1 = s \quad (3)$$

and

$$s_{n+1} = T(n, s_n, x_n) \quad , \quad n = 1, 2, 3, \ldots, N. \quad (4)$$

Let $X(s)$ denote the set of all feasible solutions to Problem $Q(s)$ and $X^*(s)$ the set of all optimal solutions to this problem. We shall assume that for each $s \in S_1$ the set $X^*(s)$ is not empty, namely, that for each initial state $s \in S_1$, Problem $Q(s)$ has at least one optimal solution. ∎

The objects figuring in this formulation are:

(1) A *stage variable* $1 \leq n \leq N$.
(2) A *state space* S.
(3) A function D prescribing the *feasible decisions* pertaining to each stage-state pair.
(4) A *transition function* T embodying the dynamics of the decision-making process in so far as the state variable is concerned.
(5) A real-valued *objective function* g on $S \times \mathbb{D}^N$, where $\mathbb{D}$ denotes the image of the function D.
(6) A subset S_1 of S specifying the admissible *initial states*.

Stated more formally, D is a function on $\{(n,s): 1 \leq n \leq N, s \in S\}$ such that to each pair (n,s) it assigns a subset of some set $\mathbb{D}$ and T is a function on $\{(n,s,x): 1 \leq n \leq N, s \in S, x \in D(n,s)\}$ with values in S. We shall refer to the set $\mathbb{D}$ as the *decision space*. Also, for simplicity, we shall assume that the problem is truncated, namely that N is finite.

Now, to set the stage for a definition of a separable objective function, define

$$S_{n+1} := \{T(n,s,x): s \in S_n, x \in D(n,s)\} \quad , \quad n=1,2,3,\ldots,N, \tag{5}$$

bearing in mind that the set S_1 is a given in the formulation of Problem $Q(s)$. In this case,

Definition 1. The objective function g is said to be *separable* if there exist a sequence of real-valued functions $G=\{g_n : 1 \leq n \leq N\}$ on $S_n \times \mathbb{D}^{N-n+1}$ respectively and a real-valued function ρ on $\{(n,s,x,a): 1 \leq n \leq N, s \in S_n, x \in D(n,s), a \in \mathbb{R}\}$, where $\mathbb{R}$ denotes the real line, such that

$$g_1(s,z) = g(s,z) \quad , \quad \forall s \in S_1, z \in X(s) \tag{6}$$

and

$$g_n(s,x,z) = \rho(n,s,x,g_{n+1}(T(n,s,x),z)) \tag{7}$$

for all $1 \leq n < N$, $s \in S_n$, $x \in D(n,s)$ and $z \in X_{n+1}(s')$, $s'=T(n,s,x)$. Here $X_n(s)$ denotes the set of all sequences $(x_n, x_{n+1}, \ldots, x_N)$ such that

$$x_k \in D(k,s_k) \quad , \quad k=n, n+1, \ldots, N, \tag{8}$$

with

$$s_n = s \tag{9}$$

and

$$s_{k+1} = T(k, s_k, x_k) \quad , \quad k=n, n+1, \ldots, N. \tag{10}$$

We shall refer to (G,ρ) as a *decomposition scheme*, to ρ as a *composition function*, and to g_n as the *modified objective function* pertaining to stage n. For more details see [7, pp. 34-38]. ∎

To cut a long story short, the idea here is to devise a decomposition scheme (G,ρ) such that would satisfy the functional equation

$$f_n(s) = \underset{x \in D(n,s)}{\text{opt}} \rho(n,s,x,f_{n+1}(T(ns,x))) \quad , \quad \forall 1 \leq n < N, s \in S_n, \tag{11}$$

where $f_n(s)$ denotes the optimal value of the modified objective function pertaining to stage n, given that $s_n = s$, that is,

$$f_n(s) := \underset{(x_n, \ldots, x_N)}{\text{opt}} g_n(s, x_n, x_{n+1}, \ldots, x_N) \quad , \quad 1 \leq n \leq N, s \in S_n \tag{12}$$

subject to (8)-(10).

Although the construction of a decomposition scheme satisfying the dynamic programming functional equation defined by (11) is not always an easy task, the existence of such a scheme is nonetheless assured as a matter of principle:

Theorem 1 ([8]-[9]). Irrespective of the structures of the constituent elements of Problem $Q(s)$, the problem in question can, as a rule, be formulated so as to yield a decomposition scheme satisfying the dynamic programming functional equation. ∎

The trouble is, however, that in certain cases a reformulation will be required such that might result in an expanded state space, whereupon the problem in question will be left exposed to the Curse of Dimensionality. The following is a case in point.

Example 1

Consider the case where

$$g(s,x_1,x_2,\ldots,x_N) = \frac{\sum_{n=1}^{N} A_n(x_n)}{\sum_{n=1}^{N} B_n(x_n)} . \tag{13}$$

Since no decomposition scheme seems to be available for this function, it is rendered non-separable.

As a way out therefore, the state space can be made to absorb two auxiliary variables, in which case the now expanded state variable would have the form

$$s'_n = (s_n, a_n, b_n) \quad , \quad n=1,2,3,\ldots,N+1, \tag{14}$$

where s_n denotes the original state variable,

$$a_n := \sum_{k=1}^{n-1} A(x_k) \quad , \quad n=2,3,4,\ldots,N+1 \tag{15}$$

$$b_n := \sum_{k=1}^{n-1} B(x_k) \quad , \quad n=2,3,4,\ldots,N+1 \tag{16}$$

and

$$a_1 := b_1 := 0. \tag{17}$$

Note that this manoeuvre will produce the following state space:

$$S' := \bigcup_{n=1}^{N+1} S'_n := S_n \times A'_n \times B'_n, \tag{18}$$

where A'_n and B'_n denote the set of feasible values that the variables a_n and b_n respectively, can take.

This in turn will require the substitution of D, T and g by

$$D'(n,s') := D(n,s) \quad , \quad s'=(s,a,b) \in S'_n \tag{19}$$

$$T'(n,s',x) := (T(n,s,x), a+A_n(x), b+B_n(x)) \quad , \quad s'=(s,a,b) \in S'_n \tag{20}$$

and

$$g'(s',x_1,x_2,\ldots,x_N) = g(s,x_1,x_2,\ldots,x_N) \ , \ s'=(s,a,b)\in S'_n \tag{21}$$

respectively. Consider now the following modified objective functions:

$$g'_n(s',x_n,x_{n+1},\ldots,x_N) := \frac{a + \sum_{k=n}^{N} A_k(x_k)}{b + \sum_{k=n}^{N} B_k(x_k)} \ , \ s'=(s,a,b). \tag{22}$$

Since it is clear that the function

$$\rho'(n,s',x,a) := a \ , \ a \in \mathbb{R} \tag{23}$$

constitutes a valid composition function for this case, it follows that the functional equations would be of the form

$$f'_n(s') = \operatorname*{opt}_{x \in D'(n,s')} f'_{n+1}(T'(n,s',x)) \ , \ 1 \leq n < N, \ s' \in S'_n \ , \tag{24}$$

observing that

$$f'_N(s') := \operatorname*{opt}_{x \in D'(N,s')} \frac{a + A_N(x)}{b + B_N(x)} \ , \ s' \in S'_N. \tag{25}$$

In short then, the incorporation of the additional state variables a_n and b_n in the formulation has the effect of producing a prohibitively large state space. ■

A similar situation may arise in cases where the objective function is of the form

$$g(s,x_1,x_2,\ldots,x_N) = \sum_{n=1}^{N} [A_n(x_n) - \frac{1}{N} \sum_{k=1}^{N} A_k(x_k)]^2 \tag{26}$$

or

$$g(s,x_1,x_2,\ldots,x_N) = \sum_{n=1}^{N} A_n(x_n) + \alpha [\sum_{n=1}^{N} B_n(x_n)]^\beta . \tag{27}$$

Now, this, however, is no indication that problems with objective functions of this type are doomed to remain unsolved. Indeed, it is our contention that in many cases - given the proper strategy - problems of this kind would lend themselves to a solution. For instance, the problem featured in Example 1 would lend itself to a solution strategy combining fractional programming ([10]-[11]) and dynamic programming techniques, and problems possessing variance-type objective functions of the kind defined by (26) would lend themselves to a solution strategy combining dynamic programming and variance separation techniques ([12]-[13]).

The common denominator of such strategies is in the two-pronged solution procedure that they lay out. To separate the objective function, they bring into play a parametric problem involving an auxiliary parameter and assign it this task. In turn, the parametric problem is solved with the aid of a dynamic programming algorithm. And to put it in more concrete terms, let us consider the following.

Example 1 (Continued)

Observe that the objective function defined in (13) is fractional. Therefore, the problem in question can be handled by Dinkelbach's method ([10]-[11]), in which case the objective function of the associated parametric problem would be of this form:

$$g(s, x_1, x_2, \ldots, x_N; \lambda) := \sum_{n=1}^{N} A_n(x_n) - \lambda \sum_{n=1}^{N} B_n(x_n) \quad, \quad \lambda \in \mathbb{R}. \tag{28}$$

Since this function is additive, the parametric problem would lend itself to dynamic programming without expansion of the state space being required.

The rationale for linking the two problems is provided by fractional programming which relies on the assurance that, under certain regularity conditions, a $\lambda \in \mathbb{R}$ exists such that all the optimal solutions to the parametric problem induced by this λ are also optimal for the target problem. Standard line search methods are used to identify the optimal value of λ [11].

To sum up, a solution strategy merging fractional programming and dynamic programming techniques amounts to a procedure whereby dynamic programming is in charge of solving the sequence of parametric problems being generated by the fractional programming algorithm. ∎

The solution strategy that we propose in this paper adopts a similar approach. It works out a scheme for the collaboration between dynamic programming and c-programming on lines similar to the above collaboration betweem fractional programming and dynamic programming. We shall demonstrate its prowess by showing how it handles problems whose objective functions are of the following form:

$$g(s_1, x_1, \ldots, x_N) = q(s_1, x_1, \ldots, x_N) + \varphi(w(s_1, x_1, \ldots, x_N)), \tag{29}$$

where

$$q(s_1, x_1, \ldots, x_N) = \sum_{n=1}^{N} q_n(s_n, x_n) \tag{30}$$

and

$$w(s_1, x_1, \ldots, x_N) = \sum_{n=1}^{N} w_n(s_n, x_n). \tag{31}$$

Note that in the case where φ is nonlinear, g is rendered non-separable so that the separation of g would necessitate the introduction of the following auxiliary state variable:

$$z_n := \sum_{k=1}^{n-1} w_k(x_k) \quad, \quad k=2,3,\ldots,N, \tag{32}$$

with $z_1 = 1$. This manipulation, needless to say, is likely to result in an inflated state space. We shall show then that the proposed strategy tackles such problems in a manner that altogether obviates tinkering with the state space.

C-PROGRAMMING

The parametric method that we have entitled c-programming involves two interrelated problems: a target problem, namely the problem of interest, and a parametric problem, which is a linearized version of the former. These two are of the following forms:

Target Problem:

$$\text{Problem P:} \qquad p := \max_{x \in X} h(x) := \phi(u(x)), \qquad (33)$$

where u is a function on some set X with values in $\mathbb{R}^K$, and ϕ is a real-valued function on the set $u(X) := \{u(x): x \in X\}$. Let X^* denote the set of optimal solutions to this problem.

Parametric Problem:

$$\text{Problem P}(\lambda): \qquad p(\lambda) := \max_{x \in X} h_\lambda(x) := \sum_{k=1}^{K} \lambda_k u_k(x), \quad \lambda \in \mathbb{R}^K, \qquad (34)$$

where λ_k and $u_k(x)$ denote the k-th component of the vectors λ and $u(x)$ respectively. Let $X^*(\lambda)$ denote the set of optimal solutions to Problem $P(\lambda)$.

A point to be made with regard to these problems is that Problem P often remains resistant to the solution techniques of standard optimization methods, whereas for each $\lambda \in \mathbb{R}^K$ Problem $P(\lambda)$ proves amenable to one or more of these methods.

Example 2

Consider the following nonlinear optimization problem

$$p := \max_{(x_1, \ldots, x_N)} \left\{ \sum_{n=1}^{N} a_n x_n - \alpha \left[\sum_{n=1}^{N} b_n^2 x_n \right]^\beta \right\} \qquad (35)$$

subject to

$$1 \leq \sum_{n=1}^{N} x_n \leq J \qquad (36)$$

$$x_n \in \{0,1\}, \quad n=1,2,3,\ldots,N, \qquad (37)$$

where the parameters α, β, $\{a_n\}$, $\{b_n\}$ and J are positive constants. This, needless to say, is a nonlinear 0-1 knapsack problem. Now, in compliance with the c-programming format set $K=2$,

$$X = \{ x \in \mathbb{R}^N : 1 \leq \sum_{n=1}^{N} x_n \leq J, \ x_n \in \{0,1\} \} \qquad (38)$$

$$u(x) = \left(\sum_{n=1}^{N} a_n x_n, \sum_{n=1}^{N} b_n^2 x_n \right) \quad , \quad x \in X \qquad (39)$$

and

$$\phi(z) = z_1 - \alpha z_2^\beta \quad , \quad z \in \mathbb{R}^2 \qquad (40)$$

so that the parametric problem would take the following form:

$$p(\lambda) := \max_{(x_1,\ldots,x_N)} \left\{ \lambda_1 \sum_{n=1}^{N} a_n x_n + \lambda_2 \sum_{n=1}^{N} b_n^2 x_n \right\} \quad , \quad \lambda \in \mathbb{R}^2 \qquad (41)$$

$$= \max_{(x_1,\ldots,x_N)} \sum_{n=1}^{N} (\lambda_1 a_n + \lambda_2 b_n^2) x_n \qquad (42)$$

subject to (36)-(37). Clearly, for every $\lambda \in \mathbb{R}^2$ this would constitute a linear 0-1 knapsack problem. ∎

Now, c-programming's basic contention is that, granted certain conditions, there exists a $\lambda \in \mathbb{R}^K$ such that $X^*(\lambda) \subset X^*$. To become clear on what these conditions require, consider the following.

<u>Assumption 1</u>. The function ϕ is differentiable on some open convex set $U \subset \mathbb{R}^K$ such that $u(X) \subset U$. ∎

Let then ∇ denote the gradient operator, and define

$$v(x) := \nabla \phi(z) \Big|_{z=u(x)} \quad , \quad x \in X, \qquad (43)$$

that is, let $v(x)$ denote the gradient of ϕ with respect to u at $z=u(x)$. It is important to note that the differentiability condition is imposed on ϕ, not on u; hence the decision set X can be discrete.

<u>Example 2 (Continued)</u>

Set $U = \mathbb{R}^+ \times \mathbb{R}^+$, where $\mathbb{R}^+$ denotes the positive segment of the real line. Since (40) implies that

$$\nabla \phi(z) = (1, -\alpha \beta z_2^{\beta-1}) \quad , \quad z \in U, \qquad (44)$$

it follows that ϕ is differentiable on U. ∎

Let us now consider the following.

<u>Theorem 2</u> ([6], [14]). If ϕ is pseudolinear on U, then $x \in X^*$ if, and only if, $x \in X^*(v(x))$. Furthermore, $X^* = X^*(v(x))$, $\forall x \in X^*$. ∎

Algorithm A ([6],[14]):

Step 1. Select an element x^1 from X and set m=1 and $\lambda^m = v(x^m)$.
Step 2. Solve Problem $P(\lambda^m)$.
Step 3. If $x^m \in X^*(\lambda^m)$, set $x^* = x^m$ and stop. Otherwise, go to Step 4.
Step 4. Select an element x^{m+1} from $X^*(\lambda^m)$ and set $\lambda^{m+1} = v(x^{m+1})$.
Step 5. Set m=m+1 and go to Step 2. ■

Theorem 3 ([6,[14]). Assume that ϕ is pseudolinear on U. Then:
a. The sequence generated by Algorithm A is strictly increasing.
b. If Algorithm A terminates, then $x^* \in X^*$ and $X^*(v(x^*)) = X^*$.
c. Algorithm A terminates if the set u(X) is finite.
d. If $\nabla \phi$ is continuous on U and the sequence $\{u(x^m)\}$ generated by Algorithm A converges to some $u' \in U$, then $X^* = X^*(\nabla \phi(u'))$. ■

We note in passing that fractional programming problems fall within the family of pseudolinear c-programming problems [6], [14].

Theorem 4 [6]. If ϕ is pseudoconvex on U, then $X^*(v(x)) \subset X^*$, $\forall x \in X^*$. ■

Theorem 5 [6]. If ϕ is pseudoconcave on U, then $x \in X^*(v(x))$ implies that $x \in X^*$. ■

C-programming's thesis is then that subject to the above conditions optimal solutions to Problem P are recoverable from the optimal solutions of Problem $P(\lambda)$. Given that the latter is for the most part far easier to solve than the former, methodologically as well as practically, this approach is totally justified, in spite of the exogenous parameter λ that it introduces.

In the next section we examine a prototype c-programming problem whose interest to this discussion is in the consequences that it has for non-separable dynamic programming problems.

SEPARABLE CONVEX C-PROGRAMMING PROBLEMS

The problems falling under this heading are of this form:

Definition 2. The function ϕ is said to be **separable and convex** if it admits the following representation:

$$\phi(z) = \sum_{k=1}^{K} \phi_k(z_k) \quad , \quad z \in U, \tag{45}$$

where for each $1 \leq k \leq K$, ϕ_k is a differentiable convex function on some open convex set $U_k \subset \mathbb{R}$ such that $u_k(x) \in U_k$, $\forall x \in X$. ■

In this case, the c-programming problem possesses the following property.

Theorem 6 [6]. Assume that ϕ is separable and convex, and for each pair $(\lambda, x) \in \mathbb{R}^K \times X$ define

$$E_\beta(\lambda, x) := \{z \in \mathbb{R}^K : z_k = \beta \alpha_k \lambda_k + (1-\alpha_k) v_k(x), \ 0 < \alpha_k \leq 1, \ 1 \leq k \leq K\}, \quad \beta \geq 0 \qquad (46)$$

and

$$E(\lambda, x) := \bigcup_{\beta \geq 0} E_\beta(\lambda, x). \qquad (47)$$

Then,

$$(\lambda \in \mathbb{R}^K, \ x \in X^*(\lambda), \ y \in X, \ v(y) \in E(\lambda, x)) \implies h(x) \geq h(y). \ \blacksquare \qquad (48)$$

Note that the following is an immediate consequence of (48):

Theorem 7 [6]. Assume that ϕ is separable and convex, and let V be any subset of $\mathbb{R}^K$ such that $v(x) \in V$, $\forall x \in X$. Now, let $\{(\lambda^m, x^m): 1 \leq m \leq M\}$ be any finite sequence such that $\lambda^m \in \mathbb{R}^K$ and $x^m \in X^*(\lambda^m)$, $\forall 1 \leq m \leq M$, and

$$V \subset \bigcup_{m=1}^{M} E(\lambda^m, x^m). \qquad (49)$$

Then, $x^i \in X^*$ for any $1 \leq i \leq M$ such that

$$h(x^i) = \max_{1 \leq m \leq M} h(x^m). \ \blacksquare \qquad (50)$$

Details concerning the design of algorithms capable of generating sequences satisfying the conditions stipulated in Theorem 7 can be found in [6]. For our purposes here it will suffice to examine the degenerate case where $K=2$ and ϕ_1 is the identity function, namely $\phi_1(z)=z$, $\forall z \in U_1$. Note that in this case Problem P takes the following form:

$$p := \max_{x \in X} \{\phi_1(u_1(x)) + \phi_2(u_2(x))\} \qquad (51)$$

$$= \max_{x \in X} \{u_1(x) + \phi_2(u_2(x))\}. \qquad (52)$$

To simplify the notation, replace u_1 by q, u_2 by w and ϕ_2 by ψ in which case (52) would assume the following form:

$$\text{Problem P:} \qquad p := \max_{x \in X} \{q(x) + \psi(w(x))\}. \qquad (53)$$

Now, since ϕ_1 is the identity function, it follows that $v_1(x)=1$, $\forall x \in X$; hence the parametric problem induced by (53) would take the following simple form:

$$\text{Problem P}(\lambda): \qquad p(\lambda) := \max_{x \in X} \{q(x) + \lambda w(x)\}, \qquad \lambda \in \mathbb{R}. \qquad (54)$$

Let X^* denote the set of optimal solutions to Problem P and $X^*(\lambda)$ the set of optimal solutions to Problem $P(\lambda)$, and assume the following to be the case.

<u>Assumption 2.</u> The function φ is differentiable and convex on some open convex set $W \subset \mathbb{R}$ such that $w(x) \in W$, $\forall x \in X$. ∎

Since convexity, in conjunction with differentiability, entails pseudoconvexity [15], the following is an immediate consequence of Theorem 4.

<u>Theorem 8</u> ([2]-[6]). If Assumption 2 holds, then

$$X^*(\nabla\varphi(w(x))) \subset X^*, \quad \forall x \in X^*. \blacksquare \qquad (55)$$

Similarly, the following is an immediate consequence of Theorem 6.

<u>Theorem 9</u> [6]. Suppose that Assumption 2 holds. Let (λ, x) be any pair such that $\lambda \in \mathbb{R}$ and $x \in X^*(\lambda)$. Then, $h(x) \geq h(y)$ for any $y \in X$ such that $\nabla\varphi(u(y)) \in I(\lambda, \nabla\varphi(u(x)))$, where

$$I(a,b) := \{z \in \mathbb{R}: z = \alpha a + (1-\alpha)b, \ 0 \leq \alpha \leq 1\}. \blacksquare \qquad (56)$$

What these results add up to is that an algorithm where (56) will function as an exclusionary rule will recover optimal solutions to Prolem P by solving Problem $P(\lambda)$. Note that standard Lagrangian arguments yield the following:

<u>Theorem 10</u> [6]. Problem $P(\lambda)$ has the following properties:

a. If $x \in X^*(\lambda^1)$ and $x \in X^*(\lambda^2)$, then $x \in X^*(\lambda)$, $\forall \lambda \in \{\alpha\lambda^1 + (1-\alpha)\lambda^2: \ 0 \leq \alpha \leq 1\}$.

b. Let $(\lambda^1, x^1, \lambda^2, x^2)$ be any quadruplet such that $(\lambda^1, \lambda^2) \in \mathbb{R}^2$, $x^i \in X^*(\lambda^i)$, $i=1,2$ and $w(x^1) \neq w(x^2)$, and set

$$\lambda' = \frac{g(x^1) - g(x^2)}{w(x^1) - w(x^2)} \qquad (57)$$

If either $x^1 \in X^*(\lambda')$ and/or $x^2 \in X^*(\lambda')$, then

$$x^i \in X^*(\lambda) \quad , \quad \forall \lambda \in \{\alpha\lambda^i + (1-\alpha)\lambda': \ 0 \leq \alpha \leq 1\}. \qquad (58)$$

for $i=1,2$. ∎

More details concerning c-programming problems of the form defined by (53) can be found in [2]-[5].

NON-SEPARABILITY REVISITED

Armed with these results let us now return to the multistage decision problem defined by (1)-(4) and examine it anew. Assume that the objective function is of the form given by (29)-(31), namely

$$g(s_1,x_1,x_2,\ldots,x_N) = \sum_{n=1}^{N} q_n(s_n,x_n) + \varphi(\sum_{n=1}^{N} w_n(s_n,x_n)). \tag{59}$$

As indicated earlier, if φ is nonlinear, g would be rendered non-separable, in which case the dynamic programming functional equation for this problem will require an expansion of the original state space. Notice, however, that should (59) fall in with the requirements laid down by the c-programming format, a solution to Problem Q(s) could be obtained by solving the following parametric problem:

Problem $Q(s;\lambda)$:

$$f(s;\lambda) := \underset{(x_1,\ldots,x_N)}{\text{opt}} \{\sum_{n=1}^{N} q_n(s_n,x_n) + \lambda \sum_{n=1}^{N} w_n(s_n,x_n)\}, \ s \in S_1, \ \lambda \in \mathbb{R} \tag{60}$$

subject to (2)-(4). ∎

Observe that since, by construction,

$$f(s;\lambda) = \underset{(x_1,\ldots,x_N)}{\text{opt}} \sum_{n=1}^{N} [q_n(s_n,x_n)+\lambda w_n(s_n,x_n)], \tag{61}$$

it follows that the parametric problem is separable, in which case for each $\lambda \in \mathbb{R}$ the latter can be solved with the following additive dynamic programming functional equation:

$$f_n(s;\lambda) = \underset{x \in D(n,s)}{\text{opt}} \{q_n(s,x)+\lambda w_n(s,x)+f_{n+1}(T(n,s,x);\lambda)\}, \ 1 \leq n < N, \ s \in S_n, \tag{62}$$

starting at n=N with

$$f_N(s;\lambda) := \underset{x \in D(N,s)}{\text{opt}} \{q_N(s,x) + \lambda w_N(s,x)\}, \ s \in S_N. \tag{63}$$

And as a final point, note that much as the c-programming format proceeds on the assumption that φ is differentiable and either convex or concave depending on whether opt=max or opt=min respectively, in truth the differentiability requirement is not crucial. Indeed, all the above results also hold true if gradients are replaced by subgradients [6].

To conclude, consider the following minimun-variance type problem:

$$f(s) := \underset{(x_1,\ldots,x_N)}{\min} \sum_{n=1}^{N} [a_n(s_n,x_n) - \frac{1}{N}\sum_{i=1}^{N} a_i(s_i,x_i)]^2 \tag{64}$$

subject to (2)-(4).

Observe that expanding the objective function in (64) would yield

$$f(s) = \min_{(x_1,\ldots,x_N)} \left\{ \sum_{n=1}^{N} a_n^2(s_n, x_n) - \frac{1}{N}\left[\sum_{n=1}^{N} a_n(s_n, x_n)\right]^2 \right\}, \qquad (65)$$

hence, in view of the the c-programming format, we can set

$$q_n(s_n, x_n) := a_n^2(s_n, x_n) \qquad (66)$$

$$w_n(s_n, x_n) := a_n(s_n, x_n) \qquad (67)$$

and

$$\varphi(z) := -\frac{z^2}{N}, \qquad z \in W := \mathbb{R}. \qquad (68)$$

Since φ is concave and differentiable on W, it follows that this formulation complies with the requirements set by Thoerem 6-Theorem 10. A discussion outlining a collaboration scheme between dynamic programming and c-programming for the solution of minimum-variance type problems can be found in [5].

ILLUSTRATIVE EXAMPLE

Let us now examine how our strategy would fare in a more practical context. Suppose that on the agenda are N projects, each of which is a potential candidate for implementation, with each being characterized by the following three parameters:

B_n = expected benefit.
L_n = amount of land required.
w_n = amount of water required.

Also assume that the total amount of land available is equal to L^*. Next, suppose that, in view of economies of scale factors, the cost of producing z units of water is stipulated by the following concave function:

$$c(z) := \alpha z^\beta, \qquad z \geq 0, \qquad (69)$$

where $\alpha > 0$ and $0 < \beta < 1$.

If our goal is to select a set of projects such that will maximize the total net benefit, the problem would read as follows:

$$p := \max_{(x_1,\ldots,x_N)} \left\{ \sum_{n=1}^{N} B_n x_n - c\left(\sum_{n=1}^{N} w_n x_n\right) \right\} \qquad (70)$$

subject to

$$\sum_{n=1}^{N} L_n x_n \leq L^* \qquad (71)$$

$$x_n \in \{0,1\} \quad , \quad n=1,2,3,\ldots,N. \tag{72}$$

Note that in terms of this formulation $x_n=1$ signifies that project n is selected, whereas $x_n=0$ signifies the opposite. To simplify the exposition we shall assume that the parameters $\{L_n\}$ and L^* are positive integers. Clearly, then, the problem under consideration is a nonlinear 0-1 knapsack problem. Now, were we to separate the above objective function in the standard fashion by expanding the state space of the corresponding linear knapsack problem, the state variable would in this case be of the form

$$s_n = (s_n(1), s_n(2)), \tag{73}$$

where

$$s_n(1) := L^* - \sum_{m=1}^{n-1} L_m x_m \quad , \quad n=2,3,\ldots,N \tag{74}$$

and

$$s_n(2) := \sum_{m=1}^{n-1} w_m x_m \quad , \quad n=2,3,\ldots,N, \tag{75}$$

with $s_1(1)=L^*$ and $s_1(2)=0$. In turn, the dynamic programming functional equation would be as follows:

$$f_n(s) = \max_{x \in D(n,s)} \{B_n x + f_{n+1}(s(1)-L_n x, s(2)+w_n x)\}, \quad 1 \leq n < N, \; s \in S_n, \tag{76}$$

where

$$f_N(s) := \max_{x \in D(N,s)} \{B_N x - c(s(2)+w_N x)\} \quad , \quad s \in S_N \tag{77}$$

$$D(n,s) := \begin{cases} \{0\} & , \; s(1) < L_n \\ \{0,1\} & , \; s(1) \geq L_n \end{cases} \quad , \quad n=1,2,3,\ldots,N \tag{78}$$

and

$S_n :=$ Set of all the feasible values of s_n as defined by (74)-(75).

On the other hand, should the problem under consideration be viewed as a c-programming problem, the following could be set:

$$q(x_1,\ldots,x_N) := \sum_{n=1}^{N} B_n x_n \tag{79}$$

$$w(x_1,\ldots,x_N) := \sum_{n=1}^{N} w_n x_n \qquad (80)$$

and

$$\varphi(z) := -\alpha z^{\beta}, \qquad z \in W := \{z \in \mathbb{R}: z > 0\}. \qquad (81)$$

Note that since $\alpha > 0$ and $0 < \beta < 1$, φ is differentiable and convex on W. The parametric problem would therefore be of the following form:

$$p(\lambda) := \max_{(x_1,\ldots,x_N)} \left\{ \sum_{n=1}^{N} B_n x_n + \lambda \sum_{n=1}^{N} w_n x_n \right\}, \quad \lambda \in \mathbb{R} \qquad (82)$$

$$= \max_{(x_1,\ldots,x_N)} \sum_{n=1}^{N} [B_n x_n + \lambda w_n x_n] \qquad (83)$$

subject to (71)-(72).

Since for each λ the objective function is additive and linear, we can solve this problem employing the state space

$$S := \{0, 1, 2, 3, \ldots, L^*\} \qquad (84)$$

and the following dynamic programming functional equation:

$$f_n(s;\lambda) = \max_{x \in D(n,s)} \{B_n x + \lambda w_n x + f_{n+1}(s - L_n x; \lambda)\}, \quad 1 \leq n < N, \ s \in S, \qquad (85)$$

starting at $n=N$ with

$$f_N(s;\lambda) := \max_{x \in D(N,s)} \{B_N x + \lambda w_N x\}, \quad s \in S, \qquad (86)$$

where

$$D(n,s) := \begin{cases} \{0\}, & s < L_n \\ \{0,1\}, & s \geq L_n \end{cases}, \quad n = 1, 2, 3, \ldots, N. \qquad (87)$$

Because the state space defined by (84) is likely to be considerably smaller than that defined by (73)-(75), it is clear that (85) will be far more efficient than (76). And yet, to make a persuasive case for the former, we would still need to demonstrate that a relatively small number of parametric problems will have to be solved for the purpose of recovering an optimal solution to the problem of interest. Let us then put the c-programming method to the test.

Consider an instance of the above problem whose parameters are stipulated in Table I.

First we need to determine the range of feasible values of $\nabla\varphi(w(x))$. Since (81) implies that $\nabla\varphi(z) = -\alpha\beta z^{\beta-1}$, define

$$W := [w_{\min}, w_{\max}], \qquad (88)$$

TABLE I. Parameters of numerical example.

n	B_n	w_n	L_n	n	B_n	w_n	L_n
1	15.0	8.8	70	11	17.1	7.3	85
2	17.8	8.3	75	12	18.3	8.1	81
3	17.3	7.6	74	13	18.7	8.6	82
4	17.0	7.9	88	14	18.8	8.9	87
5	17.6	7.5	84	15	17.4	8.1	77
6	18.1	8.0	86	16	17.1	8.2	80
7	15.8	8.2	71	17	18.5	8.5	89
8	16.6	10.8	79	18	17.6	8.6	73
9	15.7	8.4	76	19	17.4	8.4	83
10	16.3	8.7	72	20	17.5	7.6	78

$N=20$; $L^*=500$; $\alpha=3.5$; $\beta=0.7$

where

$$w_{min} := \min_{(x_1,\ldots,x_N)} \sum_{n=1}^{N} w_n x_n \qquad (89)$$

and

$$w_{max} := \max_{(x_1,\ldots,x_N)} \sum_{n=1}^{N} w_n x_n \qquad (90)$$

subject to (71)-(72).

Note that with both α and β being non-negative, $\nabla \varphi(z)$ is non-decreasing with z, so that the range of feasible values of $\nabla \varphi(z)$ on W is specified by

$$\Gamma := \{\lambda: \underline{\lambda} \leq \lambda \leq \bar{\lambda}\},$$

where

$$\underline{\lambda} := \nabla \varphi(w_{min}) = -\alpha\beta w_{min}^{\beta-1}$$

and

$$\bar{\lambda} := \nabla \varphi(w_{max}) = -\alpha\beta w_{max}^{\beta-1}.$$

Clearly,
$$w_{min} = \min_{1 \le n \le N} w_n = 7.3 .$$

Thus, solving (90) with dynamic programming yields
$$w_{max} = 54.4$$

so that we can set
$$\underline{\lambda} = -\alpha\beta(7.3)^{\beta-1} = -1.3495$$

and
$$\bar{\lambda} = -\alpha\beta(54.4)^{\beta-1} = -0.7387.$$

The next step then is to generate a sequence of λ's in order to cover the line segment $[-1.3495, -0.7387]$ as prescribed by Theorem 7. Thus, we set $\lambda^{(1)} = \underline{\lambda} = -1.3495$, and solve the corresponding parametric problem, to obtain the solution

$$x^{(1)} = (0,0,0,0,1,1,0,0,0,0,1,1,1,0,0,0,0,0,0,1),$$

which yields

$$h(x^{(1)}) = \sum_{n=1}^{N} B_n x_n^{(1)} + \varphi(\sum_{n=1}^{N} w_n x_n^{(1)}) = 55.3982$$

and

$$\lambda^{(2)} = \nabla\varphi(w(x^{(1)})) = -\alpha\beta(w(x^{(1)}))^{\beta-1} = -0.7714.$$

Hence, the line segment $[-1.3495, -0.7714]$ can be disposed of. Next, solving the parametric problem for $\lambda = \lambda^{(2)}$, we obtain the solution

$$x^{(2)} = (0,0,0,0,1,1,0,0,0,0,0,1,1,0,0,0,1,0,0,1),$$

which yields

$$h(x^{(2)}) = \sum_{n=1}^{N} B_n x_n^{(2)} + \varphi(\sum_{n=1}^{N} w_n x_n^{(2)}) = 55.8761$$

and

$$\lambda^{(3)} = \nabla\varphi(w(x^{(2)})) = -\alpha\beta(w(x^{(2)})) = -0.7656.$$

This solution eliminates the line segment $[-0.7714, -0.7656]$. Next, solving the parametric problem for $\lambda^{(3)}$, we obtain the solution $x^{(3)} = x^{(2)}$. We then set $\lambda^{(4)} = \bar{\lambda} = -0.7656$ and solve the parametric problem for this value of λ to obtain the solution

$$x^{(4)} = (0,0,0,0,1,1,0,0,0,0,0,1,1,1,0,0,0,0,0,1),$$

which yields

$$h(x^{(4)}) = \sum_{n=1}^{N} B_n x_n^{(4)} + \varphi(\sum_{n=1}^{N} w_n x_n^{(4)}) = 55.8702$$

and

$$\lambda^{(5)} = \nabla\varphi(w(x^{(4)})) = -\alpha\beta(w(x^{(4)})) = -0.7637.$$

This solution eliminates then the line segment $[-0.7637, -0.7387]$. Next, solving the parametric problem for $\lambda = \lambda^{(5)}$, we obtain the solution $x^{(5)} = x^{(2)}$, which in turn, eliminates the line segment $[-0.7656, -0.7637]$. Having thus eliminated the entire interval $[\underline{\lambda}, \bar{\lambda}]$, it follows that

$$p = \max \{h(x^{(1)}), h(x^{(2)}), h(x^{(3)}), h(x^{(4)}), h(x^{(5)})\}$$

$$= \max \{55.3982, 55.8761, 55.8761, 55.8702, 55.8761\}$$

$$= 55.8761,$$

the conclusion being that the optimal solution is $x^{(2)}$. All in all, applying the c-programming algorithm required solving six problems: one to determine the value of $\bar{\lambda}$ and five to cover the interval $[\underline{\lambda}, \bar{\lambda}]$.

SUMMARY

We have shown that c-programming offers the very machinery required to handle certain types of problems whose objective functions are rendered non separable under the terms of a standard dynamic programming formulation. Considering how often in the analysis of real-world water resources problems the analyst encounters objective functions of this type, c-programming techniques clearly provide the analyst with a potent tool. Particularly because they invite a formulation that remains true to the real-world attributes of water rersources problems.

Although it is somewhat premature at this stage to pass final judgement on the efficiency of c-programming algorithms, preliminary experiments indicate that they perform surprisingly well, especially in cases where the decision space is discrete. This is perhaps due to the fact that these algorithms operate essentially on the extreme points of the convex hull of $u(X)$. Whatever the case, it should prove interesting to examine whether the performance of these algorithms can be further enhanced by allowing them to exploit, in any given case, the peculiar features of the decision space and the objective function in question.

And finally, by its very nature c-programming has the ability to join forces with any optimization method capable of handling its parametric problem. Owing to its vulnerability to the Curse of Dimensionality, dynamic programming seems to be a prime candidate for collaboration with c-programming.

REFERENCES

1. S.J. Yakowitz, Dynamic Programming Applications in Water Resources, Water Resources Research, 18(4), 673-696 (1982).
2. M. Sniedovich, C-Programming: a Class of Nonlinear Optimization Problems, Discrete Applied Mathematics, 9, 301-305 (1984).
3. M. Sniedovich, Analysis of a Class of Proxy Problems, Operations Research Letters, 3(5), 271-273 (1984).
4. M. Sniedovich, C-Programming: an Outline, Operations Research Letters, 4(1), 19-21 (1985).
5. M. Sniedovich, A Class of Nonseparable Dynamic Programming Problems, Journal of Optimization Theory and Applications, (forthcoming).
6. M. Sniedovich, C-Programming and the Minimization of Pseudoconcave functions, TWISK #437, NRIMS, CSIR, Pretoria, RSA (1986).
7. G.L. Nemhauser, Introduction to Dynamic Programming (John Wiley, New York 1966).
8. M. Sniedovich, Dynamic Programming and Principles of Optimality: a Systematic Approach, Advances in Water Resources, 1(3), 131-139 (1978).
9. M. Sniedovich, A New Look at Bellman's Principle of Optimality, Journal of Optimization Theory and Applications, 49(1), 161-167 (1986).
10. W. Dinkelbach, On Nonlinear Fractional Programming, Management Science, 13(7), 492-498 (1967).
11. S. Schaible and T. Ibaraki, Fractional Programming, European Journal of Operations Research, 12, 325-338 (1983).
12. M .Sniedovich, A Variance Constrained Reservoir Control Problem, Water Resources Research, 16(2), 271-274 (1980).
13. M. Sniedovich, A Class of Variance Constrained Problems, Operations Research, 31(2), 338-353 (1983).
14. M. Sniedovich, A New Look at Fractional Programming, Journal of Optimization Theory and Applications (forthcoming).
15. M.S. Bazaraa and C.M. Shetty, Nonlinear Programming Theory and Algorithms (John Wiley, New York 1979).

CLOSED-LOOP CONTROL, BALANCING, AND MODEL REDUCTION OF LARGE SCALE WATER RESOURCE SYSTEMS

JOSE A. RAMOS
United Technologies Optical Systems, Inc.
Optics and Applied Technology Laboratory
P.O. Box 109660
West Palm Beach, Florida 33410-9660

ABSTRACT

A new approach for dimensionally reducing large water resource systems is presented. The approach is based on the concepts of closed-loop balancing and linear-quadratic-gaussian (LQG) methodology. We start by solving a general, discrete-time, LQG optimal control problem via dynamic programming. Its solution, by virtue of the separation principle, yields a pair of Riccati equations, one for the deterministic controller and the other for the stochastic observer. By transforming this pair of equations to a new coordinate system where costs and uncertainties of individual state components are matched, a balanced model is obtained. Then by deleting the state components that are least uncertain and at the same time contribute least to the cost function, one obtains the reduced-order model. The derivations are carried out for a typical multireservoir system operating under flood conditions.

Key Words: closed-loop balancing, dynamic programming, model reduction, separation principle, reservoir management systems.

INTRODUCTION

Dynamic programming (DP) has long been recognized as a powerful optimization tool for solving a large class of sequential decision problems encountered in water resources. The literature on DP applications to water resources problems is quite extensive and has recently been surveyed by Yakowitz [19]. Other survey articles of relevant interest are given in Ramos and Rao [12] and Yeh [20]. Much of the research on DP has concentrated on developing computationally efficient algorithms to overcome the "curse of dimensionality" - a computational barrier often found in multidimensional problems (see Morin [11] for an exposition on computational advances in DP). This, however, is not a criticism of DP but often a result of difficulties inherent in the problem at hand. Nevertheless, efficient algorithms have been developed and successfully applied to a gamut of problems in water resources.

In multireservoir operation studies, for instance, the aim has been placed at finding optimal operating rules for reservoir systems arranged in arbitrary topological order, while satisfying various system, demand and operating constraints at minimum operating costs. In general, this problem can be formulated as a nonlinear stochastic optimal control problem where the stochasticity comes from the unknown streamflows and demands. In addition, the temporal evolution of the system allows a natural stagewise decomposition of the problem, which makes DP a very attractive solution strategy.

Most reservoir operation studies in the past have dealt with the problem of specifying contract levels and prices for planning the long-term operation of the system. Short-term operation, on the other hand, makes use of the latest hydrologic information about the system in order to implement decisions in real-time. Such is the case for systems operating under flood conditions, where latest weather and streamflow forecasts are instrumental in determining the immediate releases from the reservoirs. Recent studies have shown that the use of hydrologic information in real-time can improve reservoir operating policies (Becker and Yeh [2]; Bras et al. [4]; Stedinger et al. [14]; Houck [6]). However, in going from long to short-term operation, the system is more dependent on its initial conditions and on the dynamics of the hydrologic processes driving the system. Necessarily then, for the real-time short-term reservoir operation problem, the DP model has to account for these process dynamics by augmenting the state-space (Wasimi and Kitanidis [18]). The solution to this problem via stochastic dynamic programming (SDP) requires a great deal of computational effort and is, therefore, practically infeasible.

Recently, Wasimi and Kitanidis [18] have formulated the above problem as a linear-quadratic-Gaussian (LQG) optimal control problem. Here, the multireservoir system along with the driving hydrologic processes are represented by a linear state-space model driven by Gaussian noises and a performance criterion which is a quadratic function of the state and input vectors, hence, the acronym LQG. The advantage of using such approach is that the real-time short-term reservoir operation problem can be solved in closed-form. Moreover, the solution, by virtue of the separation principle (Athans [1]), is separated into a deterministic linear-quadratic (LQ) optimal control problem and a state estimation problem. The conditions under which this separation is possible for reservoir systems has been recently studied in Kitanidis [8].

The heart of the LQG problem is the solution to a pair of Riccati equations, one for the controller and the other for the estimator. However, with the increasing interest in microcomputer applications and the development of expert systems (Houck [6]), where computer storage is a constraint, the numerical solution to these equations can be computationally prohibitive for large scale systems. The solution then calls for an approximate or reduced-order model.

Reduced-order state-space models for streamflow forecasting have been recently studied by Goldstein and Larimore [5] and Wasimi and Kitanidis [18]. More recently, Ramos [13] recognized this estimation problem as a stochastic realization problem and gave directions for model reduction as well as general approaches for solution. However, the problem of simultaneously building reduced-order state estimators and controllers has not been studied in the water resources literature, although the concept is now well established in the control engineering literature (Verriest [16,17]; Jonckheere and Silverman [7]). This paper concentrates on deriving reduced-order LQG models for water resources applications. The ideas presented here are an extension from the works of Verriest [16,17] and Ramos [13].

In Section 2, the real-time short-term reservoir operation problem is formulated as a discrete-time LQG tracking problem via DP and the separation principle, which allows the solution to be implemented as a separate controller and a state estimator. Section 4 contains a brief description of the ideas behind closed-loop balancing and model reduction. Finally, in Section 5, the implications of reducing large scale water resource systems are discussed.

FORMULATION OF THE REAL-TIME SHORT-TERM RESERVOIR OPERATION PROBLEM AS AN LQG TRACKING PROBLEM

In this section we formulate the real-time short-term reservoir operation problem as a general, discrete-time, LQG tracking problem. Our objective here is to demonstrate the utility of linear systems theory in the modeling of water resources problems. The implications of this approach is that the decision model, the model used by the decision maker to implement the operating policy, relates to the actual situation. In this context, one can analyze the decision problem to uncover some of the relationships among system structure, performance, and the need for real-time information (Wasimi and Kitanidis [18]; Houck [16]).

Without loss of generality, let us consider a typical water resource system whose conceptual representation is shown in Figure 1. The behavior of the system can be described by the following pair of difference equations

$$x_{k+1} = \phi[x_k, u_k, w_k] \quad k = 0, 1, \ldots \quad (1a)$$

$$z_k = \theta[x_k, v_k] \quad k = 0, 1, \ldots \quad (1b)$$

where at time k, x_k is the (nx1) vector of system states, u_k is the (px1) vector of controlled inputs, w_k is the (nx1) vector of process disturbances driving the sysem, z_k is the (mx1) vector of measurements or outputs, v_k is the (mx1) vector of disturbances corrupting the measurements, and ϕ and θ are functionals characterizing the properties of the system. Equation (1a) is commonly referred to as the state equation and accounts for the dynamic evolution of the system, while equation (1b) is known as the measurement or output equation.

In general, one can associate the state vector with physical realizable quantities. In reservoir systems, for instance, the states can represent the storage levels in the reservoirs, while for streamflow models where w_k is usually the output of a linear dynamic system, the state representation can vary from unit hydrograph ordinates to even meaningless quantities. The vector u_k may well represent the releases from the reservoirs, while the vector z_k represents measured quantities such as streamflows, storages, etc., which are corrupted versions of the actual processes. Usually reservoir storages are measured quite accurately, however, we will allow measurement noises (v_k) to compensate for other state variables as well.

One can easily conceptualize the system in Figure 1 as a decision making process. This is shown in Figure 2, where the water resource system is now modeled as a feedback control system subject to persistent disturbances. The system is composed of four basic subsystems: (1) the process P, (2) the decision system D, (3) a measurement system M, and (4) a disturbance system W. We will now consider the modeling aspects of the individual components of the decision making process.

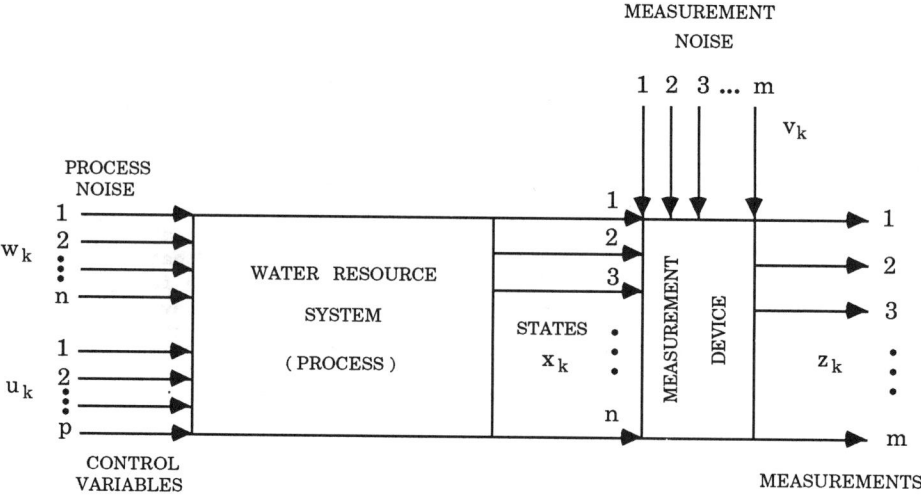

Figure 1. A typical water resource system

Process P

Suppose we have a water resource system composed of n_o reservoirs arranged in arbitrary topological order. Further suppose that $p \leq n_o$ of these reservoirs are regulated for control purposes, and that there are r streamflow measurement stations throughout the river basin. Then we can define the system components as follows: (1) states $\equiv$ storage levels - s_k ~ $(n_o \times 1)$; (2) controlled inputs $\equiv$ releases from reservoirs - u_k ~ $(p \times 1)$; and (3) process disturbances $\equiv$ streamflows - q_k ~ $(r \times 1)$. Now by invoking continuity at each reservoir, one then obtains the following state equation

$$s_{k+1} = \Theta s_k + \Phi u_k + \Gamma q_k \quad k = 0,1,\ldots \quad (2)$$

where Θ and Φ are known matrices of dimensions $(n_o \times n_o)$ and $(n_o \times p)$, respectively, and Γ is a known $(n_o \times r)$ matrix with elements 0 or 1.

Notice we have assumed that p out of n_o reservoirs are to be regulated. The reason for this is to allow linear reservoirs (conceptual) to account for the dynamics of the routing mechanism in the flood control reaches. For illustration purposes, consider the simple reservoir system with a flood control reach shown in Figure 3. The Muskingum method of streamflow routing relates the storage along the channel reach, c_k, to the inflows (r_k) and outflows (o_k) by the following relation:

$$c_k = K[xr_k + (1-x)o_k] \quad (3)$$

where K is the storage constant for the reach and x is a weighting factor varying between 0 and 0.5.

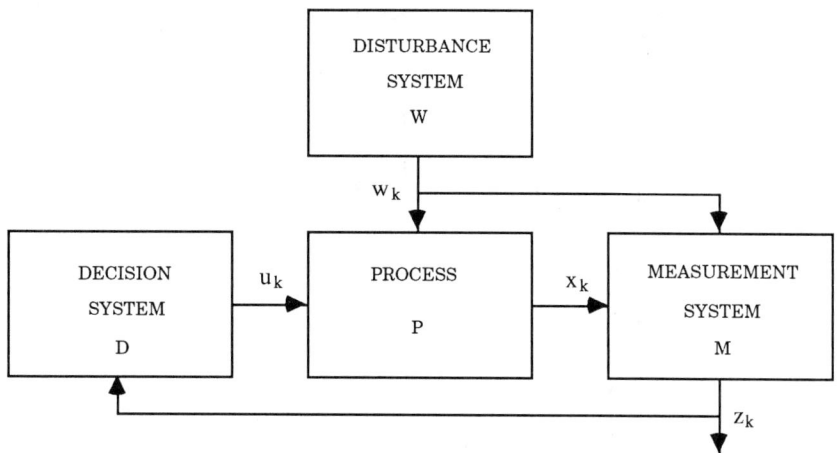

Figure 2. A general feedback control system

The continuity equation for the flood control reach is given by

$$\frac{c_{k+1} - c_k}{\Delta k} = \frac{r_k + r_{k+1}}{2} - \frac{o_k + o_{k+1}}{2} \tag{4}$$

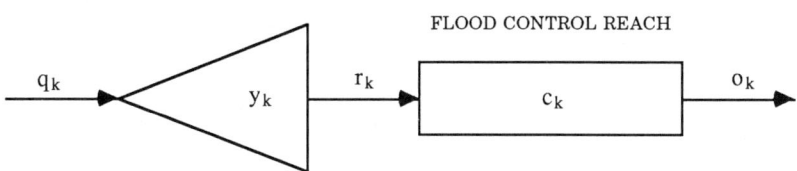

Figure 3. A simple reservoir system with flood control

By rearranging and manipulating equation (3), the outflow rate at time k can be expressed as

$$o_k = \frac{1}{K(1-x)} c_k - \frac{x}{(1-x)} r_k \tag{5}$$

Substituting equation (5) into equation (4), the state equation for the channel reach becomes

$$c_{k+1} = fc_k + g(r_k + r_{k+1}) \tag{6}$$

where the scalars f and g are appropriate functions of Δk, x, and K. Let us now combine the state equation (6) with that of the upstream reservoir, i.e.

$$\begin{bmatrix} y_{k+1} \\ c_{k+1} \end{bmatrix} = \begin{bmatrix} 1 & 0 \\ 0 & f \end{bmatrix} \begin{bmatrix} y_k \\ c_k \end{bmatrix} + \begin{bmatrix} -1 & 0 \\ g & g \end{bmatrix} \begin{bmatrix} r_k \\ r_{k+1} \end{bmatrix} + \begin{bmatrix} 1 \\ 0 \end{bmatrix} q_k \qquad (7)$$

Then if we let $s_k = [y_k \ c_k]^T$ and $u_k = [r_k \ r_{k+1}]^T$, and by appropriately defining Θ, Φ, and Γ, equation (7) reduces to equation (2) which corresponds to the model for the process P.

Disturbance System W

The disturbances, q_k, are modeled as the output of a linear system driven by white noise, i.e.

$$\theta_{k+1} = F\theta_k + \varepsilon_k \qquad k = 0,1,\ldots \qquad (8)$$

$$q_k = H\theta_k + \omega_k \qquad k = 0,1,\ldots \qquad (9)$$

where θ_k is the (n_1x1) state vector, ε_k and ω_k are respectively (n_1x1) and (rx1) white Gaussian noise vectors, and F and H are matrices of appropriate dimensions. The problem of identifying the parameters of the disturbance system is well documented in the water resources literature (Goldstein and Larimore [5]; Wasimi and Kitanidis [18]; Ramos [13]), and since it only becomes a subproblem of the overall problem, we will not discuss the details any further.

Notice that (2) and (8) can be combined to form an augmented state equation, i.e.

$$\begin{bmatrix} \theta_{k+1} \\ s_k \end{bmatrix} = \begin{bmatrix} 0 & F \\ \Theta & \Gamma H \end{bmatrix} \begin{bmatrix} \theta_k \\ s_k \end{bmatrix} + \begin{bmatrix} 0 \\ \Phi \end{bmatrix} u_k + \begin{bmatrix} \varepsilon_k \\ \Gamma\omega_k \end{bmatrix} \qquad (10)$$

or, in general,

$$x_{k+1} = Ax_k + Bu_k + w_k \ , \qquad k = 0,1,\ldots \qquad (11)$$

The new state x_k is composed of actual reservoir storages (y_k), fictitious reservoir storages (c_k), and the corresponding states from the disturbance model (θ_k). One can see that (11) describes the evolution of the general system shown in Figure 1.

Measurement System M

Let us return to our previous example of the system in Figure 3. Suppose that the measured storage level in the upper reservoir (p_k) can be obtained as a linear function of the actual storage levels (y_k) plus some noise (η_k), i.e., $p_k = hy_k + \eta_k$. Then by letting $d = [1/K(1-x)]$ and $e = [x/(1-x)]$ in (5) and making use of (9), one can represent the output of the overall measurement system M as

$$\begin{bmatrix} q_k \\ p_k \\ o_k \end{bmatrix} = \begin{bmatrix} H & 0 & 0 \\ 0 & h & 0 \\ 0 & 0 & d \end{bmatrix} \begin{bmatrix} \theta_k \\ y_k \\ c_k \end{bmatrix} + \begin{bmatrix} 0 & 0 \\ 0 & 0 \\ -e & 0 \end{bmatrix} \begin{bmatrix} r_k \\ r_{k+1} \end{bmatrix} + \begin{bmatrix} \omega_k \\ \eta_k \\ 0 \end{bmatrix} \quad (12)$$

or, in general, as

$$z_k = Cx_k + Du_k + v_k \quad , \quad k = 0,1,\ldots \quad (13)$$

where all terms are easily identified from (12).

Decision System D

The decision system develops a control strategy (policy) which is close to a set of desired target levels $\{x_{o_k}\}_{k=0}^{N}$ and $\{u_{o_k}\}_{k=0}^{N-1}$, by minimizing the cost functional given by

$$J = E\left[\frac{1}{2}\delta x_N^T P_N \delta x_N + \frac{1}{2}\sum_{k=0}^{N-1}\left[\delta x_k^T, \delta u_k^T\right]\begin{bmatrix} M & L \\ L^T & N \end{bmatrix}\begin{bmatrix} \delta x_k \\ \delta u_k \end{bmatrix}\right] \quad (14)$$

where E is the expectation operator, $\delta x_k = x_k - x_{o_k}$, $\delta u_k = u_k - u_{o_k}$, P_N is a positive definite penalty matrix, and L, M, and N are weight matrices of appropriate dimensions such that

$$\begin{bmatrix} M & L \\ L^T & N \end{bmatrix} \geq 0, \quad M \geq 0, \quad N > 0.$$

Minimization of (14) is subject to the constraints (11) and (13).

To comply with linear systems terminology we will assume the dimensions of the system to be as usual, i.e. $x_k \sim$ (nx1), $u_k \sim$ (px1), $w_k \sim$ (nx1), $z_k \sim$ (mx1), and $v_k \sim$ (mx1). The matrices A, B, C, and D are of dimensions (nxn), (nxp), (mxn), and (mxp), respectively. The second-order statistics of the system are assumed known and given by

$$E[w_k] = E[v_k] = E[x_o w_k^T] = E[x_o v_k^T] = 0, \quad k \geq 0 \quad (15a)$$

$$E\begin{bmatrix} w_k \\ v_k \end{bmatrix}[w_s^T, v_s^T] = \begin{bmatrix} Q & S \\ S^T & R \end{bmatrix}\delta_{ks} \geq 0; \quad Q \geq 0; \quad R > 0 \quad (15b)$$

$$E[x_0] = \bar{x}_0 \tag{15c}$$

$$E[(x_0 - \bar{x}_0)(x_0 - \bar{x}_0)^T] = \Sigma_0 \tag{15d}$$

The overall problem given by equations (11), (13), and (14) is known as a discrete-time LQG tracking problem. The solution to this problem is generated by the decision system D, which seeks a closed-loop policy $\{u_k\}_{k=0}^{N-1}$ as a function of $\{z_j;\ 0 \leq j \leq k\}_{k=0}^{N-1}$.

SOLUTION TO THE DISCRETE-TIME LQG TRACKING PROBLEM VIA DP

The discrete-time LQG tracking problem posed in the previous section is a closed-loop stochastic optimal control problem and can be associated with a functional minimization problem of the form

$$\min_{u(\cdot)} E_x[f(x,u(x))] \tag{16}$$

where the functions $u(\cdot)$ are defined in such a way that the expectation in (16) exists. Unfortunately, this functional minimization problem is very difficult to solve by ordinary stochastic dynamic programming. However, under certain regularity conditions (see Kitanidis [8]) (1) can be converted to a series of ordinary minimizations. The key elements here are the following identities (Meier et al. [19]).

(1) $\min\limits_{u(\cdot)} E_x[f(x,u(x))] = E_x[\min\limits_u f(x,u)] \tag{17}$

(2) $E_{x,y}[f(x,y)] = E_y[E_x[f(x,y)|y]] \tag{18}$

(3) $E_{x_k}[x_k^T P x_k | Z_k, U_{k-1}, x_0, \Sigma_0] = E_{x_k}[\hat{x}_{k|k}^T P \hat{x}_{k|k}] + \text{tr}[P\Sigma_{k|k}] \tag{19}$

where $\hat{x}_{k|k}$ is the minimum variance estimate of x_k given $Z_k = [z_0, z_1, \ldots, z_k]$, $U_{k-1} = [u_0, u_1, \ldots, u_{k-1}]$, $\bar{x}_0$ and Σ_0 are as defined in (15c) and (15d), respectively, and

$$\Sigma_{k|k} = E_{x_k}[(x_k - \hat{x}_{k|k})(x_k - \hat{x}_{k|k})^T | Z_k, U_{k-1}, \bar{x}_0, \Sigma_0] \tag{20}$$

Let us now reformulate the discrete-time LQG tracking problem to a form easily solvable by DP. Let $\pi = [\mu_0, \mu_1, \ldots, \mu_{N-1}]$ be any admissible control law and define the information vector available to the decision maker at time k as $I_k = [Z_k; U_{k-1}]$ for $k = 1, 2, \ldots, N-1$ and $I_0 = [z_0]$ for $k = 0$. Then from (1) and (14), we have

$$J_\pi = \mathop{E}_{\substack{x_0, w_k, v_k \\ k=0,1,2,\ldots,N-1}} \left[\frac{1}{2} \delta x_N^T P_N \delta x_N + \frac{1}{2} \sum_{k=0}^{N-1} [\delta x_k^T, \delta \mu_k^T(I_k)] \begin{bmatrix} M & L \\ L^T & N \end{bmatrix} \begin{bmatrix} \delta x_k \\ \delta \mu_k(I_k) \end{bmatrix} \right]$$

$$= \mathop{E}_{\substack{x_o, w_k, v_k \\ k=0,1,2,\ldots,N-1}} [g_N(x_N) + \sum_{k=0}^{N-1} g_k[x_k, \mu_k(I_k), w_k]] \tag{21}$$

If we now apply (1) and (18) for k=0,1,...,N-1 and do some manipulation, then one can convert the discrete-time LQG tracking problem into a problem with perfect state information (Bertsekas [3]), i.e.

$$J_\pi = \mathop{E}_{\substack{z_k \\ k=0,1,\ldots,N-1}} \left[\sum_{k=0}^{N-1} \widetilde{g}_k[I_k, \mu_k(I_k)] \right] \tag{22}$$

where $\widetilde{g}_k(\cdot)$ are well defined functions. One can easily show that the DP algorithm for this problem is given by (Bertsekas [3])

$$J_{N-1}(I_{N-1}) = \inf_{u_{N-1} \varepsilon \bar{U}} [\mathop{E}_{x_{N-1}, w_{N-1}} [g_N[\phi(x_{N-1}, u_{N-1}, w_{N-1})]$$

$$+ g_{N-1}[x_{N-1}, u_{N-1}, w_{N-1}] | I_{N-1}, u_{N-1}]] \tag{23}$$

$$J_k(I_k) = \inf_{u_k \varepsilon \bar{U}} [\mathop{E}_{x_k, w_k} [g_k[x_k, u_k, w_k] + J_{k+1}[I_k, \theta[\phi(x_k, u_k, w_k)], u_k, v_{k+1}] | I_k, u_k]] \tag{24}$$

where $\bar{U}$ is the admissible control region.

An optimal control law, $\pi^* = [\mu_0^*, \mu_1^*, \ldots, \mu_{N-1}^*]$, for the discrete-time LQG tracking problem can now be obtained as follows:

1) For every possible value of I_{N-1} solve (23) to obtain $\mu_{N-1}^*(I_{N-1})$ as the solution to $[\partial J_{N-1}(I_{N-1})/\partial \mu_{N-1}(I_{N-1})] = 0$. This requires use of (19) and the fact that $E[E[w_{N-1}|I_{N-1}]] = 0$. Then substitute $\mu_{N-1}^*(I_{N-1})$ in (23) to obtain $J_{N-1}^*(I_{N-1})$.

2) Use $J_{N-1}^*(I_{N-1})$ in the computation of $J_{N-2}(I_{N-2})$ via the minimization in (24), which yields $\mu_{N-2}^*(I_{N-2})$ for every possible value of I_{N-2}. Substitute $\mu_{N-2}^*(I_{N-2})$ in (24) to obtain $J_{N-2}^*(I_{N-2})$. Continue the same procedure to obtain $\mu_k^*(I_k)$ and $J_k^*(I_k)$. In this case, we have for k = 0,1,...,N-1.

$$\mu_k^*(I_k) = - [N + B^T K_{k+1} A] E[x_k | I_k] + [N + B^T K_{k+1} B]^{-1} \Theta_k \tag{25a}$$

$$J_k^*(I_k) = \frac{1}{2} [E[x_k|I_k]^T K_k E[x_k|I_k] + E[x_k|I_k]^T \pi_k + \pi_k^T E[x_k|I_k] + \sum_{i=k}^{N} \Omega_i] \tag{25b}$$

where

$$\Theta_k^T = [x_{o_k}^T L + u_{o_k}^T N - \pi_{k+1}^T B] \tag{26a}$$

$$\pi_k^T = \Theta_k^T[N + B^TK_{k+1}B]^{-1}[L^T + B^TK_{k+1}A] - \Phi_k^T \qquad (26b)$$

$$\Phi_k^T = [x_{o_k}^T M + u_{o_k}^T L^T - \pi_{k+1}^T A] \qquad (26c)$$

$$\Omega_k = tr[M\Sigma_{k|k}] + tr[K_{k+1}\Xi_k] - \Theta_k^T[N + B^TK_{k+1}B]^{-1}\Theta_k + [x_{o_k}^T, u_{o_k}^T]\begin{bmatrix} M & L \\ L^T & N \end{bmatrix}\begin{bmatrix} x_{o_k} \\ u_{o_k} \end{bmatrix} \qquad (26d)$$

$$\Omega_N = tr[P_N\Sigma_{N|N}] + x_{o_k}^T P_N x_{o_k} \qquad (26e)$$

$$\Xi_k = S[C\Sigma_{k|k-1}C^T + R]^{-1}S^T \qquad (26f)$$

One can see that the solution to the discrete time LQG tracking problem with incomplete state information can be decomposed into two separate components, one deterministic and the other stochastic. This remarkable result is known as the separation principle (Athans [1]), which states that the solution to the discrete-time LQG tracking problem is composed of a deterministic LQ regulator, except that the control is a feedback of the optimal state estimate obtained from a Kalman filter. This implies that the results derived separately for the deterministic optimal control problem and the state estimation problem are still valid. This is shown in Figure 4, where the estimator plays the role of a Kalman filter.

We now state the solution to the optimal discrete-time LQG tracking problem as follows:

Dynamic Output Feedback:

$$u_k = -G_k \hat{x}_{k|k} + F_k \qquad (27a)$$

$$\hat{x}_{k+1|k} = A\hat{x}_{k|k} + \hat{w}_{k|k} + Bu_k \quad ; \quad \hat{x}_{o|-1} = \bar{x}_o \qquad (27b)$$

$$\hat{x}_{k|k} = \hat{x}_{k|k-1} + H_k(z_k - C\hat{x}_{k|k-1}) \qquad (27c)$$

Computation of Gains:

$$G_k = [N + B^TK_{k+1}B]^{-1}[L^T + B^TK_{k+1}A] \qquad (28a)$$

$$H_k = \Sigma_{k|k-1}C^T[C\Sigma_{k|k-1}C^T + R]^{-1} \qquad (28b)$$

For k < N

$$K_k = A^TK_{k+1}A + M - [L + A^TK_{k+1}B][N + B^TK_{k+1}B]^{-1}[L + A^TK_{k+1}B]^T \quad ; \quad K_N = P_N \qquad (29)$$

For k ≥ 0

$$\Sigma_{k+1|k} = A\Sigma_{k|k}A^T + Q_k \quad ; \quad \Sigma_{0/-1} = \Sigma_o \tag{30a}$$

$$\Sigma_{k|k} = \Sigma_{k|k-1} - \Sigma_{k|k-1}C^T[C\Sigma_{k|k-1}C^T + R]^{-1}C\Sigma_{k|k-1} \tag{30b}$$

$$Q_k = Q - \Xi_k - A\Sigma_{k|k-1}C^T[C\Sigma_{k|k-1}C^T + R]^{-1}S^T - S[C\Sigma_{k|k-1}C^T + R]^{-1}C\Sigma_{k|k-1}A^T \tag{30c}$$

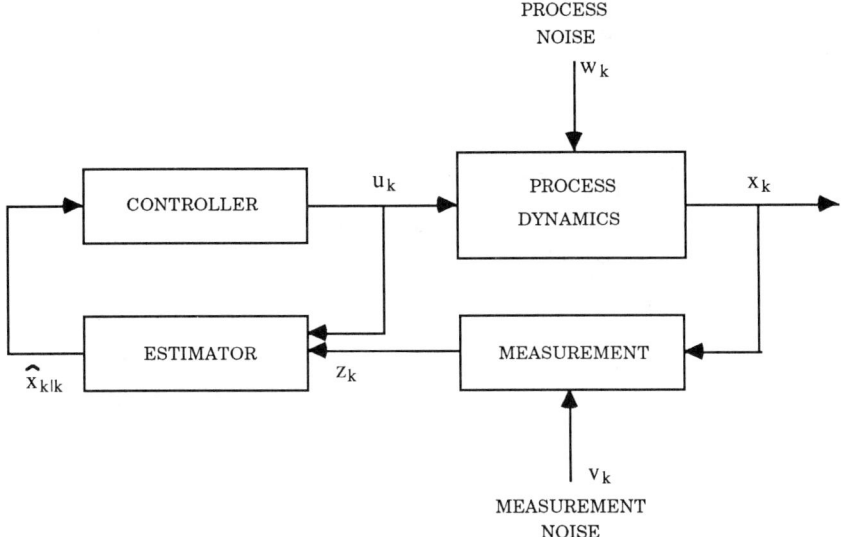

Figure 4. The separation principle

Remark 1: Equations (25) - (30) are valid only for the case where $Du_k = 0$ in equation (13). This bears no loss of generality since it implies that $x = 0$ in (3) which corresponds to the well familiar linear reservoir, i.e. $c_k = Ko_k$.

Remark 2: Notice that $F_k = [N + B^T K_{k+1}B]^{-1}\Theta_k$ and that $\hat{x}_{k|k} \equiv E[x_k|I_k]$ and $\hat{w}_{k|k} \equiv E[w_k|I_k] = S[C\Sigma_{k|k-1}C^T + R]^{-1}(z_k - C\hat{x}_{k|k-1})$ are computed from the Kalman filter.

Remark 3: If the system is stationary and the time horizon is sufficiently long, then under mild conditions, the gains G_k and H_k reach a steady-state value. This can aid significantly in the implementation of the optimal controller as well as the state estimator. One can easily show that for a steady-state system, (29) and (30a-b) satisfy the following pair of algebraic Riccati equations

$$K = A^T K A + M - [L + A^T K B][N + B^T K B]^{-1}[L + A^T K B]^T \tag{31a}$$

$$\Sigma = A\Sigma A^T + Q - [S + A\Sigma C^T][R + C\Sigma C^T]^{-1}[S + A\Sigma C^T]^T \tag{31b}$$

which clearly displays the duality between the controller and the estimator.

CLOSED-LOOP BALANCING AND MODEL REDUCTION

Since its introduction in 1978, the concept of balanced realizations has received so much attention that it is now evolving as a rich theory in its own right. Its implications in the design of digital filters, system identification, and model reduction are now well established in the control engineering literature. Open-loop balanced realizations were first introduced by Moore [10] and were later generalized by Verriest [15] for continuous-time varying systems. Closed-loop balanced realizations, on the other hand, were introduced by Verriest [16,17] and Jonckheere and Silverman [7] for continuous-time LQG systems. Here we will follow this approach for deriving the closed-loop balancing conditions for the discrete-time LQG tracking problem.

Let us first fix the ideas behind Moore's open-loop balancing in the context of the familiar linear system

$$x_{k+1} = Ax_k + Bu_k \quad ; \quad x_o = \bar{x}_o \tag{32a}$$

$$z_k = Cx_k \tag{32b}$$

where all terms are as defined previously. Here we assume that the system is stable, the triple $(A,B,C)_n$ is of minimal dimension (n), and that the pairs (A,B) and (C,A) are resepctively controllable and observable. Furthermore, if we let $W_o = O^T O$ and $W_c = CC^T$, where O and C are respectively the observability and controllability matrices, then these satisfy the following Lyapunov equations

$$W_o = A^T W_o A + BB^T \tag{33a}$$

$$W_c = AW_c A^T + C^T C \tag{33b}$$

An interpretation of these gramian matrices is as follows. Suppose we wish to find a control sequence $\{u_k : k_o \leq k \leq k_f\}$ that drives the system from $x_{k_0} = x_o$ to $x_{k_f} = 0$ in finite time. Then the minimum energy required for this control is given by (Verriest [16,17])

$$J_c = x_o^T W_c^{-1} x_o = \sum_{k=k_o}^{k_f} u_k^T u_k \tag{34}$$

Similarly, if for the autonomous system, $x_{k+1} = Ax_k$, we observe an output $z_k = Cx_k + v_k$, then the maximum output power is obtained from

$$J_o = x_o^T W_o x_o = \sum_{k=k_o}^{k_f} z_k^T z_k \tag{35}$$

Hence, the gramian matrices W_o and W_c can be interpreted as measures of energies.

Moore's idea of balancing is to transform (32) to a coordinate system where both measures of energies are matched. That is, to apply a nonsingular transformation T such as $\bar{x}_k = Tx_k$, so that under the new coordinate system $(\bar{A},\bar{B},\bar{C})_n = (TAT^{-1}, TB, CT^{-1})_n$, the transformed gramians

$\bar{W}_o = T^{-T}W_o T^{-1}$ and $\bar{W}_c = TW_c T^T$ are equal and diagonal, i.e.

$$\bar{W}_o = \bar{W}_c = \Sigma = \text{diag}[\sigma_1, \sigma_2, \ldots, \sigma_n] \qquad (36)$$

where $\sigma_1, \sigma_2, \ldots, \sigma_n$ are the eigenvalues of the product $W_o W_c$. These are strictly positive and invariant under similarity transformations.

The motivation for model reduction is then obvious. Suppose we let $x_o = e_i = (0, \ldots 0, 1, 0, \ldots, 0)^T$, where the 1 is in the ith component, then from (34) and (35) we calculate the energies as $J_c = 1/\sigma_i$ and $J_o = \sigma_i$. This tells us that if σ_i is very small, then it takes a large amount of energy to drive the system from the ith direction, while very little output energy can be observed along the same direction. Hence, the states along the ith direction are nearly uncontrollable and nearly unobservable, and thus, can be deleted.

Interestingly, Verriest [16,17] and Jonckheere and Silverman [7] interpreted the solutions to the algebraic Riccati equations in the LQG problem, κ and Σ, as being gramians in the closed-loop system. That is, κ is the observability gramian under state feedback and represents a weighting for the minimal regulation cost. Analogously, Σ represents the controllability gramian under output injection feedback and, thus, represents a measure of disturbability caused by noise. From this point of view it then seems natural to balance the LQG problem with respect to κ and Σ. This is called closed-loop balancing and amounts to finding a new coordinate system, $(\bar{A},\bar{B},\bar{C})_n = (TAT^{-1}, TB, CT^{-1})_n$, where $\bar{\kappa} = T^{-T}\kappa T^{-1}$ and $\bar{\Sigma} = T\Sigma T^T$ are equal and diagonal, i.e.

$$\bar{\kappa} = \bar{\Sigma} = \Delta = \text{diag}[\delta_1, \delta_2, \ldots, \delta_n] \qquad (37)$$

with $\delta_1 > \delta_1 > \ldots > \delta_n > 0$ being the similarity invariant eigenvalues of κΣ. The transformation $\bar{x}_k = Tx_k$ transforms the other system matrices as follows: $\bar{Q} = TQT^T$; $\bar{S} = TS$; $\bar{R} = R$; $\bar{M} = T^{-T}MT^{-1}$; $\bar{L} = T^{-T}L$; $\bar{N} = N$; $\bar{P}_N = T^{-T}P_N T^{-1}$.

In order to find the balancing conditions we proceed as follows. First compute the eigenvalue-eigenvector decomposition of Σ as

$$\Sigma = U_1 \Sigma_1^2 U_1^T \qquad (38)$$

where $U_1^T U_1 = I$. Next apply the transformation $T_1 = \Sigma_1^{-1} U_1^T$ to $\bar{\Sigma} = T_1 \Sigma T_1^T$ and $\bar{K} = T_1^{-T} K T_1^{-1}$ such that

$$\bar{\Sigma} = I \quad \text{and} \quad \bar{K} = \Sigma_1 U_1^T K U_1 \Sigma_1 \qquad (39)$$

and further decompose $\bar{K}$ as follows

$$\bar{K} = U_2^T \Delta^2 U_2 \qquad (40)$$

where $U_2^T U_2 = I$. Then the balancing transformation can be picked as

$$T = \Delta^{1/2} U_2^T \Sigma_1^{-1} U_1^T. \qquad (41)$$

Notice that Σ_1 and Δ are diagonal matrices. Furthermore, one can readily verify that $\bar{\bar{\Sigma}} = T \Sigma T^T = \bar{\bar{K}} = T^{-T} K T^{-1} = \Delta$, where Δ is also the matrix of eigenvalues of $K\Sigma$, i.e.

$$K\Sigma = T^T \Delta^2 T^{-T} = T^T \text{diag}[\delta_1^2, \delta_2^2, \ldots, \delta_n^2] T^{-T} \qquad (42)$$

Suppose we can define a threshold index r such that $\delta_r \gg \delta_{r+1}$, then one can partition the system as follows

$$\bar{x}_k = \begin{bmatrix} \bar{x}_k^1 \\ \bar{x}_k^2 \end{bmatrix} \begin{matrix} r \\ n-r \end{matrix} \quad ; \quad \Delta = \begin{bmatrix} \Delta_1 & 0 \\ 0 & \Delta_2 \end{bmatrix} \begin{matrix} r \\ n-r \end{matrix}$$

$$\bar{A} = \begin{bmatrix} \bar{A}_1 & \bar{A}_{12} \\ \bar{A}_{21} & \bar{A}_2 \end{bmatrix} \begin{matrix} r \\ n-r \end{matrix} \quad ; \quad \bar{B} = \begin{bmatrix} \bar{B}_1 \\ \bar{B}_2 \end{bmatrix} \begin{matrix} r \\ n-r \end{matrix} \quad ; \quad \bar{C} = [\bar{C}_1 \; \bar{C}_2] m$$

where the first subsystem is associated with $\Delta_1 = \text{diag}[\delta_1, \delta_2, \ldots, \delta_r]$, and the second subsystem with $\Delta_2 = \text{diag}[\delta_{r+1}, \delta_{r+2}, \ldots, \delta_n]$. The motivation for model reduction is then to delete $\bar{x}_k^2$ from the model since these are the states that contribute least to the cost criterion and at the same time are the least uncertain (i.e., $\delta_{r+1}, \delta_{r+2}, \ldots, \delta_n$ are very small). Some measures and suggestions for model reduction can be found in Verriest [16,17] and Ramos [13]. The final task then is to implement the reduced-order model. A block diagram of the implementation is shown in Figure 5. It should be noted that this implementation is suboptimal since we are deleting information from the model. Nevertheless, the degree of suboptimality in the solution when compared to the reduced computational effort, should warrant the implementation of reduced-order models.

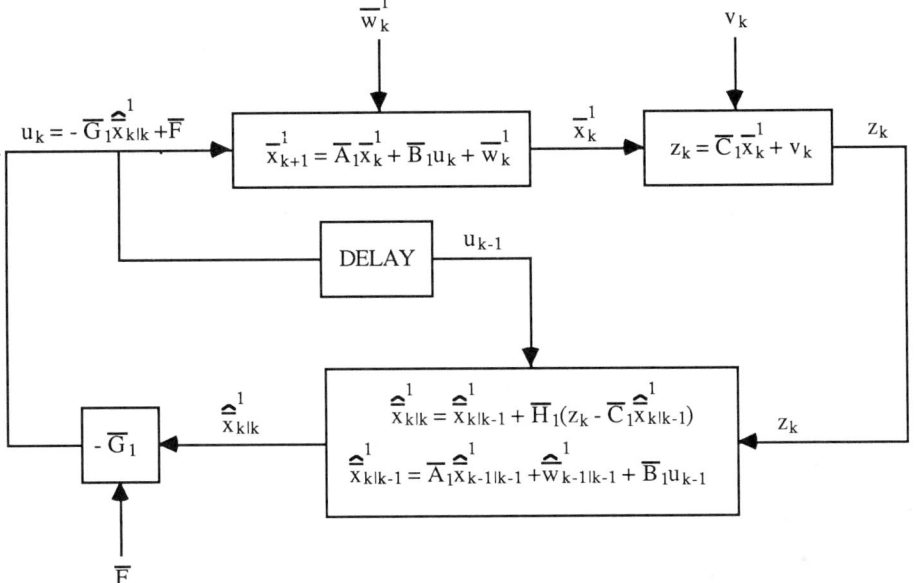

Figure 5. Implementation of the reduced-order model

DISCUSSION

The main objective of this paper has been the design of reduced-order models for short-term real-time control of large scale water resource systems. The material plays the role of a tutorial paper in the subject and should be useful to researchers in all areas of water resources where linear systems theory can be applied.

The computational effort required to implement current feedback control policies is almost proportional to the square of the state vector dimension. Thus, the application of existing design procedures to large scale water resource systems constitute an excessive burden, particularly from an implementation point-of-view. Resorting to reduced-order models, which are some form of model aggregation, will then lead to a great deal of flexibility in the design but at the expense of performing suboptimally. For a good approximation, however, the degree of suboptimality should be very small. It should be kept in mind however, that when dealing with physical systems, the balancing transformation and model reduction operation map the physical state vector into a coordinate system where the reduced state vector has no physical meaning. This bears no loss of generality since the control vector retains its physical properties. While the physical states, if needed, can be approximately recovered by applying the inverse balancing transformation. Ramos [13] has studied the forecasting aspects of reduced-order steamflow models, and found that for size reductions as large as 50%, the reduced-order models performed as good as the full-order models. The author is currently investigating the numerical implementation of reduced-order controllers, the results of which will be reported elsewhere.

REFERENCES

1. Athans, M. [1971]. "The Role and Use of the Stochastic Linear-Quadratic-Gaussian Problem in Control System Design." IEEE Transactions on Automatic Control, Vol. AC-16, No. 6, pp 529-552.
2. Becker, L., and W. W-G Yeh [1974]. "Optimization of Real-Time Operation of a Multiple-Reservoir System." Water Resources Research, Vol. 10, No. 6, pp. 1107-1112.
3. Bertsekas, D. [1976]. "Dynamic Programming and Stochastic Control." Academic Press, New York.
4. Bras, R. L., R. Buchanan, and K. C. Curry [1983]. "Real-Time Adaptive Closed Loop Control of Reservoirs with the High Aswan Dam as a Case Study." Water Resources Research, Vol. 19, No. 1, pp. 33-52.
5. Goldstein, J. D., and W. E. Larimore [1980]. "Applications of Kalman Filtering and Maximum Likelihood Parameter Identification to Hydrologic Forecasting." Technical Report TR-1480-I, TASC, Reading, Mass.
6. Houck, M. [1985]. "Designing an Expert System for Real-Time Reservoir System Operation." Civil Engineering Systems, Vol. 2, No. 1, pp. 30-37.
7. Jonckheere, E. A., and L. M. Silverman [1981]. "A New Set of Invariants for Linear Systems: Application to Approximation." Presented at the 1981 Int. Symp. Math. Theory Networks and Syst., Santa Monica, CA, pp. 129-133.
8. Kitanidis, P. K. [1983]. "Real-Time Forecasting of River Flows and Stochastic Optimal Control of Multireservoir Systems." IIHR Rep. 258, Inst. Hydraul. Res., Univ. Iowa, Iowa City.
9. Meier, L., R. E. Larson, and A. J. Tether [1971]. "Dynamic Programming for Stochastic Control of Discrete Systems." IEEE Transactions of Automatic Control, Vol. AC-16, No. 6, pp. 767-775.
10. Moore, B. [1978]. "Principal Component Analysis in Linear Systems: Controllability, Observability, and Model reduction." IEEE Transactions of Automatic Control, Vol. AC-26, No. 1, pp. 17-32.
11. Morin, T. [1978]. "Computational Advances in Dynamic Programming." In: Dynamic Programming and its Applications. M. L. Putterman, ed., pp. 53-90, Academic Press, New York.
12. Ramos, J. A. and S. G. Rao [1982]. "State of the Art Review of Optimization Techniques Used in the Operation and Planning of Water Resource Systems." Unpublished Report, Dept. of Civil Eng., Georgia Institute of Technology.
13. Ramos, J. A. [1985]. "A Stochastic Realization and Model Reduction Approach to Streamflow Modeling." Ph.D. dissertation, Dept. of Civil Eng., Georgia Institute of Technology, Atlanta, GA.
14. Stedinger, J. R., B. F. Sule, and D. P. Loucks [1984]. "Stochastic Dynamic Programming Models for Reservoir Operation Optimization." Water Resources Research, Vol. 20, No. 11, pp. 1499-1505.
15. Verriest, E. I. [1980]. "On Balanced Realizations for Time Variant Linear Systems." Ph.D. dissertation, Dept. of Elect. Eng., Stanford University, Stanford, CA.
16. Verriest, E. I. [1981a]. "Low Sensitivity Design and Optimal Order Reduction for the LQG Problem." 24th Midwest Symposium on Circuits and Systems, June 1981, Albuqerque, New Mexico, pp. 365-369.

17. Verriest, E. I. [1981b]. "Suboptimal LQG-Design via Balanced Realizations." Proceedings of the 20th IEEE Conference on Decision and Control, San Diego, CA, pp. 686-687.
18. Wasimi, S. and P. K. Kitanidis [1983]. "Operation of a System of Reservoirs Under Flood Conditions Using Linear Quadratic Gaussian Control," IIHR Rep. 268, Inst. Hydraul. Res., University of Iowa, Iowa City, Iowa.
19. Yakowitz, S. [1982]. "Dynamic Programming Applications in Water Resources." Water Resources Research, Vol. 18, No. 4, pp. 673-696.
20. Yeh, W. W.-G. [1982]. "State of the Art Review: Theories and Applications of Systems Analysis Techniques to the Optimal Management and Operation of Reservoir System." Report, School Eng. Appl. Sci., University of California, Los Angeles, CA.

PART THREE

WATER SUPPLY AND DISTRIBUTION MANAGEMENT

IDENTIFICATION OF DEMAND MODELS FROM NOISY OBSERVATIONS--AN APPLICATION TO WATER RESOURCES

LOV KUMAR KHER,* AND SOROOSH SOROOSHIAN**
*AT&T Bell Laboratories, 600 Mountain Avenue, Murray Hill, New Jersey 07974; **Dept. of Hydrology and Water Resources, University of Arizona, Tucson, Arizona 85721

ABSTRACT

One of the major problems in systems planning is the identification of a demand model when all the independent and dependent variables are noisy. The existing methods generally used for noisy variables are associated with various problems, for example, nonidentifiability and unboundedness. An alternative model, the noise-in-variable model (NVM), is presented, which explicitly considers noise in the data for all the variables of the system. A mathematical programming-based solution algorithm is used to identify the NVM model. This algorithm gives bounds on both the model parameters and the noise covariance matrix. Here a solution range is obtained instead of the single point estimate from the classical estimation theory. The problem of unboundedness is addressed by using prior information about the noise. The procedure is implemented for the water resources using real data. The NVM approach gives a more realistic picture of future water demands. Various demand values, obtained using the identified parameters' space, will correspond to a set of demand scenarios in which each scenario will have a different noise variance. Multiple step-ahead point forecasts for demand can be computed using the minimum noise variance system of the identified demand models.
Further, a suitable cost model can be identified from noisy data using the same approach as outlined earlier. Finally, the demand and cost models can be implemented in an integrated fashion to solve capacity expansion decision-making problems using dynamic programming procedures as given in Freidenfelds [1].

Keywords: Identification; linear systems; modeling; nonlinear programming; noise; optimization; realization theory; statistics; economics; water resources.

INTRODUCTION

In the real world, the future is uncertain, but not to a degree to obviate the benefits of long short-range planning or the use of mathematical models. Can we cope with it better than before with the new and improved modeling techniques that are available to us now? The obvious answer is yes. Dreyfus and Sen [2] comment: "If the wisdom of professional decision makers, forecasters, and planners indeed takes the analytical form, the computer implementation of mathematical models embodying their knowledge should be an invaluable aid to their activities." Many researchers have addressed the methodology for the analysis of empirical data for mathematical modeling. However, among these researchers, Kalman's recent work (Kalman [3], [4], [5]) presents a new improved direction towards the identification of mathematical models from real data. Kalman [3] raises an important question:

"The dilemma is that we don't know whether there is a way, which mathematicians call natural or canonical, of defining models so that they depend only on the data and not on any external biases. Such biases may be introduced, often unintentionally and unknowingly, by the special procedures or algorithms employed in constructing a realization. In the vast majority of models which have arisen historically, such biases are indeed present and cannot be justified away. The problem is subtle. The solution requires a mathematical point of view which cannot be rendered into ordinary language with precision or intuitive meaning."

In particular, we focus our attention toward mathematical modeling problems for water resources systems. There are several important problems in planning water resources systems, such as how much water should be used, where it will be needed, what purposes it will serve, when the demand for water will occur, and how much budget should be allocated to meet these demands. The actual water demand will depend on such time-dependent variables as population levels and distribution, per capita income, price of water, rainfall and temperature, technological development, consumer habits, and preferences. Developing relations between these variables and using them to estimate water demands under various conditions requires analytical approaches.

In most water management and/or planning studies, some best-guess demand function is specified by some growth rate rather than detailed year-by-year estimates (Dandy et al. [6]; Kindler and Russel [7]). In cases where detailed estimates are given, generally multiple regression or time-series methods are used (see e.g., Agthe and Billings [8]; Foster and Beattie [9]; Hansen and Narayanan [10]; and Maidment and Parzen [11]). Models based on these methods are known as error-in-equation models (EEM) which have an error term at the end of the equation, and the EEM define an inexact relationship among observable variables. Demand models developed using any of the above methods are based on the assumption that the data for all the independent variables are "exact" (i.e., data have no error). In real life, however, data are always "inexact" (i.e., noisy, uncertain), and lack of their consideration in the modeling process limits the usefulness of the results. Hanke and de Maré [12] have also made a similar observation about the noise in the data.

Without proper consideration of noise in independent variables, we can expect the resulting estimates to be biased and inconsistent and, therefore, produce inaccurate forecasts of future demands. For further discussion of the implications of some of these problems, see Hoodges and Moore [13]; Bard [14]; and Garber and Klepper [15].

To overcome the drawbacks of EEM models, some researchers (see e.g., Bard [14]; Mehra [16]; Garber and Klepper [15]; Kalman [4]; and Kher [17]) have examined the feasibility of developing models which consider noise in all the variables. In this case, the resulting water demand model will not be unique in the classical sense. Due to the consideration of noise in all the variables, a range for each model parameter is obtained rather than a point estimate by the classical estimation theory.

One class of existing models which incorporates errors in the independent variables is known as error-in-variable models (EVM). Unfortunately, the EVM are not very useful because, according to Kalman [4], all the model parameters cannot be estimated from the given behavioral relations. To overcome such a limitation, Kalman [4] has recently proposed an alternative to EVM which is based on "noisy realization theory". This procedure gives a bounded solution set when the data covariance matrix (Σ) of all the variables is inverse positive (i.e., all entries of Σ^{-1} are positive real numbers). However, when Σ is not inverse positive, then the solution set will be unbounded if a single relationship for the demand is to be identified. Klepper and Leamer [18] have discussed determining a bounded solution set for the unbounded case by making the assumption that all the independent variables will have the same percentage of noise variance.

In this paper, the general background of a procedure is discussed for identifying a model which considers the noise structure in the inputs and produces a bounded solution set for a water demand model. The proposed approach extends the application of noisy realization theory introduced by Kalman [4, 5]. The name used to identify this class of model is the noise-in-variable model (NVM). A simple first-order lag dynamical model is used to incorporate time-dependent information. The NVM model explicitly considers noise in the data for all the variables of the system. The solution procedure is developed so that the identification problem is transformed, using Cholesky-type factorization for the "true" component of the data covariance matrix, into a nonlinear programming algorithm which gives maximum and minimum bounds on both the model parameters and the noise covariance matrix. So, here we obtain a nonunique solution instead of a unique point estimate as obtained by classical estimation theory. For details about the mathematical problem formulation and solution procedure, see Kher [17] and Kher and Sorooshian [19].

The proposed procedure gives a bounded solution set for the identification of a linear relationship from the noisy data when the data covariance matrix is inverse positive. However, when it is not inverse positive, then the solution set for this will be unbounded. In this case, Kher [17] has assumed prior information about the noise covariance matrix (such as its upper/lower bound) for obtaining a bounded solution set. The use of such prior information also enables us to obtain a tighter parameter range for the NVM.

Finally, an example is presented using real data to identify a water demand model for the city of Tucson, Arizona. The solutions obtained by the proposed NVM approach are compared with the results when noise is not considered in the independent variables. Results are presented to demonstrate the usefulness and practical applicability of the proposed procedures.

BACKGROUND: NOISE-IN-VARIABLE MODELS (NVM)

Problem Formulation

According to the procedure provided by noisy realization theory, behavioral data may be translated into a "model", which is defined as a collection of all systems (or "family of models") compatible with the given data (Kalman [4]). The next step is to parameterize the model as an abstract object. If the realization problem has a nonunique solution, the "model" will consist of a family of inequivalent models. Each model belonging to the "family of models" is then described by two sets of parameters. One set of "model parameters" specifies a particular model in the family of models. The other set of "noise parameters" describes the place of this model within the family constituting the models. The objective, therefore, is to identify a single linear demand model among "true" (unobservable) variables such that the solution set (both in parameter and noise spaces) is bounded. The various means of tackling the problem of unboundedness in the solution set are discussed in detail in Kher [17].

Here, we define "noise" as any deviation of the observed variable from an unobserved "true" variable, due to some unknown factors such as measurement error, disturbances, departures from linearity, etc. This definition of noise is more useful and less complicated mathematically than it would be if we were to use measurement error, disturbance error, model error, etc., individually.

The NVM modeling approach is described briefly below. It is assumed that each measured or observable variable (Z_i, $\forall i=1,\ldots,k+2$) consists of an unknown "true" component ($\hat{Z}_i$) and an unobservable "noise" component ($\tilde{Z}_i$).

Further, it is assumed that these variables are zero mean realizations of an unknown random process, the "true" variables and their noises are independent, and also noise is assumed to be uncorrelated among variables. Because of the assumption of zero mean, the available data can be summarized by the sample covariance matrix of the observations, i.e.,

$$\Sigma = [\sigma_{ij}], \quad \Sigma \text{ is a } (k+2 \times k+2) \text{ symmetric matrix.} \qquad (1)$$

$$\sigma_{ij} = \text{Cov}\left(\hat{Z}_i \; \hat{Z}_j\right); i=1,\ldots,(k+2), j \neq i \qquad (2)$$

Here, Z is an (n x k + 2) matrix of n observations for k+2 variables. We can now state the problem as:

Given a symmetric positive data covariance matrix ($\Sigma = \Sigma' > 0$), find all possible solutions in the noise-parameter space such that:

$$\hat{Z} A = 0 \qquad (3a)$$

where

$$\hat{Z} = \begin{vmatrix} \hat{Y}_{1,1} & \hat{Y}_{0,1} & \hat{X}_{1,1} & \cdots & \hat{X}_{1,k} \\ \hat{Y}_{2,1} & \hat{Y}_{1,1} & \hat{X}_{2,1} & \cdots & \hat{X}_{2,k} \\ \vdots & \vdots & \vdots & & \vdots \\ \hat{Y}_{t,1} & \hat{Y}_{(t-1),1} & \hat{X}_{t,1} & \cdots & \hat{X}_{t,k} \\ \vdots & \vdots & \vdots & & \vdots \\ \hat{Y}_{n,1} & \hat{Y}_{(n-1),1} & \hat{X}_{n,1} & \cdots & \hat{X}_{n,k} \end{vmatrix} \quad [n \times (k+2)] \text{ matrix}$$

$Y_{t,1}$ = observed value of output;

$\hat{Y}_{t,1}$ = "true" unknown value of output;

$\tilde{Y}_{t,1}$ = "unobservable" noise in output;

$X_{t,i}$ = observed value of ith input;

$\hat{X}_{t,i}$ = "true" unknown value of ith input;

$\tilde{X}_{t,i}$ = "unobservable" noise in i^{th} input; and

β_{1i} = model parameters, $i = 0, 1, \ldots k$

$A = (1, -\beta_{10}, -\beta_{11}, \ldots, -\beta_{1k})'$ [(k+2)x 1] vector, (prime indicates transpose), and the following conditions are satisfied:

$$\hat{\Sigma} A = 0 \tag{3b}$$

$$\hat{\Sigma} \geqslant 0, \text{ corank } \hat{\Sigma} = m \geqslant 1 \tag{4}$$

$$\tilde{\Sigma} = \Sigma - \hat{\Sigma}, \text{ diagonal, and} \tag{5}$$

$$\tilde{\Sigma} \geqslant 0 \tag{6}$$

where m represents the maximum number of independent relationships among the variables that can be identified from the noisy data. $\hat{\Sigma}$ and $\tilde{\Sigma}$ are, respectively, the "true" and "noise" component of the data covariance matrix Σ, A is a (k+2 x1) vector of model parameters.

Solution Procedure

The success of such a modeling procedure is highly dependent on the availability of a proper solution methodology compatible with the underlying assumptions. Kalman [4] has given a solution procedure for this noisy problem which follows an algebraic geometric procedure. The limitations of this procedure are discussed in Kher [17]. The objective here is to develop a solution procedure so that a bounded solution set can be obtained and demand scenarios into the future can be generated.

It should be noted that for a given Σ and a given m, $\hat{\Sigma}$ (and equivalently, $\tilde{\Sigma}$) is a solution to the noisy identification problem (in the noise space) only if the above-mentioned conditions (3) through (6) are met. Here constraints on $\hat{\Sigma}$ are held implicitly by means of a Cholesky-type factorization (number here) of matrix $\hat{\Sigma}$, such as:

$$\sum_{\ell=1}^{r=(k+2)-m} t_{i\ell} t_{j\ell} = \sigma_{ij}, \quad 1 \leqslant i < j \leqslant (k+2) \tag{8}$$

$$\sum_{\ell=1}^{r=(k+2)-m} t_{i\ell}^2 \leqslant \sigma_{ii}, \quad 1 \leqslant i \leqslant (k+2) \tag{9}$$

The solution procedure consists of two phases. In the first phase, the solution is obtained with the maximum feasible corank $\hat{\Sigma}$; and in the second phase, with constraints linking noise solutions to parameter solutions, the range of each model parameter is obtained.

It should be noted that if Σ is inverse positive, m will be equal to one, implying that there exists only one linear relationship. Otherwise, m $\geqslant$ 2, implying that there will be at least two linear relationships among the variables. In other words, when Σ is not inverse positive, then the solution set for a single linear relationship will be unbounded. Therefore, some prior information about the noise covariance matrix ($\tilde{\Sigma}$ will be required to obtain a bounded solution set for a single linear relation. The details of solution procedure can be found in Kher [17] and Kher and Sorooshian [19].

Once the solution to the NVM model is obtained, it can be used for predicting the future scenarios of demand. It should be noted that here we will obtain a range (with minimum and maximum bounds) of predicted values due to the range of the parameter space. A simple prediction scheme is given in the following:

$$\hat{Y}_{1,n+h} \in (\alpha_\ell + \underline{\beta}'_\ell \hat{\underline{x}}_{n+h}), (\alpha_u + \underline{\beta}'_u \hat{\underline{x}}_{n+h}) \qquad (10)$$

Or, it can be simplified to:

$$Y_{1,n+h} = \underline{R}_\alpha + \underline{R}'_\beta \underline{x}_{n+h} + R_{\tilde{\sigma}_{11}} - \underline{R}'_\beta \text{ diag } (R_{\tilde{\sigma}_{22}}, \ldots, R_{\tilde{\sigma}_{kk}}) \qquad (11)$$

where:

$\underline{R}_\alpha$ is the vector of range (including lower and upper bounds) for constant parameter α. α can be obtained as:

$$\alpha = E(Y_{1,n}) - \sum_{i=1}^{k} \beta_i \, E(x_{i,n}) \qquad (12)$$

Lower and upper bounds of α can be obtained by substituting the corresponding lower and upper bounds of β_i in Equation (12).

$\underline{R}_\beta$ = (kx1) vector of ranges (including lower and upper bounds) for each model parameter β_i, i = 0,1,...,k.

$R_{\tilde{\sigma}_{ii}}$ = range (including minimum and maximum bounds) of one standard deviation for each noise element $\tilde{\sigma}_{ii}$, $\forall i = 1,k$.

$Y_{1,n+h}$ = the predicted output at time n+h.

$\underline{x}_{n+h}$ = vector of inputs (known) at time n+h.

Note that the NVM model is identified using data up to time n and predictions for the output are made at time n+1, n+2, ..., n+h.

The above scheme can be further improved by updating the solution space in a recursive way as new information (data) is available at different time steps.

Application

The proposed NVM approach has been successfully applied in different areas, such as identification of a water demand model for the city of Tucson, Arizona (Kher and Sorooshian [19]), identification of models and predictions for energy/electricity available for the United States (Kher, Sioshansi, and Sorooshian [20]), and identification of an industrial turnover demand model (Kher [21]).

For the purpose of practical applicability of the proposed approach here, we present an example from a water resources case study. The detailed results of these case studies can be seen in the above references.

Case Study I--Water Demand Model

Several demand models have been identified for planning the water supply system for the city of Tucson. Here, only one model is shown as given in Table 1. Domestic water demand (Q, in 1 ccf per month per household, 1 ccf is 100 cubic feet and it is equal to 2831.60 L) is an output (Y_1), and average income of a family (I, in real dollars per month per household), average price of water (P, in real cents per 100 cubic feet), monthly average temperature (T, in ^{0}F), are respectively considered as inputs (X_i, i=1,···,3). For the purpose of this case study, we have included the average price of water as an input; however, some researchers including Taylor

TABLE 1. Water demand model identified.

Model	Variables*	Data Set** (Monthly)	Remarks
WRT	1,3,4, and 6	Jan.1974-Dec.1976	Static, split data set

* 1, Y_{t1} is the water demand (Q) at time t; 3, X_{t1} is the average income of a family (I) at time t; 4, X_{t2} is the average price of water (P) at time t; 6, X_{t3} is the monthly average temperature (T) at time t.

** Data for variables Q, I, and P were obtained from the Division of Economic and Business Research, University of Arizona, Tucson, Arizona. Data for variable T were collected from climatological data, Arizona. Interested readers can obtain entire data set from from the authors.

[22] have suggested that proper modeling of decreasing block tariffs requires inclusion of both a marginal and average price as predictors in the demand function. The results of the case study are obtained by employing the computational algorithm as shown in Figure 1. This algorithm has been coded in FORTRAN; interested readers can obtain details of the programs from the authors.

Model WTR

The data covariance matrix is given as:

$$\Sigma = \begin{matrix} 43.336 & & & \\ -11.207 & 609.171 & & \\ -9.100 & 47.532 & 54.816 & \\ 74.081 & -73.993 & 2.027 & 165.332 \end{matrix}$$

The corresponding elementary regression vectors (ERV) of Σ are:

$$a_1^{(1)} = \begin{matrix} 1.00000 \\ 0.05775 \\ -0.23372 \\ 0.47679 \end{matrix} \qquad a_1^{(2)} = \begin{matrix} 1.00000 \\ 0.26996 \\ -0.42133 \\ 0.57406 \end{matrix}$$

$$a_1^{(3)} = \begin{matrix} 1.00000 \\ 0.10411 \\ -0.78501 \\ 0.50429 \end{matrix} \qquad a_1^{(4)} = \begin{matrix} 1.00000 \\ 0.06953 \\ -0.24721 \\ 0.56513 \end{matrix} \qquad (13)$$

In this case, Σ is inverse positive, and the ERV of Σ lie in the same orthant. Therefore, the solution to this problem is bounded and is shown in Figure 2. The model to be identified is given as:

$$\hat{Y}_{t1} = a_{1I} \hat{X}_{t1} + a_{1P} \hat{X}_{t2} + a_{1T} \hat{X}_{t3} \qquad (14)$$

The tighter range for each parameter is obtained by assuming that the variables I, P, and T have noise equal to 25, 15, and 20 percent,

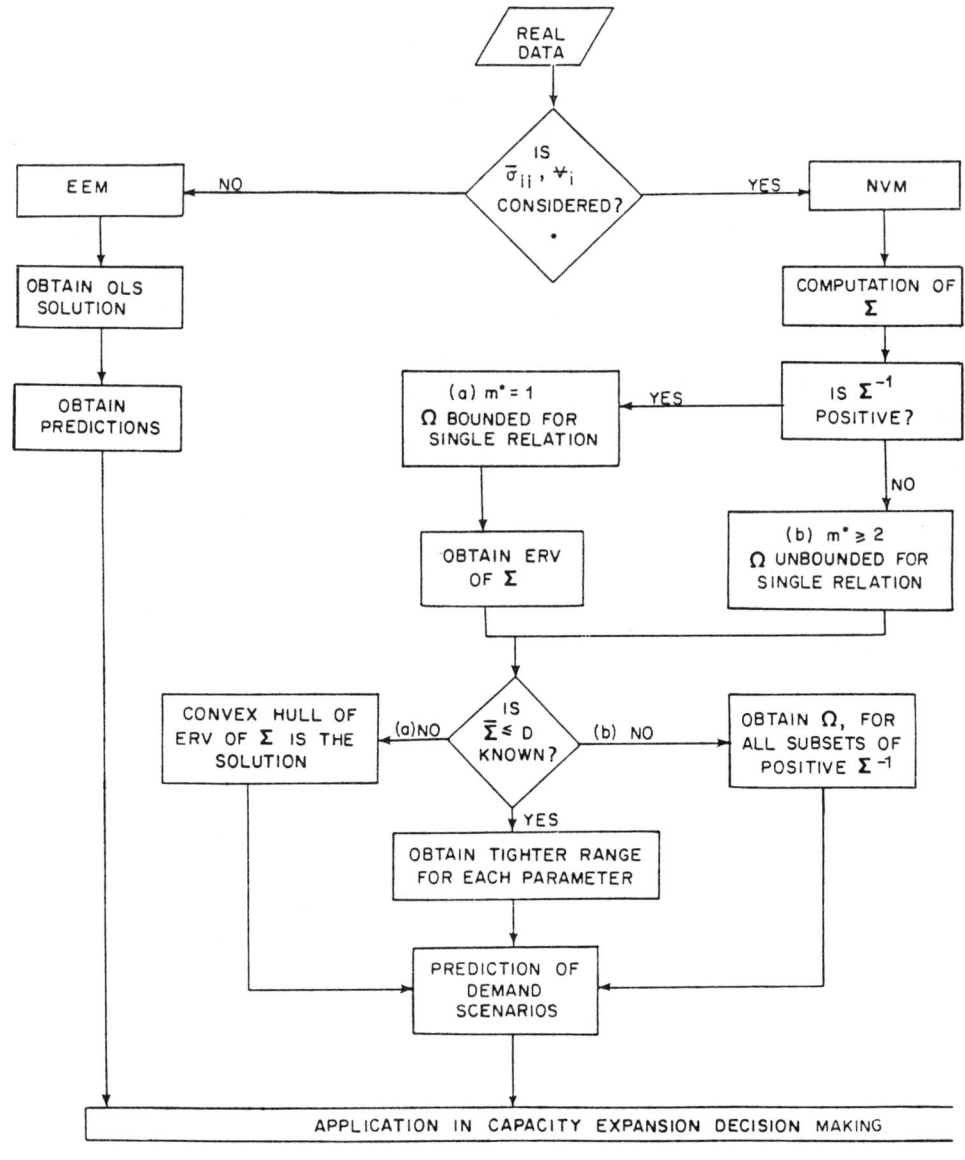

Figure 1. Block diagram of the computational algorithm.

*IS NOISE CONSIDERED IN ALL THE INPUT AND OUTPUT VARIABLES?

respectively. Therefore, the upper bound on the noise covariance matrix is given as:

$$\tilde{\Sigma} \leq D_u = \text{diag}(0, 38.073, 1.233, 6.613) \qquad (15)$$

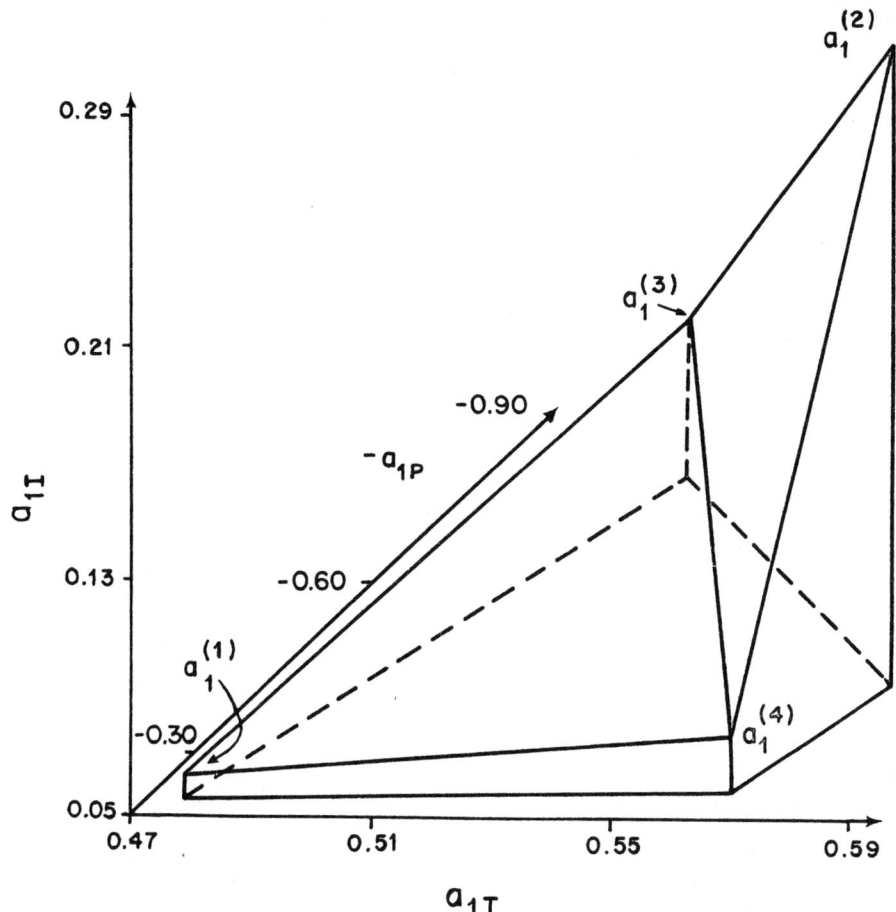

Figure 2. ERV of Σ--Model WTR 2.

Using the same prior information as given above, the summary of results (both in parameter and noise spaces) for the programming solution procedure is given in Table 2.

It is observed that in model WTR, the range of parameters a_{1I} and a_{1T} is small as compared to that in another model, whereas the range of parameters a_{1P} and σ_{1I} is higher in this case.

CONCLUSIONS AND RECOMMENDATIONS

Identification of demand models for water resources systems planning is the focus of this paper. A modeling procedure is presented which explicitly considers noise in all of the variables included in a chosen demand model.

A first-order lag dynamic noise-in-variable model (NVM) is formulated. This formulation attempts to identify a "true" relationship between the

TABLE 2. Summary of Results for Tighter Bounds of Parameters: Model WTR2

$\tilde{\Sigma}$ Corresponding to Minimum Parameter Bound	Minimum and Maximum Parameter Bounds	$\tilde{\Sigma}$ Corresponding to Maximum Parameter Bound
(0.0, 38.073, 1.233, 6.613)	$0.05775 \leq a_{1I} \leq 0.0658$	(4.741, 0.0, 0.0, 0.0)
(6.466, 38.073, 0.0, 6.613)	$-0.23954 \leq a_{1P} \leq -0.23372$	(0.0, 38.073, 1.233, 6.613)
(6.535, 38.073, 1.233, 6.613)	$0.47679 \leq a_{1T} \leq 0.49832$	(4.895, 38.073, 0.0, 0.0)

inputs and the outputs of the system. The procedure is such that we obtain both the model parameters and the noise variances from the known data statistic (i.e., data covariance matrix of all the variables) and some given behavioral relations. A mathematical programming solution procedure is developed in such a way that the original linear identification problem is transformed into a nonlinear programming problem. This procedure identifies a bounded solution set for a single linear demand model by making use of prior information about the noise covariance matrix. It should be noted that in the NVM, a range for each model parameter is obtained rather than a unique point estimate as obtained by classical estimation procedures. This implies that the "true" solution to a noisy identification problem lies within the convex polygon formed by the elementary regression vectors (ERV) of the data covariance matrix (Σ). In other words, any point within the range of each model parameter is a possible solution. At this point, it is necessary to point out that the range of each model parameter is also not unique because the bounds on each model parameter can vary with different lengths of data sets. Therefore, the selection of a proper data set also becomes an important factor.

These procedures are implemented to develop a water demand model or a "family of models" for the city of Tucson. These models are based on different combinations of variables, different lengths of data, and whether the model is static or dynamic. The variables used in these models are: water demand, income of a family, price of water, rainfall, temperature, and effective evapotranspiration.

In the first step of this procedure, all the possible models are analyzed, assuming that the data set is noise-free. This analysis gives ordinary least squares estimates and other statistical results. The second step consists of identifying a "family of models" for the given example by assuming that the data set is noisy. The "family of models" obtained by the model WTR is identified for the city of Tucson for the given data set. In this case, the data covariance matrix is inverse positive and hence the solution set is bounded. The tighter range for each model parameter is obtained by assuming that the variables income, price, and temperature have noise (e.g., ±measurement error) equal to 25, 15, and 20 percent, respectively. The identified model, once used for forecasting, will give us a range (a minimum and a maximum bound) where the future demand is expected to lie.

The mathematical programming-based solution procedure is found to be computationally efficient and fast. Another advantage of this proposed approach is that any dimensional problem can be addressed for the identification of a model for which real data are available. Some of the problems which were faced in this study are: (1) the model parameters are optimized individually and not simultaneously, thereby causing the problem of obtaining an optimum solution in a real sense for the noise-parameter space; (2) the assumption of an upper bound on the noise covariance matrix is quite significant for obtaining a bounded solution set; and (3) the selection of a proper length of data set is also important for obtaining the solution set.

In this paper, several aspects of demand modeling are discussed in the context of noisy data. However, some work is underway to improve the decision-making processes. With further improvements in the above items, the NVM approach to demand forecasting should be seriously considered as a tool in water resources systems planning.

ACKNOWLEDGMENT

The authors wish to thank Ms. Corla Thies for her editorial and typing of the drafts and the final version of the manuscript.

REFERENCES

1. J. Freidenfelds, "Capacity Expansion: Analysis of Simple Models with Applications, Elsevier, New York (1981).

2. S.E. Dreyfuss and S. Sen, Mathematical modeling and long range planning. In X.J.R. Avula (Ed.), Proc. International Conference on Mathematical Modeling, 4th, 1983, Pergamon Press, London, pp. 37-43 (1984).

3. R.E. Kalman, A System-theoretic critique of dynamic economic models. J. of Policy Analysis and Information Systems, 41(1), 3-22 (1980).

4. R.E. Kalman, Identifiability and Problems of Model Selection in Econometrics, in Advances in Econometrics, edited by W. Hildenbrand, pp. 169-207, Cambridge, New York (1982a).

5. R.E. Kalman, System Identification from Noisy Data, in Proceedings International Symposium on Dynamical Systems, edited by A. Bednarek and L. Cesari, pp. 135-164, Academic Press (1982b).

6. G.C. Dandy, E.A. McBean, and B.G. Hutchinson, A Model for Constrained Optimum Water Pricing and Capacity Expansion, Water Resour. Res., 20(5), 511-510 (1984).

7. J. Kindler and C.S. Russell, Modeling Water Demands, Academic Press, London (1984).

8. D.E. Agthe and R. Bruce Billings, Dynamic Models of Residential Water Demand, Water Resour. Res., 16(3), 476-480 (1980).

9. H.S. Foster and B.R. Beattie, Urban Residential Demand for Water in the United States, Land Economics, 55(1), 43-58 (1979).

10. R.D. Hansen and R. Narayanan, A Monthly Time Series Model of Municipal Water Demand, Water Resour. Bull., 17(4), 578-585 (1981).

11. D.R. Maidment and E. Parzen, Cascade Model of Monthly Municipal Water Use, Water Resour. Res., 20(1), 15-23 (1984).

12. S.H. Hanke and L. de Mare´, Municipal Water Demands in Modeling Water Demands, edited by J. Kindler and C.S. Russell, pp. 149-169, Academic Press, London (1984).

13. S.D. Hoodges and P.G. Moore, Data Uncertainties and Least Squares Regression, Applied Statistics, 21, 185-195 (1972).

14. Y. Bard, Nonlinear Parameter Estimation, Academic Press, New York (1974).

15. S. Garber and S. Klepper, Extending the Classical Normal Errors-In-Variable Model, Econometrica, 48(6), 1541-1546 (1980).

16. R.K. Mehra, Identification and Estimation of the Error-In-Variables Model (EVM), in Mathematical Programming Study, 5, p. 191-210, North Holland (1976).

17. L.K. Kher, Demand Modeling for Systems Planning Using Noisy Realization Theory, Ph.D. dissertation, 303 pp., Case Western Reserve University, Cleveland, Ohio (1985).

18. S. Klepper and E.E. Leamer, Consistent Sets of Estimates for Regressions with Errors in All Variables, Econometrica, 52(1), 163-183 (1984).

19. L.K. Kher and S. Sorooshian. Identification of water demand models from noisy data. Water Resour. Res., 22(3), 322-330 (1986).

20. L. Kher, F.P. Sioshansi, and S. Sorooshian. Energy demand modeling with noisy input-output variables. Paper accepted for publication in The Energy Journal (December 1986).

21. L.K. Kher, A New Look Into Demand Modelling Using Noisy Realization Theory. In Proc. 4th IFAC/IFORS Symp on Large Scale Systems: Theory and Application, edited by M. Mansour and H.P. Geering, Pergamon Press, England (August 1986).

22. L.D. Taylor, The Demand for Electricity: A Survey, The Bell Journal of Economics, 6(1), 74-110 (Spring 1975).

ON OPTIMAL PUMPING POLICIES FOR GROUND WATER MANAGEMENT

TOSHIO ODANAKA
Tokyo Metropolitan Institute of Technology, 6-6 Asahigaoka,
Hino-city, Tokyo, Japan

ABSTRACT

This paper is concerned with the optimal allocation over time of the resources which are in supply only partially renewable at a point in time. A functional equation is obtained from a dynamic programming formulation of the problem. This functional equation is used to derive an optimal decision rule for resource use as a function of current supply. The results are applied to ground water storage control separately and conjunctively. They are then tested empirically by comparison with a decision rule obtained by a detailed numerical method. A more general problem is next discussed. Finally, we solve a continuous version of the model in complete detail. The advantage of having a complete solution to the problem is that it is possible to determine turnpike horizon policies and to develop a practical method for ground water system operation. The stochastic case is also discussed.

INTRODUCTION

Water management, under conditions of uncertain supply and controlled demand, may be viewed as an inventory problem. The objective is to control the demand (withdrawal) in such a way that the expected value of net benefits at present value is maximized. This inventory control is of extreme importance for ground water systems management because its quantity in storage is often large in relation to the annual use rate. The inventory problem is similar to the models of surface reservoirs systems.

The supply of water which is available for possible capture is a random variable and thus the solution of the inventory problem requires an estimate of the probability distribution for this water supply. It is assumed that the probability distribution is known or that a good estimate is available from sources such as time series data.

The origin of the sequential decision model used dates back to Massé [1]. Some of the early pioneering works in the theory of inventory control of interest are those by Arrow et al. [2]. Bellman [1957] generalized the concepts of sequential decision processes and coined the term dynamic programming [3]. Odanaka studied the multistage inventory control problem using the technique of dynamic programming [7]. One of the first applications of modern sequential decision theory to water storage problems was, however, by Little [4].

In our study, the simplest water management problem is first considered; that is, the case of only underground water storage. A mathematical model is constructed which permits maximization of present value of expected net benefits for any length of planning horizon, under specified physical conditions of recharge, storage capacity, etc. (See O.R. Burt [6] for example).

The case of two storage facilities is next considered; that is, a single surface reservoir and one underground reservoir. The mathematical model formulated as a basis for quantitative analysis of this situation can be quite complex. The need for some simplification is correctly recognized by Buras and Hall [5]. Some approximation procedures are suggested. Most of the analysis will be readily understandable by first considering the single storage facility case as background.

In the case of two storage facilities, we want to give a completely rigorous mathematical analysis to support our development. A functional equation is obtained from a dynamic programming formulation of the problem. This functional equation is used to derive an optimal decision rule for resource use as a function of current supply. The results are applied to ground water and surface reservoir storage control and then tested empirically by comparison with a decision rule obtained via some detailed numerical methods.

Finally, we solve in complete detail a continuous version of the model. The advantage of having a complete solution to the problem is that it is possible to determine turnpike horizon points or policies and to develop a practical ground water management protocol system.

OPTIMAL PUMPING POLICIES FOR GROUND WATER

The economical use of ground water is becoming an important consideration in the world. As the demand for water continues to grow, it will become increasingly important to "economically integrate" the use of the two basic sources of water, namely surface and ground water.

In this study the net function viewpoint is taken. The question is, "What are the characteristics and determining factors of the water utilization policy which maximizes the net of operating a ground water system?"

In each of the following sections assumptions are made about net factor, random flow, demand for ground water, and the limitations of the ground water facilities. From these assumptions salient features of the most economical water utilization policy are deduced. This is the main objective of the study.

In this paper, it is assumed that the net incurred in the ith subinterval is a concave function of the amount of ground water produced during that subinterval. It is assumed that all water produced can be sold at some price. If the price of ground water is a decreasing function of the amount of water supplied, this fact can be recognized through the use of concave net functions which increase as x_i increase for sufficiently large values of x_i (x_i is the volume of water produced during the ith subinterval, i= $1, 2, \ldots, T$).

To make nets incurred during one subinterval comparable to those incurred in another subinterval, a discount factor a is introduced.

The following assumptions are inessential to the basic result of this paper. They have been made for the sake of mathematical simplicity.

1. The well is assumed to be of unlimited capacity.

2. The amount of ground water which may be produced during a subinterval is assumed not to be limited by a pump capacity restriction.

3. The water flowing into the well during the ith subinterval, w_1 is not available for pumping generation until the next subinterval.

4. There is assumed to be no depletion of the water through bottom leakage. In practice, such leakage could be recognized in this model by an appropriate adjustment of the probability distribution.

5. The cost function to be used in measuring the net incurred during the ith subinterval will be denoted by $G_i(x_i, s_i)$ that are K-convex, differentiable and nondecreasing in x.

Attention is focused on maximizing the expected net incurred during a certain time period [0,T]. This time period is divided into T subintervals of equal length. The volume of water which flows into the well during the ith subinterval is denoted by w_i (i=1,2,......,T). This variable w_i is assumed to be a continuous random variable with a probability density $h(w_i, s_i)$ (i=1,2,......,T).

The following elementary relations and notation are used in this proof:

1. The T subintervals are numbered in the reverse order of their occurrence.

2. The volume of water utilized during the ith subinterval for the pumping of ground water is x_i (i=1,2,......,T). The value of x_i that minimize the discounted expected cost incurring during the ith and all subsequent subintervals is denoted by x_i^* (i=1,2,......,T).

3. The cost function $G(x,s)$ that may be incurred during the ith subinterval, if x_i units are produced during that subinterval (i=1,2,......,T) is K-convex, differentiable and nondecreasing in x. Where $R(x)$ is the net function which is approximated by the quadratic equation in x and $c(s,x)$ is pumping cost.

$$c(s,x) = (c_0 - cs)(x - E(w)) + k, \qquad x > 0$$
$$= 0 \qquad x = 0 \qquad (1)$$

$$G(x,s) = c(s,x) - R(x)$$

4. The volume of water on the basin at the beginning of the ith subinterval which can be used to pump is denoted by s_i (i=1,2,......,T). For the model, the following recurrence relation is valid:

$$s_i = s_{i+1} + w_{i+1} - x_{i+1} \qquad (i=1,2,......,T-1) \qquad (2)$$

5. The discounted expected net if an optimal water storage policy is followed during i consecutive subintervals, and the ground water level at the beginning of these i subintervals is s_i, is denoted by $f_i(s_i)$, for i=1,2, ...,T. The following basic recurrence relation can be deduced from this definition (and using equation 1 and 2):

$$f_i(s_i) = \min_{0 \le x_i \le s_i} \left\{ G_i(x_i, s_i) + \alpha \int_0^\infty f_{i-1}(s_i + w_i - x_i) h_i(w_i) dw_i \right\} \qquad (3)$$

$$(i=2,3,......,T)$$

If the planning horizon is infinite, we will be interested in the limiting form of (1). In the limit as $n \to \infty$, $f_{n+1}(s) = f_n(s) = f(s)$, and we have

$$f(s) = \min_x \{G(x,s) + \alpha \int f(s+w-x)h(w)dw\} \quad (4)$$

This is the equation which will provide a means for obtaining decision rules.

The following two definitions are crucial to the subsequent development.

Definition. A (s_n, S_n) policy is a policy for period n that takes no action if state variable $s_n \geq s \geq S_n$ and takes action with state reduction to S_n if $s > s_n$.

Definition. A function $f(s)$ is said to be K-convex if, for any three points $0 \leq r \leq b \leq s < \infty$, $f(r) + K \geq [f(b) - f(s)] \cdot (b-r)/(s-b) + f(b)$. If $f(s)$ is differentiable, then $K \geq -f'(s) \cdot (s-r) + f(s) = f(b)$.

Theorem. Subject to assumptions, (i) (s_n, S_n) policy is optimal for period i, and (ii) $f_i(s)$ is a continuous, differentiable, nondecreasing and K-convex function of s.

CONJUNCTIVE SURFACE AND GROUND WATER

Let us now consider the multi-dimensional version of the problem. Here we have N water resources whose storage levels will be denoted by $s_1, s_2, \ldots, S_n$, and whose probable inflow $(w_1, w_2, \ldots, w_n)$ at any time is subject to a joint distribution function whose density is $h(w_1, w_2, \ldots, w_n)$.

In formulating the functional equation for the function $f(s_1, s_2, \ldots, s_n)$, the maximum expected overall discounted return, let us, for the sake of simplicity, consider only the two-dimensional case, conjunctive use of surface and ground water.

Let us put no set up cost, then the optimal pumping levels for the ith reservoir are $\bar{x}_i$, $i=1,2$.

AN EMPIRICAL APPLICATION

This application was worked previously using numerical methods to obtain the functional equation. The region of study is an area serviced by the Tama River region, a suburban region of Tokyo Metropolis [10].

TURNPIKE HORIZONS

5.1 Deterministic case

In this section, we solve a continuous version of the model in complete detail. The reason we are able to obtain a complete solution is that the linear decision rule, which is optional here as in other quadratic models, permits the elimination of the adjoint function from the state variable equation after one differentiation of the latter. Thus the difficult two point boundary value problem which usually arises in control problems is converted into an ordinary second order differential equation, which is readily solved [11].

Consider the optimal pumping policies for ground water at a well. Define the following quantities:

$s(t)$ = the volume of water at the basin at time t.

$x(t)$ = the volume of water utilized at time t for the pumping of ground water.

$w(t)$ = the volume of water flowing into the well at time t.

T = length of planning period, the net incurred at time t.

$G(x,s)$ = the convex cost function to be used in measuring and no set up cost.

We now state the conditions of the model. The first condition is the stockflow equation:

$$\dot{s} = w - x, \quad s(0) = s_0$$

The second is the objective function:

$$\text{Minimize } \{f = \int_0^T G(x,s)\,dt\}$$

Introducing the Lagrange multipliers $\lambda(t)$, we then consider the new criterion functional

$$\int_0^T \{G(x,s) + (\lambda(t), \dot{s} - w + x)\}\,dt$$

Consider, for example,

$$G(x,s) = a_1 x^2 + a_2 x + a_3 + a_4 sx + a_5 s$$

The Euler equations and boundary conditions are

$$\dot{s} = w - x, \quad s(0) = s_0$$

$$2a_1 x + a_2 + a_4 s + \lambda = 0, \quad \lambda(T) = 0$$

$$a_4 x + a_5 - \dot{\lambda} = 0$$

Solving for x yield the so-called optimal decision rule

$$x^* = -\frac{a_2 + a_4 s + \lambda}{2a_1}$$

Substituting (8) into (5) gives

$$\dot{s} = w + \frac{a_2 + a_4 s + \lambda}{2a_1}, \quad s(0) = s_0$$

Note that if $\frac{a_2 + a_4 s + \lambda}{2a_1} < 0$ then x^* in (8) becomes negative. When pumping water is constrained nonnegative the form of the optimal control to be changes from (28) to $x^* = \text{mas}\left[\dfrac{a_2+a_4 s+\lambda}{2a_1},\ 0\right]$.

Next we drive the equation for the Lagrange multipliers,

$$\dot{\lambda}=a_4 x+a_5,\quad \lambda(T)=0$$

Differentiating (29) with respect to t and substituting in $\dot{\lambda}$ from (30) and $\dot{s}$ from (25).

$$\ddot{s}=\dot{w}+\dfrac{a_4\dot{s}+\dot{\lambda}}{2a_1} = \dot{w}+\dfrac{a_4(w-x)+(a_4 x+a_5)}{2a_1}$$

$$=\dot{w}+\dfrac{a_4 w+a_4}{2a_1} = \dot{w}+\dfrac{a_4}{2a_1}w+\dfrac{a_5}{2a_1}$$

From the elementary theory of differential equations the solution to (31) is

$$s=c_1+c_2 t+Q(t),\quad s(0)=s_0$$

where $Q(t)$ is the special particular integral of (31). Differentiation of (32) and substitution into (29) gives

$$\lambda = 2a_1[c_2+\dot{Q}(t)-w] = a_2 - a_4 s$$

$$= 2a_1[c_2+\dot{Q}(t)-w] - a_2 - a_4)c_1 + c_2 t + c_1(t)$$

$$= 2a_1 c_2 - a_2 - a_4 c_1 + 2a_1[\dot{Q}(t)-w-\dfrac{a_1}{2a_1}Q(t)] - a_4 c_2 t,\ \lambda(T) = 0$$

We now discuss the imposition of the boundary conditions in (32) and (33). Defining

$$b_1 = s_0 - Q(0)$$
$$b = \dot{0}(T) - w(T) - \dfrac{a_4}{2a_1}Q(T)$$

imposing the condition t=0 in (32) and t=T is (33) and solving the resulting equations gives

$$c_1 = s_0 - Q(0) = b_1,$$

$$c_2 = \dfrac{a_2+a_4 b_1+2a_1 b_2}{2a_1-a_4 T}$$

Using these results we can write expressions s, t and λ. Note that we break each expression into three parts: the first expression labelled "Starting Correction" is a term that can be nonzero for all t; the second expression labelled "Turnpike Expression" can be nonzero for all t in the range $0\leq t\leq T$; and the last expression labelled "Ending Correction" can be nonzero only when t is close to T.

	Starting Correction	Turnpike Expression	Ending Correction
$s =$	b_1	$+ \quad (\dot{Q}(t))$	$+ (c_2 t)$
$t =$		$(w - \dot{Q}(t))$	$+ (-c_2)$
$\lambda =$	$(2a_1 c_2 - c_2 - a_2 - a_4 c_4)$	$- 2a_1 [\dot{Q}(t) - w(t) - \dfrac{a_4}{2a_1} Q(t)]$	$- a_4 c_2 t$

5.2 Stochastic case

In this section, we develop some asymptotic properties of the optimal pumping processes that consider Eq.(4) in Section 2. In particular, a turnpike planning horizon theorem is presented for this model when there is discounting and nondiscounting.

Roughly speaking, the statement of this theorem is, there exists an N* such that for all $n \geq N^*$, the optimal immediate decisions when there are n periods remaining in the planning horizon is one of the decisions which is optimal when the horizon is infinite or unbounded. Thus, if there is a unique optimal stationary strategy over the infinite horizon then the theorem suggests the following procedure when the planning horizon is a finite number $n > N^*$. For the initial (n-N*) periods, follow the unique stationary strategy which is optimal when the horizon is infinite. For the remaining N* periods, terminate with an optimal transient strategy which can be found by the use of dynamic programming recursions. The optimal turnpike is the optimal stationary strategy, and the theorem states that an optimal finite horizon strategy is to follow the turnpike for n-N* periods, and (possibly) to leave it during the terminal N* periods [12],[13].

DISCUSSION

At first, using techniques of dynamic programming, the sequence of optimal policies is determined in terms of these expected future utilities. First the policy for the ground water, then the optimal policies for the surface and the ground water, can be determined. The advantage of using dynamic programming is that the structure of the optimal policy is understood. That is, in place of determining the optimal sequence of decisions from some fixed state of the system, we wish to determine the optimal decision to be made at any state of the system.

Secondly, we solved a continuous version of the model in complete detail, using the nonlinear boundary value problem. One advantage of having a complete solution to the problem is that it is possible to determine turnpike horizon points. These correspond to zeros of the adjoint function, and have the property that if they are known exactly, then pumping plan which is optimal up to the next horizon point also forms parts of the overall optimal plan. A second advantage of having the complete solution available is that it is possible to develop a practical ground water system which intermingles a prediction procedure with the solution procedure so that a comparison between predicted and actual ground water levels can be made continuously. Whenever the discrepancy between these two becomes sufficiently large, the model suggests proper actions to be taken.

ACKNOWLEDGEMENTS

The author would like to thank the guest editor and the referees for their constructive inputs to the original draft and editorial assistance.

REFERENCES

1. P. Massé, Application des probabilités en chaine à l'hydrologie statistique et au jeu des réservoirs. Rept. to the Société de Statistique de Paris, June 21, 1944, Berger-Levrault, Paris. (1945).
2. K.J. Arrow, T.E. Harris, and J. Marsnak, Optimal inventory policy. Econometrica, 19 (3). (1951).
3. R. Bellman, Dynamic Programming, Princeton Univ. Press. (1957).
   - " -   Adaptive Control Processes: A guided tour. Princeton Univ. Press. (1961).
4. J.D.C. Little, The use of storage water in a hydroelectric system. J. Operations Research Soc. of America, 3 (3), May. (1955).
5. N. Buras, and W.A. Hall, An analysis of reservoir capacity requirements for conjunctive use of surface and ground water storage. International Ass'n. for Scientific Hydrology, Publ. 57: 556-61. (1961).
6. Oscar R. Burt, Optimal resource use over time with an application to ground water. Management Science, 11 (1). (1964).
7. (a) T. Odanaka, A Study of Multistage Inventory Control, TIT Ph.D. Thesis. (1964).
   (b) - " -   Optimal Inventory Processes, Katakura Libri. (1983).
8. D.L. Jaquette, A discrete-time population-control model with set up cost. Operations Research, 22, No. 2. (1974).
9. Z.A. Saleem, and C.E. Jacob, Optimal use of coupled leaky aquifers. Water Resources Research, 7, No. 2. (1971).
10. Y. Kitabataka, T. Miyazaki, and M. Takahashi, Regional multiobjective planning of water supply and the disposal of residuals with due regard to intraregional population distribution. Environment and Planning A., 12. (1980).
11. G.L. Thompson, and S.P. Sethi, Turnpike horizons for production planning, Management Science, 24, No. 3. (1980).
12. J.F. Shapiro, Turnpike Planning Horizons for a Markovian Decision Model, Management Science, 14, No. 5. (1968).
13. D.J. White, A Multi-Objective Version of Bellman's Inventory Problem, JMAA, 87, 219-227. (1982).

DYNAMIC PROGRAMMING FOR OPTIMIZED CONTROL OF WATER SUPPLY AND DISTRIBUTION SYSTEMS.

B.COULBECK.

Water Control Unit, School of Electronic and Electrical Engineering, Leicester Polytechnic, P.O. Box 143, Leicester LE1 9BH, England.

ABSTRACT

The article introduces the general problems of least cost control of water supply and distribution systems, and their solutions based on dynamic programming techniques. A description is given of the hydraulic operation of typical systems and components together with an evaluation of operating cost factors. The resulting mathematical formulation is then presented as a general dynamic optimization problem and forward dynamic programming solutions are developed for specific single reservoir systems. Simplifications are incorporated to cater for practical requirements and a resulting scheme is illustrated by evaluation of optimal schedules for an actual system. finally, the solution methods are extended to cover compatible multireservoir systems.

INTRODUCTION

In common with executives of other public utilities, managers of water supply and distribution systems are now seeking to implement overall automatic control in order to achieve more efficient operation of systems of ever-increasing complexities and costs. Existing technology has long been capable of providing computerized hardware for measurements and control; however, computer software, in the form of program algorithms, is not in such an advanced state that effective on-line control can be achieved, and additional research is required in this area.

As an initial step towards improved control, many authorities are now completing installation of systems for computerized monitoring with limited control features. The current work forms part of continuing collaboration with various UK water authorities to devise and present computer algorithms suitable for on-line control of complete water distribution systems.

Of major importance in the control schemes is the concept of optimization of operation, which attempts to achieve lowest operating costs consistent with providing a satisfactory service to customers. It has been shown [1, 2, 3] that the project is very complex involving control of large-scale non-linear dynamic systems subject to unknown disturbances.

The optimization methods must cater for high state and control dimensionality, with further complications of highly non-linear performance indices, and must incorporate both continuous and discrete controls.

Distribution networks consist of large numbers of interconnecting pipes with occasional control valves, both of which have a non-linear relationship between flow and head loss. Reservoirs are connected at strategic points throughout the network to provide storage capability and maintain required pressure levels. Individual consumer demands occur at distributed points throughout the network but, since there is usually minimal monitoring it may only be possible to estimate the total demand from measurements of pump flows and reservoir levels. In many regions boreholes are a typical source of water supply, with pumping to the network via pump stations using parallel combinations of fixed and variable speed pumps. Booster pumps, together with control valves, are normally used for transer of water between reservoirs of differing pressure zones. In both cases the pump flows are influenced by the reservoir levels and the resultant costs are dependent upoon electricity unit and demand charges.

Since water networks contain storage, the optimization problem reduces to minimization of electricity charges and associated costs for the complete network over the entire control period. This requires the control of pumping and storage while catering for consumer demands and maintaining desired reservoir levels. Consequently the successful application of optimization methods depends significantly upon the formulation of simplified dynamic models for rapid and repeated evaluation of the effects of control strategies upon the network operation. In addition, a prediction scheme is required which will provide a forecast of consumer demands for the complete control period.

It is essential to ensure that theoretical developments are applicable to actual systems and meet all operational constraints. However, to avoid unnecessary experimental manipulation of operational systems, system analysis and initial validation of results must rely upon accurate simulation methods. In this article particular consideration is given to development of operation and costs in a suitable form for treatment by forward dynamic programming. This optimization technique is illustrated by application to one single reservoir system supplied by fixed speed pumps and another single reservoir system supplied by a combination of fixed and variable speed pumps. For extension of the basic technique to cover multi-reservoir systems it is necessary to devise methods to alleviate the attendant dimensionality problems. Two restricted classes of multi-reservoir systems are considered with a solution for one case obtained using <u>successive approximations</u> [4] and a proposal for the other case using <u>state increment dynamic programming</u> [5].

SYSTEM ANALYSIS AND SIMULATION

Figure 1 shows a very general system configuration which can be adapted to represent the practical situations under consideration in this article. The following sections cover the analysis and simulation of such general systems.

Pump Analysis and Simulation

As far as control of operation and costs are concerned, pumps are the most important system components. Pump characteristics are typically non-linear and the flow and input power are both dependent upon pump operation and instantaneous operating conditions within the network [6].

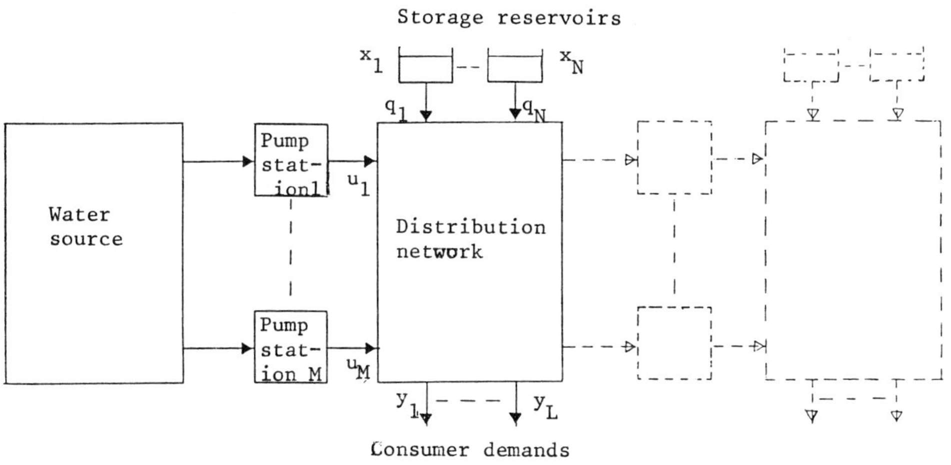

Fig.1 General system configuration.

Within each pump station sufficiently identical parallel pumps can be considered on a group basis. For general groups of fixed or variable speed pumps the total flow, q_j can be represented in terms of the pump head increase Δh_j, together with two independent pump control parameters, as:

$$q_j = f_j(\Delta h_j, r_j, s_j) \qquad (1)$$

where f_j is a non-linear flow function for the pump group j network link.

The control parameter, r_j, represents the number, or proportion, of total pumps in use, while the parameter s_j represents the speed. Figures 2 and 3 show typical characteristics which illustrate the effects of these parameters and the resulting flows for specified network interaction.

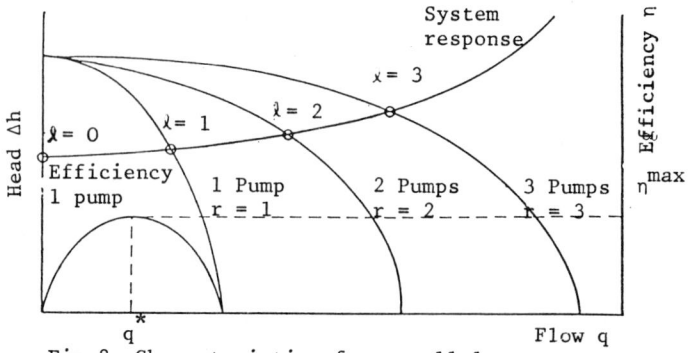

Fig.2 Characteristics for parallel pump combinations at a fixed speed.

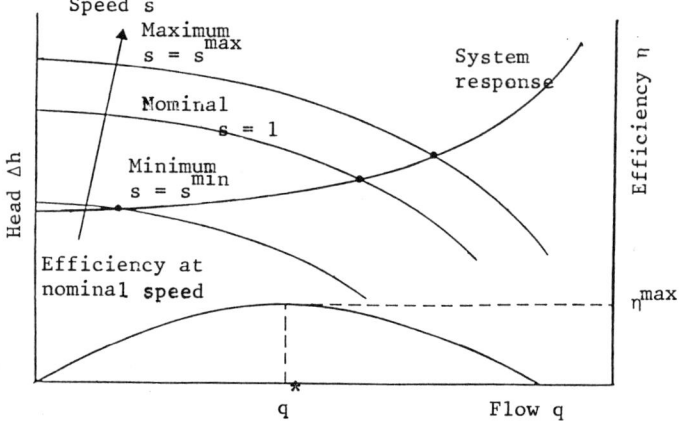

Fig. 3. Characteristics for a parallel pump combination with varying speed

The exact form of the function can be obtained by fitting a curve to the manufacturer's characteristics; typically, this could be given by:

$$\Delta h_j / s_j^2 = A_j (q_j / r_j s_j)^2 + B_j (q_j / r_j s_j) + C_j \qquad (2)$$

subject to $r_j \in R_j$ and $s_j^{min} \leq s_j \leq s_j^{max}$

where R_j is the maximum number of pumps in group j, and s_j^{min}, s_j^{max} are the lower and upper speed limits, respectively.

A_j, B_j and C_j are the curve coefficients which apply to a single pump of a group ($r_j = 1$) running at nominal speed ($s_j = 1$).

The following general non-linear electrical power function represents pump group input power, p_j, allowing for variation of all relevant variables:

$$p_j = g_j(\Delta h_j, r_j, s_j) \qquad (3)$$

where g_j is a non-linear power functional for pump group j.

For each pump group the electrical power can be calculated from:

$$p_j = g\, \Delta h_j\, q_j/\eta_j \qquad (4)$$

where g is the conversion factor appropriate to the density of water and associated units. Figures 2 and 3 show typical efficiency characteristics which can be used to derive the approximate group pumping efficiency as:

$$\eta_j = \eta_j^{max} \{1-(q_j/r_j s_j q_j^*-1)^2\} \qquad (5)$$

where η_j^{max} and q_j^* are maximum efficiency and corresponding flow respectively, for a single pump running at nominal speed.

A computer program is available [7] which uses the developed equation to simulate pump operations, and can give results as shown in Figs. 2 and 3. This facility provides an independent assessment for control and optimization of overall systems.

<u>Network Analysis and Simulation</u>

Distribution networks consist of nodes connected in pairs by pump, pipe and valve elements. Any node can be connected to a reservoir and can also supply a consumer determined demand flow.

Pipes and control valves can be represented by a general element flow relationship of:

$$q_j = f_j(\Delta h_j, r_j) \qquad (6)$$

A simplified expression for the flow is

$$q_j = |\Delta h_j/r_j|^{0.54} \qquad (7)$$

with the flow direction determined by the sign of the head drop.

The parameter r_j now represents hydraulic resistance, which can be fixed for pipes at a value R_j or continuously variable for valves with a value of v_j. Thus, in addition to pumps, valves can be used as control devices in order to influence the operation and efficiency of the system.

Because of the storage capability, the levels, x_i, of individual reservoirs located at specific nodes will vary with time according to:

$$\dot{x}_i = q_i/A_i \tag{8}$$

subject to $x_i^{min} \leq x_i \leq x_i^{max}$ where q_i is the reservoir flow, and A_i is the cross-sectional area of the reservoir at the current level. x_i^{min} and x_i^{max} are the fixed levels which correspond to the reservoir being empty or full, respectively.

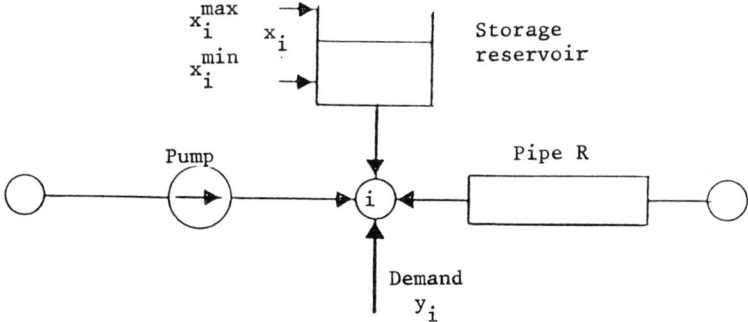

Fig.4. General network node configuration.

For any general node of the network, as illustrated in Fig.4, the sum of: connected element flows, f_i, reservoir flows, q_i, and demand flows, y_j, must balance to give:

$$f_i + q_i + y_i = 0 \tag{9}$$

where f_i is the total element flow into node i for the j elements connected to the node which can be expressed as:

$$f_i = \sum_j f_j \tag{10}$$

The network is considered to be solved when all heads and flows are known at one time. This requires the evaluation of: reservoir flows, q_i, for nodes with heads determined by an associated reservoir level, and pressure heads, h_i, for nodes with outflows determined by an associated consumption.

Considering the instantaneous operation, generalized sets of equations can be formulated, for any interconnected network, as:

$$\underline{f}(\underline{x},\underline{h},\underline{u},\underline{v}) + \underline{q} + \underline{y} = \underline{0} \tag{11}$$

where the new variables are the vector equivalents of the previously defined ones. The solution of Eq.(11) will give the unknown values of reservoir flows and pressure heads which can be interpreted as obeying numerical relationships of the form:

$$\underline{q} = \underline{\Psi}_q(\underline{x},\underline{u},\underline{y},\underline{v}) \qquad (12)$$

$$\underline{h} = \underline{\Psi}_h(\underline{x},\underline{u},\underline{y},\underline{v}) \qquad (13)$$

The time varying reservoir flow equations can be generalized to:

$$\underline{\dot{x}} = \underline{F}\,\underline{q} \qquad (14)$$

where $\underline{F}$ is a diagonal reservoir coefficient matrix whose elements relate change of level to flow.

A complete network simulation involves a calculation of future reservoir levels, over an extended time period, for known pump and valve controls and consumer demands.

This can be achieved by repeated forward solution of the static and dynamic network equations (11) and (14). Computer programs are available for this purpose [8, 9] which, however, require the use of time-consuming iterative techniques to solve the non-linear equations.

These programs provide a realistic way of analyzing the system operation under varying conditions and also cater for validation of control schemes, but are unsuitable for direct use in optimized control schemes.

Simulation can also give a guide to derivation of simplified dynamic models for on-line control purposes. One solution is to express changes in operating conditions, from Eqs. (12) and (13), as [10]:

$$d\underline{q} = \underline{A}_q\,d\underline{x} + \underline{B}_q\,d\underline{u} + \underline{C}_q\,d\underline{y} + \underline{D}_q\,d\underline{v} \qquad (15)$$

$$d\underline{h} = \underline{A}_h\,d\underline{x} + \underline{B}_h\,d\underline{u} + \underline{C}_h\,d\underline{y} + \underline{D}_h\,d\underline{v} \qquad (16)$$

The matrix coefficients $\underline{A}$, $\underline{B}$, $\underline{C}$ and $\underline{D}$ can be determined from the simulation program [9]; they permit the development of a model which is linear around average operating conditions for small changes in the control variables.

Demand Analysis and Prediction

The combined effect of large numbers of consumers gives rise to overall consumption profiles with certain well defined features. These are illustrated in Fig.5 which shows a typical weekly pattern of measured demands. In general, this indicates a daily periodic profile with a slowly varying component and added random variations. Processing to remove any corrupted data and minimize the random effects will leave a good estimate of the underlying consumption behaviour.

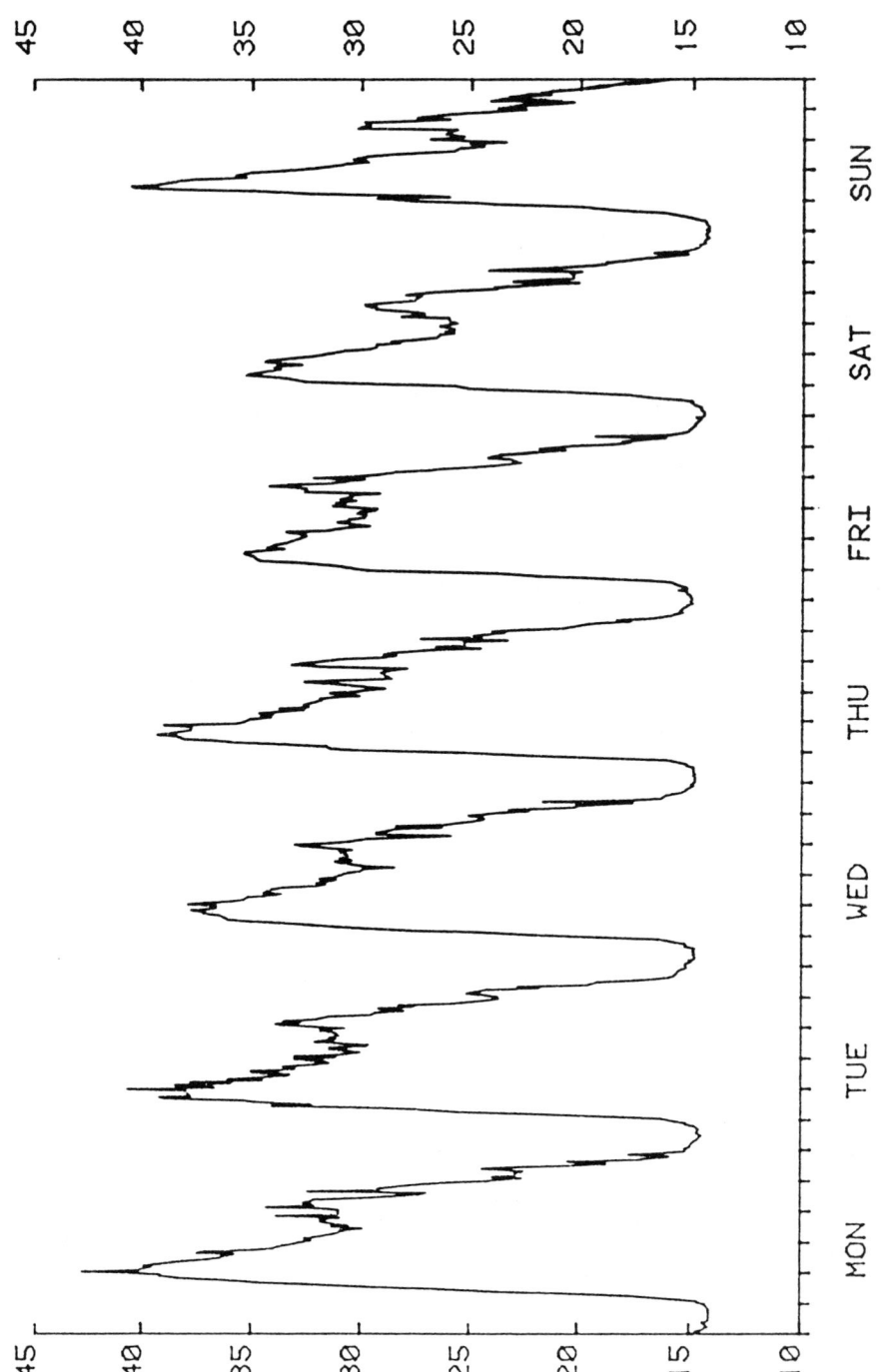

Fig.5. Typical weekly profile of measured demands.

A preliminary analysis is required to determine specific quantifiable features of typical demand patterns for each geographical area for which demands are to be monitored and predicted. These features can then be used to provide coarse screening and fine smoothing of the data values so that; the processed measurements represent the best available estimate of the actual demands, and the present demand data can be used for extrapolation of future demands.

For operational control of water systems a special requirement is to generate forecasts of complete consumption patterns at the beginning of each defined period.

One simple means of forecasting is known as exponential smoothing [11]. The technique takes the average of previous demand patterns with increased weighting for the most recent. Trends and error corrections are automatically incorporated so that the predicted patterns reflect the smoothed history of previous periods. As a consequence of the smoothing and correction process the predicted patterns can track normal consumption variations over periods of several days and provide acceptable results for optimized control purposes [12]. A description of a computer program for demand analysis and prediction is given [13].

SYSTEM CONTROL MODEL

Optimized control requires calculation of pump and valve schedules so as to minimise overall costs while providing predicted demands and operating within system constraints. This involves definition of appropriate system and cost equations which can be used in conjunction with the initial conditions to yield an optimal control sequence solution. The required optimization computations can only be performed for sufficiently simplified systems. In this respect, introduction of the stage variable, k, is advantageous and allows formulation of discrete time equations throughout, which are suitable for efficient optimized control schemes.

System Equations

An appropriate simplification occurs for a single reservoir system used under the assumption that input pump station flows, $u_m(k)$, and valve flows, $v_r(k)$, are independent of heads and can be controlled at given values. Under these conditions the concept of total volumetric balance applies and the linear one-dimensional equation describing the system dynamic operation becomes:

$$x(k+1) = x(k) + (\Delta t/A)\sum_{m=1}^{M} u_m(k) + (\Delta t/A)\sum_{\ell=1}^{L} y_\ell(k) + (\Delta t/A)\sum_{r=1}^{R} v_r(k) \quad (17)$$

under the constraints:

$$x^{min} \leq x(k) \leq x^{max} \quad (18)$$

for $k = 0, 1, \ldots, K-1$

where Δt is the time interval for each stage increment, m is the pump station index with a maximum value of M, ℓ is the demand index with a maximum value of L, and r is the valve index with a maximum value of R.

For a given initial value of $x(0)$, predicted values of $y(k)$, and specified values of $u(k)$ and $v(k)$, Eqs.(17) and (18) describe the reservoir level variation over a total control period of K Δt.

Equation (17) applies even for non-linear head-flow relationships within the network, but excludes direct incorporation of independent pump and valve control parameters. To cater for this, the section on <u>Optimization of Single Reservoir Systems</u> invokes additional hydraulic relationships and uses Eq.(17) as the basis for optimization of a single reservoir system.

The same principle can be extended to cover the case of mutiple zones consisting of one reservoir for each zone, under the same assumption but allowing for controllable inter-zonal flow, to give the following linear multivariable formulation:

$$\underline{x}(k+1) = \underline{A}\ \underline{x}(k) + \underline{B}\ \underline{u}(k) + \underline{C}\ \underline{y}(k) + \underline{D}\ \underline{v}(k) \tag{19}$$

In this case $\underline{A}$ will be a unit diagonal matrix and $\underline{B}$, $\underline{C}$ and $\underline{D}$ will have constant or zero elements dependent upon system configuration. This formulation is exploited for the optimization of a class of multireservoir systems as described in later sections.

More general systems consist of several reservoirs interconnected by non-linear head-dependent elements such as pipes, pumps and control valves. Network inflows are typically from multiple pump stations where the number and speed of the pumps on-line can be controlled and the actual pump flows are also dependent on the network head values. Consumption flows are distributed throughout the network and are assumed to be independent of heads and available as known functions of time. While total volumetric balance still applies, this will now determine the total volume of water in all reservoirs and will not indicate the relative volumes in individual reservoirs. A simple linear formulation as for Eq.(19) is, therefore, not feasible. However, a simplified model can be obtained from Eqs.(15) and (16), to cover deviations from average operating conditions as [10]:

$$\underline{dx}(k+1) = \underline{Ax}\ \underline{dx}(k) + \underline{Bx}\ \underline{du}(k) + \underline{Cx}\ \underline{dy}(k) + \underline{Dx}\ \underline{dv}(k) \tag{20}$$

$$\underline{dz}(k) = \underline{Az}\ \underline{dx}(k) + \underline{Bz}\ \underline{du}(k) + \underline{Cz}\ \underline{dy}(k)\ \underline{Dz}\ \underline{dv}(k) \tag{21}$$

In this case the network model consists of a set of linear equations in terms of matrix coefficients which relate changes in reservoir levels, $\underline{dx}(k)$, and nodal heads, $\underline{dz}(k)$, to changes in control and disturbance parameters. The model now treats the case of head-dependent pump and valve flow by means of independent parameters which can be used as continuously variable or discrete controls.

Cost Factors

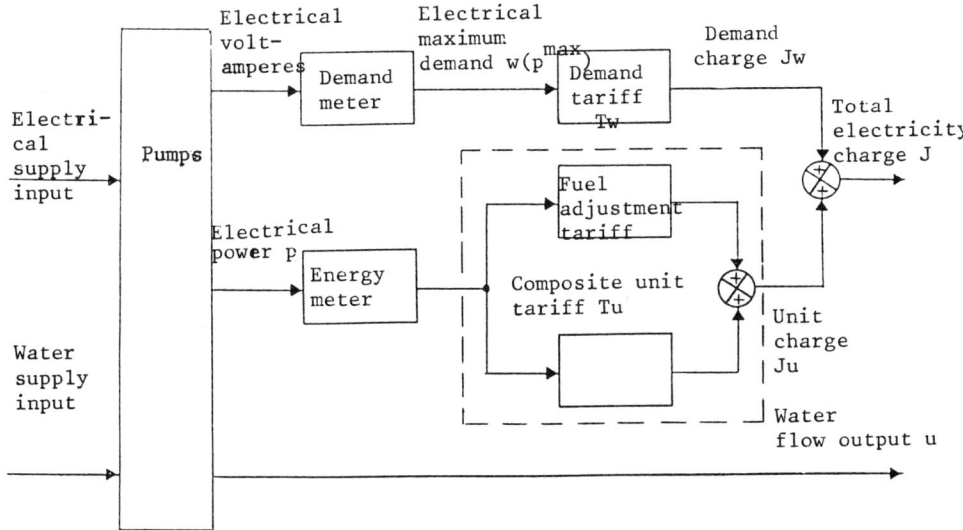

Fig.6. General electricity cost model of a pumping station

Major operating costs are due to electricity charges for pumping. Figure 6 shows a general cost model of a pumping station which takes account of typical UK electricity tariffs. Charges are made both for electrical units used and maximum demand achieved over the whole of the tariff period. The unit charge is based upon a unit tariff which can also be a function of the electricity maximum demand. A reasonably accurate representation of these effects can be obtained by forming a composite time-varying unit tariff, $Tu(k)$, for each optimization period, which can then include night unit rebates and fuel adjustment charges. The demand tariff, Tw, can vary throughout the year but will be constant over each tariff period. Under these conditions the unit and demand charges for each pump station, over the given optimisation period, will be:

$$Ju = \sum_{k=0}^{K-1} Tu(k) \, p(k) \, \Delta t \qquad (22)$$

$$Jw = Tw \, p(k)^{max} \qquad (23)$$

where $p(k)$ is the pump station electrical power consumption at stage k.

Pump maintenance requirements are related to the number of times pumps are switched on and off. A factor can be derived to penalise excessive switching, which takes the form:

$$Js = \sum_{k=0}^{K-1} Qs \, t(k) \qquad (24)$$

where Qs is the pump switch penalty factor and $t(k)$ is the number of pumps switched at stage k. This approach is not necessarily optimal but satisfies the practical requirements.

A cost factor also arises by virtue of the permitted range of reservoir levels. Allowing the terminal level to be free would cause the reservoir to be emptied at the end of the optimization period. Any previously stored water would already have been paid for and the optimization policy would dictate use of this water rather than incur additional pumping costs. Because of the necessarily limited optimization period, it is essential to provide a sufficient reserve of stored water in order to meet future anticipated demands without incurring heavy electricity maximum demand charges. For short term optimization periods a realisable, but non-optimal, solution is to specify a desirable terminal quantity and impose a cost penalty for any deviation from this value.

Depending on the system, this factor can also be applied to penalise reservoir level deviation from desirable values throughout the optimization period. The desired levels may be consistent with providing satisfactory service pressures while minimising pressure-related leakage; alternatively there may be a preferred level corresponding to most efficient pumping conditions.

The general cost factor is:

$$Jx = \sum_{k=1}^{K} Qx(k)[x(k)-x(k)^{des}]^2 \qquad (25)$$

where Qx is the reservoir level penalty and x^{des} is the reservoir level desired value.

Optimized Control

With the model simplifications, general systems now correspond to discrete time multivariable dynamic types, with non-linear cost factors and constraints on states and controls, where the controls can also take continuous and/or discrete values. This type of problem is ideally suited for optimization by dynamic programming which, in principle, can handle all of the complexities and obtain a global solution by evaluation and comparison of the cost of all feasible controls [4]. In practice the conventional dynamic programming procedure is restricted to low-dimensional problems and modifications to the standard procedure must be sought to allow solution of higher-order systems.

There are two basic types of dynamic programming: backward, in which optimal trajectories are calculated from all initial and intermediate states leading to a single terminal state; and forward, in which the trajectories are calculated from a single initial state leading to all intermediate and final states.

For a practical implementation the optimization method must cater for disturbance effects which cause the system to deviate from its optimal trajectory. The use of backward dynamic programming is most suitable in this situation but forward dynamic programming has certain advantages for the present application and can still be applied by recalculation using the disturbed state as a new initial value. The following sections illustrate the application of forward dynamic programming to single reservoir systems and particular classes of multireservoir systems.

OPTIMIZATION OF SINGLE RESERVOIR SYSTEMS

Single Reservoir Systems with Fixed Speed Pumps

Figure 7 shows the basic classes of supply systems which are considered [14]. These consist of one or more pump stations, containing only fixed speed pumps, which supply a single reservoir via a simple network of trunk mains. For simplicity any consumptions en route are transferred to the reservoir demand. It is assumed that pump station suction heads remain at constant values for each selected combination of pumps. Also typical storage reservoirs operate with a fixed inlet level which is independent of the reservoir stored quantity. Under the above conditions the combined pump station flows and the power consumptions will have constant values for each selected combination of pumps. For practical systems the results are likely to be approximately true irrespective of the above assumptions.

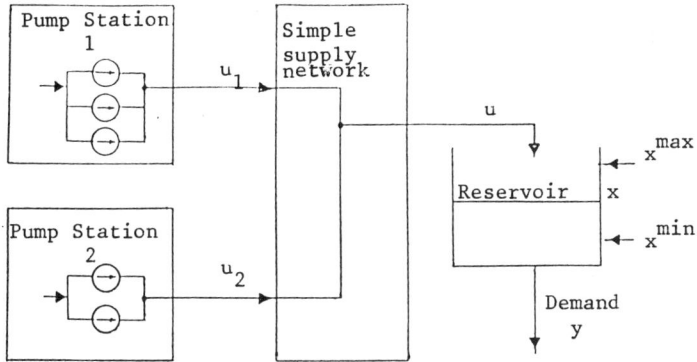

Fig. 7. Single reservoir system with fixed speed pumps

The reservoir level variation will be governed by:

$$x(k+1) = x(k) + u(k) \Delta t/A - y(k) \Delta t/A \qquad (26)$$

subject to $x^{min} \leq x(k) \leq x^{max}$ for $k = 0, 1, \ldots K-1$ with $x(0)$ and $y(k)$ given.

For the calculation of costs over the control period the factors which are considered are those for direct pumping, based on electricity unit and demand charges, and for indirect costs based on pump switching and desirable intermediate and terminal reservoir levels. This can be expressed as:

$$J = \sum_{k=0}^{K-1} Tu(k) p(k) \Delta t + Tw \, p(k)^{max} + \sum_{k=0}^{K-1} Qs \, t(k) +$$
$$+ \sum_{k=1}^{K} Qx(k)[x(k) - x(k)^{des}]^2 \qquad (27)$$

The pump station flow and power can be represented as:

$$u(k) \in U(\ell) \qquad (28)$$

$$p(k) \in P(\ell) \qquad (29)$$

where $U(\ell)$ and $P(\ell)$ are the fixed values for each selected pump combination, ℓ, where $\ell = 1,...L$. In practice these values can be obtained by direct measurement or by off-line simulation.

It is now required to determine the values of $x(k)$ and $u(k)$ which satisfy the system state equation (26) within the constraints of Eqs. (28) and (29), and which result in a minimization of the total operating cost defined in Eq. (27).

Calculation of optimized schedules. In minimizing the total operating cost, the optimization method has to cope with discrete controls, non-linear system equations, and constraints on a one-dimensional state. Dynamic programming [4] can be relied upon to give a robust solution with appropriately discretised variables. In particular, forward dynamic programming offers computational advantages for simplified incorporation of pump switching costs and maximum demand charges.

The technique is well known and only a summary of the algorithm features significant to this section will be given.

(i) Select a state versus stage grid, appropriate to the requirements of Eq.(26), and insert the initial conditions $x(k)$ for $k = 0$.

(ii) For each pump combination use Eq.(28) to generate all $u(k)$ and Eq.(29) to generate all $p(k)$.

(iii) Starting from each $x(k)$; use Eq.(26) with each $u(k)$ to generate all valid $x(k+1)$ state values (not necessarily at grid points) and use Eq.(27) with each $p(k)$ to evaluate each corresponding trajectory cost.

(iv) Interpolate the state costs to give values at each closest k+1 grid point; retain least cost trajectories and store with actual state values at closest k+1 grid point.

(v) If $k < K$, increment k and return to (ii).

(vi) If $k = K$, select closest trajectory to desired terminal state and recover the optimal controls which generate $x^*(k)$ and $u^*(k)$ for $k = 0,1,...K-1$.

A typical operational application would require a minimum control period of 24 hours, with 2 hourly control decisions to give a 12 stage process. The reservoir level could require partitioning into 100 state increments and one pump station, with 3 identical pumps, would imply 4 discrete pump combinations. An evaluation of the corresponding time and memory for a computer solution would indicate that the above method is only appropriate for one-dimensional systems.

Single Reservoir Systems with Fixed and Variable Speed Pumps

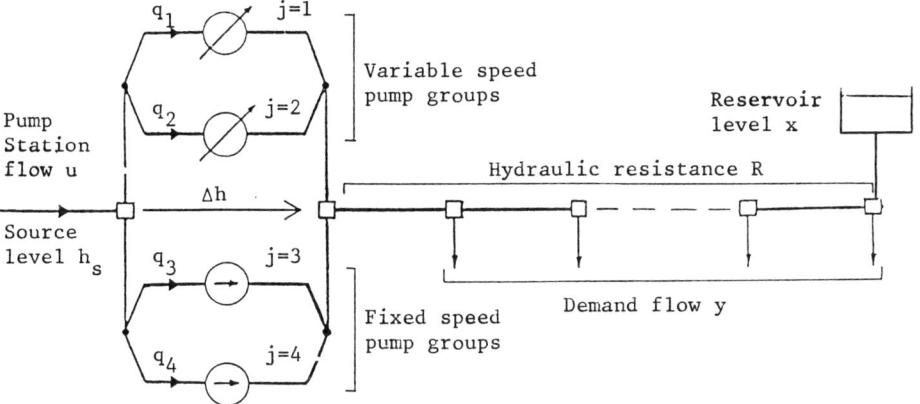

Fig.8. Single reservoir system with fixed and variable speed pumps.

Figure 8 shows the generalized one-dimensional system on which this section is based. The pumps may be supplied from a borehole, the reservoir of another zone, or a large impounding reservoir, in each case the source level can be assumed to vary with predictable values. Within the pump station both fixed and variable speed pump groups are now designated, each group containing pumps of sufficiently identical characteristics. Fixed speed pump groups are controllable by selection of numbers of parallel pumps in use while the variable speed pump groups are also controllable over continuous, but limited, ranges of speeds. To simplify the analysis any consumptions en route are again transferred to the reservoir demand and only one group of variable speed pumps is considered. Analysis of more general systems is covered in [15].

The reservoir level variation will again be governed by:

$$x(k+1) = x(k) + u(k)\Delta t/A - y(k)\Delta t/A \tag{30}$$

subject to $x^{min} \leq x(k) \leq x^{max}$ for $k = 0, 1, \ldots, K-1$ with $x(0)$ and $y(k)$ given.

For each group of parallel pumps the head and flow are related by

$$\Delta h_j / s_j^2 = A_j (q_j / r_j s_j)^2 + B_j (q_j / r_j s_j) + C_j \tag{31}$$

for $r_j = 1,...R_j$ and subject to $s_j^{min} \leqslant s_j \leqslant s_j^{max}$ for variable speed

Allowing for all pump groups the station flow will be:

$$u = \sum_{j=1}^{J} q_j \qquad (32)$$

Head relationships for the connected system can be expressed as:

$$\Delta h_j = x - h_s + R u^{1.852} \qquad (33)$$

where h_s is the system source head and R is the equivalent pipe resistance.

For given reservoir and source levels together with speed and combination of each pump group, Eqs.(31) to (33) could be solved iteratively to yield the corresponding total and individual pump flows, and common pump head increase.

However, when some variable speed pumps are available for use a more efficient computational procedure can be devised. This involves specifying the reservoir level at the next stage and using Eq.(30) to evaluate the required total pump station flow. For each selected pump combination Eqs.(31) to (33) can now be solved directly to yield the corresponding pump group heads, flows and speeds.

Once the pump group heads, flows and speeds have been calculated, Eqs.(4) and (5) can be combined to give:

$$P_j = g \, \Delta h_j r_j s_j q_j^{*} / \{\eta_j^{max}(2-q_j/r_j s_j q_j)\} \qquad (34)$$

and used to determine each corresponding group operating power. The total pump station power will then be:

$$P = \sum_{j=1}^{J} P_j \qquad (35)$$

For the same cost factors as given in <u>Single Reservoir Systems with Fixed Speed Pumps</u>, the total system cost will also correspond to Eq.(27).

<u>Calculation of optimized schedules</u>. Forward dynamic programming is again an appropriate method of determining optimized pump schedules, for the reasons already given for single reservoir systems with fixed speed pumps.

The algorithm for systems containing some variable speed pumps is now as follows:

(i)   Select a state versus stage grid, appropriate to the requirements of Eq.(30), and insert the initial conditions $x(k)$ for $k = 0$.

(ii)   Use Eq.(30) to evaluate u(k) for all state trajectories x(k) leading to all grid points at k + 1.

(iii) For all u(k) apply all available pump combinations and use Eqs.(31) to (33) to evaluate hydraulic conditions, Eqs.(34) and (35) to evaluate power conditions and Eq.(27) to evaluate each trajectory cost, at the k + 1 grid points.

(iv)   Retain least-cost state trajectories and store at each k + 1 grid point.

(v)   If k < K, increment k and return to (ii).

(vi)   If k = K, select closest trajectory to desired terminal state and recover the optimal controls which generate $x^*(k)$ and $u^*(k)$ for k = 0,1,...K-1.

For typical practical systems with variable speed pumping capability it has been observed [15] that there is little loss of optimality in selecting a limited number of stage increments which coincide with the times of electricity tariff change. This is practically beneficial since it considerably reduces the dynamic programming computation and restricts the number of pump switches and speed changes which can occur.

Selection of a limited number of state increments will also have a significant effect on the dynamic programming computation and memory requirements. Since this will result in a coarsely defined optimal state trajectory the desired solution accuracy must be obtained by repeated application of the complete algorithm. Each iteration will then take place with a progressively reduced, but refined, state corridor around each computed state trajectory.

Application to a supply system. The method is illustrated by application to an actual supply system [16].

Operational requirements are for a control period of 24 hours, starting from midday, together with initial and desirable final reservoir quantities of 70% of full capacity. No penalties are imposed for deviation from the desired level at intermediate times. System constraints include maximum and minimum reservoir quantities of 90% and 50% respectively, and an upper limit on the electricity maximum demand. When used in conjunction with the water demands, Fig.9(a), the optimized pump schedules, Fig.9(b), cause the reservoir level to vary as shown in Fig.9(c). Electricity charges are made for units and maximum demands, and night rebates are given for units used between 0000-0800 hours.

Discussion of results. The solution is in terms of both discrete and continusous control variables which relate directly to the practical system requirements for on-off pump combinations and pump speeds. A detailed review of the generated schedules shows that they are entirely consistent with expectations based on the electricity tariff structure and operational considerations. Specific points noted are: use of most efficient pumps, increased overnight pumping, and limitations on maximum power demanded and pump switching.

The particular formulation has allowed the sytem non-linearities to be incorporated directly and efficiently within the optimization routine, hence yielding results sufficiently accurate for open-loop control. The success of the scheme depends primarily on the prediction accuracy of water demands. Preliminary trials on actual systems, with manual implementation of the generated schedules, have given satisfactory results and indicated that cost savings of 5-10% can be expected.

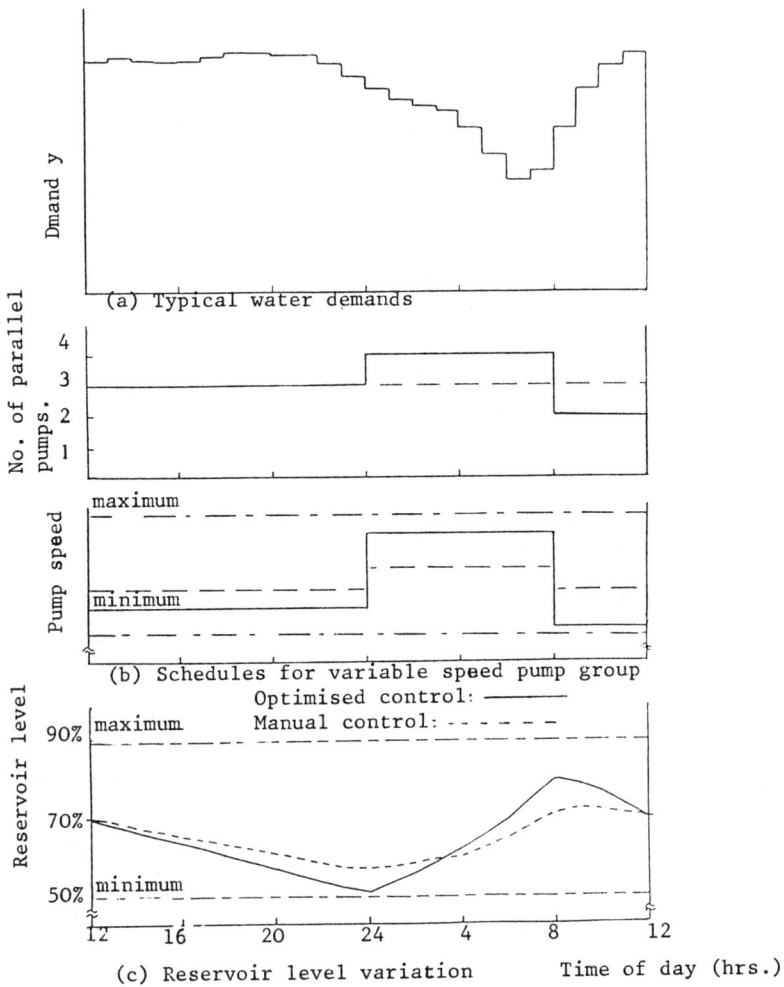

Fig.9. Typical operation of a single reservoir system

OPTIMIZATION OF MULTIRESERVOIR SYSTEMS

Parallel-connected Subsystems with Fixed Speed Pumps

This section considers the possibility of an optimized control scheme for a multi-reservoir system having a particular parallel structure and containing some automatic control facilities [17, 18]. Figure 10 shows the type of system to be controlled, which consists of a number of source inputs supplying a distribution network. The source inputs are of the type shown in Fig.7, which have been independently optimized in the corresponding section.

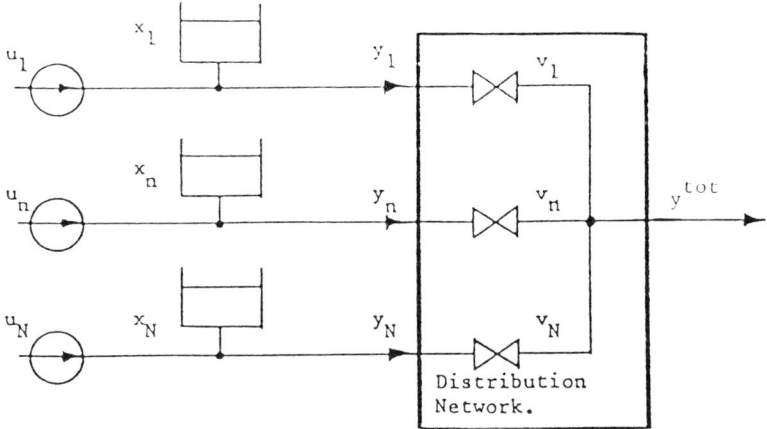

Fig.10. Parallel connected sub-systems with fixed speed pumps.

While the complete distribution network supplies various water demands, via numerous pipes and valves, only a simplified representation is shown in terms of a set of dominant valves and the total system demands.

The overall optimal control problem reduces to control of pumps and valves to ensure that the most economical sources supply the total demands on a time-varying basis. Constraints are set by the required maintenance of reservoir levels within specified ranges and an acceptable profile of pressures at network nodes. In order to guarantee meeting all required operating conditions and achieve maximum economy of operation, it would be necessary to incorporate an integrated scheme allowing for full interaction between all system variables. This is rarely possible for large scale non-linear systems. The composite system must usually be divided into smaller sub-systems with limited interaction in order to allow a solution to be obtained. For the particular structure shown a major partitioning can be obtained by considering the sources and the network separately. The solution procedure will consist of initially optimizing the source pumping schedules so that each source can supply the optimum proportion of the total demand. This will be followed by control of the network valves to realise and maintain the optimum demand distribution. Figure 11 shows the proposed scheme and the following sections discuss the optimization and control requirements and procedures in greater detail.

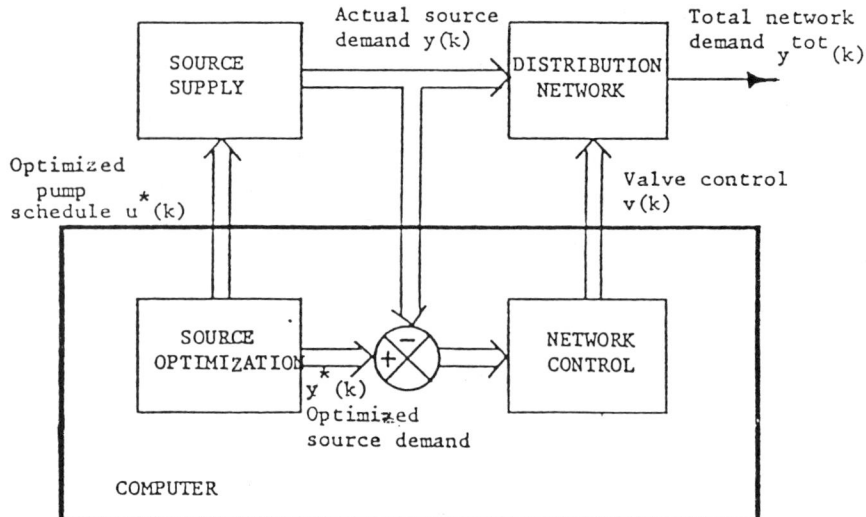

Fig.11 Proposed optimization and control scheme.

<u>Source optimization</u>. The source optimization problem is to determine a least-cost schedule of pumping at all source inputs which will provide the total predicted consumer demand while maintaining reservoir levels within limits and meeting any other operational constraints. The individual sources cannot be considered in isolation since they are coupled by the total water demand requirement.

For a practical case of parallel combinations of on-off pumps with several reservoirs, an integrated optimal solution is likely to prove impossible due to the dimensionality and complexity of the sytem. However, a procedure is described which converges to an optimal result.

In effect, the procedure allows variation of only one source reservoir at a time with all others held constant. The application is repeated for variation of each reservoir in turn until no further reduction in operating cost is achieved. For the given total demand the procedure results in optimal reservoir level trajectories and pumping schedules, together with the time-varying proportion of total demand required from each supply source.

Since the $\underline{A}$, $\underline{B}$ and $\underline{C}$ matrices of Eq.(19) are now diagonal, then variation of each of the N source reservoirs over the control period [0,K] can be represented as:

$$x_n(k+1) = x_n(k) + u_n(k) \Delta t/A_n - y_n(k) \Delta t/A_n \qquad (36)$$

where n takes values of 1...N.

corresponding constraints take the form:

$$x_n^{min} \leq x_n(k) \leq x_n^{max} \qquad (37)$$

$$y_n^{min} \leq y_n(k) \leq y_n^{max} \qquad (38)$$

where the y_n bounds correspond to the feasible supply flow range for each source.

For fixed speed pumps, each pump station flow and power can be generally represented as:

$$u_n(k) \in U_n(\ell_n) \qquad (39)$$

$$p_n(k) \in P_n(\ell_n) \qquad (40)$$

where ℓ_n represents the pump combination for each source and takes values of $1 \ldots L_n$.

Also the total demand is assumed to be known and must be satisfied for each control increment by:

$$y^{tot}(k) = \sum_{n=1}^{N} y_n(k) \qquad (41)$$

The total operating cost over the control period is defined as:

$$J = \sum_{n=1}^{N} J_n \qquad (42)$$

where J_n is the cost of supply from each source and can be written as:

$$J_n = \sum_{k=0}^{K-1} Tu_n(k) \, p_n(k) \, \Delta t + Tw_n \, p_n(k)^{max} + \sum_{k=0}^{K-1} Qs_n \, t_n(\kappa)$$

$$+ \sum_{k=1}^{K} Qx_n(k)[x_n(k) - x_n^{des}(k)]^2 \qquad (43)$$

It is now required to determine the values of $x(k)$, $u(k)$ and $y(k)$ which satisfy the system Eqs.(36) and (41) within the constraints of Eqs.(37) to (40) and which result in a minimization of the operating costs defined in Eqs.(42) and (43). The solution in one dimension for fixed $y(k)$ has already been achieved. However, for more than one dimension the computing requirements in terms of memory and speed can become excessive. Under the given conditions this problem can be overcome by applying the one-dimensional optimization method to each source supply in turn while all other reservoir trajectories are held constant at their previously computed values. This is known as the Successive Approximation Dynamic Programming Technique [4], which guarantees at least a local optimum and can be applied in this case as follows:

If m represents any of the N-1 sources, with pre-determined fixed levels of $x_m(k)$, Eq.(36) can be solved for the permitted source flows, $y_m(k)$, arising from application of all the pump combinations, $u_m(k)$, to give:

$$y_m(k) = x_m(k) A_m/\Delta t - x_m(k+1) A_m/\Delta t + u_m(k) \qquad (44)$$

with $y_m(k)$ subject to the constraints of Eq.(38) and for $m = 1...N$, $m \neq n$.

With n representing the source with allowed level variation, Eq.(41) can be solved using all the permitted values of $y(k)$ to give the range of predetermined demand flows, $y_m(k)$, as:

$$y_n(k) = y(k)^{tot} - \sum_{\substack{m=1 \\ m \neq n}}^{N} y_m(k) \qquad (45)$$

with $y_n(k)$ subject to the constraints of Eq.(38).

Using the standard dynamic programming method described in the section for <u>Single Reservoir Systems with Fixed Speed Pumps</u>, all combinations of $u_n(k)$ can now be applied to Eq.(36) to determine the optimal trajectories of the reservoir level, $x_n^*(k)$, for $k = 0, 1,...,K-1$, in terms of $u_n^*(k)$ and $y_n^*(k)$. This process is completed for $n=1...N$, and then continued until two consecutive iterations yield the same minimum overall cost. When this occurs an overall (possibly local) minimum will have been achieved and the values of $x_n^*(k)$, $u_n^*(k)$ and $y_n^*(k)$, for $n=1...N$ and $k=0,1,...,K-1$, define a source optimization policy.

<u>Network control</u>. The network control problem involves adjustment of the network parameters to automatically achieve the preferred proportion of the total demand to be supplied by each source. One way of maintaining the source supplies at their optimal values could be by means of output feedback; in this method the reservoir outflows would be monitored and the errors between the desired and monitored flows used to control a set of dominant network valves. Since there is no storage within the network, such valves could only control the distribution of water and could not affect the total flow balance. Consequently, any major discrepancy between the predicted and actual total demand could lead to unacceptable nodal pressures. The proposed control scheme should thus allow for demand prediction errors and them in principle, the same method could be used to control a set of nodal pressures at desired values.

The total problem corresponds to continuous control of a non-linear system with multiple inputs and outputs. In order to apply satisfactory control it is desirable to obtain a set of linear system equations. It has been shown that the time-varying reservoir flows, $\underline{q}$, and the nodal pressures, $\underline{h}$, can be represented by the following set of linear matrix equations [10]:

$$d\underline{q}(k) = \underline{A}_q \, d\underline{x}(k) + \underline{B}_q \, d\underline{u}(k) + \underline{C}_q \, d\underline{y}(k) + \underline{D}_q \, d\underline{v}(k) \qquad (46)$$

$$d\underline{h}(k) = \underline{A}_h \, d\underline{x}(k) + \underline{B}_h \, d\underline{u}(k) + \underline{C}_h \, d\underline{y}(k) + \underline{D}_h \, d\underline{v}(k) \qquad (47)$$

Use of Eqs.(46) and (47) for design of a non-interactive control system would depend upon choice of a set of valves which have major effects on the flows and heads to be controlled. One possibility would be for valves to be selected to form variable-demand zones located around each source input.

More elegant optimization methods could be used for multiple valve adjustment to meet the conflicting requirements of flow and pressure control (e.g. generalised reduced gradients [19]).

However, this would require more accurate information on the network and detailed knowledge of the distributed demand flows.

Series-connected Subsystems with Fixed and Variable Speed Pumps

The type of system for consideration can consist of a branch type structure containing a number of series and/or parallel connected subsystems each with some variable speed pumps.

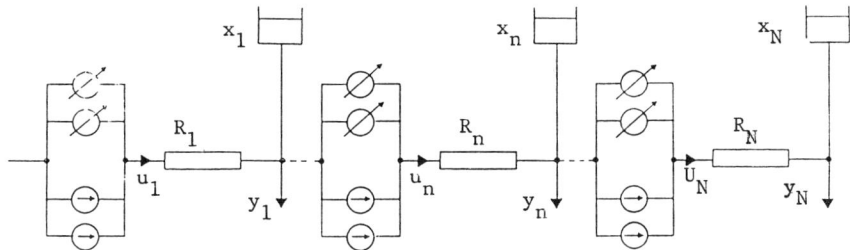

Fig.12. Series connected subsystems with fixed and variable speed pumps.

For simplicity only series connections are included as shown in Fig.12. For this system the general state equations will be:

$$x_n(k+1) = x_n(k) + u_n(k) \Delta t/A_n - y_n(k) \Delta t/A_n - u_{n+1}(k) \Delta t/A_n \qquad (48)$$

where n takes values of $1...N$, and $u_{N+1}(k) = 0$.

East state is subject to constraints of the form of Eq.(37) and the operating costs will be as given by Eqs.(42) and (43).

Joalland and Cohen [20] have proposed a decomposition and co-ordination method for optimizing systems of this type. Their method involves dynamic programming solutions for each sub-system and the method would appear to be applicable for systems with variable speed or fixed speed pumps. However, for systems with variable speed pumping capability the following proposals represent an extension of the method developed and proven for independent sub-systems.

For optimization of this type of multireservoir system the necessary reduction in computer memory and time can be achieved by using State Increment Dynamic Programming (SIDP) [5,21]. This method can be applied to an invertible system with continuous controls and guarantees at least a local optimum. SIDP is an iterative technique in which the recursive equation of dynamic programming is used to search for an improved trajectory among discrete states adjacent to an initial feasible trajectory. Figure 13 illustrates the technique and demonstrates that only 3^N states of $\underline{x}$ and $\underline{x} \pm \Delta\underline{x}$ need be explored at each intermediate stage of an N-dimensional process. The selection of values of $\Delta\underline{x}$ and other computational features are covered in [5]. The algorithm for this case follows the one described for <u>single reservoir systems with fixed and variable speed pumps</u>, except that the trajectories are now defined in the N-dimensional state-space. For each prescribed state, Eqs.(48) must be solved in reverse order, to yield $u_N(k)...u_1(k)$, and the costs evaluated from Eqs.(42) and (43). The optimal schedules of pump combinations and speeds which generate $x_n^*(k)$ and $u_n^*(k)$, for $n = 1,...N$ and $k = 0, 1,...K-1$, will define an optimization policy for the complete system.

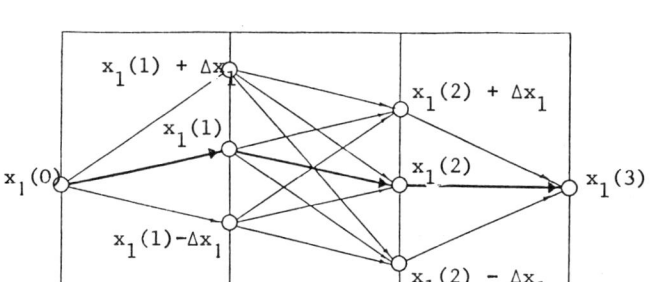

Stage k

Fig.13. Illustration of state increment dynamic programming procedure.

SUMMARY

Optimization of water distribution systems presents a very complex problem, when all operating factors have to be taken into account, and no entirely satisfactory solution methods are currently available. In order to determine possible solution techniques which will cater for some of the requirements, it is necessary to simplify the problem. By adopting a compromise between accuracy and feasibility, hydraulic and cost models can be formulated which are sufficiently representative for operational control purposes but which allow for optimization by efficient computational methods. These particular formulations have allowed development of two flexible computational modules for optimizing some typical one-dimensional water supply systems. Both the algorithms use dynamic programming techniques and one of them has been programmed for interactive use on a minicomputer. The validity of the results has been demonstrated by applying the generated least-cost pumping schedules to actual operating systems.

Additional simplifications are required for multidimensional systems which usually rely upon decomposition methods. For a parallel connected class of systems useful results can be obtained by structural decomposition, which leads to a dynamic source optimization problem and a static network control problem. Further decomposition, using the <u>dynamic programming method of successive approximations</u>, then allows each of the source supplies to be optimized using one of the previously developed computational modules. A series connected class of systems with continuous controls can take advantage of the computational efficiencies offered by <u>state increment dynamic programming</u>. This method extends the applicability of the corresponding one-dimensional module to series-connected multidimensional systems.

ACKNOWLEGEMENTS

Anglian Water Authority, Cambridge Division and Severn Trent Water Authority, Tame Division, have supplied the data on which some of the results are based.

The programs were written and the computed results obtained by Dr C H Orr of the Leicester Polytechnic Water Control Unit. The support of the above parties is gratefully acknowledged.

REFERENCES

1. B. Coulbeck, and M.J.H. Sterling 'Optimized control of water Distgribution systems', Proc IEE, $\underline{125}$(9), 1039-1044 (1978)

2. R. De Moyer, and L.B. Horwitz, 'A system approach to water distribution modelling and control',(Lexington Books 1973).

3. F. Fallside, and P.F. Perry, 'Hierarchical optimization of a water supply network', Proc. IEE, $\underline{122}$, 202-208 (1975).

4. R. E. Larson, 'State increment dynamic programming', (American Elsevier, New York 1967).

5. M. Heidari, V.T. Chow, P.V. Kokotovic, and D.D. Meredith, 'Discrete differential dynamic programming approach to water resources systems optimization', Water Resources Research, $\underline{7}$, 273-282 (1971).

6. M.J.H. Sterling, and B. Coulbeck, 'Analysis of pump interaction and efficiency in control of water supply', Trans. Inst. MC, $\underline{1}$(2), 79-84 (1979).

7. C.H. Orr and B. Coulbeck, 'Analysis and Simulation of Pump characeristics in the control of Water Distribution Systems', Leicester Polytechnic Water Control Unit, Research Report No. 30 (1983).

8. H.S. Rao, and D.W. Bree, 'Extended period simulation of water systems; Part A'. J. Hydraul. Div. ASCE, HY2(103), 97-108 (1977).

9. B. Coulbeck, and C.H. Orr, 'A Network Analysis and Simulation Program for Water Distribution Systems', Civil Engineering Systems, 1, 139-144 (1984).

10. B. Coulbeck 'Dynamic simulation of water distribution systems', Maths and Computers in Simulation, XXII, 222-230 (1980).

11. N.T. Thomopoulos, 'Applied forecasting methods', (Prentice Hall 1980).

12. S.M. Moss 'On-line optimal control of a water supply network', Cambridge University Engineering Detartment, PhD Thesis (1975).

13. B. Coulbeck, S.T. Tennant, and C.H. Orr, 'Development of a Demand Prediction Program for use in Optimal Control of Water Supply', Proceeding of 4th International Conference on Systems Engineering, Coventry Lanchester Polytechnic, 10-12 September, 176-183 (1985).

14. M.J.H. Sterling, and B. Coulbeck, 'A Dynamic Programming Solution to Optimization of Pumping Costs'. Proc. ICE, 59(2), 813-818 (1975).

15. B. Coulbeck and C.H. Orr. 'Optimized Pumping in Water Supply Systems', IFAC IX Triennial World Congress, Budapest, Hungary, July 2-6, IV, 158-163 (1984).

16. B. Coulbeck, and C.H. Orr, 'Optimized pump scheduling for water supply systems', 3rd IFAC Symposium on Control of Distributed Parameter Systems, Toulouse, France, 29 June - 2 July, XVIII, 29-33 (1982).

17. B. Coulbeck, 'A proposed optimization and control procedure for a class of water supply systems', Proc. International Conference on Systems Engineering', Lanchester Polytechnic, Coventry, 9-11 September, 558-567 (1980).

18. B. Coulbeck. 'Optimization of Water Networks', Trans, Inst. Measurement and Control, 6 271-280 (1984).

19. J. Abadie, Integer and Non-Linear programming. (North-Holland Publishing Company (1970).

20. G. Joalland and G. Cohen, 'Optimal Control of a water distribution network by two multilevel methods', Automatica, 16, 83-88 (1980).

21. W.J. Trott, and W.W.G. Yeh, 'Optimization of a multiple reservoir system', J.Hydraul. Div. ASCE, 381-391 (1972)

NOTATION

A   quadratic coefficient for pump head-flow characteristics; also reservoir cross-sectional area

$\underline{A}$ reservoir coefficient matrix

B   linear coefficient for pump head-flow characeristics

$\underline{B}$ pump coefficient matrix

C   constant coefficient for pump head-flow characteristics

$\underline{C}$ demand coefficient matrix

$\underline{D}$ valve coefficient matrix

f   generalized function for individual and total connected network link flows

$\underline{f}$ vector functional with elements corresponding to summation of network link flow relationship for each node

$\underline{F}$ diagonal reservoir coefficient matrix whose elements relate change of water level to flow

g   generalized function; alsos coefficient for pump electrical input power

h   general network node pressure head with Δh for network link head increase or head drop

$\underline{h}$ pressure node vector

J   cost factor defining operating costs; also total number of pump groups

k   stage variable (0,1,....,K-1) correspondiong to time increments Δt

K   total control period

L   total consumer demands; also total number of pump combinations

M   total pump stations

N   total reservoirs

p   pump electrical power consumption

P   fixed value of pump station power consumption

q   general network link flow and reservoir node flow

$\underline{q}$ reservoir node flow vector

Q   cost weighting factor

r	general network link parameter for pipes, pumps, and control valves
R	total pump/valve controls; also fixed pipe resistance
s	pump speed control parameter as a ratio of actual to nominal speed
t	number of pumps switched
T	pump station electricity unit or maximum demand tariff values
$\underline{u}$	pump control vector corresponding to pump station flow or pump combinations and speeds
U	fixed value of pump station flow
$\underline{v}$	valve control vector corresponding to valve flow or valve resistance
$\underline{x}$	reservoir level vector
$\underline{y}$	consumer demand vector
$\underline{z}$	general pressure node vector corresponding to pressure head or head drop
ψ	network solution functional
η	pump efficiency

Subscript

i	general network node identification index
j	general network link identification index; also general index for pump groups $1,\ldots J$.
ℓ	general index for consumer demands $1,\ldots,L$; also general index for pump combinations $1,\ldots L$
m	general index for pump stations $1,\ldots,M$
n	general index for reservoirs $1,\ldots,N$
r	general index for pump/valve controls $1,\ldots,R$

Superscript

des	desired value
min	minimum value
max	maximum value
tot	total value
*	Optimal value

Suffix

q,h,x,z   values related to respective previously specified variables

u,w   values related to electricity units and maximum demands

s   values related to pump switching

DYNAMIC PROGRAMMING FOR OPTIMIZATION OF PUMP SELECTION AND SCHEDULING IN WATER SUPPLY SYSTEMS

B. COULBECK and C.H. ORR
Water Control Unit, School of Electronic and Electronic Engineering, Leicester Polytechnic, P.O. Box 143, Leicester, LE1 9BH, England

ABSTRACT

This article considers the problem of optimal control of water systems which consist of reservoirs and pump stations coupled by direct pipe lines. Under specified conditions, the overall problem of optimized pumping can be divided into the two main areas of optimized pump selection and optimized pump scheduling. Optimized pump selection is a static optimization problem to search for a suitable pump combination, with the least power consumption, to satisfy the instantaneous system operating conditions. Optimized pump scheduling is a dynamic optimization problem to take account of system storage and time varying electricity tariffs in order to minimize the overall operating costs over a particular control period. A forward dynamic programming procedure has been formulated which incorporates both aspects to provide optimal operation of single reservoir sub-systems. Data is provided which completely defines an actual supply system and the procedure is used to optimize the system operation. The computer program developed as part of the study is now in operational use.

INTRODUCTION

Typical water supply systems consist of large numbers of interconnecting pipes together with valves and pumps. The networks are often provided with storage capacity in the form of service reservoirs. Determination of both instantaneous pumping conditions and longer term operational policies has long been a major concern in the control of these systems. Results from the water industry show that a significant proportion of the total operational expense is due to electricity costs for pumping. This provides a strong motivation for optimizing the pumping operations. Locally, the optimization objective is to select the pump combination to minimize the resulting power consumption. Globally, the main objective is to determine pumping policies which minimize the overall costs.

Improved control schemes require accurate models of the distribution networks. However, the strong interaction between the non-linear network components and the large number of system variables imposes severe theoretical and computational restrictions in the application of optimization techniques for system control. Various proposals have been made for optimization of particular systems, [1,2,3,4,5]. In most cases, restrictive assumptions have been made which either, simplify the optimization considerations while retaining the overall system model, or attempt complete optimization of a simplified system.

In this article, the fundamental optimization problems of both pump selection and pump scheduling are identified and examined. Pump stations usually contain parallel combinations of pumps where each pump can have different control characteristics. This can result in either: a discrete type control, or a mixed type control with both discrete and continuous variables. The primary objective in pump optimization is to determine appropriate combinations of these variables. The resulting formulation then forms the basis for development of optimized scheduling procedures; these can be are used in conjunction with the forward dynamic programming technique for the optimization of a class of water supply systems.

For demonstration purposes the developed method is applied to an actual supply system. A comprehensive set of data is included which completely defines the system structure and the characteristics of the components. The resulting optimized schedules correspond to a given set of practical operating conditions.

OPTIMIZED PUMP SELECTION

The major objective in optimized pump selection is to minimize the pump power consumption while maintaining given system conditions. With assigned pump controls, the power consumption depends on both the pump head and flow values. Since these pump hydraulic values interact with the network response, solutions for pump heads and flows involve solutions for the entire network. However, under certain circumstances, the pump head and flow values will be prescribed for a given pre-defined network response. Provided sufficient flexibility is available from the pump characteristics, the corresponding pump controls can then be determined directly from the pump hydraulic properties. The pump operating conditions can thus be obtained without further consideration of the network response.

For convenience parallel pumps of sufficiently identical characteristics are combined in groups. The pump group characteristics for various pump categories have previously been analyzed [6]. In summary, the hydraulic head-flow characteristics for groups of fixed speed pumps, variable speed pumps, and variable throttle pumps can be represented by:

$$\Delta h_f = A(q/r)^2 + B(q/r) + C \qquad (1)$$

$$\Delta h_s/s^2 = A(q/rs)^2 + B(q/rs) + C \qquad (2)$$

$$\Delta h_v = A(q/rv)^2 + B(q/rv) + C \qquad (3)$$

where:

Δh_f, Δh_s, and Δh_v are the head increases across the pump groups for fixed speed, variable speed, and variable throttle pumps, respectively,

q is the total flow for all active pumps in the particular pump group,

A, B, and C are the coefficients obtained from the hydraulic curves for a single pump (r = 1) running at nominal speed (s = 1.0) with no throttling (v = 1.0),

r is the discrete control factor, representing the number of pumps of similar characteristics in use, with values from r^{min} to r^{max}

s is the continuous speed ratio factor (ratio of actual speed n to nominal speed N with constraints on speed from n^{min} to n^{max}), v is the throttle factor, valid values ranging from 0.0 to 1.0, generally v^{min} to v^{max}, for throttle valve fully shut to completely open.

The general pump group efficiency equation is derived as:

$$\eta = \eta^{max}\{1 - (q/rsq^* - 1)^2\} \qquad (4)$$

with s = 1.0 for fixed speed and variable throttle pumps.

where:

η is the operating efficiency, and η^{max} and q^* are the maximum efficiency and corresponding maximum efficiency flow, respectively, for a single pump (r = 1) running at nominal speed (s = 1.0).

The true power consumption for all pump types can be expressed as

$$p = g \, \Delta h \, q/\eta \qquad (5)$$

where p is the pump group electrical input power,
g is the unit conversion factor, and
Δh is the pump group head increase.

g will have a value of 0.0098066 for p in kilowatts, Δh in metres, q in litres/second, η in per unit values, and density of water taken to be unity.

If the power factor of the electrical pump drive is close to unity, Eq.(5) can also be used for the apparent power demanded in kVA. For evaluation of system operation under normal circumstances, the pump controls can have pre-specified values. This includes pump combinations for all types of pumps. For variable speed pumps and variable throttle pumps, there are also corresponding speed and throttle factors. The pump station is then considered in conjunction with the entire network to obtain the overall operating condition. Because of the system non-linearities, iterative numerical procedures must be applied for the solution of the entire system to obtain pump head and flow values, and the corresponding power consumptions. Various combinations of available pumps can be tested, under the same operating condition, and a comparison of their respective power consumptions used as a basis for the optimization.

When both the head and flow values of the pump can be pre-defined by the system operational conditions, the pump station will be effectively decoupled from the rest of the system. The pump operating condition and corresponding power consumption can then be obtained directly from the pump characteristics. However, the particular requirement to satisfy the given pump head and flow values demands continuous variation of the pump characteristics, which can only be obtained with variable speed or variable throttle pumps. The basic approach involves selection of appropriate speed ratio or throttle factor to satisfy the operating condition. Optimized pump selection is then made for various pump combinations based on their respective power consumptions. Pump throttling is an inefficient way of operating pumps, due to the additional power dissipation at the throttle valves, and their use is only considered to overcome the inflexibility of the fixed speed pump characteristics.

For parallel combinations of fixed and variable pump groups, the same station head will apply, and the fixed speed pump flows can be computed directly from the fixed speed pump group head-flow relationships. The difference between the total pump flow and the fixed speed pump flows then prescribes the variable speed or variable throttle pump flows.

Considering a combination of two variable speed pump groups with different characteristics. In this case, there are 2 sets of continuous variables s_1, and s_2. When only a total station flow u is given, a primary optimization procedure is required to sub-divide this total station flow into individual pump group flows q_1 and q_2 in such a way that the sum of the individual pump group powers p_1 and p_2 are minimized and the resulting system operation is optimal. It is convenient to divide the station flow u according to

$$q_1 = \alpha u \quad \text{and} \quad q_2 = (1 - \alpha)u \tag{6}$$

For a convex optimization problem the unique minimum can be found from:

$$dp/d\alpha = dp_1/d\alpha + dp_2/d\alpha = 0 \tag{7}$$

With the simplifying assumptions of $\eta_1^{max} = \eta_2^{max}$ and $s_1 = s_2 = 1.0$ the minimizing value of α can be shown to be:

$$\alpha = r_1 q_1^* / (r_1 q_1^* + r_2 q_2^*) \tag{8}$$

Simulation tests with practical data have demonstrated that the above requirements and assumptions are valid for the systems studied.

Equation 8 can be generalised for J variable speed pump groups to:

$$\alpha_j = r_j q_j^* / (\sum_{j=1}^{J} r_j q_j^*) \tag{9}$$

subject to $\sum_{j=1}^{J} \alpha_j = 1.0$

where each individual pump group flow is given by:

$$q_j = \alpha_j u = \alpha_j \sum_{j=1}^{J} q_j \tag{10}$$

Once the individual pump group flows are determined for each of the variable speed pump groups, the optimization selection for minimum pump power can be carried out. For multiple groups of variable throttle pumps, throttle variation of only one group is required to obtain the imposed operating condition with the remaining groups being used as normal fixed speed pumps. This will ensure optimal usage of all available pumps.

It follows that this latter approach allows direct manipulation of the available pumping resources without requiring iterative numerical solutions for the entire network. This results in a considerable reduction in the overall solution procedure and provides a direct and efficient computational approach to system control. In the following sections, the optimized control considerations will be based on this particular procedure whenever sufficient degrees of freedom are available from the selected pumps.

OPTIMIZED PUMP SCHEDULING

Optimized pump scheduling is a dynamic optimization problem. The main objective is to minimize the total operational cost over a given control period. A large part of the operational cost is due to electricity charges for pumping. These can be offset by taking advantage of system reservoir storage and the time-varying peak/off-peak electricity tariff structure, subject to the system operational constraints. However, minimization of total system operational costs does not necessarily imply minimization of total power consumption because savings can be made with the rebate rate even though the operation may demand higher overall power.

Well established optimization techniques are covered by a range of literature [7,8]. Because of : the complexity of general water networks, the discrete and continuous control characteristics, and the time varying nature of the cost function; an integrated optimal approach for the overall solution has yet to be realised. Among the existing techniques, Dynamic Programming has been found to be a useful approach [1,2]. In the following sections a forward dynamic programming procedure is defined for evaluation of optimized pump schedules in a water supply system.

Further details of the hydraulic and cost models and the optimization algorithms are given in the accompanying article entitled: **Dynamic Programming for Optimized Control of Water Supply and Distribution Systems**.

System Formulation

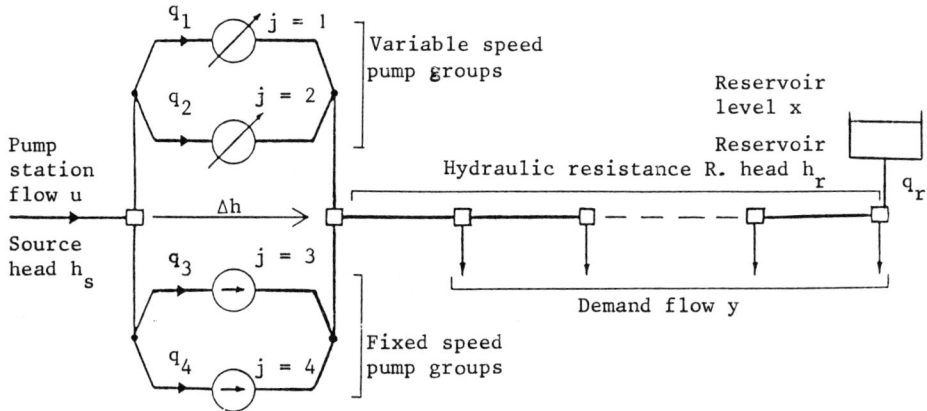

Fig.1. Basic water supply system.

Figure 1 shows a generalized one-dimensional supply system typical of bulk supply schemes. The system consists of a pump station which pumps water from a source along a direct equivalent pipe line, containing a number of demand nodes, to a terminal reservoir.

While the source reservoir level is assumed to vary in a predictable manner, the controlled reservoir level variation is governed by the system demand and station pump flows according to:

$$q_r = u - y \qquad (11)$$

where

q_r is the reservoir inflow

u is the total pump station outflow defined as $\sum_{j=1}^{J} q_j$ for J pump groups,

y is the total network demand flow, defined as $\sum_{\ell=1}^{L} y_\ell$ for L consumer demands.

Because of abstraction licenses and treatment works restrictions, the station flow can be subject to minimum and maximum levels, u^{min} and u^{max}, respectively, such that $u^{min} \leq u(k) \leq u^{max}$.

In discrete time form the incremented reservoir level x(k + 1) is given by:

$$x(k + 1) = x(k) + (K_r D \, \Delta t/V) q_r(k) \qquad (12)$$

where:
k is the discrete time index varying over 0, 1, ..., K-1, for K stages,
D is the reservoir depth,
V is the reservoir volume,
Δt is the time interval for each stage, and
K_r is the reservoir conversion factor appropriate to the units.

K_r will have a value of 0.0036 for x(k) in metres, $q_r(k)$ in litres/second, D in metres, V in megalitres, and Δt in hours.

The variation of the reservoir level is subject to the minimum and maximum levels x^{min} and x^{max}, respectively, such that $x^{min} \leq x(k) \leq x^{max}$. The initial reservoir level x(0) is generally known whilst there is usually a desirable final reservoir level $x(K)^{des}$ at the end of the control period.

For the pipe-line the head drops can be computed from the Hazen-Williams relationship of:

$$\Delta h = R|q|^{1.852} = \{(K_\ell L)/(D^{4.87} C^{1.852})\} |q|^{1.852} \tag{13}$$

where:

Δh is the head drop with sign determined by the direction of the flow, q,
R is the pipe hydraulic resistance
K_ℓ is the pipe conversion factor appropriate to the units,
L is the pipe length,
D is the pipe diameter,
C is the Hazen-Williams friction coefficient,

K_ℓ will have a value of 1.21216×10^{10} for Δh in metres, q in litres/second L in metres, and D in millimetres.

The control objective is to schedule the pump groups for least cost operation while satisfying the demand flows, the station flow constraints, and the reservoir level constraints. Individual systems may have additional operational constraints on parameters such as, node pressures and pipe velocities.

Cost of maximum power demanded, Jw, is given by:

$$Jw = Tw\, p(k)^{max} \tag{14}$$

subject to $\quad w(0) \leq p(k)^{max} \leq w^{max}$

where:

$p(k)^{max}$ is the maximum demand attained during the control period,
Tw is the corresponding tariff rate,
$w(0)$ is the initial value of the maximum demand (to account for any previously incurred demand within the same tariff charging period), and
w^{max} is the available maximum demand (to account for electricity supply limitations).

Cost of units consumed, Ju, is given by:

$$Ju = \sum_{k=0}^{K-1} Tu(k)\, p(k)\, \Delta t \tag{15}$$

where:

$p(k)$ is the total pump power at stage k, and
$Tu(k)$ is the unit tariff rate at stage k which allows for peak and off-peak rates.

Indirect pump operating costs include those for pump maintenance which is related to the frequency of pump switching. Satisfactory results, for the systems studied, have been obtained by use of:

$$Js = \sum_{j=1}^{J} Js_j = \sum_{j=1}^{J} \sum_{k=0}^{K-1} Qs_j \, t_j(k) \qquad (16)$$

where:

$t_j(k)$ is the number of pump state changes for pump set j at stage k with initial pump state $r_j(0)$ given, and

Qs_j is the estimated maintenance cost for each change of pump state for pump set j.

Another indirect cost factor is due to reservoir level fluctuations (e.g. to account for pressure related leakage) which can be expressed as:

$$Jx = \sum_{k=1}^{K} Qx(k)[x(k) - x(k)^{des}]^2 \qquad (17)$$

where:

$Qx(k)$ is the empirical time varying cost penalty for level variation, and

$x(k)^{des}$ is the desired reservoir level at stage k.

Normally $Qx(k)$ and $x(k)^{des}$ would be assigned values to influence the operation over the intermediate stages of $1, \ldots, K-1$, and $Q_x(K)$, $x(K)^{des}$ would be assigned values to dictate the operation at the final stage K.

The overall cost function J_o can then be formulated as:

$$Jo = Jw + Ju + Js + Jx \qquad (18)$$

which is to be minimized over the control period, subject to the system constraints, to yield the optimal pump schedule.

OPTIMIZATION PROCEDURE

At any time stage k, the individual demand quantities y_ℓ for all pipe sections are obtained by prediction. The pre-defined change in reservoir levels defines the reservoir flow q_r. The pump station flow can then be decided from Eqs. (11) and (12). This station flow and the individual demand flows cause a head drop along the pipe-line, which can be computed from the pipe characteristics and the flow quantities according to Eq.(13). With the given heads for the source and controlled reservoirs, and the head drop values along the pipe-line, the pump station head increase $\Delta h(k)$ can be obtained as:

$$\Delta h(k) = h_r(k) - h_s(k) + \sum_{\ell=1}^{L} \Delta h_\ell(k) \qquad (19)$$

where:

        $h_T(k)$ is the total reservoir head,
        $h_S(k)$ is the source reservoir head, and
        $\Delta h_\ell(k)$ is the head drop across the ℓth section of the pipe-line.

Depending upon the reservoir configuration the total head can vary according to $x(k) + H_b$ (where H_b is the reservoir bottom level) or can be fixed at H_i (where H_i is the reservoir inlet level).

Eqs.(11) and (19) determine the operating condition of the pump station. The resulting situation is identical to the decoupled pump operation described under <u>Optimised Pump Selection</u>.

The technique of Dynamic Programming is well documented in other literature [9,10] and only brief details are included here. For application to the particular supply scheme, the computational algorithm is based on forward dynamic programming as summarized in Fig.2. The reservoir operational range is first divided into a number of discrete levels. These levels then form the basis for selection of reservoir level trajectories and hence define corresponding reservoir flows at any stage k. The required pump flow being divided between fixed speed pump groups and variable speed pump groups. The individual and optimized variable speed pump group flows are obtained using the primary pump flow optimization procedure. The pump power consumption is obtained from the station head increase and the individual pump group flows. The optimized pump selection principle is applied to retain the minimum cost combination for that particular operating condition. The same procedure is applied until the end of the control period. At the terminal stage, penalty costs are implemented according to the final reservoir level specification. The optimal trajectory for the control period is then recovered to obtain the respective reservoir level variation and pump selection at each stage.

The accuracy of the Dynamic Programming technique in this application depends on the number of reservoir level divisions. In order to achieve the required accuracy, without a significant increase in computing resources, certain refinements are necessary. The basic characteristics of reservoir level control and pump station decoupling from the network have allowed the adaptation of iterative procedures. This alternative computation scheme is based on a coarse reservoir level grid to establish an initial reservoir level variation throughout the control duration. Refined calculations are then carried out using a reduced reservoir range to form a corridor around the established reservoir level variation. This, in effect, improves the accuracy of the reservoir level control and hence the overall system results. The same refinement procedure can be repeated until the iterative calculations converge within a given limit. In this manner, the corresponding computing time increases only linearly with the repeated calculations and there are no additional computer memory requirements.

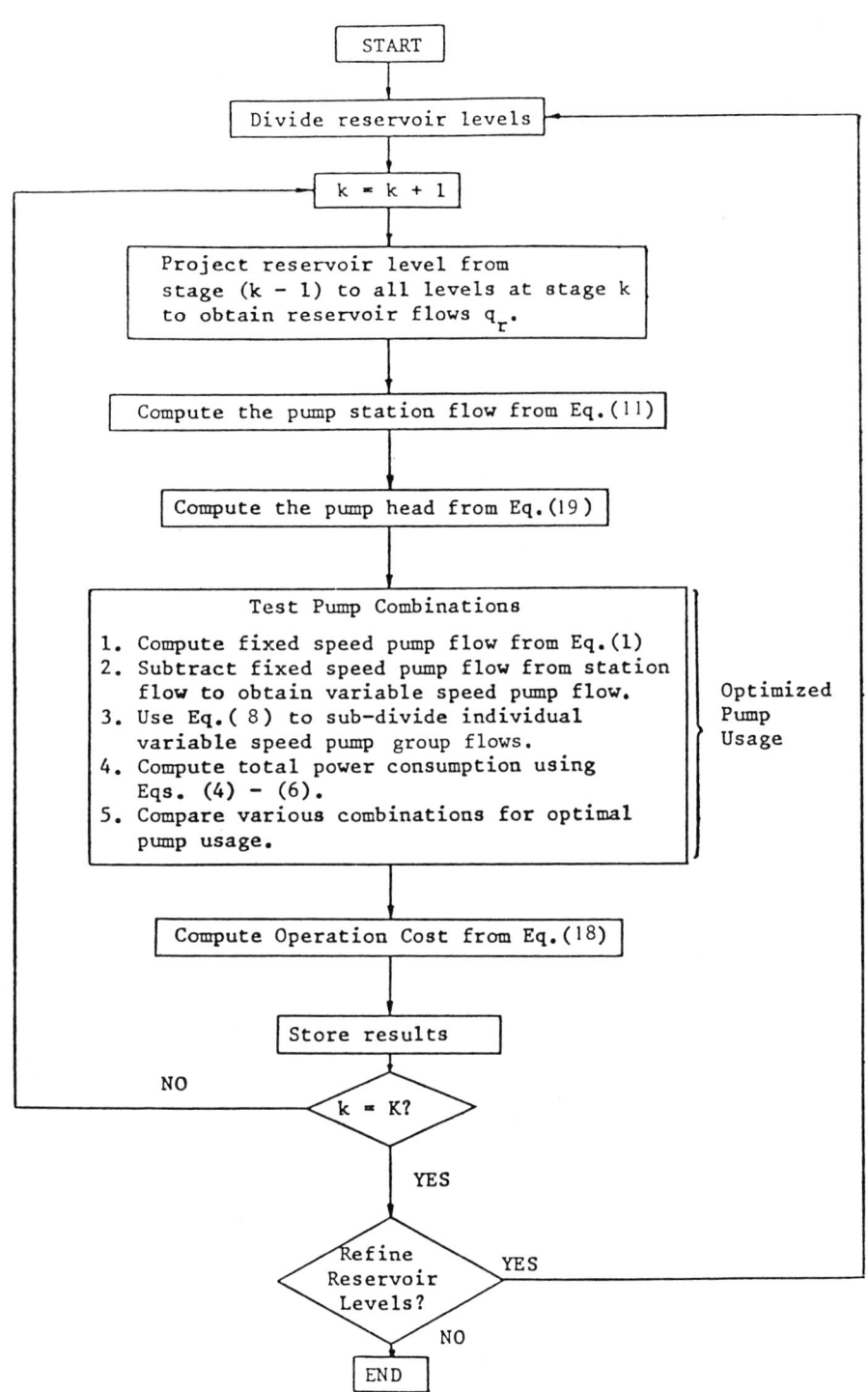

Fig.2. Flow Diagram of the Optimization Procedure.

APPLICATION TO A SUPPLY SYSTEM.

A graphical and interactive computer program has been written to perform the combined calculations for optimized pump selection and optimized pump scheduling [3,11]. Various versions of the program have been under application to actual supply systems since early 1982. Details of further applications and validations of the results are covered in [12]. The selected subsystem for demonstration purposes consists of 2 groups of variable speed pumps (VSP1 and VSP2) and 1 group of fixed speed pumps (FSP1). The pipe network has 7 sections, each with its own terminating node and take-off demand. The terminal reservoir of the present system is connected as the source reservoir of the following subsystem (which forms the major demand in the present system). Figure 3 shows the annotated system diagram with the complete set of system data being given in Table 1.

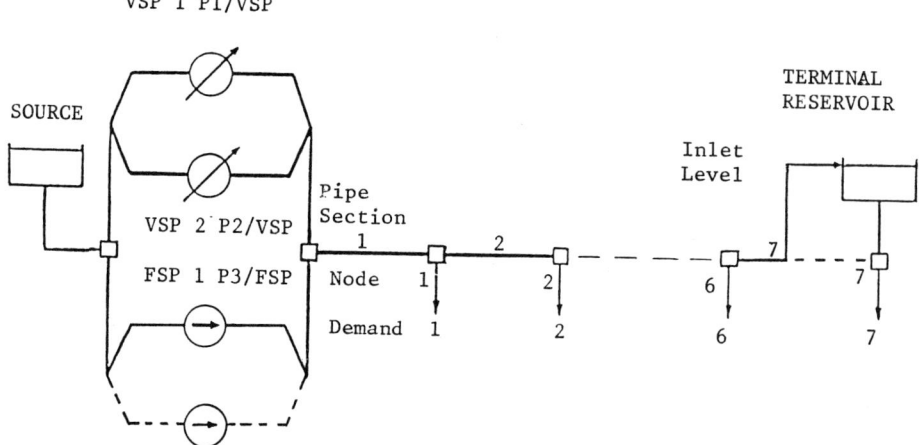

Fig.3. Actual supply system.

For convenience the consumer demand period is defined, independently of the optimization period, to always start at 00.00 hours. The present set of demand data take constant values over each 1 hour interval for a 24 hour period, typically as shown on Fig.4. Under these conditions the optimization also covers a 00.00 to 24.00 hour period but, to improve the computational efficiency, the optimization calculations are only performed at times of tariff changes. These changes occur at 01.00 and 08.00 hours which correspond to the beginning and end of the unit rebate period, respectively. The results of the optimization are shown in Table 2, with typical graphical results on Figs.5 and 6.

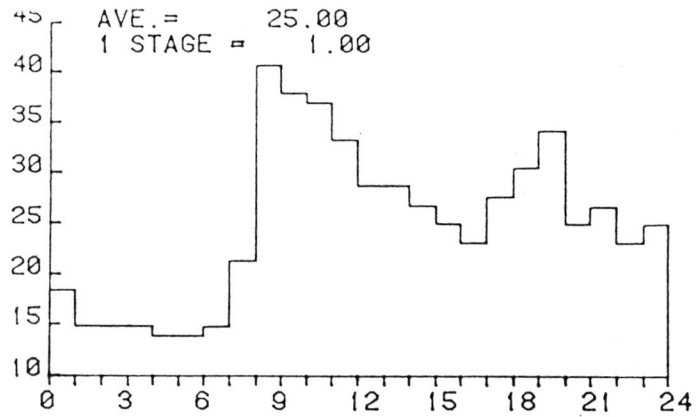

Fig. 4 Typical Prescribed Demand Variation

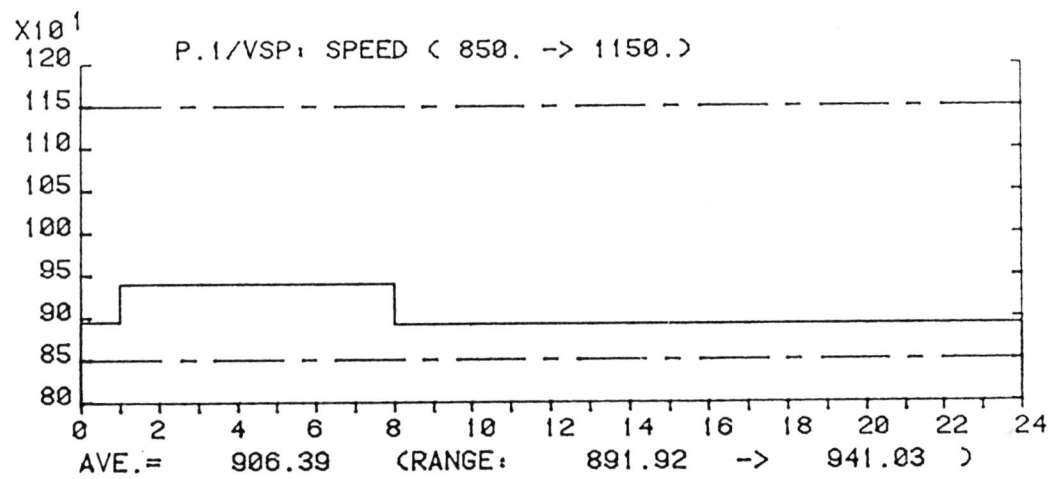

Fig. 5 Typical Optimized Pump Speed Variation

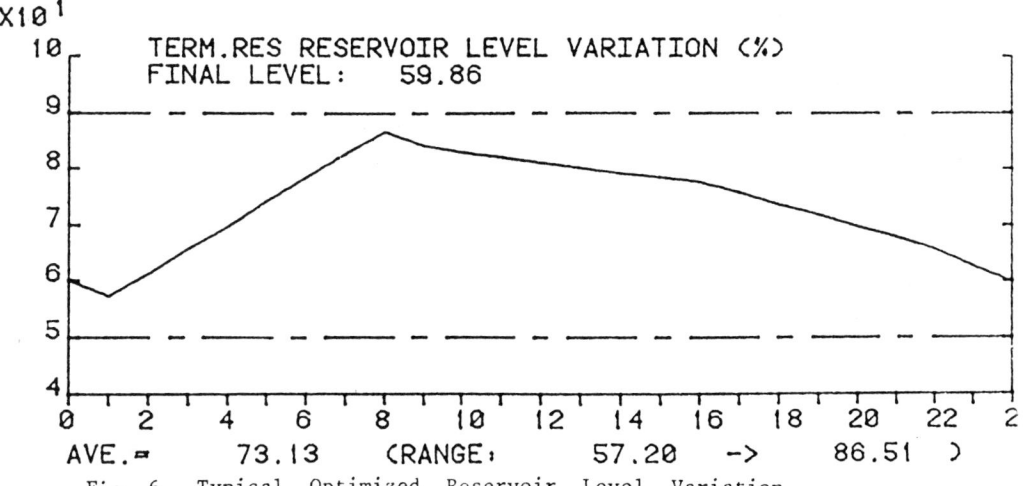

Fig. 6 Typical Optimized Reservoir Level Variation.

TABLE I System Data

** NETWORK: CONTROL PARAMETERS **

NETWORK PIPE SECTIONS: L	7
LOAD STEP (h): ΔT	1
LOAD STAGE: K	24
RESERVOIR CALCULATION LEVEL:	11
RESERVOIR CALCULATION RANGE (%):	70
INTERMEDIATE LEVEL (%): $x(k)^{des}$	70.00
INTERMEDIATE LEVEL PENALTY (£): $Q_x(k)$	0.00
ITERATIVE SOLUTION LIMIT:	100
ITERATIVE HEAD TOLERANCE (m):	0.100

** TIME CONTROL DATA **

START TIME (h):	0
END TIME (h):	24

** RESERVOIR LEVEL CONTROL DATA **

RESERVOIR NAME:	TERM.RES
OPERATING RANGE (%): $x^{min} \rightarrow x^{max}$	50.00 $\rightarrow$ 90.00
INITIAL LEVEL (%): $x(o)$	60.00
FINAL LEVEL (%): $x(K)^{des}$	60.00
FINAL LEVEL PENALTY (£): $Q_x(K)$	100.00

** STATION DATA **

MAXIMUM STATION DEMAND (kVA): w^{max}	5000.00
INITIAL MAXIMUM DEMAND (kVA): $w(o)$	1725.00
STATION FLOW RANGE (l/s): $u^{min} \rightarrow u^{max}$	500.00 $\rightarrow$ 5000.00

** STATION PUMP DATA **

VSP 1 NAME:	P.1/VSP
OPERATION CONTROL RANGE: $r_1^{min} \rightarrow r_1^{max}$	0 $\rightarrow$ 1
INITIAL STATE: $r_1(o)$	0
SWITCHING COST (£): Q_{s1}	10.00
SPEED RANGE (RPM): $n_1^{min} \rightarrow n_1^{max}$	850.00 $\rightarrow$ 1150.00
VSP 2 NAME:	P.2/VSP
OPERATION CONTROL RANGE: $r_2^{min} \rightarrow r_2^{max}$	1 $\rightarrow$ 2
INITIAL STATE: $r_2(o)$	2
SWITCHING COST (£): Q_{s2}	10.00
SPEED RANGE (RPM): $n_2^{min} \rightarrow n_2^{max}$	800.00 $\rightarrow$ 1150.00
FSP 1 NAME:	P.3/FSP
OPERATIONAL CONTROL RANGE: $r_3^{min} \rightarrow r_3^{max}$	0 $\rightarrow$ 2
INITIAL STATE: $r_3(o)$	0
SWITCHING COST (£): Q_{s3}	0.00

TABLE I Continued

** SOURCE HEAD CURVE **

SOURCE NAME: SOURCE
HEAD VALUES (m): $h_S(k)$

49.57	49.57	49.57	49.57	49.57	49.57
49.57	49.57	49.57	49.57	49.57	49.57
49.57	49.57	49.57	49.57	49.57	49.57
49.57	49.57	49.57	49.57	49.57	49.57

** DEMAND DATA ** (l/s):y

NODE 1: DEM.ND-1 y_1 AVERAGE DEMAND: 25.00

DEMAND CURVE VALUES:

18.49	14.79	14.79	14.79	13.87	13.87
14.79	21.26	40.68	37.91	36.98	33.28
28.66	28.66	26.81	24.96	23.11	27.74
30.51	34.21	24.96	26.81	23.11	24.96

NODE 2: DEM.ND-2 y_2 AVERAGE DEMAND: 20.00

DEMAND CURVE VALUES:

24.22	16.51	16.51	16.51	16.51	16.51
16.51	20.91	24.22	24.22	23.12	22.01
20.91	19.81	18.71	18.71	19.81	20.91
20.91	22.01	20.91	19.81	18.71	20.91

NODE 3: DEM.ND-3 y_3 AVERAGE DEMAND: 136.00

DEMAND CURVE VALUES:

144.45	139.43	140.43	138.43	138.43	139.43
136.42	136.42	134.42	134.42	133.41	134.42
135.42	135.42	133.41	135.42	134.42	132.41
133.41	136.42	133.41	133.41	135.42	135.42

NODE 4: DEM.ND-4 y_4 AVERAGE DEMAND: 25.00

DEMAND CURVE VALUES:

21.63	19.46	18.38	18.38	17.30	18.38
18.38	24.87	31.36	31.36	31.36	30.28
30.28	27.03	27.03	23.79	25.95	27.03
27.03	29.20	25.95	24.87	25.95	24.87

NODE 5: DEM.ND-5 y_5 AVERAGE DEMAND: 12.00

DEMAND CURVE VALUES:

12.78	10.81	10.81	10.81	10.81	10.81
10.81	12.78	12.78	12.78	12.78	12.78
12.78	11.79	11.79	11.79	11.79	12.78
12.78	12.78	11.79	11.79	12.78	11.79

NODE 6: DEM.ND-6 y_6 AVERAGE DEMAND: 18.00

DEMAND CURVE VALUES:

13.87	12.02	12.02	11.10	11.10	10.17
12.02	18.50	25.90	24.97	25.90	23.12
21.27	21.27	19.42	17.57	16.65	19.42
19.42	22.20	18.50	19.42	18.50	17.57

TABLE I Continued

NODE 7: DEM.ND-7 y_7 AVERAGE DEMAND: 800.00
DEMAND CURVE VALUES:

835.98	990.33	976.35	995.32	980.35	987.34
984.34	999.32	764.71	646.91	642.92	638.92
646.91	642.92	642.92	642.92	752.73	748.73
752.73	744.75	740.75	752.73	850.56	838.59

** MAXIMUM DEMAND TARIFF **

TIME (h)	RATE (p)
0 → 24	20.00

** UNIT TARIFF **

TIME (h)	RATE (p)
0 → 1	3.50
1 → 8	1.50
8 → 24	3.50

** NETWORK: RESERVOIR DATA **

CAPACITY (Ml): V	32.50
BOTTOM LEVEL (m): H_b	125.00
DEPTH (m): D	6.70
INLET LEVEL (m): H_i	125.50

** NETWORK: PIPE SECTION DATA **

	LENGTH (m)	DIAMETER (mm)	C-VALUE
STATION → DEM.ND-1	7000.00	1410.00	130.00
DEM.NO-1 → DEM.ND-2	6760.00	1410.00	130.00
DEM.NO-2 → DEM.ND-3	2250.00	1410.00	130.00
DEM.NO-3 → DEM.ND-4	2900.00	1375.00	130.00
DEM.NO-4 → DEM.ND-5	4350.00	1375.00	130.00
DEM.NO-5 → DEM.ND-6	4500.00	1375.00	130.00
DEM.NO-6 → DEM.ND-7	2740.00	1375.00	130.00

** NETWORK: PUMP CHARACTERISTICS DATA **

VSP 1:   P.1/VSP
HYDRAULIC COEFFICIENT A: $m/(l/s)^2$ -0.000100
HYDRAULIC COEFFICIENT B: $m/(l/s)$ -0.185000
HYDRAULIC COEFFICIENT C: m   205.00
PEAK EFFICIENCY VALUE (%): η_1^{max} 77.50
PEAK EFFICIENCY FLOW (l/s): q_1^* 465.00
NOMINAL SPEED (RPM): N_1 975.00

TABLE I Continued

VSP 2:	P.2/VSP
HYDRAULIC COEFFICIENT A: m/(l/s)^2	-0.000700
HYDRAULIC COEFFICIENT B: m/(l/s)	-0.107000
HYDRAULIC COEFFICIENT C: m	180.00
PEAK EFFICIENCY VALUE (%): η_2^{max}	75.00
PEAK EFFICIENCY FLOW (l/s): q_2^*	205.00
NOMINAL SPEED (RPM): N_2	1000.00
FSP 1:	P.3/FSP
HYDRAULIC COEFFICIENT A: m/(l/s)^2	-0.000130
HYDRAULIC COEFFICIENT B: m/(l/s)	-0.000200
HYDRAULIC COEFFICIENT C: m	165.00
PEAK EFFICIENCY VALUE (%): η_3^{max}	80.00
PEAK EFFICIENCY FLOW (l/s): q_3^*	605.00

TABLE II Results for Optimized Scheduling

HR.	TERM.RES % LEVEL (x)	P.1/VSP NO. (r_1)	P.1/VSP SPEED (n_1)	P.2/VSP NO. (r_2)	P.2/VSP SPEED (n_2)	P.3/FSP NO.(r_3)	STATION HEAD(Δh)	STATION FLOW(u)	TOTAL DEMAND
0	60.00	1	896	2	826		80.13	818.35	1071.42
1	57.20	1	941	2	873	1	92.29	1579.55	1201.46
8	86.51	1	892	2	823		80.00	811.01	961.40
24	59.86								

TIMES (h): HOURLY % RESERVOIR LEVEL VARIATION (x)

0 → 5:	60.00	57.20	61.36	65.68	69.83	74.15
6 → 11:	78.39	82.66	86.51	84.05	82.94	81.89
12 → 17:	80.97	80.02	79.18	78.41	77.70	75.77
18 → 23:	73.79	71.74	69.63	67.79	65.82	62.78
24 → 24:	59.86					

TOTAL DIRECT COST (£):   710.87

  MAXIMUM DEMAND CHARGE (£): Jw   12.57
  MAXIMUM DEMAND POWER (kVA):$p(k)^{max}$ 1885.82

  UNIT CHARGE (£): Ju   K-1   698.30
  ELECTRICITY UNIT (kWh): $\sum_{k=0} p(k)\Delta t$ 27493.82

TOTAL INDIRECT COST (£):   10.00

  PUMP SWITCHING COST:
    P.1/VSP (£): Js_1 10.00
    P.2/VSP (£): Js_2 0.00
    P.3/FSP (£): Js_3 0.00

CONCLUSIONS

In this article, the essential features of optimized pump selection and optimized pump scheduling are identified. Optimized pump selection can be regarded as a static optimization problem, in which the specific aim is to select least power pump combinations to satisfy pre-specified system operating conditions. The minimum power condition will then ensure minimum electricity cost at that particular time instant.

In contrast, optimized pump scheduling is a dynamic optimization problem to minimize the operating cost over a given control period. In order to achieve this, the dynamic characteristics of both the electricity tariffs and the reservoir storage must be considered. At each time step, an average operating condition is assumed. The resulting system behaviour is then similar to the instantaneous operating conditions during optimal pump selection. Therefore, in optimized pump scheduling, the static optimization procedure is extended to take advantage of the dynamic nature of the system behaviour to derive cost effective operating policies for the overall system.

A flexible computer program has been written to perform both of these optimization tasks. The program is based on a general one-dimensional supply system and provides an accurate model of all non-linearities and system interactions. Forward dynamic programming is used as the basic pump scheduling optimization procedure with pump combination selection based on pump flow and power considerations. This particular formulation has provided an accurate and efficient computational algorithm for optimized system control. The solution gives both discrete and continuous controls corresponding to on-off pump combinations and pump speeds. Obtained results for the given set of system data have demonstrated the expected operation of the pump sources, notably

(1) an increase in overnight pumping,
(2) the use of the most efficient pumps,
(3) reduction of pump switching,
(4) limitations on variations of reservoir levels,
(5) limitations on maximum power demanded, and
(6) limitations on source supply flows.

The achievable 5-10 percent saving can be significant since typical water supply schemes have annual operating costs exceeding one million pounds (Sterling).

ACKNOWLEDGEMENTS

The work presented here has been developed with the collaboration of a number of U.K. Water Authorities. The authors would like to thank Mr.B.R.J.Perry, Mr.R.Minns, and Mr.R.Glass of Anglian Water Authority, Cambridge Division, for their cooperation and provision of system data. Thanks are also due to Mr.M.Williams, Mrs.S.O'Neill (Tame Division), and Mr.G.H.Snell (Avon Division) of Severn Trent Water Authority for many suggestions. The financial support of the UK Science and Engineering Research Council is also gratefully acknowledged.

REFERENCES

1. G. Cohen, Optimal Control of Water Supply Networks, in: Optimization and Control of Dynamic Operational Research Models, S.G. Tzafestas (ed.), (North Holland, Amsterdam, 1982)

2. B. Coulbeck, and C.H. Orr., Optimized Pump Scheduling for Water Supply Systems, Proc. of 3rd IFAC Symposium on the Control of Distributed Parameter Systems, Toulouse, France, 29th June - 2nd July, XVIII 29-33 (1982)

3. B. Coulbeck, and C.H. Orr, Development of an Interactive Pump Scheduling Program for Optimized Control of Bulk Water Supply, Proc. of 2nd International Conference on Systems Engineering, Lanchester Polytechnic, Coventry, England, 14th-16th, Sept., 176-186 (1982)

4. F. Fallside, and P.F. Perry, Hierarchical Optimization of a Water Supply Network, Proc. IEE, 122, No.2, 202-208 (1975)

5. M.J.H. Sterling, and B. Coulbeck, A Dynamic Programming Solution to Optimization of Pumping Costs, Proc. ICE, Part 2, 59, 813-818 (1975)

6. C.H. Orr, and B. Coulbeck, Analysis and Simulation of Pump Characteristics in the Control of Water Distribution Systems, Research Report No.30, School of Electronic and Electrical Engineering, Leicester Polytechnic, England (1983)

7. L.S. Lasdon, (ed.), Optimization Theory for Large Systems, (MacMillan, 1970)

8. D.A. Wismer, (ed.), Optimization Methods for Large Scale Systems... With Applications, (McGraw Hill, 1971)

9. R. Bellman, and S. Dreyfus., Applied Dynamic Programming (Princeton University Press, 1962)

10. R.E. Larson, State Increment Dynamic Programming, (Elsevier Publishing Co., 1968)

11. C.H. Orr, B. Coulbeck, J.P. Rance, and R.E. Korzeniewski, Computer-Aided Pump Scheduling for Efficient Operation of Water Supply Systems using the Program GIPOS3. Research Report No.42 School of Electronic and Electrical Engineering, Leicester Polytechnic, England, March (1986)

12. G.H. Snell, Practical Application of the Water Distribution System Simulation and Optimization Programs GINAS and GIPOS, Jnl. Inst. Water Pollution Control, 84, No.4, 459-469 (1984)

OPTIMAL DESIGN OF LARGE COMPLEX WATER RESOURCES CONVEYENCE SYSTEMS
VIA NONSERIAL DYNAMIC PROGRAMMING ***

AUGUSTINE O. ESOGBUE,* AND CHAE YOUNG LEE**
*School of Industrial and Systems Engineering
Georgia Institute of Technology,
Atlanta, Georgia 30332-0205, U.S.A.;
**Korea Institute of Technology
Taejun, Korea

1. INTRODUCTION

    Water resources planning is a multi-faceted multi-staged, continuous process which occurs at several levels and in various locations. At the state, regional and even local levels the systems under consideration are more often than not large scale in nature. That is, planning whether for design, operation or maintenance relates to a system of units rather than a single unit. Thus, cost effectiveness considerations require the treatment of all systems units as a whole. In general, in such large scale systems, the number of variables and alternatives that must be considered forces the planner or analyst using classical approaches favored by practicing engineers to eliminate a large number of possible alternatives. This is done in order to focus on the few that are considered most promising. Only very few experienced engineers can use such a trial and error approach in combination with good judgment to produce cost-effective designs, most of the time, in the usual time and resource constrained design environment. The use of mathematically reliable models, especially those that can be automated has tended to minimize the problems inherent in traditional practices.

    As documented in the literature, systems and optimization based approaches have become a useful tool of the modern design engineer. Since its development by Bellman and later by numerous authors [5], dynamic programming has become a very attractive modeling and design tool. However, because of the well known but perhaps somewhat exaggerated problem of dimensionality, its utility to the practicing engineer has been quite limited. Various authors and model developers have sought approaches to circumvent this problem. Unfortunately, the casualty is usually the problem. Oversimplification and sometimes sensible decomposition methods have been advocated. We have erstwhile postulated that these problems can only be eliminated or more realistically ameliorated when large computers of the super variety, efficient algorithms geared towards memory reduction, parallel computing and above all adroit problem formulation which ab initio requires a minimal number of state variables are efficiently utilized to address a given problem. Some of the foregoing perquisites are beginning to be made available to the systems designer.

*** This paper has been accepted for publication in the Journal of Computers and Mathematics with Applications. It is specially dedicated to Professor Warren A. Hall, who introduced the first author to water resources engineering, and the memory of Professor Richard Bellman, who introduced both Professor Hall and the first author to dynamic programming.

Our principal contribution is in the area of modeling and
computational technology, but specifically nonserial dynamic programming.
The purpose is to show how a problem which naturally occurs as a nonserial
system but which has hitherto been approximated as a classical serial
dynamic program can be directly assaulted via nonserial dynamic
programming. This approach naturally minimizes approximation errors and,
ipso facto, increases accuracy of results. Most of all, it benefits from
the global optimality characteristic of classical dynamic programming.

2. THE PROBLEM

A gamut of problems in water resources but particularly those
involving water conveyance systems design, multi-basin capacity expansion
projects, optimal operation of reservoirs located on different streams in
a region, and storm water systems design are large scale in nature.
Further, they occur as nonserial, branched systems problems. Most of
them, however, have hitherto been treated as serial ones. When nonserial
methods have been used only very simple classical structures such as
diverging and converging systems have been utilized - and even then the
algorithms have been inefficient and computationally unattractive. We
have recently developed efficient and even high level computing
algorithms for processing various complex nonserial systems [10] and
naturally wish to apply them to water resources problems where their use
will prove beneficial.

We consider a few examples. Larson and Keckler [12] present a state
increment dynamic programming analysis of a four reservoir operation
problem (Fig. 1) which is an example of a highly simple converging branch
system. They however do not use nonserial dynamic programming to treat
their problem. Instead they solve a four state variable dynamic program
with the level of water in each reservoir considered a state variable.
Hall and Shephard [11] treat a more realistic reservoir operation problem
in the California Central Valley Project consisting of at least four
reservoirs with power generation capacities and an assortment of dams,
lakes, etc. located on separate streams but converging at a node in the
delta. Again, this is a good example of a multi-converging branch system
which was analyzed by a combination of linear and dynamic programming by
Hall and Shephard, Becker and Yeh [2], and Yeh et al. [3]. Real-time
hourly operation problems for the same complex multi purpose reservoir
system were considered by Yeh et al. [8,9]. In each of the foregoing,
however, decomposition and approximation methods were used in conjunction
with serial dynamic programming - in some cases differential dynamic
programming as the optimization procedure. Although useful solutions were
obtained, computational complexities and inefficiencies characterize the
approach and thus make their use costly.

The approaches proposed in the Central Valley Project by Yeh et al.
(Fig. 3) have been modified and applied to other important basins in
various parts of the United States. For example, the Tennessee Valley
Authority implemented Yeh et al. [21] incremental dynamic programming and
successive approximation model for their real-time optimal scheduling of
water releases for flood control and hydro-electric power generation
problem. The Texas Water Development Board has considered the use of
optimization and computerized models for various water resources problems.
A classic problem of concern is the optimal capacity expansion model for
surface water resources systems in multi basins. Some examples of these
multi basin networks are given in Figures 4 and 5 for the White

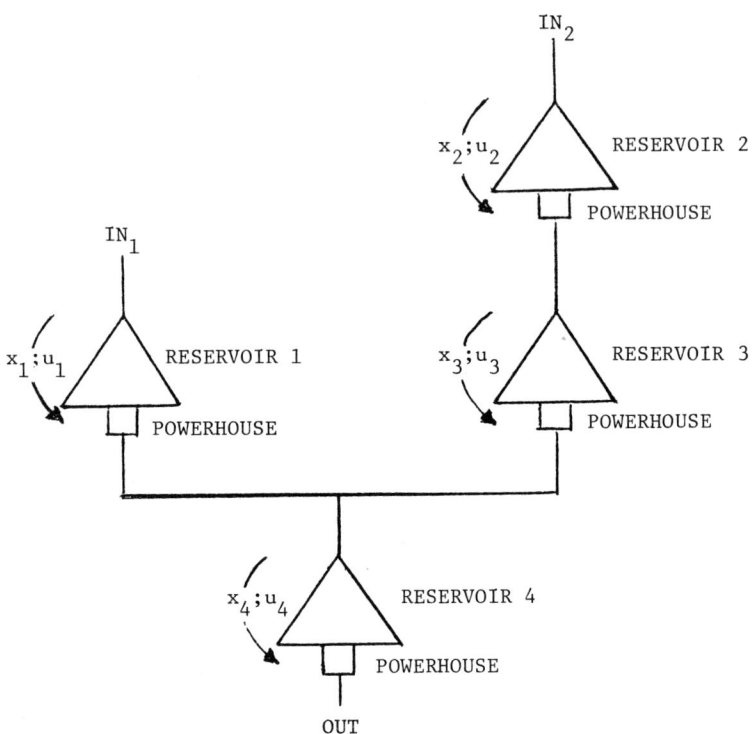

Figure 1. Network Configuration of Four-Reservoir Problem Considered by Larson and Keckler [12]

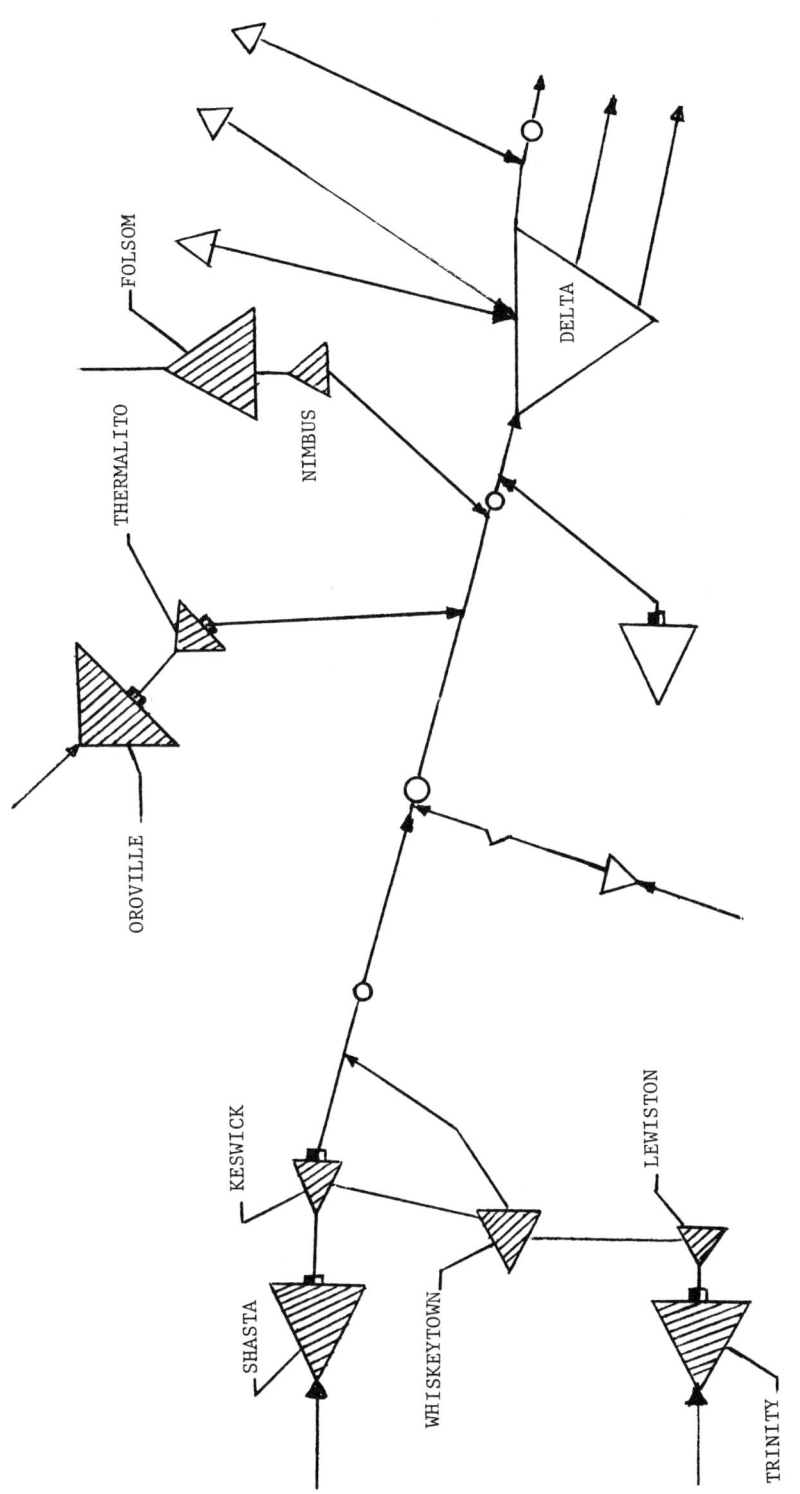

Figure 2. Operation Studies of the California CVP studied by Hall and Shephard [11]

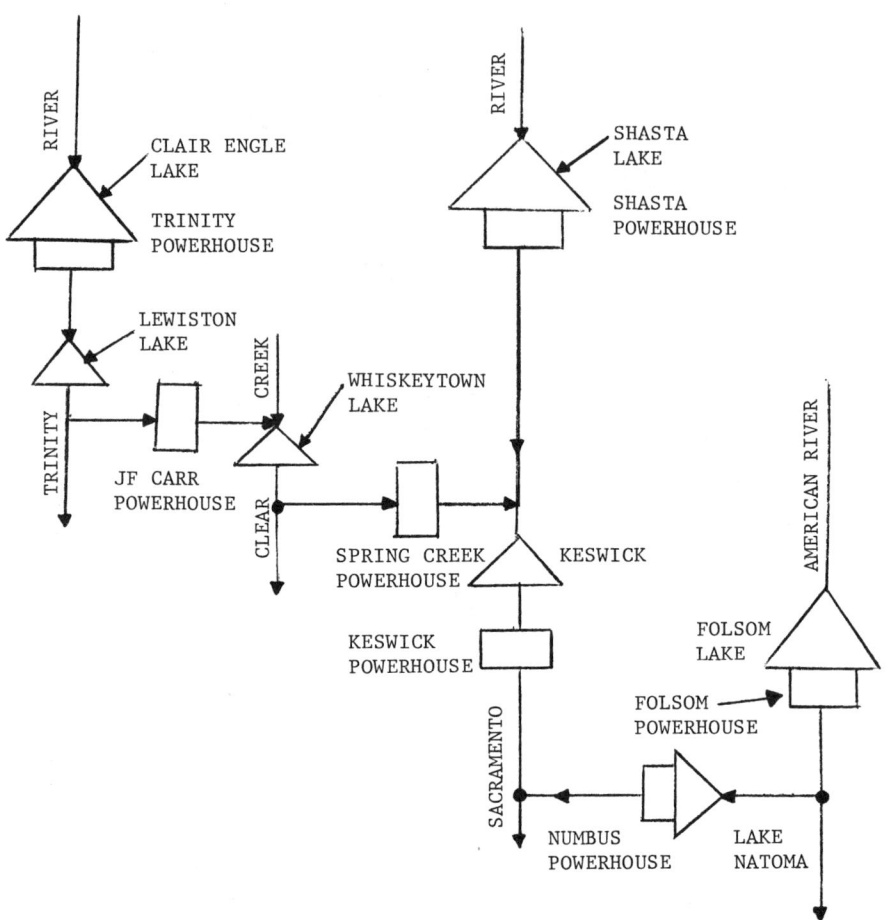

Figure 3. The California Valley Project System (Northern Portion) considered by Yeh, Becker and Chu [21]

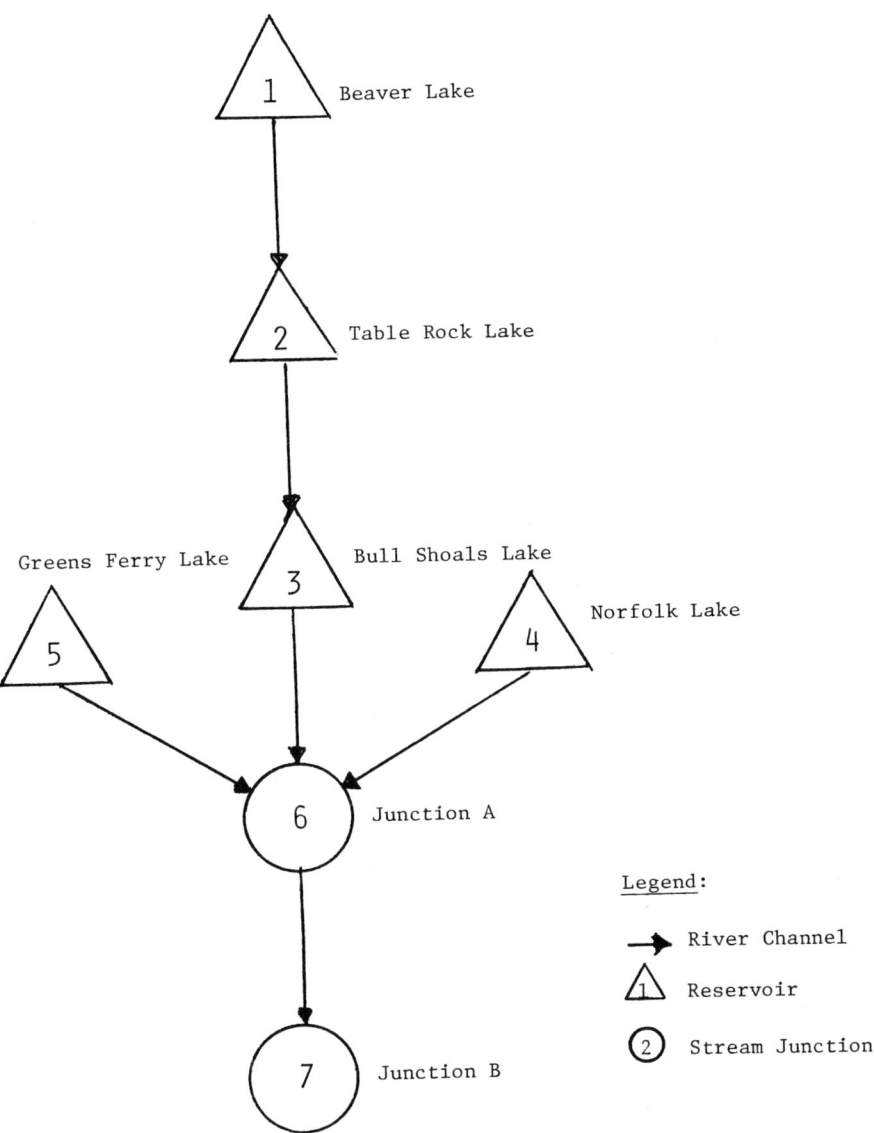

Figure 4. Network Representation of White River System Considered by Martin [14]

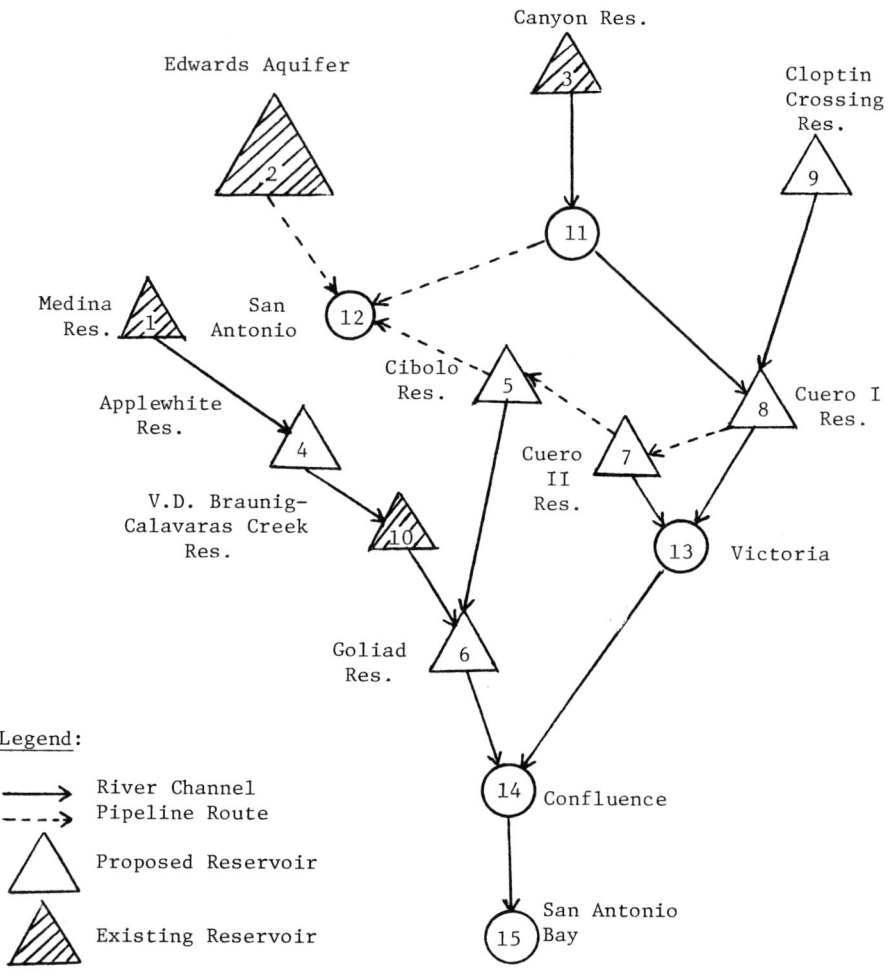

Figure 5. Network Representation of the San Antonio - Guadalupe River System Studied by Martin [14]

River Basin and the San Antonio-Guadalupe River System respectively. The first is an example of a simple converging branch system while the second illustrates a more complex multi converging branch network.

In different problem areas such as sewer systems design, but particularly least cost design of urban storm sewers, the basic problem is a classic example of a complex nonserial network. Modern optimality based approaches, a departure from simulation and heuristic methods, employ dynamic programming but usually discrete differential dynamic programming – but again of the serial variety. Tang, Mays, and Yen [8] for example, describe the use of dynamic programming in the design.

A paper of great interest and one that testifies to the need of our approach is due to Mays and Yen [15]. After pointing out the advantages of dynamic programming over other methods, they treat a multi-converging branch sewer system rightfully using a nonserial dynamic programming approach. Their method, however, has been known to be grossly inefficient [8] and thus not surprisingly was dropped in favor of a discrete differential dynamic programming (DDDP) method. Although DDDP proved to be computationally more attractive – savings in computer time, for the example treated, the accuracy and memory requirements were approximately the same. However, Maidment et al. [13] show that for large systems DDDP is superior both from the memory and time standpoints. Mays and Yen clearly point out the deficiencies of DDDP including its sometimes local optimality and other computational difficulties, especially when incorporating realistic hydraulic characteristics and configurations. They conclude that "ramification and improvement of the optimization techniques to other branched systems is highly desirable". This dilemma is partly responsible for our work in nonserial dynamic programming and the application reported here.

3. METHODOLOGY FOR OPTIMAL ANALYSIS OF CONVERGING BRANCH NONSERIAL DYNAMIC PROGRAMMING

In a converging branch system as illustrated in Figure 6, the transition and return at the junction stage, s are triple input functions given by

$$x_{s-1} = t_s(x_{01}, x_s, d_s) \tag{1}$$

$$r_s = r_s(x_{01}, x_s, d_s). \tag{2}$$

At all other stages, the usual transformations of serial dynamic programming are used for the main and branches:

$$x_{n-1} = t_n(x_n, d_n), \qquad n = 1,\ldots, N, n \neq s \tag{3}$$

$$x_{m\theta 1,1} = t_{m1}(x_{m1}, d_{m1}), \qquad m = 1,\ldots, M \tag{4}$$

In addition, the stage returns are defined as

$$r_n = r_n(x_n, d_n), \qquad n = 1,\ldots, N, n \neq s \tag{5}$$

$$r_{m1} = r_{m1}(x_{m1}, d_{m1}), \qquad m = 1,\ldots, M \tag{6}$$

Thus the optimization of a single converging branch system can be

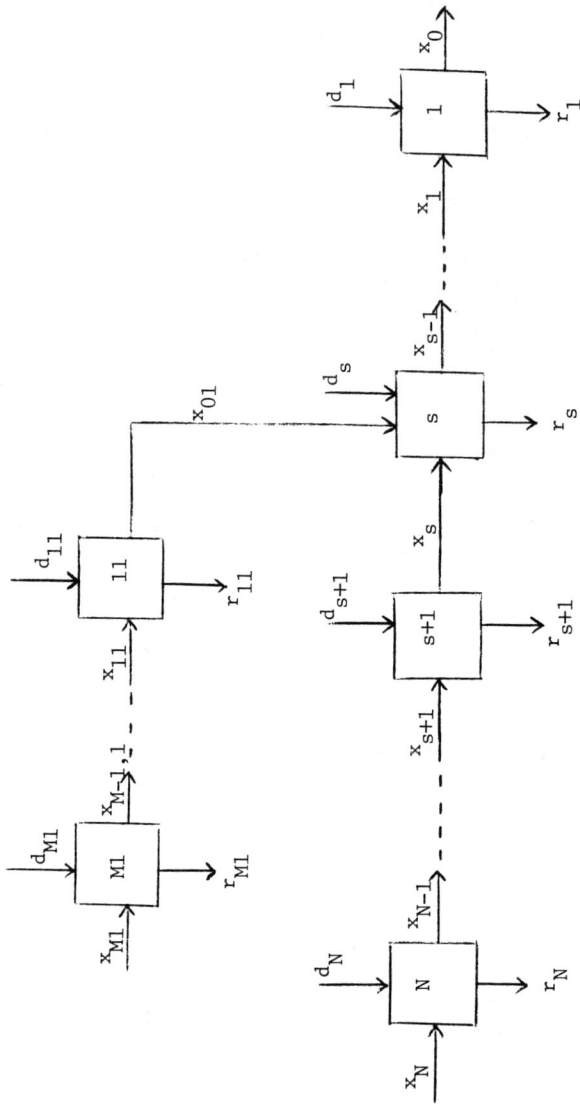

Figure 6. A Single Converging Branch System

formulated as follows:

$$\max_{\substack{d_1,\ldots,d_N \neq s \\ d_{11},\ldots,d_{M1}}} \sum_{n=1}^{N} r_n(x_n, d_n) + r_s(x_{01}, x_s, d_s) + \sum_{m=1}^{M} r_{m1}(x_{m1}, d_{m1}) \quad (7)$$

s.t.
$$x_{n-1} = t_n(x_n, d_n), \quad n = 1,\ldots, N, \; n \neq S \quad (8)$$

$$x_{s-1} = t_s(x_{01}, x_s, d_s) \quad (9)$$

$$x_{m-1,1} = t_{m1}(x_{m1}, d_{m1}), \quad m = 1,\ldots, M \quad (10)$$

$$x_n \in X_n, \; d_n \in D_n, \; x_{m1} \in X_{m1}, \; d_{m1} \in D_{m1}, \; \forall \; n, m \quad (11)$$

Note that in the foregoing formulation, at the junction stage s the stage return r_s is a function of input variables x_{01}, x_s and the stage decision d_s. Since the branch output x_{01} is a function of the branch input x_{M1} and the decisions $(d_{11},\ldots, d_{M1})$, the decision at the converging branch affects the return from the main serial process. Thus, the converging branch can not be optimized independently of the main serial process as in the case of the diverging branch nonserial network. In general, the converging branch is treated as an initial-final value problem (often termed a two-point boundary value problem); this therefore results in a two dimensional optimization problem.

The main serial system is optimized as the usual serial dynamic programming process up to stage s-1. At the junction stage s, however, the s-stage return combines the branch return $f_{M1}(x_{01}, x_{M1})$ with the (s-1)-stage optimal return $f_{s-1}(x_{s-1})$ and the return at stage s. Thus, the s-stage return function is given by

$$f_s(x_s, x_{M1}) = \max_{x_{01}, d_s} [r_s(x_{01}, x_s, d_s) + f_{s-1}(t_s(x_{01}, x_s, d_s)) + f_{M1}(x_{01}, x_{M1})] \quad (12)$$

Aris, Nemhauser and Wilde [1] suggest that x_{01} be treated as a "cut state". They choose a particular value for the branch output x_{01}, and use the boundary value optimization to obtain the optimal branch output $f_{M1}(x_{01}, x_{M1})$. Simultaneous selection of d_s would, for a given value of x_s, determine a total return in Equation (12). Once this quantity has been stored, together with the corresponding decision, new values of x_{01} and d_s can be chosen by a direct search method. Repetition of this guided search eventually gives $x_{01}^*(x_s)$ and $d_s^*(x_s)$ the optimal values of x_{01} and d_s for every value of x_s. This is not difficult to do if the branch input x_{M1} is a constant. Notice, however, that if x_{M1} is not a constant, then the branch input x_{M1} also has to be treated as another "cut state" which results in the following three-decision optimization problem:

$$f_s(x_s) = \max_{\substack{x_{01}, x_{M1} \\ d_s}} [r_s(x_{01}, x_s, d_s) + f_{s-1}(t_s(x_{01}, x_s, d_s)) + f_{M1}(x_{01}, x_{M1})] \quad (13)$$

This is clearly a difficult problem from the standpoint of computational complexity. We have nevertheless developed an algorithm in which this three-decision optimization problem has been reduced to three one-decision problems which are much easier to solve. The details of a high level computing version of this algorithm are presented in Esogbue and Warsi [10].

We now illustrate the above with an example, and then focus our attention to the development of a methodology for the analysis and design of complex multi-converging branch systems which may prove helpful in analyzing real life water resource systems.

Example 1 To illustrate consider a single converging branch system as shown in Figure 7 below with the following transition and return functions and restrictions on the variables:

$$t_n = x_n + d_n, \quad n = 1,2,4,5$$

$$t_3 = x_3 + x_{01} + d_3$$

$$t_{m1} = x_{m1} + d_{m1} \quad m = 1,2$$

$$r_n = x_n + d_n^2 \quad n = 1,2,4,5$$

$$r_3 = x_3 + x_{01} + d_3$$

$$r_{m1} = x_{m1} + d_{m1}^2 \quad m = 1,2$$

$$29 \leq x_1 \leq 38, \quad 27 \leq x_2 \leq 36, \quad 5 \leq x_3, x_4 \leq 14, \quad 0 \leq x_5 \leq 9$$

$$16 \leq x_{01} \leq 20, \quad 10 \leq x_{11}, x_{21} \leq 19$$

$$0 \leq d_n, \quad d_{m1} \leq 2$$

$$x_n, x_{m1}, d_n, d_{m1}: \text{integer}$$

Note that this simple system contains only one converging branch which consists of two stages, i.e. M = 2 while the main chain contains five stages, i.e. N = 5. The converging node s = 3. The algorithm receives input data either in terms of functions or tables.

Table 1 illustrates the computational results with the optimal input, decision and return at each stage of the system, as well as the input tables.

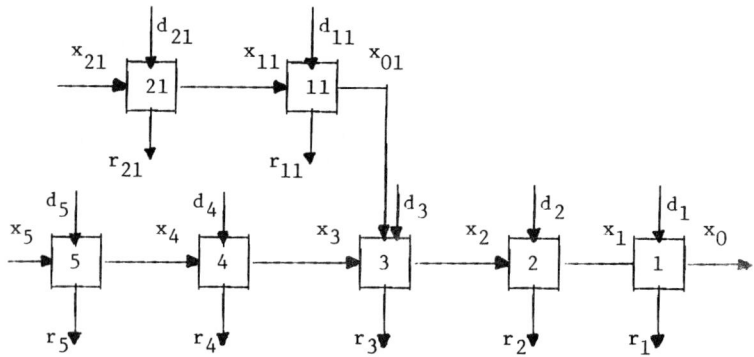

Figure 7. A Converging Branch System for Example 1

Table 1. Computer Output for the Converging Branch System of Example 1

```
DECISION TABLE FOR BRANCH 1
****************************
****************************
****************************
****************************
      2. *********************
   1.    2. ****************
   0.    1.    2. **********
****      0.    1.    2. *****
***********      0.    1.    2.
*****************      0.    1.
      2.    2.    2.    2.    2.
      2.    2.    2.    2.    2.
      2.    2.    2.    2.    2.
      2.    2.    2.    2.    2.
      2.    2.    2.    2.    2.
      1.    2.    2.    2.    2.
      0.    1.    2.    2.    2.
      2.    0.    1.    2.    2.
      2.    2.    0.    1.    0.
      2.    2.    2.    0.    0.
OPTIMAL BRANCH INPUT & RETURN OF BRANCH1
   X01       INPUT     RETURN
    16        14.        34.
    17        15.        36.
    18        16.        38.
    19        17.        40.
    20        16.        42.
"RETURN TABLE FOR MAIN PROCESS"
 33.  34.  35.  36.  37.  38.  39.  40.  41.  42.
 64.  66.  68.  70.  72.  74.  76.  78.  80.  82.
133. 136. 139. 142. 145. 148. 151. 154. 157. 160.
148. 152. 156. 160. 164. 168. 172. 176.   0.   0.
  4.   5.   6. 155. 160. 165. 170. 175. 180. 185.
"DECISION TABLE FOR MAIN PROCESS"
  2.   2.   2.   2.   2.   2.   2.   2.   2.   2.
  2.   2.   2.   2.   2.   2.   2.   2.   2.   2.
  2.   2.   2.   2.   2.   2.   2.   2.   2.   2.
  2.   2.   2.   2.   2.   2.   2.   2.   0.   0.
  2.   2.   2.   2.   2.   2.   2.   2.   2.   2.

OPTIMAL INPUT X01 FROM BRANCH 1  TO JUNCTION  #
 20.  20.  20.  20.  20.  20.  20.  20.  20.  20.

"OPTIMAL SOLUTION OF THE SYSTEM"
"STAGE  INPUT  DECIS   RETURN"
"MAIN SERIAL PROCESS"
    5       9        2      13.
    4      11        2      15.
    3      13        2      35.
    3      20
    2      35        2      39.
    1      37        2      41.

BRANCH   1  FROM JUNCTION 3
    2      16        2      20.
    1      18        2      22.
TOTAL OPTIMAL RETURN IS    185.
THE CPU TIME IN SECONDS IS  .235
```

Our algorithm is exceedingly efficient and overcomes the problems associated with previously reported ones as for example Mays and Yen. Its efficiency is exemplified when we consider a multi-converging branch system.

4. AN ALGORITHM FOR MULTI-CONVERGING BRANCH SYSTEMS

We define a multi-converging branch system as a system with more than one converging branch as shown in Figure 8. Each branch in the Figure has M_i stages and converges to a stage s_i in the main system.

When a system has multi-converging branches, it is important to carry out optimizations over branch input $x_{M_i i}$ at each corresponding stage s_i. Otherwise, $x_{M_i i}$ would have to be carried as state variables to be examined exhaustively during the optimization at some stages in the main serial system. More specifically, at each junction s_i of the i-th branch the branch input $x_{M_i i}$ has to be kept as a state variable. Thus, at the last junction s_D, the stage return function may be expressed as

$$f_{s_D}(x_{s_D}, x_{M_1 1}, \ldots, x_{M_D D}) = \max_{x_{OD}, d_{s_D}} [r_{s_D}(x_{s_D}, d_{s_D}) + f_{M_D D}(x_{M_D D}, x_{OD})$$

$$+ f_{s_D - 1}(t_{s_D}(x_{s_D}, d_{s_D}), x_{M_1 1}, \ldots, x_{M_{D-1} D-1})]. \tag{14}$$

However, when we apply the procedure developed for converging branch systems, the D+1 dimensional optimization problem can be reduced to a sequence of one dimensional problems for the main serial chain. This is a tremendous reduction of dimensionality which is very useful in applications.

In analyzing the computational demands of the multi-converging branch system, we notice that they are highly affected by the optimization order of the converging branches. If we optimize the branches first, each branch calls for the memory space to store the optimal branch return. However, by employing the method that optimizes and combines branch i when the procedure has reached the corresponding converging stage s_i in the main serial process, we can save the memory space for the optimal branch returns. The reduction effect in this multi-converging branch system is greater than in the multi-diverging branch system, since the optimization of a single converging branch is two dimensional.

Now, to generalize our analysis, it is instructive to consider a system with branches of the following two classes:

1) Class I; Branches that converge at the main serial system.

2) Class II; Branches that do not converge at the main serial system.

Figures 8 and 9 illustrate the branches of these two classes. Notice that in Figure 8, each branch converges separately and directly at the main chain. In Figure 9, however, one branch converges at C1 - a node in another branch. Also, two branches converge simultaneously at node S2 of the main chain. When a node in the main chain has branches of the two

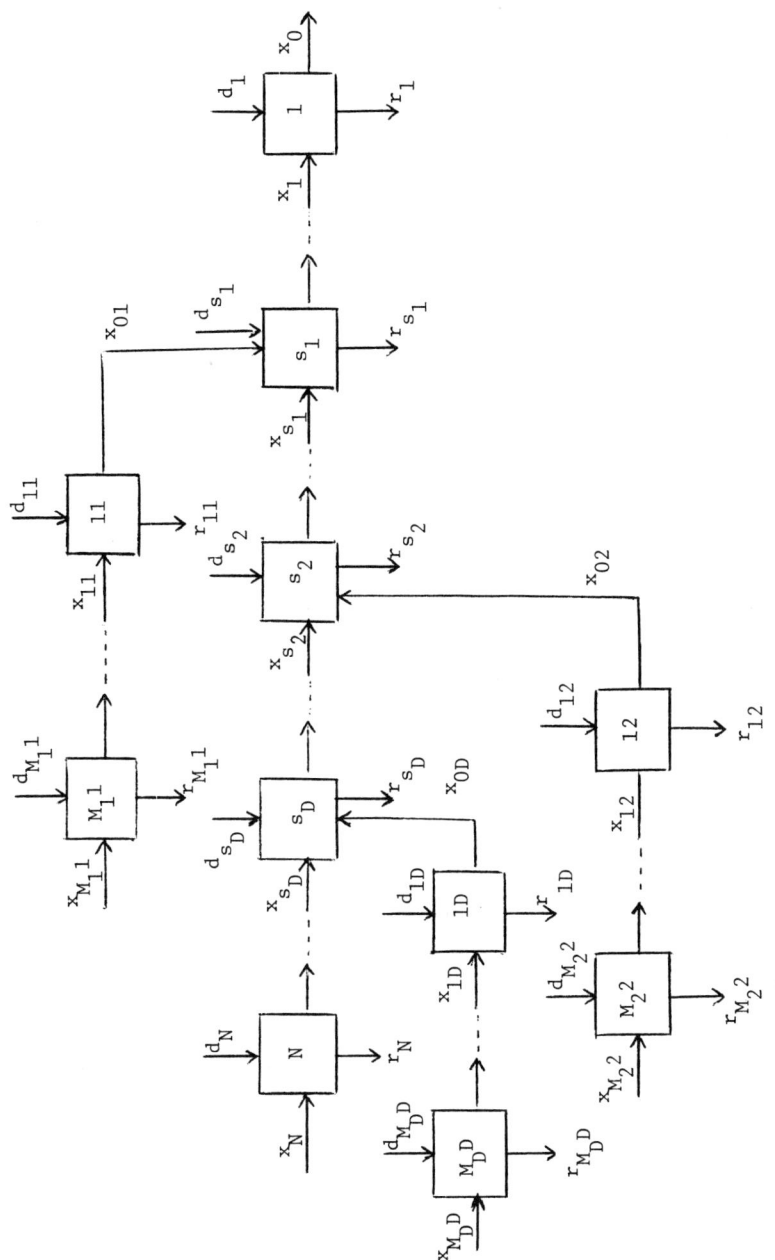

Figure 8. A Multi-converging Branch System

247

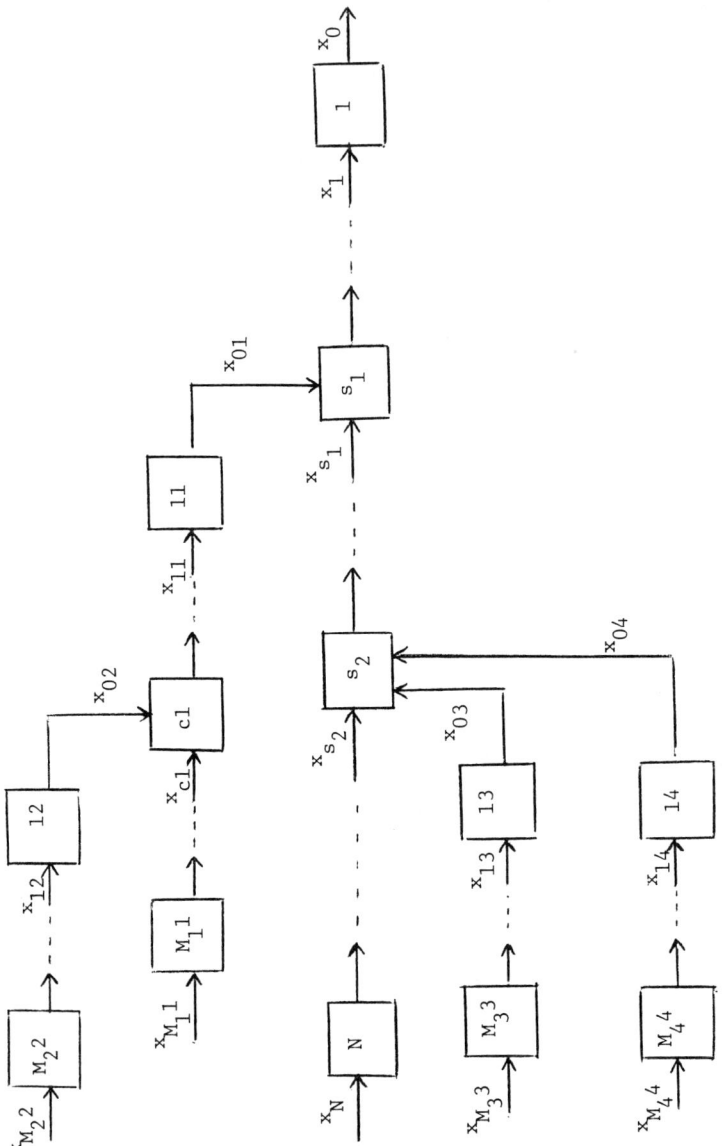

Figure 9. A Complex Multi-converging Branch System

classes, the branch of class II needs to be optimized prior to that of class I in order to reduce the computational storage demand.

We wish to present a detailed optimization procedure for processing a complex multi-converging branch system of the type illustrated in Figure 9. The flow chart of the algorithm is also presented in Figure 10. To aid in understanding our presentation we first define U_n and V_n respectively as the

> number of branches of class I that converge into stage n of the main serial process and the

> number of branches of class II that are connected to stage n of the main serial process.

4.1 Optimization of the Multi-Converging Branch System

Initialization: Let $n = 0$ and $f_n(x_n) = 0$ and go to Step 1.

Step 1: Replace n with n+1. If $n = N+1$, then stop. The optimal system return $f_N(x_N)$ is obtained. Otherwise, if stage n has any converging branch i of class I, i.e., $u_n \neq 0$, then go to Step 2. If $u_n = 0$, let $f_{0i}(x_{0i}) = 0$ and go to Step 4.

Step 2: If stage n has any branch j of class II, i.e., $v_n \neq 0$, then optimize the branches and obtain the optimal branch returns

$$f_{0j}(x_{0j}, x_{M_j i}). \text{ Let } f_{0j}(x_{0j}) = \max_{x_{M_j j}} f_{0j}(x_{0j}, x_{M_j i}).$$

Store the optimal branch inputs $x_{M_j j}^*(x_{0j})$ and go to Step 3. Otherwise, if $v_n = 0$, then let $f_{0j}(x_{0j}) = 0$ and go to Step 3.

Step 3: Optimize branches of class I by absorbing $f_{0j}(x_{0j})$ at the corresponding converging stages. Store the optimal branch output x_{0j}^*. Obtain the optimal branch returns $f_{0i}(x_{0i}, x_{M_i i})$. Let

$$f_{0i}(x_{0i}) = \max_{x_{M_i i}} f_{0i}(x_{0i}, x_{M_i i}).$$

Store the optimal branch inputs $x_{M_i i}^*(x_{0i})$ and go to Step 4.

Step 4: Obtain the optimal n-stage return of the main serial chain by absorbing all the branches as

$$f_n(x_n) = \max_{x_{0i}} [\Sigma_i f_{0i}(x_{0i}) + \max_{d_n} [r_n(x_{0i}, x_n, d_n)$$
$$+ f_{n-1}(t_n(x_{0i}, x_n, d_n))]].$$

Store the optimal branch outputs $x_{0i}^*(x_n)$ and go to Step 1.

This concludes the optimization phase of the multi-converging branch system. In the next section we show how this algorithm can be utilized to solve a complex water resources problem.

Figure 10. Flow Chart of the Multi-Converging Branch Algorithm

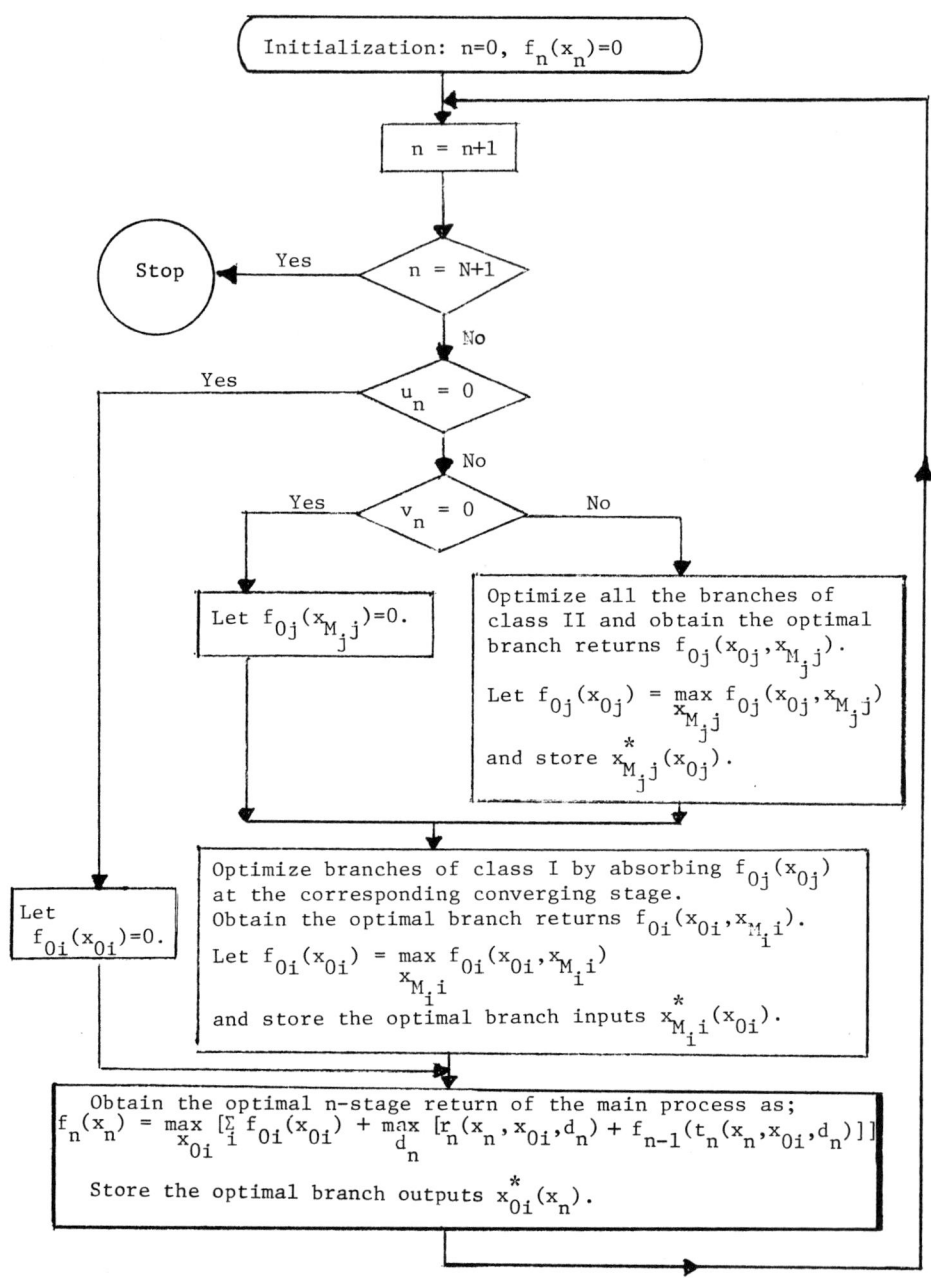

5. OPTIMAL ANALYSIS OF BRANCHED SEWER SYSTEMS VIA THE MULTI CONVERGING BRANCH ALGORITHM

The problems of designing the size and slope of sewer pipes in a drainage system can be solved by employing the optimization procedure for the multi-converging branch systems. Mays and Yen [15] consider a sewer system illustrated in Figure 11 where many branches converge into a main stream. The main stream consists of 9 sewer pipes and 10 manholes. Connected to the main system are four branches that converge into the main stream at manholes 3, 4, 5 and 6 respectively.

The problem is to determine the optimum pipe size and the elevation of upstream and downstream of each pipe such that the installation cost for the sewer system is minimized under several constraints. Mays and Yen solved this problem by applying discrete differential dynamic programming. They change the state space in each iteration by improving the trial trajectory (the sequence of states for different stages). The conventional dynamic programming approach is used within the neighborhood states of the previous trial trajectory. The procedure terminates when the increment of the state is less than a predetermined value.

Following Mays and Yen's problem formulation, we now solve the branched sewer system by applying the optimization algorithm developed in Section 4. First, the following assumptions and constraints are used.

1. Any size of the diameter of the sewer pipe is available that is computed by using the Manning's formula (see equation (11) of [15].)

2. The diameter of a pipe can not be less than that of the pipe preceding it.

3. The upstream elevation of a pipe is equal to the lowest downstream of the pipes preceding it.

4. A minimum soil cover depth of 8 feet above the crown of each pipe is assumed.

To solve the problem we also define the following dynamic programming variables:

<u>Stages</u> : The stages are analogous to the ordering of manholes in the main stream and the branches, i.e.,

$$n = 1,\ldots, 9$$

$$m = 1,\ldots, M_i, \quad i = 1,\ldots, 4$$

where $M_1 = M_2 = M_3 = 3$, and $M_4 = 2$.

<u>States</u> : The state x_n (or x_{mi}) at manhole n (or mi) is analogous to the crown elevation of the pipes connected to the manhole.

<u>Decisions</u> : The decision d_n (or d_{mi}) is the drop of the pipe across the stage.

<u>Returns</u> : The return r_n (or r_{mi}) is the cost of the installation of the manhole n (or mi) and the pipe connecting it to the manhole n-1 (or m-1, i).

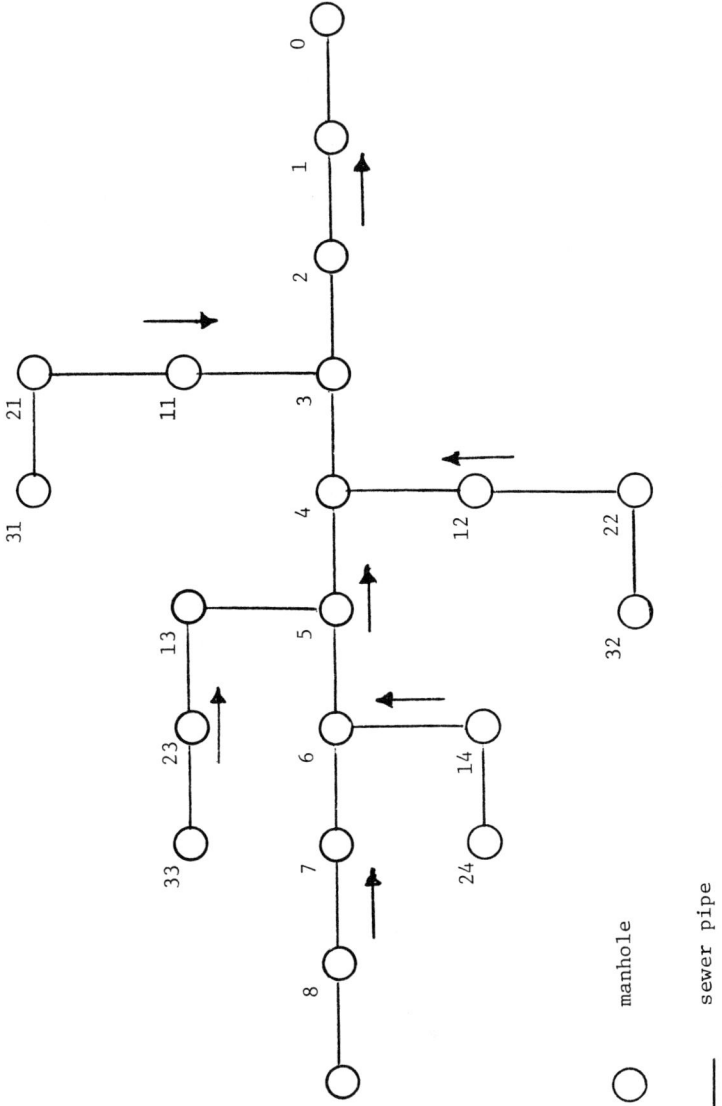

Figure 11. An Example of Sewer Systems

252

Transitions : The transition function is defined as

$$x_{n-1} = x_n - d_n, \quad n = 1, \ldots, 9$$

$$x_{m-1,i} = x_{mi} - d_{mi}, \quad m = 1, \ldots, M_i, \; i = 1, \ldots, 4$$

6. DISCUSSION OF RESULTS

Now by applying the multi-converging branch algorithm developed in Section 4 (see Figure 10 for the functional equation at each stage) we optimize the system from manhole 1 through manhole 9 in the main stream. The optimum return of each branch is computed and combined to the main system at the corresponding junction manhole. Table 2 illustrates the computational results for the problem. At each stage, 11 discretizations were used for the input elevations of each pipe. The physical data and cost functions given in [15] were used to determine the optimal diameter and the slope of each sewer pipe. The optimal solution is obtained with the outlet elevation of the system to be at 435 feet. Due to the several different assumptions and constraints used, the upstream and downstream elevations of the pipes are slightly different from the solution given by Mays and Yen.

The computational complexity (both space and time complexity) of the two approaches, however, differed by more than 75%. The multi-converging branch algorithm required 19723 elementary operations with 11 discretizations of the state variables. See Esogbue and Warsi [10] for computational complexity analysis of converging branch systems. For comparative purposes, consider the discrete differential dynamic programming approach which uses 5 discretizations at each stage in each iteration. The recursive equations would require three additions and one comparison at each junction and two additions and one comparison at all other stages. The total number of operations results in 34925, an astronomically higher number than our nonserial dynamic programming approach. The computer time requirement of the two approaches were also examined. The multi-converging branch approach required a total processing time (compilation time + execution time) 20.5 CPU seconds (CYBER 855) while the discrete differential dynamic programming approach required 28.2 ∼ 43.3 CPU seconds (IBM 360.75). The minimum cost solution indeed involved eleven iterations and a total processing time of 43.3 seconds for the DDDP approach while the inefficient DP approach took 113.7 seconds.

From the above results we conclude that the computational demands of the multi-converging branch algorithm is much less than the discrete differential dynamic programming approach, which is currently used in practice. Further, the computational superiority of our algorithm becomes more impressive when solving higher dimensional (more branches and nodes per branch and main chain) and more complex (the structure of convergence as illustrated in Figure 9) systems. Finally, global optimality is assured in all cases and the application is not restricted to special cost functions nor specially structured hydraulic systems as in the discrete deterministic dynamic programming model.

Table 2 . Computer Output of the Problem for the Sewer System

```
OPTIMAL BRANCH INPUT & RETURN OF BRANCH 1

   XO1         INPUT        RETURN
  447.00       460.00        15837.
  446.80       460.00        15825.
  446.60       460.00        15815.
  446.40       460.00        15806.
  446.20       460.00        15799.
  446.00       460.00        15794.
  445.80       460.00        15790.
  445.60       460.00        15786.
  445.40       460.00        15804.
  445.20       460.00        15836.
  445.00       460.00        15868.

OPTIMAL BRANCH INPUT & RETURN OF BRANCH 2

   XO2         INPUT        RETURN
  457.00       477.00        22911.
  456.80       477.00        22885.
  456.60       477.00        22861.
  456.40       477.00        22861.
  456.20       477.00        22881.
  456.00       477.00        22901.
  455.80       477.00        22924.
  455.60       477.00        22948.
  455.40       477.00        22973.
  455.20       477.00        23000.
  455.00       477.00        23028.

OPTIMAL BRANCH INPUT & RETURN OF BRANCH 3

   XO3         INPUT        RETURN
  462.00       482.00        23441.
  461.80       482.00        23419.
  461.60       482.00        23400.
  461.40       482.00        23382.
  461.20       482.00        23367.
  461.00       482.00        23389.
  460.80       482.00        23413.
  460.60       482.00        23439.
  460.40       482.00        23466.
  460.20       482.00        23495.
  460.00       482.00        23524.

OPTIMAL BRANCH INPUT & RETURN OF BRANCH 4

   XO4         INPUT        RETURN
  472.00       482.00        12367.
  471.80       482.00        12353.
  471.60       482.00        12341.
  471.40       482.00        12331.
  471.20       482.00        12322.
  471.00       482.00        12316.
  470.80       482.00        12310.
  470.60       482.00        12307.
  470.40       482.00        12326.
  470.20       482.00        12365.
  470.00       482.00        12405.

"STAGE RETURN AT EACH STAGE OF MAIN PROCESS"
```

```
AT STAGE 1
   37528.74    38394.02    39288.22    40214.15    41175.04    42174.63    43217.29    44308.11
   45453.16    46659.64    47936.30
AT STAGE 2
   68695.39    69779.50    70937.79    72183.35    73532.93    74891.78    76279.62    77699.27
   79153.97    80629.76    82123.32
AT STAGE 3
  104441.86   105076.27   105741.41   106440.77   107178.56   107959.90   108791.09   109679.95
  110638.28   111675.75   112809.01
AT STAGE 4
  148489.61   148710.24   148936.27   149188.91   149444.80   149704.09   149966.94   150233.52
  150504.03   150778.64   151057.59
AT STAGE 5
  188045.28   188285.84   188536.82   188798.85   189072.67   189372.88   189685.26   190011.25
  190352.55   190711.20   191089.64
AT STAGE 6
  212424.59   212500.84   212580.43   212726.15   212887.26   213051.97   213220.26   213392.09
  213569.66   213749.50   213931.72
AT STAGE 7
  218012.27   218071.40   218131.60   218192.94   218255.49   218319.34   218384.58   218451.31
  218557.32   218666.88   218778.85
AT STAGE 8
  223826.38   223886.95   223948.38   224010.72   224074.01   224138.33   224203.72   224270.25
  224337.99   224440.48   224552.56
AT STAGE 9
  228315.70   228377.86   228441.67   228507.29   228574.89   228644.67   228716.87   228791.77
  228869.71   228951.07   229060.20

"DECISION TABLE FOR MAIN PROCESS"

AT STAGE 1
    5.00         4.80        4.60        4.40        4.20        4.00        3.80        3.60
    3.40         3.20        3.00

AT STAGE 2
    3.00         2.80        2.60        2.40        2.20        2.20        2.20        2.20
    2.20         2.00        2.00

AT STAGE 3
    4.00         3.80        3.60        3.40        3.20        3.00        2.80        2.60
    2.40         2.20        2.00

AT STAGE 4
   10.00         9.80        9.60        9.40        9.20        9.00        8.80        8.60
    8.40         8.20        8.00

AT STAGE 5
    5.00         4.80        4.60        4.40        4.20        4.00        3.80        3.60
    3.40         3.20        3.00

AT STAGE 6
   10.00         9.80        9.60        9.40        9.20        9.00        8.80        8.60
    8.40         8.20        8.00

AT STAGE 7
```

```
              7.00      6.80      6.60      6.40      6.20      6.00      5.80      5.60
              5.40      5.20      5.00

AT STAGE 8
              8.00      7.80      7.60      7.40      7.20      7.00      6.80      6.60
              6.40      6.20      6.00

AT STAGE 9
              5.00      4.80      4.60      4.40      4.20      4.00      3.80      3.60
              3.40      3.20      3.00

OPTIMAL INPUT  XO1   FROM BRANCH  1   TO JUNCTION  3

            447.00    446.80    446.60    446.40    446.20    446.00    445.80    445.60
            445.60    445.60    445.60

OPTIMAL INPUT  XO2   FROM BRANCH  2   TO JUNCTION  4

            457.00    456.80    456.60    456.60    456.60    456.60    456.60    456.60
            456.60    456.60    456.60

OPTIMAL INPUT  XO3   FROM BRANCH  3   TO JUNCTION  5

            462.00    461.80    461.60    461.40    461.20    461.20    461.20    461.20
            461.20    461.20    461.20

OPTIMAL INPUT  XO4   FROM BRANCH  4   TO JUNCTION  6

            472.00    471.80    471.60    471.40    471.20    471.00    470.80    470.60
            470.60    470.60    470.60

"OPTIMAL SOLUTION OF THE SYSTEM"

"STAGE  PIPE-DIAM  UP-ELEV  DECIS  DOWN-ELEV"

"MAIN SERIAL PROCESS"
    9       .98      492.00    5.00    487.00
    8      1.13      487.00    8.00    479.00
    7      1.24      479.00    7.00    472.00
    6      1.77      472.00   10.00    462.00
    5      2.57      462.00    5.00    457.00
    4      2.79      457.00   10.00    447.00
    3      3.31      447.00    4.00    443.00
    2      3.68      443.00    3.00    440.00
    1      3.68      440.00    5.00    435.00

BRANCH  1   TO JUNCTION   3
    3      1.04      460.00    4.00    456.00
    2      1.15      456.00    4.00    452.00
    1      1.32      452.00    5.00    447.00

BRANCH  2   TO JUNCTION   4
    3      1.24      477.00   10.00    467.00
    2      1.64      467.00    5.00    462.00
    1      1.79      462.00    5.00    457.00

BRANCH  3   TO JUNCTION   5
    3      1.35      482.00    5.00    477.00
    2      1.36      477.00   10.00    467.00
    1      1.64      467.00    5.00    462.00

BRANCH  4   TO JUNCTION   6
    2      1.00      482.00    5.00    477.00
    1      1.32      477.00    5.00    472.00



TOTAL OPTIMAL RETURN IS    231418.

THE CPU TIME IN SECONDS IS      8.659
```

REFERENCES

1. Aris, R., G.L. Nemhauser and D.J. Wilde, "Optimization of Multistage Cycle and Branching Systems by Serial Procedures," J. Am. Inst. Chem. Eng., Vol. 10, No. 6, 1964, pp. 913-919.
2. Becker, L. and W.G. Yeh, "Optimization of Real Time Operation of a Multiple Reservoir System," Water Resources Research, Vol. 10, 1974, pp. 1107-1112.
3. Becker, L., W.G. Yeh, D. Fults, and D. Sparks, "Operations Models for Central Valley Project," ASCE J. Water Resources Planning and Management Division, Vol. 102, 1976, pp. 101-115.
4. Beightler, C.S. and William Meier, "Branch Compression and Absorption in Nonserial Multistage Systems," Journal of Mathematical Analysis and Applications, Vol. 21, 1968, pp. 426-430.
5. Bellman, R.E. and S.E. Dreyfus, Applied Dynamic Programming, Princeton Univ. Press, New Jersey, 1962.
6. Bertele, Umberto and Francesco Brioschi, Nonserial Dynamic Programming, Academic Press, New York, 1973.
7. Chow, V.T., D.R. Maidment, and G.W. Tauxe, "Computer Time and Memory Requirements for DP and DDDP in Water Resources Systems Analysis," Water Resources Research, Vol. 11, No. 5, 1975, pp. 621-628.
8. Esogbue, A.O. and Barry Marks, "Nonserial Dynamic Programming - A Survey," Operational Research Quarterly, Vol. 25, No. 2, 1974.
9. Esogbue, A.O. and Barry Marks, "Dynamic Programming Models of the Nonserial Critical Path-Cost Problem, Management Science, Vol. 24, No. 2, 1977, pp. 200-209.
10. Esogbue, A.O., "Dynamic Programming Algorithms and Analyses for Nonserial Networks," Technical Report, J-83-3, School of Industrial and Systems Engineering, Georgia Institute of Technology, Atlanta, Georgia.
11. Esogbue, A.O. and N.A. Warsi, "A High-Level Algorithm for Diverging and Converging Branch Nonserial Dynamic Programming Systems," Journal of Computers and Mathematics with Applications, Vol. 12A, No. 6, 1986, pp. 719-732.
12. Hall, W.A. and R.W.L. Shepard, "Optimum Operations for Planning of a Complex Water Resources System," Contribution No. 122, UCLA Water Resources Center, Los Angeles, 1967.
13. Larson, R.E. and W.G. Keckler, "Applications of Dynamic Programming to Water Resources Systems," Proceedings of IFAC Haifa Conference on Computer Control of Natural Resources and Public Utilities, Haifa, Israel, 1967.
14. Martin, Q., "Minimum Cost Capacity Expansion of a Linear Water Conveyance Pipeline," ORSA/TIMS Bulletin, No. 3, San Francisco, California, May 1977.
15. Mays, L.W. and B.C. Yen, "Optimal Cost Design of Branched Sewer Systems," Water Resources Research, Vol. 11, No. 1, 1975, pp. 37-47.
16. Mays, L. and H. Wenzel, "Optimal Design of Multilevel Branching Sewer Systems," Water Resources Research, Vol. 12, No. 5, 1976, pp. 913-917.
17. Nemhauser, G.L., Introduction to Dynamic Programming, John Wiley and Sons, New York, 1967.
18. Parker, M.W. and R.M. Crisp, "Decomposition of Converging Branch Multistage Systems," AIIE Transactions, Vol. 2, 1970, pp. 185-190.
19. Tang, W.H., L.W. Mays, and B.C. Yen, "Optimal Risk Based Design of Stormed Sewer Networks," Paper presented at 45th Joint National Meeting of TIMS and ORSA, Boston, Mass., April 22-24, 1974.

20. Wong, Peter and E. Larson, "Optimization of Tree Structured Natural Gas Transmission Networks," <u>Journal of Mathematical Analysis and Applications</u>, Vol. 24, 1968, pp. 613-626.
21. Yakowitz, S., "Dynamic Programming Applications in Water Resources," <u>Water Resources Research</u>, Vol. 18, No. 4, 1982, pp. 673-696.
22. Yeh, W.G., L. Becker and W.S. Chu, "Optimization of Real-Time Hourly Operations of a Complex, Multiple Purpose Reservoir System," UCLA Engineering Report No. UCLA-ENG. 7807, Los Angeles, California, 1978.
23. Yeh, W.G., L. Becker and W.S. Chu, "Real Time Hourly Reservoir Operation," <u>ASCE, J. Water Resources Planning and Management Division</u>, Vol. 105, 1978, pp 187-203.

PART FOUR

WATER QUALITY, WASTE TREATMENT AND FLOOD CONTROL SYSTEMS

IMPROVING WATER QUALITY BY OPTIMAL AERATION CONTROL VIA DYNAMIC PROGRAMMING

HIROSHI SUGIYAMA
Professor Emeritus of Mathematical Engineering
Osaka University
Osaka, Japan

ABSTRACT

The problem of improving water quality via aeration control at certain fixed points of a slow river stream is addressed in this paper. The biochemical oxygen demand (BOD) and dissolved oxygen (DO) at time t are denoted by B(t) and D(t) respectively and we consider (B(t), D(t)) as the system state. Criterion functional J(u) is defined by an integral over a specified time interval, where the integrand is weighted sum of squares of B(t), DO deficit, and control u(t). By minimizing J(u) via dynamic programming (DP) subject to a set of differential equations relating the system state and control u(t), we wish to obtain the optimal aeration control $u^*(t)$ numerically, starting from a specified state (B(o), D(o)). By the mathematical formulation stated above, we are able to achieve objectives of decreasing BOD level and increasing DO level simultaneously, taking aeration cost into our account. In order to solve this problem in a feasible way, we discretized our process allowing t to take on discrete values $0, \Delta, 2\Delta, \ldots$ Then, replacing u(t) by $u_0, u_1, u_2, \ldots$ and solving our differential equations successively by step-by-step method, our discretized DP formulation is set up for minimizing the value of discretized criterion function J_n corresponding to the criterion functional J(u). An efficient and feasible policy is devised for solving these DP equations and practical solution algorithm is obtained. Thus computations required can be readily performed even via hand held programmable calculators. Finally, application of Box's hill-climbing method is first exemplified in our simple examples for obtaining approximate minimum value of our DP return functions. This approach in general, together with extended Kiefer-Wolfowitz stochastic approximation, will be useful for locating global minimum or maximum of DP return functions fairly acculately, reducing the dimension of DP computations effectively. The author considers that all the methods proposed in this paper are viable and conducive to our practical purposes.

[Keywords] BOD, DO, Water Quality, Optimal Aeration Control, Criterion functional, Dynamic Programming, System State, Discretized State Space, Box's Hill-climbing Method.

1. THE WATER QUALITY PROBLEM

A considerable portion of the water resources literature has been devoted to water quality problems and the remedial strategies for improving water quality of polluted river water, from the environmental, technical, and mathematical perspectives.

The problem of improving polluted water quality by artificially induced aeration, introducing aerobiotic micro-organisms into the polluted water and thus activating oxidation functioning of micro-organisms, became of interest to a number of researchers including the auther himself.

Since the construction of aeration devices in certain estuary portion of a river inside Osaka City area in 1979, observations have been made of water quality based upon the measurement data of *BOD* and *DO* at this point of the river, and comparisons have been made intensively between the states of water quality before and after the installation of aeration devices. It has been reported that striking improvement has been realized by the application of aeration.

Specifically, it was observed that the *BOD* level decreased, the *DO* level increased and much of the fish population that had dwindled witnessed a considerable growth. [12]

This report is rather impressive with the result that the optimal aeration control for improving water quality became a very important problem requiring the mathematical methods such as the ones suggested in this paper for solving such aeration control problems.

Our contribution is the development of an efficient and practical mathematical method for solving such problems.

2. SYSTEM STATE AND CONTROL

We consider the problem of controlling the aeration rate at a very short local portion of a polluted river of slow stream, e.g. the river stream which is artificially damed between tidal cycles like the case inside Osaka City area, mentioned earlier.

Actually, we denote the *BOD* and *DO* of a river or its estuary as $B(t, x)$ and $D(t, x)$, taking x (the location parameter along the river stream) into our account, as in the mathematical model due to Hullett. [11]

In this paper, however, we are concerned with *BOD* and *DO* at $x = x_0$ (actually, certain neighborhood of fixed point x_0), and thus we denote *BOD* and *DO* by $B(t)$ and $D(t)$ respectively, dropping x_0 out of $B(t, x_0)$ and $D(t, x_0)$ in the subsequent considerations of this paper.

Now, we assume the state $(B(t), D(t))$ is governed by the following set of differential equations at the location $x = x_0$ of a river, i.e.

$$(2.1) \quad \frac{dB(t)}{dt} = -k_1 B(t) - r D(t) + G,$$

$$(2.2) \quad \frac{dD(t)}{dt} = k_2 (D_s - D(t)) - k_1 B(t) + u(t),$$

where r is *DO* decay rate in (2.1) and D_s is assumed to be a fixed saturation level of dissolved oxygen for specified environmental conditions, and G is assumed to be a positive constant which denote the instream pollutant discharge, and $u(t)$ is our aeration rate control.

In the foregoing, k_1 is the *BOD* decay rate and k_2 is the atmospheric aeration rate.

If we put $r = 0$, this model, described by (2.1) and (2.2), coincides whith the Hullett model (1974) when $x = x_0$ [3]

3. CRITERION FUNCTIONAL AND FORMULATION OF THE AERATION CONTROL PROBLEM

Keeping the context which have been explained in the foregoing sections, we assume the following quadratic type of criterion functional for our optimal aeration control; i.e.

$$(3.1) \quad J\{u(t)\} = \int_0^T [b\{B(t)\}^2 + \{D_s - D(t)\}^2 + \lambda \{u(t)\}^2] dt,$$

where b and λ are positive weight parameters, and we write as,

$$(3.2) \quad J\{u(t)\} = b \int_0^T B^2(t) dt + \int_0^T \{D_s - D(t)\}^2 dt + \lambda \int_0^T u^2(t) dt.$$

If we put $b = 0$, the above criterion functional coincides with that of Hullett again, at $x = x_0$.

Specifically, λ denotes the weight of aeration energy cost, and the integrand of the second term of (3.2) is the squared value of oxygen deficit.

Selecting b and λ in an appropriate way, we are able to optimize our objectives of decreasing *BOD*, increasing *DO*, with our aeration cost, by minimizing criterion functional (3.1), over the time interal $[0, T]$. In addition, $B(t)$, $D(t)$, and $u(t)$ are assumed to be continuous scalar functions in this continuous modelling, and thus the integrand of (3.1) is the function of $L^2[0, T]$.

From the mathematical model established so for, our optimal aeration control is reduced to the problem of deriving $u(t)$ which minimizes $J\{u(t)\}$ in (3.1), subject to the differential equations (2.1) and (2.2), starting from a specified initial state $(B(o), D(o))$.

To this end, we wish to solve the differential equations (2.1) and (2.2) by step-by-step method, using an appropriate constant mesh size $\Delta t \triangleq \Delta$, starting from the initial state $(B(o), D(o))$, replacing $u(t)$ by $u_0, u_1, \ldots, u_{n-1}$. For this purpose, we have to replace the differential equations (2.1) and (2.2) by the corresponding differential difference equations. These differential difference equations are the following.

$$(3.3) \quad B(n\Delta) = B(n\Delta - \Delta) + \{G - k_1 B(n\Delta - \Delta) - rD(n\Delta - \Delta)\}\Delta,$$

$$(3.4) \quad D(n\Delta) = D(n\Delta - \Delta) + \{k_2(D_s - D(n\Delta - \Delta)) - k_1 B(n\Delta - \Delta) + u_{n-1}\}\Delta.$$

On the other hand, our criterion functional is also required to be discretized. We assume $T = N\Delta$. Then, $J\{u(t)\}$ is decomposed as,

(3.5) $\int_0^T [bB^2(t) + \{D_s - D(t)\}^2 + \lambda u^2(t)] dt$

$= \sum_{n=1}^{N} \int_{(n-1)\Delta}^{n\Delta} [\ldots] dt$

$= b \sum_{n=1}^{N} \int_{(n-1)\Delta}^{n\Delta} B^2(t) dt + \sum_{n=1}^{N} \int_{(n-1)\Delta}^{n\Delta} \{D_s - D(t)\}^2 dt$

$+ \lambda \sum_{n=1}^{N} \int_{(n-1)\Delta}^{n\Delta} u^2(t) dt.$

Now, t is allowed to take only discrete values $0, \Delta, 2\Delta, \ldots, N\Delta$. Then, our optimal control problem is reduced to choosing a sequence of scalar values $u_n \triangleq u(n\Delta)$, $n = 0, 1, \ldots, N - 1$, which minimizes the following function $J_N(\{u_n\})$, instead of choosing a continuous function $u(t)$ which minimizes the previous criterion functional (3.1), starting from $(B(0), D(0))$. Now, we define as follows, for $n = 1, 2, \ldots, N$.

(3.6) $J_n(\{u_k\}) \triangleq \sum_{k=1}^{n} \Delta_k J(u_{k-1}|u_0, u_1, \ldots, u_{k-2})$, where $u_k \triangleq 0$ if $k < 0$, and

(3.7) $\Delta_k J(u_{k-1}|u_0, u_1, \ldots, u_{k-2})$

$\triangleq [b\{B(k\Delta)\}^2 + \{D_s - D(k\Delta)\}^2 + \lambda \{u_{k-1}\}^2]\Delta.$

Note that the initial state $(B(0), D(0))$ is steered to the subsequent state $(B(\Delta), D(\Delta))$ by the first choice of aeration rate u_0 operated upon the initial state $(B(0), D(0))$, and the state $(B(\Delta), D(\Delta))$ is steered to the next state $(B(2\Delta), D(2\Delta))$ by the second choice of aeration rate u_1 operated upon $(B(\Delta), D(\Delta))$, and so on, and, in general, the state $(B((k-1)\Delta), D((k-1)\Delta))$ is steered to the next state $(B(k\Delta), D(k\Delta))$ by the aeration rate u_{n-1} operated upon the prior state $(B((k-1)\Delta), D((k-1)\Delta))$, until at last the state $(B((N-1)\Delta), D((N-1)\Delta))$ is steered to the terminal state $(B(N\Delta), D(N\Delta))$ by u_{n-1}, governed by the set of differential difference equations (3.3) and (3.4).

Now, we wish to determine the optimal sequence of aeration rates $\{u_k^*\}$, $k = 0, 1, 2, \ldots, n-1$, which minimizes $J_N(\{u_k\})$ via dynamic programming in the following way. First, we have to notice the sequence of $B(n\Delta)$ and $D(n\Delta)$ are concretely expressed by the linear combination of $u_0, u_1, \ldots$, as in the following, due to the recurrence relationships described by (3.3) and (3.4), successively. For example, in the case where $\Delta = 1$ for simplicity, we have the following expressions.

(3.8) $B(1) = B_0(1)$
$D(1) = D_0(1) + u_0,$

(3.9) $B(2) = B_0(2) - ru_0$
$D(2) = D_0(2) + (1-k_2)u_0 + u_1$

(3.10) $B(3) = B_0(3) - u(2-k_1-k_2)u_0 - ru_1$
$D(3) = D_0(3) + \{(1-k_2)^2 + rk_1\}u_0 + (1-k_2)u_1 + u_2,$

and so on. In addition, the procedure to rule out the respective coefficient is simple and thus these expressions can be easily obtained by the computer computation machinery. Here, $B_0(n)$ and $D_0(n)$ denote the values of $B(n)$ and $D(n)$ when $u_n \equiv 0$ is applied for all n.

Now, let us define

(3.11) $\quad f_n(C_1, C_2) \triangleq \min_{\{u_0,\ldots,u_{n-1}\}} J_n(\{u_k\})$, starting from state (C_1, C_2).

Then, by the principle of optimality, there hold the following relations.

(3.12) $\quad f_N(B(0), D(0)) = \min_{u_0} [\Delta_1 J(u_0) + f_{N-1}(B(\Delta), D(\Delta)|u_0)]$

$= \min_{\{u_0, u_1\}} [\Delta_1 J(u_0) + \Delta_2 J(u_1|u_0) + f_{N-2}(B(2\Delta), D(2\Delta)|u_0, u_1)]$,

..., and so on.

If we write here, using notations $J_n(\ldots)$

(3.13) $\quad \Delta_1 J(u_0) = J_1(u_0)$.

(3.14) $\quad \Delta_1 J(u_0) + \Delta_2 J(u_1|u_0) = J_2(u_0, u_1), \ldots$

(3.15) $\quad \Delta_1 J(u_0) + \Delta_2 J(u_1|u_0) + \Delta_3 J(u_2|u_0, u_1) + \ldots$

$+ \Delta_n J(u_{n-1}|u_0, u_1, \ldots, u_{n-2}) = J_n(u_0, u_1, \ldots, u_{n-1})$,

in general, assuming $n \geq 2$, (3.12) is rewritten as,

(3.16) $\quad f_N(B(o), D(o)) = \min_{u_0} [J_1(u_0) + f_{N-1}(B(\Delta), D(\Delta)|u_0)]$

$= \min_{\{u_0, u_1\}} [J_2(u_0, u_1) + f_{N-2}(B(2\Delta), D(2\Delta)|u_0, u_1)], \ldots$,

(3.17) $\quad f_N(B(o), D(o))$

$= \min_{\{u_0,\ldots,u_{N-2}\}} [J_{N-1}(u_0, u_1, \ldots, u_{N-2}) + f_1(B((N-1)\Delta), D((N-1)\Delta))]$, where

$f_1(B((N-1)\Delta), D((N-1)\Delta))$

$= f_1(B((N-1)\Delta), D((N-1)\Delta))|u_0, u_1, \ldots, u_{N-2})$

$= \min_{u_{N-1}} \Delta_N J(u_{N-1}|u_0, u_1, \ldots, u_{N-2})$

$= \min_{u_{N-1}} [b\{B((N)\Delta)\}^2 + \{D_s - D((N)\Delta)\}^2 + \lambda u_{N-1}^2]$

(3.18) $\quad = \min_{u_{N-1}} [(\lambda+1) u_{N-1}^2 - R_{1, N-1}(u_0, u_1, \ldots, u_{N-2}) u_{N-1}$

$+ R_{2, N-1}(u_0, u_1, \ldots, u_{N-2})]$,

where $R_{1, N-1}(u_0, u_1, \ldots u_{N-2})$ and $R_{2, N-1}(u_0, u_1, \ldots, u_{N-2})$ are quadratic functions in respective u_k, $k = 0, 1, \ldots, N-2$, but free from u_{N-1}.

In addition $R_{1, N-1}$ and $R_{2, N-1}$ involves parameters $k_1, k_2, D_s, r, \lambda$. Anyway, by the above minimization of the value inside the bracket with respect to u_{N-1}, we obtain the following results immediately.

(3.19) $\quad f_1(B((N-1)\Delta), D((N-1)\Delta)$

$= R_{2, N-1}(u_0, u_1, \ldots, u_{N-2}) - \dfrac{\{R_{1, N-1}(u_0, u_1, \ldots, u_{N-2})\}^2}{4(\lambda+1)}$ together

with the optimal aeration rate at the last stage,

(3.20) $\quad u_{N-1}^* \equiv u^*((N-1)\Delta) = \dfrac{R_{1, N-1}(u_0, u_1, \ldots, u_{N-2})}{2(\lambda+1)}$

Thus, the minimized value inside the bracket in (3.18) with respect to u_{N-1} is given by (3.19) as the function of remaining variables $\{u_0, u_1, \ldots, u_{N-2}\}$, free from u_{N-1}.

Let us call this operation as "the minimizing elimination of u_{N-1}". Then, the result (3.20) obtained is again a quadratic polynomial in u_{N-2} and $J_{N-1}((u_0, u_1, \ldots, u_{N-2}))$ is also a quadratic polynomial in u_{N-2}, the value in the bracket of (3.18) is again a quadratic polynomial in u_{N-2}, and thus we repeat the minimizing elimination of control variable u_{N-2}, until the last variable u_0 is eliminated.

Thus, we are able to obtain the value of $f_N (B(o), D(o))$ as a function of paramenters, $\lambda, r, k_1, k_2, D_s$ and $B(o), D(o)$.

This is a striking feature of our quadratic criterion function.

In addition, we note that $B(o)$ and $D(o)$ are arbitrary, and thus for any value of $C_1 = B(o)$ and $C_2 = D(o)$, we are able to compute the value of $f_N (C_1, C_2)$, together with the sequence of optimal aeration rates $u_0^*, u_1^*, \ldots, u_{N-1}^*$.

4. APPLICATION TO REAL LIFE PROBLEMS: SOME NUMERICAL EXAMPLES

[Example 1] Computation of $f_3(10, 3)$, together with the optimal aeration rates $\{u_0^*, u_1^*, u_2^*\}$, when $\lambda = 10, b = 1, r = 0.2,$ and $k_1 = 0.16, k_2 = 0.66, D_s = 8$.

(Solution) Self-made HP 67 Program Card was used. Assuming $\Delta = 1$, the specified parameter values were stored to start with. Programs stored in this tiny piece of card enable us to compute state values $\{(B(n), D(n))\}$ and the corresponding values of $\Delta_n J (u_{n-1}| u_0, u_1, \ldots, u_{n-2})$. They are displayed by pressing keys A, B, and E in succesion, starting from the prestored initial state values of $B(o)$ and $D(o)$ at the location number 7 and 8 in our case, if we key in successive values $u_0, u_1, u_2, \ldots$ into location 1 in this case.

Now, if we continue to store 0 at this location 1 at each stage, we obtain the set of values

$(B_0(1), D_0(1)) = (9.8, 4.7)$

$(B_0(2), D_0(2)) = (9{,}292, 5.310)$

$(B_0(3), D_0(3)) = (8.7433, 5.5987)$,

where $(B_0(n), D_0(n)), n = 1, 2, 3$ are the values described in our formula (3.8), (3.9), and (3.10), with explanation.

Thus, $J_3(u_0, u_1, u_2) = \Delta_1 J(u_0) + \Delta_2 J(u_1|u_0) + \Delta_3 J(u_2|u_0, u_1)$

$= 9.8^2 + (3.3 - u_0)^2 + 10u_0^2$

$+ (9{,}292 - 0.2u_0)^2 + (2.69 - 0.34u_0 - u_1)^2 + 10u_1^2 + (8.7433 - 0.2360u_0 - 0.2u_1)^2$

$+ (2.4013 - 0.1476u_0 - 0.34u_1 - u_{\underline{2}})^2 + 10u_{\underline{2}}^2$

$\underset{u_{\underline{2}}}{\min} \ 9.8_2^2 + \ldots + (8.7433 - 0.2360u_0 - 0.2u_1)^2$

$+ \dfrac{10}{11}(2.4013 - 0.1476u_0 - 0.34u_1)^2$

$= 11.1451 u_{\underline{1}}^2$

$- 2 \times \{(2.69 - 0.34u_0) + 0.2 \times (8.743 - 0.2360u_0)$

$+ \dfrac{10}{11} \times 0.34 \times (2.4013 - 0.1476u_0)\} u_{\underline{1}}$

$+ 9.8^2 + \ldots + \dfrac{10}{11} \times (2.4013 - 0.1476u_0)^2$

$\underset{u_1}{\min} \ - \dfrac{(5.1809 - 0.4328u_0)^2}{11.1451}$

$+ 9.8^2 + (3.3 - u_0)^2 + 10u_0^2$

$+ (9{,}292 - 0.2u_0)^2 + (2.69 - 0.34u_0)^2$

$+ (8.7433 - 0.2360u_0)^2$

$+ \dfrac{10}{11} \times (2.43013 - 0.1476u_0)^2$

$= 11.2143 u_{\underline{0}}^2 - 16.5149 u_{\underline{0}} + 279.7863 \xrightarrow{\underset{u_0}{\min}} 279.7863 - \dfrac{16.5149^2}{4 \times 11.2143}$

$\doteq 273.7061$, and $u_0^* = 0.7363$, etc.

[Remark 1] The notations, $\frac{\min}{u_n}$ means "by the minimizing elimination with respect to u_n, it is equal to".

[Remark 2] Each time when we perform the operation of the minimizing elimination with respect to u_n, relationships between the optimal aeration controls u_0^*, u_1^*, u_2^* are obtained, and u_n^*'s are obtained therefrom.

In the above [Example 1], the optimal aeration rates were obtained to be $u_1^* = 0.4363$, $u_2^* = 0.1949$, successively, after $u_0^* = 0.7363$ was obtained.

In addition, in this example and also in the subsequent examples in this paper, we have assumed $\Delta = 1$, $G = 2$ (mg/l), and $b = 1$.

[Remark 3] Using aeration rates obtained as above, our self-made HP 67 program card was used again, starting from the initial state $(B(0), D(0)) = (10, 3)$ i.e. $B(0) = 10$ (mg/l), $D(0) = 3$ (mg/l). The following results were thus obtained in a few minutes: i.e.

$(B(1), D(1)) = (9.8, 5.4363)$, and $\Delta_1 J(u_0^*) = 108.0339$

$(B(2), D(2)) = (9.1447, 5.9966)$, and $\Delta_2 J(u_1^*|u_0^*) = 89.5433$

$(B(3), D(3)) = (8.4823, 6.0506)$, and $\Delta_3 J(u_2^*|u_0^*, u_1^*) = 76.1289$,

and, $f_3(10, 3) = \Delta_1 J(u_0^*) + \Delta_2 J(u_1^*|u_0) + \Delta_3 J(u_2^*|u_0^*, u_1^*) = 273.7058$.

[Example 2] Derivation of the general formula for obtaining the value of $f_2(C_1, C_2)$ with the incidental optimal aeration rates u_0^* and u_1^*, where $(C_1, C_2) = (B(\tilde{0}), D(\tilde{0}))$ is any specified initial state. Note that $\Delta = 1$, $G = 2$, and $b = 1$ are assumed for simplicity.

(Solutions) Using (3.8) and (3.9), and from the above conditions,

$\Delta_1 J(u_0) + \Delta_2 J(u_1|u_0)$

$= (B_0(\tilde{1}))^2 + \{D_s - D_0(\tilde{1}) - u_0\}^2 + \lambda u_0^2$
$+ \{B_0(\tilde{2}) - ru_0\}^2 + \{D_s - D_0(\tilde{2}) - (1-k_2)u_0 - u_1\}^2 + \lambda u_1^2$

$= (\lambda + 1)u_1^2 - 2\{D_s - D_0(\tilde{2}) - (1-k_2)u_0\}u_1$
$+ \{D_s - D_0(\tilde{2}) - (1-k_2)u_0\}^2 + \ldots$

$\frac{\min}{u_1} - \frac{\{D_s - D_0(\tilde{2}) - (1-k_2)u_0\}^2}{\lambda + 1} + \{D_s - D_0(\tilde{2}) - (1-k_2)u_0\}^2$
$+ (B_0(\tilde{1}))^2 + \{D_s - D_0(\tilde{1}) - u_0\}^2 + \lambda u_0^2$
$+ \{B_0(2) - ru_0\}^2$

$= \{(\lambda + 1) + r^2 + \frac{\lambda}{\lambda+1}(1-k_2)^2\}u_0^2$
$- 2\{(D_s - D_0(\tilde{1})) + rB_0(\tilde{2}) + \frac{\lambda}{\lambda+1}(D_s - D_0(\tilde{2}))(1-k_2)\}u_0 + \ldots$

$\frac{\min}{u_0} - \frac{B^2}{A} + (B_0(\tilde{1}))^2 + (B_0(\tilde{2}))^2 + \{D_s - D_0(\tilde{1})\}^2 + (D_s - D_0(\tilde{2}))\}^2 - \frac{(D_s - D_0(\tilde{2}))^2}{\lambda+1}$

$= f_2(C_1, C_2)$, where

$A \triangleq (\lambda + 1) + r^2 + \frac{\lambda}{\lambda+1}(1-k_2)^2$,

$B \triangleq (D_s - D_0(\tilde{1})) + rB_0(\tilde{2}) + (D_s - D_0(\tilde{2}))(1-k_2) - \frac{1}{\lambda+1}(D_s - D_0(\tilde{2}))(1-k_2)$,

and $u_0^* = \frac{B}{A}$, $u_1^* = \frac{D_s - D_0(\tilde{2}) - (1-k_2)u_0^*}{\lambda+1}$.

[Remark] Using these results, HP 67 program card is easily made for obtaining the values of $f_2(C_1, C_2)$ and u_0^*, u_1^*.

Actually, by a piece of HP 67 self-made program card for calculating these values of the above [Example 2], we wish to obtain the value of u_1^* by the card used for [Example 1], cosidering (9.8 5.4363) = $(B(\tilde{0}), D(\tilde{0}))$ is our initial state, and we wish to derive two stages optimal aeration rates. Note that the above initial state was obtained as the state $(B(1), D(1))$ steered by $u_0^* = 0.7363$ from $B(0), D(0)) = (10, 3)$

Now, we applied the above program card, assuming (9.8, 5.4363) as our initial state, and $u_0^* = 0.4363$ was obtained immediately. This value of u_0^* now obtained is naturally identical with $u_1^* = 0.4363$ obtained in [Example 1], as the solution to the three stages optimal aeration problem via dynamic programming.

[**Example 3**] <u>Derivation of the general formula for obtaining the value of $f_3(C_1, C_2)$ together with the incidental optimal aeration rates $\{u_0^*, u_1^*, u_2^*\}$, starting from any specified initial state $(C_1, C_2) = (B(\tilde{0}), D(\tilde{0}))$, with the same conditions described in [Example 2].</u>

(Solution) Using (3.8), (3.9), and (3.10), and following precisely in the same way as in previous [Example 2], we obtain the following formula.

(4.1) $f_3(C_1, C_2) \equiv f_3(B(\tilde{0}), D(\tilde{0})) = C - \dfrac{B^2}{A}$, where

$C = (B_0(\tilde{1}))^2 + (B_0(\tilde{2}))^2 + (B_0(\tilde{3}))^2 + (D_s - D_0(\tilde{1}))^2 + (D_s - D_0(\tilde{2}))^2$
$\quad + \dfrac{\lambda}{\lambda+1}(D_s - D_0(\tilde{3}))^2 - \dfrac{b_0^2}{a_0}$,

$B = (D_s - D_0(\tilde{1})) + rB_0(\tilde{2}) + (1-k_2)(D_s - D_0(\tilde{2})) + r(2-k_1-k_2)B_0(\tilde{3})$
$\quad + \dfrac{\lambda}{\lambda+1}\{(1-k_2)^2 + rk_1\}(D_s - D_0(\tilde{3})) - \dfrac{b_0^2}{a_0}$,

$A = (\lambda+1) + r^2 + (1-k_2)^2 + r^2(2-k_1-k_2)^2 + \dfrac{\lambda}{\lambda+1}\{(1-k_2)^2 + rk_1\}^2 - \dfrac{c_0^2}{a_0}$,

where $a_0 = (\lambda+1) + r^2 + \dfrac{\lambda}{\lambda+1}(1-k_2)^2$,

$b_0 = (D_s - D_0(\tilde{2})) + rB_0(\tilde{3}) + \dfrac{\lambda}{\lambda+1}(1-k_2)(D_s - D_0(\tilde{3}))$,

$c_0 = (1-k_2) + r^2(2-k_1-k_2) + \dfrac{\lambda}{\lambda+1}(1-k_2)\{(1-k_2)^2 + rk_1\}$, together with

(4.2) $u_0^* = \dfrac{B}{A}$, $u_1^* = \dfrac{b_0 - c_0 u_0^*}{a_0}$, and

$u_2^* = \dfrac{D_s - D_0(\tilde{3}) - \{(1-k_2)^2 + rk_1\}u_0^* - (1-k_2)u_1^*}{\lambda+1}$.

[Remark 1] These results obtained above are apparently complicated and tedious for numerical computations, but here also a few pieces of *HP 67* self-made program cards are easily available.

By a few pieces of *HP 67* program cards prepared by the present authour, based upon the formula (4.1) and (4.2), the solutions to the optimiaztion problem treated in our [Example 1] are obtained by the following steps. Assuming no control case i.e. $u_0 = u_1 = u_2 = 0$, we obtain the following states, to begin with.

$(B_0(1), D_0(1)) = (9.8, 4.7)$

$(B_0(2), D_0(2)) = (9.292, 5.310)$

$(B_0(3), D_0(3)) = (8.7433, 5.5987)$

We store these data in the secondary memories, $B_0(1), B_0(2), B_0(3)$ at the locations 4, 5, 6 and $D_0(1), D_0(2), D_0(3)$ at the locations 1, 2, 3. Then, using another card, $a_0 = 11.1451$, $b_0 = 5.1809$, $c_0 = 0.4328$ are instantly obtained and stored in the secondary memories at the locations 7, 8, 9.

After these preparations, one more card is used. The values of $B, C,$ and A in (4.1) are furnised by the strokes of *HP 67* keys in succession. Thus, the values $B = 8.2574, C = 279.7863, A = 11.2142, f_3(10, 3) = 273.7062$, and $u_0^* = 0.7363$ are displayed by the above key strokes of HP 67 instantly. Then, $u_1^* = 0.4363$, and $u_2^* = 0.1949$ are obtained; exactly indentical with the results obtained by the calculations in [Example 1].

[Remark 2] It was the author's intension that the values of $f_1(B(0), D(0))$, $f_2(B(0), D(0))$, $f_3(B(0), D(0))$ are easily obtainable even by a programmable palm calculator like HP 67, in a few minutes, together with the incidental aeration control rates via dynamic programming.

5. FEASIBLE AERATION CONTROL EFFICIENT FOR APPLICATION

In the preceding section we have shown the general formula for obtaining the values of $f_2(C_1, C_2)$ and $f_3(C_1, C_2)$, with the incidental aeration rates u_0^*, u_1^*, and u_2^*.

We are able to extend the same way of computation for obtaining solutions to the optimal aeration control via dynamic programming to larger scaled problems but using large scale computing machinery. However such extensions may be highly expensive.

An approach to avoid the curse of dimensionality usually associated with such large dynamic programming problems is to use the aeration rate u_n at the n the aeration stage which will minimize the value of criterion function $\Delta_{n+1} J(u_n)$ at the next stage, i.e. the value u_n which minimizes the following,

(5.1) $\{(1 - k_1) B(n) - rD(n) + 2\}^2$
$+ \{D_s - (1 - k_2) D(n) + k_1 B(n) - k_2 D_s - u_n\}^2 + \lambda u_n^2$, assuming $\Delta = 1$ and $G = 2$.

Such value of u_n is obviously given by

(5.2) $\tilde{u}_n = \dfrac{(1 - k_2) \{D_s - D(n)\} + k_1 B(n)}{\lambda + 1}$.

Let us call this u_n by "one stage optimal aeration rate", which is free from the curse of dimensionality problem. From the author's experiences of simulation studies concerning the present work, aeration control rates at the beginning stages seems to be important and thus should be estimated elaborately. So, applying aeration rates via dynamic programming of manageable dimensionality at the beginning stages of our aeration, starting from our initial state $(B(0), D(0))$, and then using perhaps one stage optimal aeration rate u_n at the subsequent stages respectively. Actually, it is rather difficult to see what extent of aeration rate will be efficient to start with. So the feasible aeration control policy stated above must be certainly reasonable.

In the following, we show how efficient such feasible control can be for practical purposes. Some simulated numerical examples are given.

[Example 4] We consider aeration rate control processes with the following parameters, starting from an initial state $(B(0), D(0)) = (10, 3)$. The parameter values are, $r = 0.3, \lambda = 10, k_1 = 0.16, k_2 = 0.66, D_s = 8$ (mg/l) $G = 2, \Delta = 1$.

(Solution) To solve this problem we first consider a three stage of dynamic programming version. By our procedures for computing $f_3(10, 3)$ with u_0^*, u_1^*, u_2^*, we list the values obtained in the following.

$(B_0(1), D_0(1)) = (9.5, 4.7)$

$(B_0(2), D_0(2)) = (8.57, 5.3580)$

$(B_0(3), D_0(3)) = (7.5914, 5.7305)$

$a_0 = 11.1951$ $\qquad B = 9.5448$

$b_0 = 5.6209$ $\qquad C = 241.0546$

$c_0 = 0.4968$ $\qquad A = 11.332$

$f_3(10, 3) = 233.0160$, with

$u_0^* = 0.8422$, $u_1^* = 0.4647$, u_2^* coincides with $\tilde{u}_2 = 0.1794$. Then, we applied (5.2) for our subsequent aeration control.

Table 5.1 Resulting processes obtained by the feasible aeration control for [**Example 4**] are listed in the following table.

n	u_n	$B(n)$	$D(n)$	$\Delta_n J(u_{n-1}\vert\ldots)$
0	0.8422*	10 (mg/l)	3 (mg/l)	No Account here.
1	0.4647* (no control)	9.5000 (9.5000)	5.5422 (4.7000)	103.3838 (101.1400)
2	0.1794*	8.3173 (8.5700)	6.1090 (5.3580)	74.9133 (80.4251)
3	0.1595	7.1539 (7.5914)	6.2057 (5.7305)	54.7189 (62.7799)
4	0.1387	6.1475 (6.6576)	6.4048 (6.0138)	40.5909 (48.2691)
5	0.1191	5.2425 (5.7883)	6.6128 (6.2595)	29.6002 (36.5336)
6	0.1011	4.4198 (4.9843)	6.8087 (6.4821)	21.0961 (27.1474)
7	0.0846	3.6701 (4.2422)	6.9889 (6.6864)	14.5939 (19.7217)
8	0.0696	2.9863 (3.5575)	7.1536 (6.8746)	9.7052 (13.9224)

[Remark] The corresponding values for the case of no control are listed in the respective parenthesis.

From **Table 5.1**, it can be seen that our feasible aeration rate control seems to be considerably efficient.

[Example 5] Parameter values in this example are $\lambda = 1$, $b = 1$, $r = 0.2$, $k_1 = 0.2$, $k_2 = 0.6$, $D_s = 8$, together with $G = 2$, and $\Delta = 1$.

Two stages DP policies were used to start with. The subsequent aeration rates are u_n's. The resulting data are shown in **Table 5.2**.

Once again it can be seen that our feasible aeration control policies are quite efficient, resulting in dramatic improvement of water quality and appreciable reduction of the total cost different from the case of no control.

Table 5.2 Resulting processes obtained by the feasible aeration control for [Example 5], where $u_0^* = 3{,}0377$, $B(0) = 10$, $D(0) = 3$, are listed in the following table.

n	u_n	$B(n)$	$D(n)$	$\Delta_n J$
$n = 1$	$u_1^* = 1.1325$ (no control)	9.4000 (9.4000)	7.0377 (4.000)	98.5136 (104.3600)
$n = 2$	1.0377	8.1125 (8.7200)	6.8675 (4.5200)	68.3769 (88.1488)
$n = 3$	0.9192	7.1165 (8.0720)	6.9623 (4.8640)	52.7978 (74.9917)
$n = 4$	0.8139	6.3007 (7.4848)	7.0808 (5.1312)	41.3889 (64.2522)
$n = 5$	0.7252	5.6244 (6.9616)	7.1861 (5.3555)	32.9589 (55.4571)
$n = 6$	0.6513	5.0623 (6.4982)	7.2748 (5.5499)	26.6789 (48.2293)
$n = 7$	0.5897	4.5949 (6.0886)	7.3487 (5.7203)	21.9614 (42.2675)
$n = 8$	0.5386	4.2062 (5.7268)	7.4103 (5.8704)	18.3875 (37.3312)
$n = 9$	0.4960	3.8829 (5.4073)	7.4614 (6.0028)	15.6569 (33.2282)
$n = 10$	0.4606	3.6140 (5.1253)	7.5040 (6.1197)	13.5532 (29.8046)
$n = 11$	0.4312	3.3904 (4.8763)	7.5394 (6.2228)	11.9192 (26.9370)
$n = 12$	0.4067	3.2045 (4.6565)	7.5688 (6.3139)	10.6403 (24.5261)
$n = 13$	0.3863	3.0498 (4.4624)	7.5933 (6.3942)	9.6320 (22.4917)
$n = 14$	0.3694	2.9212 (4.2911)	7.6137 (6.4652)	8.8317 (20.7591)
$n = 15$	0.3553	2.8142 (4.1398)	7.6306 (6.5279)	8.1926 (19.3054)

6. APPLICATION OF BOX'S HILLCLIMBING METHOD FOR COMPUTING SOLUTIONS TO DYNAMIC PROGRAMMING

[Example 6] Let us explain the idea of this method by the following example. We wish to derive the values of $f_4(10,3)$ together with the incidental optimal aeration control $u_0^*, u_1^*, u_2^*, u_3^*$, when $r = 0.2$, $\lambda = 10$, $k_1 = 0.16$, $k_2 = 0.66$, and $D_S = 8$ with $G = 2$ and $\Delta = 1$.

Of course, it is possible to compute the above solutions directly in the same way as we did in the preceding section 4, likewise the case of $f_2(10,3)$, $f_3(10,3)$ together with the incidental aeration control rates.

Instead of proceeding in such a manner, we wish to show a new approach for computing

$f_4(10, 3)$ and u_i^*'s, using the search technique due to G.E.P. Box, i.e. Box's hillclimbing method using the values of $f_2(C_1, C_2)$ which are obtainable directly by our method explained in section 4.

The Box's hillclimbing method is well known in statistical design and analysis of experiments for searching the location of maximum or minimum of response surface. This method of hillclimbing is based upon the first and the second strategies. We wish to use here the first strategy which is the statistical version of steepest ascent (or descent).

Now, let us explain by our example. Look at the following relation due to the principle of optimality:

(6.1) $f_4(10, 3) = \min_{\{u_0, u_1\}} [J_2(u_0, u_1) + f_2(B(2), D(2)|u_0, u_1)]$, where

(6.2) $J_2(u_0, u_1) = \Delta_1 J(u_0) + \Delta_2 J(u_1|u_0)$, by our notations.

The minimization in (6.1) with respect to $\{u_0, u_1\}$ is a search problem in two-dimensional (u_0, u_1) plane. So, we set 4 grid points with coordinates (1.5, 1.5), (1.5, 0.5), (0.5, 1.5), and (0.5, 0.5). The coordinate of the center of square is obviously (1.0, 1.0). If we transform our coordinate system from (u_0, u_1) to (x_1, x_2) by putting $x_1 = \frac{u_0}{0.5}, x_2 = \frac{u_1}{0.5}$, the four grid points are denoted by (1,1), (1, −1), (−1, 1), and (−1, −1), i.e. the most simple case of 2^n type orthogonal design in $n = 2$ dimensions, and therefrom, if we know the values of $J_2(u_0, u_1) + f_2(B(2), D(2)|u_0, u_1)$, at the four grid points in (u_0, u_1) plane, or equivalently the above four grid points in (x_1, x_2) plane, we are able to fit a plane in the neighborhood of the center of the square as, $f(x_1, x_2) = \beta_0 + \beta_1 x_1 + \beta_2 x_2$. Thus, [Example 6] is the simplest case of the Box's hillclimbing method.

Note that

(6.3) $f_4(10, 3) = \min_{\{x_1, x_2\}} f(x_1, x_2)$.

Starting from our initial state (10, 3), the following four sets of results were obtained.

(6.4) At $(u_0, u_1) = (1.5, 1.5)$,
  (1) $(B(1), D(1)) = (9.800, 6.200)$, with $\Delta_1 J = 121.780$, after applying aeration control $u_0 = 1.5$ upon $(B(0), D(0))$.
  (2) Likewise by $u_1 = 1.5$, we have $(B(2), D(2)) = (8.992, 7.320)$, with $\Delta_2 J = 103.818$.
  (3) Starting from $(B(0), D(0)) \stackrel{i.e.}{=} (B(2), D(2))$, $f_2(8.992, 7.320)$ was computed instantly by our method in section 4, and we obtained $f_2(8.992, 7.320) = 126.800$, together with $u_0^* = 0.337$ & $u_1^* = 0.159$.
  (4) Thus, at the grid point (1.5, 1.5), $J_2(1.5, 1.5) + f_2(8.992, 7.320) = 352.398$.

(6.5) At $(u_0, u_1) = (1.5, 0.5)$,
$J_2(1.5, 0.5) + f_2(8.992, 6.320) = 343.105$, with $u_0^* = 0.375$ & $u_1^* = 0.171$,

(6.6) At $(u_0, u_1) = (0.5, 1.5)$,
$J_2(0.5, 1.5) + f_2(9.192, 6.980) = 349.236$, with $u_0^* = 0.356$ & $u_1^* = 0.166$, in the same way as we followed in (6.4).

(6.7) At $(u_0, u_1) = (0.5, 0.5)$,
$J_2(0.5, 0.5) + f_2(9.192, 5.980) = 340.931$, with $u_0^* = 0.395$ & $u_1^* = 0.178$.

Note that, in the corresponding (x_1, x_2)-plane, our experimental design can be described by

$$x = \begin{bmatrix} 1 & 1 & 1 \\ 1 & 1 & -1 \\ 1 & -1 & 1 \\ 1 & -1 & -1 \end{bmatrix}, \text{ with observation vector } y = \begin{bmatrix} 352.398 \\ 343.105 \\ 349.236 \\ 340.931 \end{bmatrix},$$

assuming such structure for our calculated value as $y_\alpha = \beta_0 + \beta_1 x_{1\alpha} + \beta_2 x_{2\alpha} + \epsilon_\alpha (\alpha = 1, 2, 3, 4)$, where ϵ_α is mutually independent Gaussian noise ($\sim$ iid $N(0, \sigma^2)$).

Then, we have

(6.8)
$$x'y = \begin{pmatrix} 1 & 1 & 1 & 1 \\ 1 & 1 & -1 & -1 \\ 1 & -1 & 1 & -1 \end{pmatrix} \begin{pmatrix} 352.398 \\ 343.105 \\ 349.236 \\ 340.931 \end{pmatrix}$$

(6.9)
$$= \begin{pmatrix} 1395.670 \\ 5.336 \\ 17,598 \end{pmatrix}, \quad \text{and finally}$$

(6.10)
$$\hat{\beta} = \begin{bmatrix} \hat{\beta}_0 \\ \hat{\beta}_1 \\ \hat{\beta}_2 \end{bmatrix} = (x'x)^{-1} x'y = \begin{bmatrix} 346.418 \\ 1.334 \\ 4.400 \end{bmatrix}.$$

Thus, we have obtained the fitted plane as

(6.11) $f(x_1, x_2) \cong 316.418 + 1.334 x_1 + 4.4 x_2$.

Now, from (6.11), the direction of steepest descent is estimated by the coefficient 1.334 and 4.4, and thereby the point where the value of $f(x_1, x_2)$ is minimum was estimated to be around $(0.85, 0.5)$ in the (u_0, u_1) plane, where the value of $f(x_1, x_2) = f(2u_0, 2u_1)$ is 339.125.

Further on, the more elaborate search can be possible, though not listed here, around the point $(u_0, u_1) = (0.85, 0.5)$, and once (u_0^*, u_1^*) is estimated, $f_2(B(2), D(2)|u_0^*, u_1^*)$ can be derived imediately by our self-made *HP* 67 program card mentioned in the previous section 4.

[Remark] Using the same set of parameters and assumptions, $f_4(10, 3)$ was computed by the repetition of the author's minimizing elimination procedure of control variables, and $f_4(10, 3)$ tured out to be 338.9509, together with $u_0^* = 0.8988$, $u_1^* = 0.6122$, $u_2^* = 0.3830$, and $u_3^* = 0.1742$. These results are certainly consistent with the graphical results aided by the Box's hillclimbing method plus $f_2(C_1, C_2)$ by programmed computations, as explained in this example.

[Example 7] Estimation of $f_5(10, 3)$ under the same set of parameter values as the values in [Example 6], with the aid by Box's method. Now,

(6.11) $f_5(10,3) = \min_{\{u_0, u_1\}} [f_2(u_0, u_1) + f_3(B(2), D(2)|u_0, u_1)]$ is our basis of this example.

Based upon our method and program explained in section 4 for computing the value of $f_3(C_1, C_2)$, where $(C_1, C_2) = (B(\tilde{0}), D(\tilde{0}))$ is arbitrary, the same way of computations were performed. After performing the following computations,

$$x'y = \begin{pmatrix} 1 & 1 & 1 & 1 \\ 1 & 1 & -1 & -1 \\ 1 & -1 & 1 & -1 \end{pmatrix} \begin{pmatrix} 403.3050 \\ 397.2871 \\ 397.7022 \\ 398.2288 \end{pmatrix} = \begin{pmatrix} 1596.5231 \\ 4.6611 \\ 5.4913 \end{pmatrix},$$

$$\hat{\beta} = (x'x)^{-1} x'y = \begin{pmatrix} 1/4 & 0 & 0 \\ 0 & 1/4 & 0 \\ 0 & 0 & 1/4 \end{pmatrix} \begin{pmatrix} 1596.5231 \\ 4.6611 \\ 5.4913 \end{pmatrix}$$

$$= \begin{pmatrix} 399.1308 \\ 1.1653 \\ 1.3728 \end{pmatrix} \quad \text{was obtained.}$$

Thus, $f(x_1, x_2) \cong 399.131 + 1.165 x_1 + 1.373 x_2$ was fitted, and our estimate, $f_5(10, 3) \cong 397$ was obtained.

7. DISCUSSIONS AND FURTHER REMARKS

I. It is certain that there are a number of ways of computing the values of $f_n(B(o), D(o))$, together with the incidental aeration control rates $\{u_i^*\} (i = 0, 1, 2, \ldots, n)$, for the larger n, combining the values of $f_2(\ldots), f_3(\ldots), f_4(\ldots), \ldots$, i.e. the values of $f_n(C_1, C_2)$ for the smaller n, easily obtainable by our methods shown in this paper, using many ways of recurrence relationships due to the principle of optimality of dynamic-programming. The present author considers that dynamic programming formulation is easy in general by the principle of optimality discovered by Richard Bellman as a natural law, but solving is difficult, easily involved in the difficulty socalled the curse of dimensionality even with a large scale computing machinery. But, in this paper, the author devised feasible solving scheme free from the curse of dimensionality and yet extremely efficient for the practical purposes, at least for the optimal aeration problem treated in this paper.

Dr. Richand Bellman emphasized, throughout his life, the increasing importance of using the large scale of digital computers in solving dynamic programming problems numerically. After his epochmaking discovery of the principle of optimality, he thought of storing numerical values of return functions at a considerable number of discretized grid points in order to find out the optimal path i.e. the maximum or the minimum values of return functions at the various stages. We consider it important to solve dynamic programming problems in a practical way, even by approximate solutions, by minicomputers, or even by programmable hand held calculators, like *HP 67* used in this paper.

The form of our criterion functional was quadratic. This fact was the incentive of overcoming the difficulties by our approaches discussed above.

II. Our feasible solutions, which might be called "suboptimal solutions" to our dynamic programming formulation, turned out to be very efficient, considering the numerical examples shown in this paper.

As mentioned in the forgoing sections, *DP* solutions are very useful for setting the aeration rates at the beginning stages of aeration control to start with, based upon the results of $f_2(10, 3)$, or $f_3(10, 3)$. Though, these aeration rates seem would otherwise be difficult to assume intuitively.

As Dr. Richard Bellman stressed in his paper [3], "most functionals are fairly flat" in nature learning from many years of Rayleigh-Ritz-Galerkin methods, and thus, the author considers that the subsequent aeration rates are not required to be so precise.

Based on our experiences it is not necessary to compute $f_n(10, 3)$ for large n, say $f_{20}(10, 3)$ as well as the associated aeration rates $\{u_i^*\}$, $i = 0, 1, 2, \ldots, 19$, for instance.

III. Dr. Bellman emphasized the future importance of applying the methods of stochastic approximation for facilitating the computation of solutions to dynamic programming problems. We have shown and suggested the usefulness of hillclimbing method due to G.E.P. Box in this paper for computing numerical solutions to the optimal control problems formulated via dynamic programming.

We emphasized the usefulness and the importance of applying the Kiefer-Wolfowitz stochastic approximation with added artificial noise for locating the global maximum or global minimum in our search region [6] [14]. Thus, we equally must emphasize the combined use of Box's type hillclimbing method and the above type of stochastic approximation methods to be more useful for the numerical computation of *DP* solutions, because of the slow convergence of stochstic approximation processes.

IV. As described in Hullett's paper [11], the minimization problem of quadratic criterion functional subject to a set of differential equations is already solved.

This problem is known as an optimization problem in Hilbert space, and the optimal feedback control is known to be expressed by linear operator which is the solution of infinite dimensional Ricatti equation with certain terminal condition. In this context, the optimal control is already obtained, but on the contrary tremendous effort is required in order to obtain concrete numerical solutions for practical applications, starting from apparently simple expression of the optimal feedback control stated above.

Pontryagin's maximum (or, minimum) principles, the necessary conditions thereof, are also considered to be too mathematical and rather complex for practical applications, particularly for aeration control problems.

Thus, we feel that dynamic programming is the most practical and useful technique for our aeration control problems, because it is simple in its formulation and useful for our aeration control of water quality.

V. In this paper, we discussed aeration control approach for improving water quality based upon the number of parameter values specified precisely. In the actual situations however, such parameter values are not known in advance. We know that there exist no such precise values nor constant values for an actural river water, and thus, only rough estimates of parameter values are sufficient for the efficient water quality aeration control, since any mathematical methods of optimization offers us only good suggestions or ideas for practical purposes. Thus our "optimal solutions" must be robust for the set of parameter values.

Finally, it is most important for us to be aware of all the available technical strategies, local and global, for improving water quality if needed. One should thus not be confined to an aeration technique with introduced aerobiotic micro-organisms.

8. REFERENCES

1. BELLMAN R., Dynamic Programming, Princeton University, Princeton, N.J., 1957.
2. BELLMAN R., Dynamic Programming and Stochastic Control Processes, Information and Control 1, No 3, (September 1958), 228-239.
3. BELLMAN R., Some Directions of Research in Dynamic Programming, Unternehmensforschung 7, No 3, (1963), 97-102.
4. BELLMAN R., KAGIWADA H. and KALABA R., Dynamic Programming and an Inverse Problem in Transport Theory, Computing 2, (1967), 5-16.
5. BELLMAN R., Functional Equations in the theory of Dynamic Programming: Minimum Convolutions and Green Functions, J. Math. Anal. Appl. 33, No 3, (March 1971), 497-499.
6. BELLMAN R., Mathematical Methods in Medicine, World Scientific, 1983.
7. BOX G.E.P. and WILSON K.B., On the experimental attainment of optimum conditions, J. of Stat. Soc. B, 13, (1951), 1-45.
8. BOX G.P.E. and HUNTER, Experimental Designs for the Exploration and Exploitation of Response Surfaces, Biometrics 10, (1954), 502-536.
9. COCHRAN W.G. and COX G.M., Experimental Design, John Wiley, 1957.
10. HULLETT W., Applications of Distributed Control Theory in Optimal Estuary Aeration, Ph.D. in Engineering, University of California, Los Angeles, 1972.
11. HULLETT W., Optimal Estuary Aeration: An application of distributed control theory, J. Appl. Math and Optim. 1, No 1, (1974), 20-63.
12. NINOMIYA T., MATSUKAWA A., HAMADA K., HIRAGAKIUCHI A. and IMANISHI E., Purification of Dohtonbori River and Higashi-Yokobori River, Reports of the City of Osaka, Public Engineering Works Foundation 3, (1980), 101-112. (In Japanese).
13. REDDIEN G.W., Collocation at Gauss points as a discretation in optimal Control, SIAM J. Control, and Optimization 17, No 2, (March 1979), 298-306.
14. SUGIYAMA H., The Principle of Dynamic Programming as a natural law discovered by R. Bellman, J. Math. Anal. Appl. 119, Nos. 1/2, October/November, 1986, 55-71.
15. WATANABE Y., HOSODA K., HIRAGAKIUCHI A., IMANICHI E., KIDA G. and KOZIMA J., Reports of the City of Osaka, Public Engineering Works Foundation 4, (1983), 792-814. (In Japanese).

[Appendices]

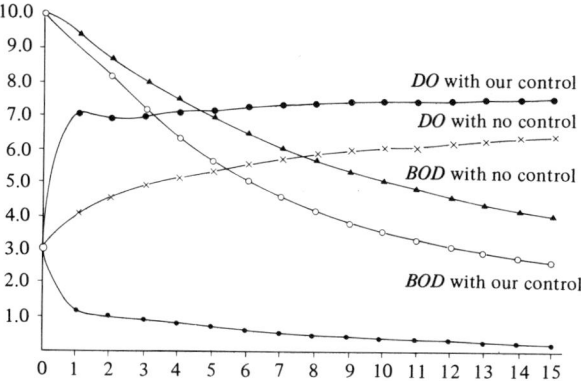

FIG. 1

Graphical Representation of the Results in Table 5.2
$\lambda = 1, b = 1, r = 0.2, k_1 = 0.2, k_2 = 0.6, D_s = 8, G = 2, \Delta = 1$.

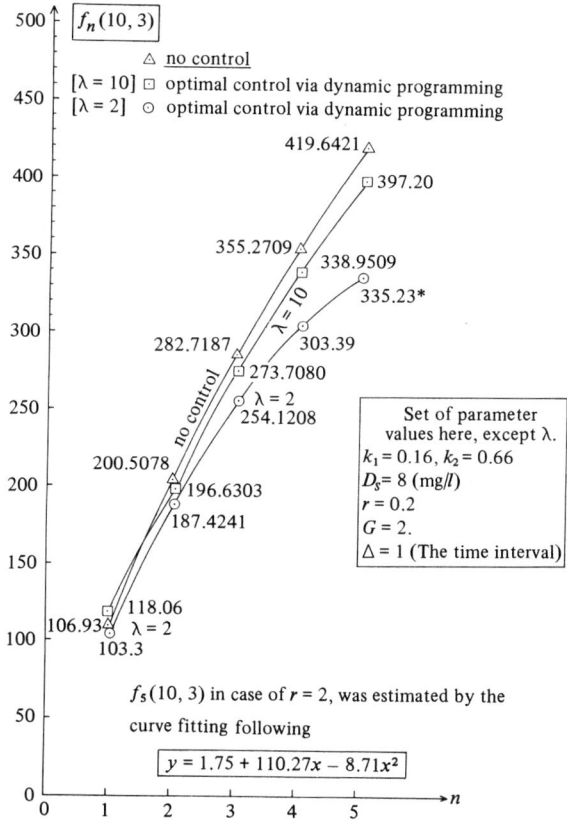

FIG. 2

Values of $f_n(10, 3)$ are plotted in the cases mentioned above.

A VARIABLE STATE-SPACE DYNAMIC PROGRAMMING MODEL FOR OPTIMIZING
INDUSTRIAL WASTE TREATMENT SEQUENCES

J. HUGH ELLIS
Department of Geography and Environmental Engineering
The Johns Hopkins University, Baltimore, Maryland 21218

ABSTRACT

A stochastic methodology is described for identifying cost-effective treatment sequences for a centralized liquid industrial waste treatment facility. The dynamic programming (DP) optimization model delineates those sequences of unit treatment processes which will produce an acceptable effluent quality, given probabilistically-generated influent waste regimes. Considerable flexibility is embedded in the methodology through allowing user-determined options such as waste characterizations, unit processes for consideration, performance functions for the processes, probabilistic descriptions for the influent wastes, and others. From a DP viewpoint the model is somewhat atypical in that it possesses a state-space which varies by stage and furthermore, the number of stages required in any given application is unknown a'priori.

INTRODUCTION

The design of centralized hazardous waste treatment systems can be improved through proper consideration of system uncertainties. Important elements of uncertainty are related to the prediction of influent waste flows and contaminant strengths which in turn impact on the sequencing of unit treatment processes. Additional uncertainty is associated with the assessment of treatment efficiencies of individual processes as functions of influent waste characteristics, (i.e., composition, strength and volume). Criteria to be used to select individual (unit) treatment processes and their sequencing for a centralized facility are far from obvious. Successful treatment can be accomplished through the use of more or less conventional means of selecting processes and process sequences, but this may not yield an optimally cost-effective design [1]. Moreover, optimization of individual process design does not guarantee optimality of the entire process sequence. These considerations become even more complicated if variability in influent waste strengths, compositions, volumes, etc., is addressed. Although the selection of optimal treatment sequences has been investigated previously (e.g., see [1], [2], [3], [4], [5]), the above-noted concerns remain unresolved.

The purpose of this paper is to develop a methodology for stochastic optimization of LIW treatment sequences which, in view of the previously-discussed design considerations, possesses the following attributes:

  i) objective selection of unit treatment process sequences based on cost-effectiveness, for the purpose of generating least-cost configurations;

  ii) variability in influent contaminant concentrations incorporated into the model through the use of probabilistic representations of waste strength, (as opposed to simply using expected-values); and,

iii) ability to accommodate the joint treatment of several waste streams with stream-specific compositions.

OVERVIEW OF THE DYNAMIC PROGRAMMING MODEL

The central component of the optimization process involves Dynamic Programming [6]. The function of the model is to delineate least-cost treatment sequences which will produce an acceptable effluent stream quality, given a probabilistically-generated influent waste regime. The model is prescriptive and hence, "screens out" processes and combinations of processes with desirable cost-effectiveness attributes. It should be noted that multi-stage treatment comprised of even a limited number of processes involves an extensive array of possible treatment sequences. Thus, the screening procedure is essential in identifying a manageable number of potentially cost-effective sequences which may be subsequently subject to detailed simulation studies for the purpose of establishing final designs. Conventional means of selecting process sequences are generally incapable of evaluating extensive arrays of potentially desirable sequences.

The optimization model is structured to permit the following user-determined options:

i) waste types and respective volumes in the waste inventory,

ii) specific contaminants within each waste type,

iii) contaminant-specific probability density functions for waste strength,

iv) unit treatment processes, including performance efficiencies and related costs; and,

v) individual contaminant effluent standards.

Of general concern in the analyses is whether acceptable cost-effectiveness criteria can be met through the use of a single process sequence for treatment of a variety of influent waste types. This scheme possesses the obvious advantage of simplicity over a scheme which employs waste stream-specific treatment sequences.

PREVIOUS APPLICATIONS

Shih and Krishnan, [5], developed a cost-minimizing DP model for optimization of pulp and paper liquid waste treatment. The waste stream contained two constituents: suspended solids, (SS), and biochemical oxygen demand, (BOD). Influent data was expressed probabilistically with design based upon a ten percent exceedence level, (90th percentile). The annual least-cost design involved: primary clarification with an overflow rate of 400 gal/day/ft^2; 30% BOD removal by trickling filter; 80% BOD removal via activated sludge treatment and tertiary treatment by carbon adsorption. This DP model, while admittedly intended for individual process design purposes, nonetheless fails to meet the required criteria as established earlier on the grounds that pre-arranged sequences were used for treatment of a limited number of waste types, (one), containing a limited number of pollutants, (two). Adams and Panagiotakopoulos [2] developed a Network Analysis Model for waste treatment optimization. The test application was that used by Shih and Krishnan, hence this model also does not meet the criteria described earlier.

MECHANICS OF THE STOCHASTIC OPTIMIZATION MODEL

Input to the stochastic DP model consists of contaminants (pollutants) within a waste stream expressed as contaminant-specific probability density functions, (pdf's). A Monte Carlo technique [7], [8] is then used to generate specific realizations of pollutant concentrations. One complete set of concentration realizations therefore constitutes an influent waste "regime." The DP algorithm then determines the least-cost treatment sequence for that particular regime. Successive stages of treatment are evaluated by the algorithm until such time as the effluent stream from a unit treatment process (i.e., a state) meets a prescribed effluent standard. Therefore, the number of stages of treatment required for a particular waste stream realization is unknown a'priori. This procedure is repeated using new input realizations in an attempt to identify convergence towards the selection of some particular treatment sequence.

MATHEMATICAL DESCRIPTION

A mathematical description of the DP model begins with the establishment of certain definitions, requisite functions and boundary conditions. The waste inventory is comprised of v waste types. Each type contains m, (m=m(v)), contaminants. Each type also has a specified volume to be treated, Q, (Q=Q(v)). For a given waste type v, the influent concentration of contaminant m to unit process j in stage i can be written as $w=w(i,j,k,m)$, with k representing the unit treatment process, (state variable), selected at previous stage i-1. A transition function T_j is defined as $T_j=T_j(D_i)$. The D_i are the decision variables, (i.e., choice of unit treatment process j at state i). The effluent concentration of contaminant m, in waste type v, from unit process j in stage i, is denoted $\widetilde{w}$ and can be expressed as,

$$\widetilde{w}(i, j, k, m) = T_j(D_i, w(i, j, k, m))$$

The transition function T_j embodies two parameterized functions. Firstly, the removal efficiency for the m'th contaminant of waste type v by unit process j is defined as R, where $R = R(j, m, Q)$. Secondly, the cost of unit process treatment, which in this study is expressed on a "per-volume-treated" basis, is C, where $C = C(j, m, q, R)$.

The set S represents the waste constituents that have met or surpassed the effluent standard set, E_m, after stage i treatment. It is written $S = W(\widetilde{w}(i, j, k, m) \leq E_m)$, for all m. The set of unit processes eligible for selection at stage i is denoted by P_i, where, $P_i = P_i(w(i,j,k,m) \leq I(j,m))$ for all m. Eligibility is determined through the use of "influent screens," denoted by $I(j,m)$. More specifically, $I(j,m)$ is the set of upper bounds on contaminant concentrations. These bounds are specific to each unit treatment process and are intended to ensure satisfactory unit process operation, (i.e., satisfactory in the sense of maintaining process removal efficiencies and costs within design guidelines). It should be noted that the use of influent screens creates a DP model which is a departure from the norm. The model now possesses a state-space which is variable (i.e., stage-specific). State-spaces for initial stages are typically small due to high raw influent waste strengths. That is, certain unit processes in, say, the first one or two stages of treatment may be excluded from consideration due to their presumed inability to successfully treat certain relatively high strength wastes. State-spaces subsequently enlarge in later stages as waste stream strength diminishes.

Sets S and P_i govern the state transition decision variable, D_i, i.e., $D_i = D_i(S, P_i)$. The treatment cost of waste stream v by unit process j is expressed on a marginal basis. A "cost-per-removal" function is defined as,

$$MC(j, m) = \frac{C(j, m, Q, R)}{R(j, m, Q)}, \qquad j \in Q, P \qquad (1)$$

The cost of state transition, $Y = Y(i, j, v, Q)$ can be defined as,

$$Y(i, j, v, Q) = \sum_m \{MC(j, m) \cdot F(i, j, m)\} \qquad (2)$$

where, $F(i, j, m)$ is a dichotomous, (dirac delta) variable whose value is given by,

$$F(i, j, m) = 0, \text{ if } w(i, j, k, m) \leq E_m, \text{ for all } m \qquad (3)$$

and

$$F(i, j, m) = 1, \text{ if } w(i, j, k, m) \geq E_m, \text{ for all } m. \qquad (4)$$

Through this device, only influent contaminant concentrations exceeding the effluent standard and hence requiring treatment, are included in determining the "cost" of state transition.

The foregoing definitions and functions can now be coupled to yield the mathematical mechanics of the DP automaton itself. Nomenclature used is similar to that of Shih and Krishnan, [5]. For notational convenience, $\hat{w}$ is defined as the set of w_i, for all j, k, m.

The overall system return, $\overline{Y}$, (minimum cost), constrained by the conditions $j \in Q, P$, is shown in recursive transition function form as,

$$\overline{Y} = \sum_{i=1}^{N} Y_N \Big\{ T_{N-1} \Big\{\Big\{ T_{N-2}, \ldots,$$

$$T_{i+1}\Big\{\Big\{D_{i+1}\Big\{T_i(D_i)\Big\}\Big\}, \ldots, D_{N-1}\Big\} D_N \Big\} \qquad (5)$$

with, $\overline{Y} = \overline{Y}\Big[\hat{w}_1, D_1, D_2, \ldots, D_N\Big]$, where (6)

$\hat{w}_1$ represents the initial, (given) waste regime. Beginning at $T_1(D_1)$, successive decisions are made until such time as the effluent stream meets the prescribed standard. As was noted earlier, the number of terms in the summation in equation 5, (i.e. "N"), is unknown a'priori and depends explicitly upon both the characteristics of the raw input waste regime, $\hat{w}_1$ and the removal efficiencies of the chosen unit processes, D_i.

An important characteristic of the model is described by initially considering a subset q of N stages with partial system return denoted B_q, where,

$$B_q = \sum_{i=1}^{q} Y(i, j, v, Q), \text{ with} \qquad (7)$$

$$B_q = B_q(\hat{w}_1, D_1, D_2, \ldots, D_q), \qquad (8)$$

Noting the use of forward recursion,

$$B_q = B_{q-1}(\hat{w}_1, D_1, D_2, \ldots, D_{q-1}) + Y(q, j, v, Q). \qquad (9)$$

In accord with the principles of DP, a system which is optimal to stage q-1, is optimal to stage q if the q'th state transition is itself optimal. To develop this concept mathematically, let some system return, $G_q(w_q)$, which is analogous to B_q, be defined by,

$$G_q(\hat{w}_q) \leq B_q(\hat{w}_1, D_1, D_2, \ldots, D_q), \text{ and}, \qquad (10)$$

$$G_q \; (\hat{w}_{q-1}) \leq B_{q-1} \; (\hat{w}_1, \; D_1, \; D_2, \; \ldots, \; D_{q-1}). \tag{11}$$

For the set of all D, and all $q = 1, 2, \ldots, N$;

$$G_q \; (\hat{w}_q) = \underset{D_q}{\text{MIN}} \left\{ G_{q-1} \; (\hat{w}_{q-1} + Y \; (q, \; j, \; v, \; Q) \right\} \tag{12}$$

Expressed in terms of transition functions, this relation becomes,

$$G_q \; (\hat{w}_q) = \underset{D_q}{\text{MIN}} \left\{ G_{q-1} \; [T_q \; (D_q)] + Y(q, \; j, \; v, \; Q) \right\} \tag{13}$$

Finally, for the entire system, the optimal, (least cost) return, Y can be written,

$$\overline{Y}^* = \underset{D_N}{\text{MIN}} \left\{ G_{N-1} \; [T_N \; (D_N)] + G_N \; (D_N) \right\} \tag{14}$$

Relations 7-14 describe a more or less traditional DP shortest path algorithm. Given a choice of paths to evaluate, such an algorithm always chooses the shortest, (i.e., least number of stages). A difficulty then arises however, because in this application a shorter path (sequence) may not possess a lesser cumulative cost. Modifying the algorithm to consider <u>all</u> sequences which produce acceptable effluent streams, regardless of sequence length, negates of course the considerable computational benefits of a shortest path DP approach. Although several heuristic modifications to the algorithm were considered, it was decided that the shortest path approach should be retained subject to the caveat that the algorithm cannot guarantee globally optimal solutions. This complication proved to be of little pragmatic consequence in a limited testing exercise. Analyses conducted on several DP-generated sequences showed those paths to be optimal with respect to their respective waste input realizations.

TEST APPLICATION

A literature review was conducted in order to obtain the input data needed to test the stochastic DP model. Space limitations preclude a detailed description of the data collection/manipulation process, hence we simply note that the following data were utilized: probability density functions and covariance estimates for a total of twelve waste constituents in three waste types; removal efficiency and treatment cost estimates for nine commonly-used unit processes; unit process-specific restrictions (i.e., upper bounds) on influent concentrations; and an effluent standard. The interested reader is referred to [9] for details of the data used in the test application.

As a means of examining the utility of the methodology, a Fortran computer program of the DP model was run one-hundred and fifty times, (fifty for each of the three waste types: "Metal," "Chemical," and "Tannery"). This sample size is obviously arbitrary but was computationally tractable and considered reasonably representative. Each of the fifty runs represented a randomly generated specific combination of influent contaminant concentrations. Hence, for each waste type, fifty "optimal" treatment sequences were delineated by the DP model.

MODELLING RESULTS - DP

Within each waste type, the fifty derived sequences were analyzed for the purpose of identifying convergence towards the selection of a particular process or process sequence. The results of the analysis showed however, that repeated and consistent selection of any one particular sequence did not occur. Variability in influent strength was therefore shown to have a significant impact upon treatment sequence delineation for all waste types examined. The results were then analyzed to determine which treatment sequences were selected most often by the DP model. In this context, a sequence is defined as a combination of processes which yielded an acceptable effluent. The procedure employed was to calculate the probability of each unit process being selected (by stage) and subsequently calculate the overall conditional probability of process selection over all stages and hence establish a "sequence" selection probability. The treatment sequences selected most frequently were therefore:

Waste Type	Sequence	Frequency of Selection
Metal	RO	16.7%
Chemical	CA (2 in series)	15.7%
Tannery	AS	26.0%

Detailed results are given in Table 1 below.

TABLE 1 - PART 1 - METAL WASTE

METAL WASTE

UNIT PROCESS SELECTION (number of instances each unit process was selected in each stage)

STAGE	UF	IE	AS	RO	CP	CA	OR	CL	FL
1	0	15	1	15	6	9	1	0	4
2	0	5	0	4	2	12	0	0	8
3	0	5	0	0	1	7	0	0	1
4	0	2	0	0	0	3	0	0	1
5	0	1	0	0	0	1	0	0	1

UTILIZATION RANKING (number of instances each process was selected for all stages)

CA	IE	RO	FL	CP	AS	OR
30	28	19	15	9	1	1

(i.e., CA was selected 9 times in stage 1, 12 in stage 2, 7 in stage 3, 3 in stage 4 and once in stage 5; this gives a total number of selections = 30)

MOST FREQUENTLY SELECTED SEQUENCES
(percent of time the given sequence was selected)

RO	-	16.7
IE/CA	-	9.3
IE/FL	-	9.3

TABLE 1 - PART 2 - CHEMICAL WASTE

CHEMICAL WASTE

UNIT PROCESS SELECTION

STAGE	UF	IE	AS	RO	CP	CA	OR	CL	FL
1	0	10	6	4	0	21	9	0	1
2	1	2	4	1	0	41	2	0	1
3	0	2	1	1	0	11	0	0	1
4	0	0	0	0	0	2	0	0	1
5	0	0	0	0	0	0	0	0	1

UTILIZATION RANKING

$$\frac{CA}{75} \quad \frac{IE}{12} \quad \frac{AS}{11} \quad \frac{OR}{11} \quad \frac{RO}{6} \quad \frac{FL}{6} \quad \frac{UF}{1}$$

MOST FREQUENTLY SELECTED SEQUENCES
(percent of time the given sequence was selected)

CA/CA - 15.7

TABLE 1 - PART 3 - TANNERY WASTE

TANNERY WASTE
UNIT PROCESS SELECTION

STAGE	UF	IE	AS	RO	CP	CA	OR	CL	FL
1	0	3	17	0	0	0	0	0	6
2	0	0	2	1	0	0	6	0	1
3	0	0	0	6	0	0	0	0	1
4	0	0	0	0	0	0	0	0	1
5	0	0	0	0	0	0	0	0	1

UTILIZATION RANKING

$$\frac{AS}{19} \quad \frac{FL}{10} \quad \frac{RO}{7} \quad \frac{OR}{6} \quad \frac{IE}{3}$$

MOST FREQUENTLY SELECTED SEQUENCES
(percent of time the given sequence was selected)

AS   - 26.0
NO TREATMENT - 36.0

It is interesting to note that the most frequently selected sequence, when evaluated over all three waste types, is the same as the Chemical waste-specific path. This is attributable to frequent selection of the CA unit process.

To put these results into proper perspective, it is important to point out that in this type of prescriptive modelling exercise, two basic types of results are possible, namely:

A - repeated selection or convergence toward a single sequence

B - absence of a predominant sequence suggesting that sequence selection is sensitive to influent waste stream strength.

With a strong caveat that we are dealing with a limited sample size, the DP results shown previously appear to be type B. They do not represent acceptable final designs. For all waste types, the aforementioned sequences would produce excessive numbers of treatment failures. That is, many waste streams treated by these sequences would either,

i) fail to meet the effluent standard due to insufficient treatment; or,

ii) within the DP model, fail to pass appropriate influent screens and hence also constitute treatment failure in the design exercise.

SIMULATION MODELLING

A second, post-optimization level of analysis was deemed necessary in order to establish acceptable final designs. It involves descriptive or simulation analyses. Four principal steps are involved, i.e.:

1) for the three DP-derived sequences, determine the causes of treatment failure (i.e., effluent standard or influent screen violations), and identify the offending waste constituents;

2) based on waste type and the extent of unit process selection in the DP analysis, identify the process which would reduce failure rate if included in the existing sequence;

3) expose the new sequence to fifty influent realizations, (using the Monte Carlo technique);

4) evaluate the treatment success/failure of the sequence as in step 2, and if unsatisfactory, return to step 1 using the new (enlarged) sequences.

SIMULATION MODELLING RESULTS

The simulation schedule was set up in a similar manner to that used in the DP optimization procedure, but using a total of 250 runs. Fifty influent realizations were again input to each of three types of sequences. In addition, the Metal and Tannery wastes were input to the Chemical treatment sequence, noting that the Chemical sequence and the sequence selected most frequently for all wastes combined were identical. The intent was to determine the feasibility of treating the three different types of waste with one sequence.

The results of the first simulation step are shown in Table 2.

TABLE 2 - SIMULATION RESULTS - (FIRST ITERATION)

NUMBER OF OCCURRENCES

	NO TREATMENT REQUIRED	SUCCESSFUL TREATMENT	EFFLUENT STANDARD VIOLATION	INFLUENT SCREEN VIOLATION
METAL WASTE* SPECIFIC SEQUENCE	1	12	19	18
CHEMICAL WASTE** SPECIFIC SEQUENCE	0	19	0	31
TANNERY WASTE*** SPECIFIC SEQUENCE	26	9	15	0
NON-WASTE**** SPECIFIC SEQUENCE - METAL WASTE INPUT	1	10	8	31
NON-WASTE SPECIFIC SEQUENCE - TANNERY WASTE INPUT	26	0	0	24
NON-WASTE SPECIFIC SEQUENCE - ALL WASTE INPUTS	27	29	8	86

* Treatment Sequence = RO
** Treatment Sequence = CA/CA
*** Treatment Sequence = AS
**** Same as Chemical Waste - Specific Sequence

As was predicted earlier, unsatisfactory results were obtained as shown by the high incidence of effluent standard and influent screen violations. Using the four-step procedure delineated above, an acceptable system performance was obtained in three simulation iterations. The final sequences are:

Waste Type	Sequence
Metal	FL/RO/CA(2)
Chemical	OR/FL/CA(2)
Tannery	AS/FL/RO

It is interesting to note that the influent screens apparently performed their function successfully from the point of view that the final sequences exhibited no obvious anomalies such as inserting reverse osmosis in a sequence ahead of, say, coarse filtration (FL). The treatment of all three waste types by a single sequence, (Chemical), yielded unsatisfactory results due to excessive influent screen violations. The final simulation results are presented in Table 3.

TABLE 3 - SIMULATION RESULTS - (THIRD, AND FINAL ITERATION)

	NUMBER OF OCCURRENCES			
	NO TREATMENT REQUIRED	SUCCESSFUL TREATMENT	EFFLUENT STANDARD VIOLATION	INFLUENT SCREEN VIOLATION
METAL WASTE* SPECIFIC SEQUENCE	0	31	11	8
CHEMICAL WASTE** SPECIFIC SEQUENCE	0	44	0	6
TANNERY WASTE*** SPECIFIC SEQUENCE	25	23	0	2
NON-WASTE**** SPECIFIC SEQUENCE- METAL WASTE INPUT	0	30	11	9
NON-WASTE SPECIFIC SEQUENCE - TANNERY WASTE INPUT	25	5	0	20
NON-WASTE SPECIFIC SEQUENCE - ALL WASTE INPUTS	25	79	11	35

  * Treatment Sequence = FL/RO/CA/CA
 ** Treatment Sequence = OR/FL/CA/CA
*** Treatment Sequence = AS/FL/RO
**** Same as Chemical Waste-Specific Sequence

Effluent standard violations for the Metal waste treatment sequence were consistently minimal, (less than 10 percent of the standard). Given uncertainties in other model inputs, it appeared imprudent to add another stage of treatment to reduce these failures. The influent screen violations were solely a result of the Monte Carlo generation technique. The contaminants which caused violations were consistently unrealistically high in concentration magnitude, (e.g., 10^6 mg/l). It was previously decided to permit generation of such concentrations, (i.e., use of an unbounded normal deviate), so as not to disturb the probability densities of waste contaminants.

CONCLUSIONS

The stochastic optimization/simulation methodology is a useful tool for delineating least-cost liquid industrial waste treatment sequences. An important feature of the model is the characterization of influent contaminant concentrations by lognormal probability density functions, with influent waste streams generated by a Monte Carlo technique. Interdependencies between contaminants can be preserved in the generated influent realizations given estimates of their correlation structure.

Another important aspect of the methodology is automatic, objective evaluation and selection of extensive arrays of treatment sequences. The use of influent screens accomplishes this task through the creation of a stage-specific, variable state-space DP structure.

Depending upon application-specific boundary conditions and transition functions, the stochastic optimization exercise may yield acceptable final treatment configurations or alternatively, serve as a preliminary screening device. As a screening device, the optimization analyses identify unit processes with desirable cost-effectiveness attributes. For this type of result, subsequent iterative analyses involving stochastic simulation are needed to generate acceptable treatment configurations.

Numerous opportunities exist for enhancing the utility of the methodology through more detailed, comprehensive representations of model input and certain critical model components. Of particular note is the fact that unit treatment process removal efficiencies are generally not constant, as was assumed in this study. The consideration of removal efficiency variation, perhaps as a function of waste strength and flow for example, represents a logical next step in this modelling approach. This extension of the methodology could then more realistically depict the fact that variation in removal efficiencies modifies the form of the original waste contaminant probability density functions.

REFERENCES

1. Shih, C.S. and De Filippi, J.A., "System Optimization of Waste Treatment Plant Process Design," A.S.C.E., Sanitary Engineering Division, April 1970, page 409.
2. Adams, B.J. and Panagiotakopoulos, D., "Network Approach to Optimal Wastewater Treatment System Design," University of Toronto, Publication 75-13.
3. Bertouex, P.M. and Polkowski, L.B., "Optimum Waste Treatment Plant Design Under Uncertainty," Journal WPCF, September 1970, page 1589.
4. Evenson, D.E., et al., "Preliminary Selection of Waste Treatment Systems," Journal WPCF, November 1969, page 1845.
5. Shih, C.S. and Krishnan, P., "Dynamic Optimization for Industrial Waste Treatment Design," P.I.W.C., Number 24, page 456.
6. Bellman, R.E., "Dynamic Programming," Princeton University Press, 1957.
7. Berthouex, P.M. and Brown, L.C., "Monte Carlo Simulation of Industrial Waste Discharges," A.S.C.E., Sanitary Engineering Division, October, 1969, page 887.
8. Fraser, D.A.S., "Probability and Statistics Theory and Applications," Duxbury Press, Massachusetts, 1976.
9. Ellis, J.H., E.A. McBean and G.J. Farquhar, "Stochastic Optimization/Simulation of Centralized Liquid Industrial Waste Treatment," A.S.C.E., Journal of Environmental Engineering, Vol. III, No. 6, pp. 804-821.

DYNAMIC PROGRAMMING FOR FLOOD CONTROL PLANNING: THE OPTIMAL MIX OF ADJUSTMENTS TO FLOODS[1]

T. L. MORIN,* W. L. MEIER, JR.** AND K. S. NAGARAJ***

*School of Industrial Engineering, Purdue University, West Lafayette, IN 47907;
**College of Engineering, Pennsylvania State University, University Park, PA 16180;
***College of Business Administration, University of Iowa, Iowa City, IA 52242.

ABSTRACT

Despite substantial expenditures on flood protection structures, flood damages continue to increase. It has been observed that structural measures, such as dams, levees and channel improvements, often provide a false sense of security to existing and potential floodplain occupants and, as such, may actually result in increased flood damages, contrary to their intended purpose. This realization among others has led to an increasing awareness of and interest in the role of nonstructural measures, such as floodplain zoning, land use allocation, insurance, and warning, as an important and integral part of any overall flood damage mitigation program. However, determining an "optimal" mix of adjustments is very difficult as a consequence of both the interdependence between the structural and nonstructural measures and its inherent computational complexity that is the result of the multitude of feasible combinations of structural and nonstructural measures which must be considered over time and space. This paper develops a Dynamic Programming (DP) algorithm for the solution of the optimal mix of adjustments problem by recognizing that the overall problem has an underlying sequencing nature that bears distinct similarities to the sequencing approaches used in electric generation planning problems. Specifically, we adapt Erlenkotter and Rogers [36] dynamic programming approach to the optimal mix of adjustments problem. Computational refinements, data requirements, and implementation details are also discussed and details of an application to a real-world problem are given.

1. INTRODUCTION

Floods are the most widespread geophysical hazard in the United States and they account for greater average annual property losses than any other single geophysical hazard [115]. Moreover, despite substantial expenditures for flood control measures, flood damages continue to increase [110]. Specifically, the total annual national flood damages have been increasing by about 4 percent annually in real dollars during this century and there are indications that this rate has accelerated to the 6 to 7 percent range during the last decade [89]. The dollar value of these losses is truly staggering -- the total annual flood damages were $3.4 billion in 1975 and it has been estimated [88] that even with improved flood stream management, damages will exceed $4.3 billion (measured in 1975 dollars) by the year 2000. Without such improvements in flood plain management, the damages could approach $6 billion (in 1975 dollars) by the year 2000. In Indiana alone the annual flood damages for 1980 were $128 million (measured in 1978 dollars) [103].

---

[1] This research was supported in part by Department of the Interior, Office of Water Research and Technology Grant B-109-IND and Office of Naval Research Contract N00014-86-K-0689 to Purdue University.

The magnitude of problem prompted Congress to instruct the National Science Foundation to conduct a flood hazard mitigation study during fiscal year 1980 (*House Report 96:91*). NSF concluded that [89, p.1], "Innovative approaches and increased attention to flood problems nationwide are required if the United States is to arrest, much less reverse, rising flood losses and the social and economic burden they place on the people and the nation's tax-supported flood-relief institutions". This paper discusses one such innovative approach -- the development of a Dynamic Programming (DP) algorithm for the determination of an optimal mix of adjustments to floods.

The mix of adjustments to floods involves both structural and nonstructural measures. The structural (protective) measures for flood control typically include levees, floodwalls, channel improvements, and storage reservoirs. The nonstructural measures for flood control typically include land use control and management (*i.e.*, floodplain zoning, outright purchase of portions of the floodplain, and land use conversion), flood-proofing, warning and evacuation, relief and rehabilitation, and flood insurance. A recent analysis [115, p. 103] of the net benefits of various measures as a function of the magnitude of the catastrophe is presented graphically in Figure 1 -- see also [116].

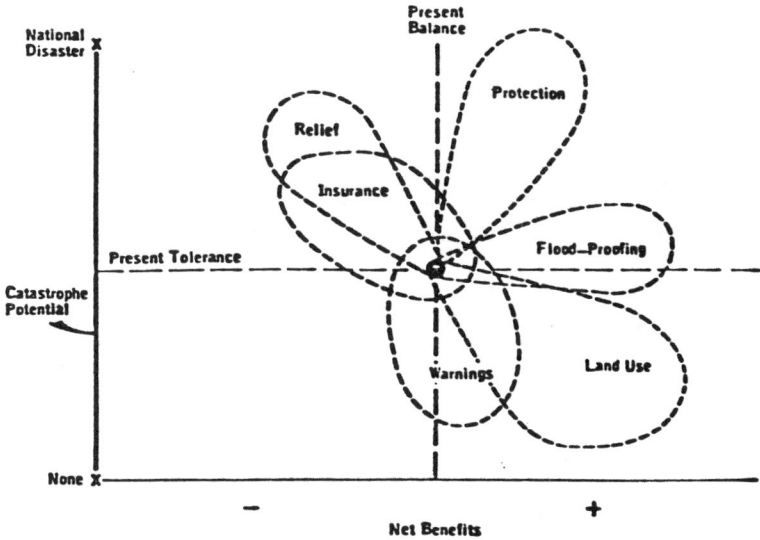

FIG. 1. Trends and limits of adjustments to floods [115].

Reliance has historically been placed on structural measures for flood control. However, it has been observed that structural measures often provide a false sense of security to existing and potential floodplain occupants and, as such, may actually result in increased flood damages, contrary to their intended purpose. That is, the potential benefits from structural flood control measures are often lost through subsequent unwise development in the areas presumed to be protected [89]. A tragic example of this was the overtopping of the Valont Dam in Veneto Province, Italy on October 9, 1963, which was caused by a massive rockslide (see Jansen [60]). It resulted in the almost total destruction of the floodplain below and the deaths of 2600 floodplain occupants. Such events and realizations have led to an increasing awareness and interest in the role of nonstructural measures as an important and integral part of any overall flood damage mitigation program -- for example see [3, 48, 116]. Furthermore, the age-old hope for complete protection from floods has given way to the realization that a more realistic and viable goal is the mitigation of flood damages [89]. In fact, the NSF study concluded [89, p. 214] that, "Flood hazard mitigation strategies can be effective only if they reflect mixes of alternative structural and nonstructural approaches appropriate to the circumstances. Much more work needs to be done on improving methods of planning for different aspects of flood hazard mitigation". This is precisely the issue addressed in the paper.

The paper is organized in the following manner. In §2 we review the relevant literature on previous approaches to flood control planning and related problems and outline our approach to the problem. The lack of viable planning methodologies is a consequence of the highly complicated nature of the problem and its inherent computational complexity that result, respectively, from the interdependence between the structural and nonstructural measures and from the multitude of feasible combinations of structural and nonstructural measures that must be considered over time and space. These factors have contributed to the fact that, although planning methods for problems involving only structural [77] and nonstructural [51] measures have been developed, the optimum mix of adjustment problem has yet to be satisfactorily resolved. We present a general formulation of the optimal mix of adjustments problem in §3 and develop a Dynamic Programming (DP) solution algorithm for its solution in §4. Computational refinements, data requirements and implementation details are also discussed and the results of an application to a real-world problem are presented in §4. The paper concludes with a discussion in §5.

2. FLOOD CONTROL PLANNING

Models of flood control planning problems cater to a variety of needs, have different emphases and considerations, and, hence, are not easily compared on the basis of their comparative strengths and weaknesses. Moreover, many of the models have been formulated with a particular application in mind and are, therefore, not readily adaptable to general models. Our approach is to first formulate a general model of the Flood Control Planning (FCP) Problem. Specifically, we first propose a general scenario of the (FCP) and then present a model which is suitable and adaptable to any particular application. The decision maker is typically a local government or legislative body that is interested in allocating funds so as to minimize the net annual damages incurred due to recurrent floods. Despite the fact that some of the funds may be allocated to measures which have benefits besides damage reduction, we shall restrict our considerations solely to damage reduction benefits of such measures. Specific structural and nonstructural measures include:

Structural	**Nonstructural**
Levees	Land Use Allocation
Reservoirs	Flood Plain Zoning
Channel Improvements	Flood-Proofing
Floodwalls	Insurance
	Warning
	Temporary Evacuation
	Emergency Procedures
	Relief and Rehabilitation

An important consideration in the evaluation of any single measure is that the measure's damage reduction capability is dependent on the presence and levels of the other measures and, hence, cannot be chosen independently. For example, the effectiveness of any nonstructural measure depends on the mix of previously built structural measures. We assume that the potential locations and levels of the structural alternatives have been specified and the decision maker then decides which of them are to be employed and when they shall be built. We also assume that the decision maker decides which of the nonstructural measures are to be used at any instant of time and their levels. Since the optimal levels of the nonstructural measures may vary over time (as a result of the damages changing over time), the decision maker can take one of two approaches: (1) allow for the possibility of the nonstructural measures to vary with time, or (2) consider the possibility of the nonstructural measures changing their levels only with the construction of a new structural measure. The next section formulates models and develops solution techniques for both of these problems. We assume that associated with each structural measure is a capital cost and an annual maintenance cost, while the nonstructural measures have their annual costs as functions of their levels. We are considering the preliminary planning stage where the decision maker is primarily concerned with how the funds should be allocated rather than determining, say, the precise detail of some land use allocation scheme.

An abstract version of the (FCP) can then be stated as follows: Find a combination (x,y) of structural (x) and nonstructural (y) measures so as to

$$\text{Max } f(x,y)$$
$$\text{subject to } (x,y) \in G(x,y) \quad \text{(FCP)}$$
$$x \in X$$
$$y \in Y,$$

in which $x = (x_1, x_2, ..., x_N)$ is the vector of structural measures, where $x_j = 1$ if structural measure j is selected and 0 if not, $y = (y_1, y_2, ..., y_K)$ is the vector of nonstructural measures, where y_k is the level of the kth nonstructural measure selected, $f(x,y)$ is the objective function; e.g., the discounted net reduction in flood damages resulting from plan (x,y), G(x,y) is the set of feasible plans (x,y), i.e., those satisfying the planning, financial, engineering, and social constraints, and X and Y, respectively, are the sets of feasible structural and nonstructural measures.

A portion of the large body of literature on flood control planning is relevant to the (FCP). Specifically, since man has historically attempted to control floods through physical structures, there have been numerous efforts, e.g., [9, 10, 18, 19, 21, 22, 40, 65, 85, 117, 118], to model the (FCP) as one of choosing from some discrete set of structural measures so as to minimize expected flood damages. Most of these models evaluate the flood routing effects of different combinations of structural measures through the use of damage functions. The structural measure (FCP) thus reduces to a capacity expansion (project sequencing) problem [2, 34, 77, 84] in which damages are minimized subject to budgetary constraints. This problem can be solved by dynamic programming [33, 34, 77, 120].

Unfortunately, structural measures have not alleviated all of the flood damages and there have even been cases documenting *increased* flood damages as the result of intensified land use in the floodplain below the structural measures (see Jansen [60]). With this there came the recognition [49, 67, 75, 98] that in order to more effectively reduce flood damages the structural measures should be complemented by nonstructural measures, such as land use planning and flood plain zoning, to control undue encroachment into the flood plain. Since then, there have been significant efforts [7, 12, 13, 23, 24, 25, 51, 64, 101] to model the damage reduction problem as a land use allocation problem for a given flood routing scheme, including the use of dynamic programming models. Specifically, Hopkins et al. [51, 52, 53] model the problem as a discrete dynamic programming problem with the reaches corresponding to the stages in the dynamic program.

The land use planning models have the objective either of minimizing damages or of maximizing economic rents subject to fixed flood routing characteristics and, hence, are only suitable for applications in which structural growth has been curtailed. A number of conceptual frameworks [91, 96, 115] for the *combined* (FCP) of determining both the structural and nonstructural measures (of which land use allocation is only one) consider the interaction among various measures -- see Figure 2 [Fig. 3-3, 116] for an example of an interaction matrix. A number of combined models [7, 8, 10, 13, 19, 21, 26, 27, 31, 54, 56, 59, 95, 103, 111] have been developed including those which consider only agricultural damage reduction [69, 94, 100]. Several models [41, 72, 74] emphasize insurance as an important nonstructural measure for distributing the flood losses more evenly rather than just reducing them.

Initial Adjustment	Control and Protection	Flood-Proofing	Land Use Planning	Warnings	Insurance	Relief and Rehabilitation
Control and Protection		O	O	O	O	O
Flood-Proofing	O		O	●	?	O
Land Use Planning	O	?		●	●	O
Warnings	O	●	●		?	O
Insurance	O	?	?	●		O
Relief and Rehabilitation	●	O	O	O	O	

(Other Adjustment Affected)

Stimulated by the initial adjustment:

● — High stimulation
O — Little or none
? — Doubtful

FIG. 2. Matrix of interaction of adjustments to floods [116].

James [56, 57, 58, 59] apparently provided the first published attempts at modelling the combined (FCP) problem. Flood-proofing is taken into account with the help of 'reduction factors', and the planning horizon is split into time blocks. The different structural possibilities are completely enumerated and the best picked. The approach of enumerating the structural measures has also been employed by Day and Weisz [26, 112, 114] where they determine the optimal land use allocation via linear programming. However, enumeration severely limits the number of combinations and sequences of structural measures that can be considered. To accommodate a greater number of structural measures, Cortes-Rivera [19] solved the combined (FCP) problem using discrete differential dynamic programming. In his dynamic programming model each stage corresponds to a time period and the land use allocation is determined using parametric linear programming with the controlling parameter being a function of the probability of flood determined by the particular mix of structural measures. Cortes-Rivera also applied his model to a real-world problem involving the Embarras river basin in South-Central Illinois. We will subsequently solve this problem to demonstrate the computational superiority of our dynamic programming approach to the (FCP).

Combined simulation/optimization models in which the control benefits are determined by simulation have also been used to solve the (FCP) [21, 30, 39, 63, 71]. These models are similar to dynamic programming models for electrical generation planning, such as [4, 11, 14, 20, 32, 36, 37, 38, 42, 61, 62, 68, 70, 80, 87, 90, 93, 97], in which the variable systems operating costs, as opposed to the flood control benefits, are determined by the simulation. We will exploit this similarity to develop our dynamic programming approach to the (FCP).

To summarize, there appears to be a lack of models and solution techniques which can satisfactorily evaluate comprehensive flood control plans involving both structural and nonstructural measures. One reason for this is the lack of suitable data [46, 47, 99, 119] on the interdependence between structural and nonstructural measures in terms of damage reductions. For example, Shabman et al. [99] state that the influence of flood plain zoning cannot be easily identified. However, White [116] has made a significant progress in enumerating the various interaction possibilities qualitatively (see Figures 1 and 2). Some examples of the interrelationships include the use of warning complementing flood-proofing in terms of damage reduction and the use of engineering projects for controlling floods in one portion of the floodplain accompanied by local regulations preventing further encroachments into other sectors of the floodplain. Depth/stage-damage functions could play a critical role in specifying the interrelationships and data for these functions are widely available [15, 17, 50, 66, 92, 104-109, 113]. Unfortunately, many of these functions are descriptive rather than prescriptive and, furthermore, are typically restricted solely to land use allocation schemes.

Another reason for the current lack of models that comprehensively consider all the nonstructural possibilities is the corresponding lack of suitable solution procedures for such a difficult optimization problem. To see this we only have to look at Problem (FCP) which is a very difficult optimization problem for a number of reasons. Firstly, the reduction in flood damages for any x_j or y_k is *dependent* not only on the value of x_j or y_k but also on all of the measures which both precede it and those that come after it. That is, the benefits which accrue from the nonstructural measures are dependent on structural measures selected and *vice versa* and these vary over the time horizon. Secondly, the problem involves both discrete and continuous decision variables -- see [44] for a discussion of the difficulties involved in such problems. That is, whereas the decisions for the structural measures involve only which measures (x) to select, the decisions for the nonstructural measures involve both which and how much (y) of each measure to select. For example, if j corresponds to a storage reservoir of a certain predetermined size, a decision is made either to build it ($x_j = 1$) or not ($x_j = 0$), whereas if k corresponds to floodproofing, a decision has to be made whether to floodproof or not ($y_k = 0$) and if so, at what level ($y_k > 0$). Finally, the problem involves the evaluation of an enormous number of feasible combinations (x,y) of structural and nonstructural measures that must be considered over time and space, and even the structural measure subproblem has been shown to be NP-complete in the strong sense [1].

In the next section we formulate a model of the optimal mix of adjustments problem that is general enough to encompass the important considerations and relationships involved in flood control planning and yet is specific enough to recognize the special structure of the (FCP).

3. OPTIMAL MIX OF ADJUSTMENTS

The Optimal Mix of Adjustments (OMA) Problem can be viewed as a capacity expansion problem involving a discrete set of structural measures accompanied by continuous set of nonstructural measures. Capacity expansion problems that involve operating level decisions which are continuous in nature are fairly common. A sampling of such discrete/continuous planning problems includes: simultaneous investment sequencing and allocation decisions for water resources systems design

[5, 55]; multi-commodity distribution systems design [45]; generation expansion of hydroelectric systems [14]; and investment sequencing with price sensitive dynamic demand [35]. We will first review the available solution methodologies for the general class of discrete/continuous planning problems in order to determine their applicability to the (OMA). These solution methodologies can be divided into two different types.

The first type of solution methodology consists of formulating the sequencing and selection of projects problem as an integer program using 0-1 variables. The continuous variable selection problem could in general be a nonlinear programming problem and is conditioned upon the presence or absence of key 0-1 variables. Thus, the overall model becomes a mixed 0-1 nonlinear integer programming problem. An example of this approach was presented by Armstrong and Willis [6] who formulate a water resources planning problem as a mixed-integer quadratic programming problem. One of the more efficient algorithmic approaches for this type of formulation appears to be Generalized Benders Decomposition [43]. Generalized Benders Decomposition exploits a special feature of the discrete/continuous formulation: when one of the two classes of variables (termed "complicating") are fixed, the problem of determining the optimal values of the variables in other class is much simpler than the overall problem. However, Generalized Benders Decomposition has the following undesirable characteristics when applied to the (OMA):

(1) The planning horizon has to be split into planning periods and capacity expansion decisions made period by period. This restricts us to discrete time finite horizon problems. More importantly, the size of the problem increases multiplicatively with the number of planning periods. In fact, we [1] have recently proven that the structural problem alone is NP-complete.

(2) Since the project selection and sequencing problem is formulated as an integer program, when Benders cuts are added to the integer program in the course of the decomposition algorithm we are essentially losing the special sequencing structure and, thus, are not able to most effectively exploit this structure.

Both of these undesirable characteristics are absent from the second solution methodology which recognizes that the sequencing problem is the key component and considers the overall problem as essentially being a sequencing problem subject to certain restrictions. Indeed, the general framework of sequencing problems defined by Erlenkotter and Rogers [36] subsumes versions of several of the planning problems referred to earlier as special cases. Their elegant and simple model exploits the sequencing nature of the problem and, furthermore, considers timings of projects as being variables, thus precluding the need to divide the planning horizon into planning periods.

The Optimal Mix of Adjustments (OMA) problem can be described in Erlenkotter and Roger's framework for sequencing competitive expansion projects [36]. There are two key observations that permit this. The first is to recognize that, whereas the structural and nonstructural measures compliment each other, the structural measures compete with one another in terms of their net effect on reducing flood damages. That is, the net reduction in flood damages from a structural measure diminishes as additional measures are added. For example, the net reduction of flood losses due to a dam diminishes as additional dams are added upstream. The second key observation is that flood damages can be interpreted as (operating) costs that vary with the level of the nonstructural measures.

We will consider the planning horizon as being a multiple of some basic unit rather than use Erlenkotter and Roger's continuous-time framework in order to more meaningfully express the expected flood damages. That is, in analyzing the effects of flood events arising as a result of natural conditions which repeat in cycles,

It would not be possible (except in trivial cases) to estimate flood damage over a partial cycle. Thus, we need to consider time in multiples of a basic cycle unit (say a year) rather than as a continuous variable. For ease of formulation, we shall consider the time unit to be a year.

We first construct a model for the Optimal Mix of Adjustments Problem in which the nonstructural measures may vary over time. The selections of the nonstructural measures are related to the mix of the structural measures in a natural way. For any year of the planning horizon and any given set of structural measures present during that year, the determination of the optimal levels, y^*, of the nonstructural measures that minimize the expected damages during the year is a relatively well defined problem. This is in accordance with the viewpoint that the nonstructural measures essentially *compliment* a given set of structural measures in terms of damage reduction. We can express this problem as

$$P(I,t) = \min_{y \in Y \cap G(I)} \{D(I,y,t) + C(y,t)\} \tag{1}$$

where $D(I,y,t)$ is the annual flood damage in the tth year for a given combination I of the structural measures when the level of nonstructural measures are y; $C(y,t)$ is the annual cost incurred in the tth year with the levels of the nonstructural measures at y; and G and Y are similar to those defined in the (FCP) problem. $P(I,t)$ denotes the minimal sum of the *net* annual flood damages and the nonstructural measure costs in the tth year for the combination I of the structural measures. We observe that the concept of considering the nonstructural measure determination as a subproblem conditioned on I has previously been used. Specifically, in [19] (1) is a linear programming problem parametized by some function of I.

We note that when the decision maker cannot differentiate between the years while constructing the damage function (as is normally the case), then $P(I,t)$ would simply reduce to $P(I)$. However, in order to not lose the general structure, we shall allow P to be a function of t. If we interpret $P(I,t)$ as the *annual operating cost* in the tth year of the set I of structural measures, then we can use (a discrete time version of) the very general sequencing framework developed by Erlenkotter and Rogers [36]. The following assumptions from [36] are relevant to our formulation of the Optimal Mix of Adjustments Problem.

(a) A finite number, m, of structural measures may be undertaken, and each project is indexed by an $i \in I^* = \{1,2,...,m\}$. The investment cost for project i is given by $c_i > 0$. This also includes an allowance for the present value of maintenance, replacement and other fixed operating costs.

(b) Let I denote an arbitrary subset of project indices and ℓ denote the power (or ground) set consisting of all the possible 2^m subsets I. The variable operating cost rate (annual net flood damage as a function of the nonstructural measures) in year t for the project set I is expressed by $P(I,t) \geq 0$. Furthermore, for each I, some project $i \in I$ must be established and added to I no later than the time $T(I) \geq 0$, where $T(I) \leq T(I \cup i)$ for all $i \in I$ and $T(I^*) = +\infty$.

(c) Costs are continuously discounted at a constant rate, $r > 0$, leading to a discount factor of e^{-rt} from time t to the initial time 0.

(d) Sequencing and timing decisions for the projects are to be selected so as to minimize the total net discounted damages over an infinite horizon.

The minimum damage sequence is selected from the m! possible orderings for the projects. To formulate this selection problem, we introduce the following additional notation from [36] : i[k] is the project index assigned to the kth position in a sequence; {i[k]} is the complete assignment of project indices for a particular sequence, where k = 1,2,...,m; S_I is the set of all permutations of project indices in I; I_k is the set of first k project indices for a particular sequence, where $I_0 = \phi$, $I_{k+1} = I_k \cup 1[k+1]$ for $k = 0,1,...,m-1$, and $I_m = I^*$; τ_k is the establishment time for the kth project in a sequence, where $\tau_0 = 0$, $\tau_k \leq \tau_{k+1}$, and $\tau_{m+1} = +\infty$; $C(I^*,\infty)$ equals the total net flood damages over the time interval $[0,\infty)$ discounted to time 0 for a minimum-damage sequencing.

Our model of the Optimal Mix of Adjustments (OTA) Problem is then

$$C(I^*,\infty) = \min_{\{i[k]\} \in S_{I^*}} \min_{\{\tau_k\}} \sum_{k=0}^{m} \left\{ \sum_{t=\tau_k}^{\tau_{k+1}-1} P(I_k,t) e^{-rt} + \sum_{k=1}^{m} c_{i[k]} e^{-r\tau_k} \right\}, \quad (2)$$

where

$$0 = \tau_0 \leq \tau_1 \leq \cdots \leq \tau_m < \tau_{m+1} \text{ and } \tau_{k+1} \leq T(I_k), k=0,1,...,m-1.$$

Note that the model (2) allows for the possibility of not establishing some projects since an establishment time equal to a very large value implies indefinite postponement, which is equivalent to eliminating that project from consideration.

We now consider the model of the (OTA) when we allow the nonstructural measures to change their level only with the construction of a new structural measure. Let y_k be the new level of nonstructural measures accompanying the construction of structural measure 1[k], where y_0 is the initial level of the measures. In the format of (2) this becomes

$$C(I*,\infty) = \min_{\{i[k]\}} \min_{\{\tau_k, y_k\}} \sum_{k=0}^{m} \sum_{t=\tau_k}^{\tau_{k+1}-1} [D(I,y_k,t) + C(y_k,t)] e^{-rt} + \sum_{k=1}^{m} c_{i[k]} e^{-r\tau_k}, \quad (3)$$

where

$$\tau_0 \leq \tau_1 \leq \cdots \leq \tau_m < \tau_{m+1} \text{ and}$$

$$\tau_k \leq T(I_{k-1}), y_k \in G(I_k) \cap Y, \quad k=1,...,m.$$

We will conclude this section with a discussion of the generality of our model. Recall that the subproblem (1) involving the optimization over the nonstructural measures is

$$P(I,t) = \min_{y \in Y \cap G(I)} \{D(I,y,t) + C(y,t)\}, \quad (4)$$

in which $P(I,t)$ is the net minimum annual flood damage as a function of the nonstructural measures in year t when the mix of structural measures is I. The key to the applicability of our model lies in the elements of expression (4) and their meaningful evaluation. Since one would not normally expect the damage effects to change with time, we will consider a simpler version of (4); namely,

$$P(I) = \min_{y \in Y \cap G(I)} \{D(I,y) + C(y)\}. \quad (5)$$

This function characterizes the effects of the nonstructural measures with respect to damage reduction. Various considerations that can be incorporated through (5) are:

- *Social and political constraints* can be expressed through the set Y or the constraint set of G(I).

- *Demographic changes over time* can be considered as affecting the damage function D(I,y,t) where now t has been appended in order to incorporate the changing considerations with time.

- *Flood Routing*: We can consider the damages that occur during the presence of different levels of structural measures to be a function of some hydrological parameters. Thus, the variable cost of routing for a given set I of structural measures could be incorporated into the damage function. Flood routing could be done using a Standard Project Flood or a distribution of floods to estimate expected hydrological parameters.

- *Damage function D(I,y)*: A simulation could be performed to determine the damage if need be. Note that this function is crucial in applying the model. Our models (2) and (3) provide for ease of incorporating different types of damage functions.

We note that determining P(I) should be kept as simple as possible (for example, via a parametized linear program) since it will be used frequently during the execution of the Dynamic Programming (DP) algorithm discussed in the following section.

4. DYNAMIC PROGRAMMING ALGORITHM

We now turn to the solution of the Optimal Mix of Adjustments Problem in models (2) and (3). The basis for our approach to model (2) is the Dynamic Programming (DP) algorithm presented by Erlenkotter and Rogers [36]. For convenience, we will rewrite model (2) as

$$C(I^*,\infty) = \min_{\{i[k] \in S_I^*\}} \min_{\{\tau_k\}} \left\{ \sum_{k=0}^{m} \sum_{t=\tau_k}^{\tau_{k+1}-1} P(I_k,t)e^{-rt} + \sum_{k=1}^{m} c_{i[k]} e^{-r\tau_k} \right\}, \qquad (6)$$

where

$$0 = \tau_0 \leq \tau_1 \leq \cdots \leq \tau_m < \tau_{m+1} = \infty, \quad 0 \leq \tau_k \leq T(I_{k-1}) \quad k=1,\ldots,m,$$

and τ_k is the *beginning* of the year in which project 1[k] is available for use. If $\tau_{k+1} - 1 < \tau_k$ we define the second summation in (6) to be 0.

Note that (6) is a very difficult optimization problem involving both discrete and continuous variables. Fortunately, however, this problem reduces to a sequencing problem when the following transformations are made. First, observe that (6) is of the form:

$$C(I^*,\infty) = \min_{\{i[k]\} \in S_I^*} \{F(1[k])\}. \qquad (7)$$

That is, the function F is optimized over all permutations of the 1's. For a general F, (7) is a difficult combinational problem. However, if F is separable into functions $g(I_k, 1[k])$, k=1, ...m, then we can exploit branch-and-bound strategies to solve (7). In particular, (7) would fit naturally into a Dynamic Programming (DP) framework and, indeed, it is this approach that we shall use to solve (6).

Initially it may not be obvious how to separate the cost (damage) function into a sum of $g(I_k, 1[k])$ functions. That is, when we consider each 1[k] to be associated with the determination of τ_k, the limits, τ_{k+1} -1 and τ_k of the summation term complicate matters in regards to its separation into functions of I_k and 1[k]. However, Erlenkotter and Rogers [36] use the following simple result to transform the cost function

$$\sum_{t=\tau_k}^{\tau_{k+1}-1} P(I_k,t)e^{-rt} = \sum_{t=\tau_0}^{\tau_{k+1}-1} P(I_k,t)e^{-rt} - \sum_{t=\tau_0}^{\tau_k-1} P(I_k,t)e^{-rt}. \tag{8}$$

Thus, (6) can be rewritten as

$$C(I^*,\infty) = \min_{\{i[k]\}\in S_{I^*}} \sum_{k=1}^{m} \min_{\tau_{k-1} \leq \tau_k \leq T(I_{k-1})} \left\{ \sum_{t=\tau_0}^{\tau_k-1} [P(I_{k-1},t)-P(I_k,t)]e^{-rt} + c_{i[k]}e^{-r\tau_k} \right\}$$

$$+ \sum_{t=\tau_0}^{\tau_{m+1}-1} P(I^*,t)e^{-rt}. \tag{9}$$

We can now interpret

$$\min_{\tau_{k-1} \leq \tau_k \leq T(I_{k-1})} \sum_{t=\tau_0}^{\tau_k-1} [P(I_{k-1},t) - P(I_k,t)]e^{-rt} + c_{i[k]}e^{-r\tau_k}, \tag{10}$$

as the cost associated with adding 1[k] to obtain the combination I_k. We now consider using DP to solve (9), with I as a state, $|I| = n$, the number of projects in subset I as the stage associated with I, and adding $l \in I$ as the last project to obtain I as the required decision at state I in stage $n = |I|$. In DP notation (see [79]) the cost function, h (ξ, y', d) can be written

$$h(\xi,I=I_k, l=1[k]\in I) = \xi + \min_{\tau_{k-1} \leq \tau_k \leq T(I-i)} \sum_{t=\tau_0}^{\tau_k-1} \left\{ [P(I-1,t) - P(I,t)]e^{-rt} + c_i e^{-r\tau_k} \right\} \tag{11}$$

where h (ξ, y', d) is the cost of reaching state t (y', d) by an initial sequence of decisions that reaches state y' from the initial state at a cost of $\xi \in \mathbf{R}$ and is then extended by the decision d. This cost function (11) does not necessarily satisfy the monotonicity condition [79] required for the validity of the functional equations of DP. That is, let ξ_1, and τ_{k-1}^1 be the total return and project timing, respectively associated with l_{k-1}^1 as the last project added to construct I-1, and let ξ_2 and τ_{k-1}^2 be those associated with l_{k-1}^2 as the last project added to construct I-1. Then if $\xi_1 < \xi_2$ and $\tau_{k-1}^1 > \tau_{k-1}^2$ it is possible that $h(\xi_1,I,1) > h(\xi_2,I,1)$. Notice that the reason that this cost function does not satisfy the monotonicity assumption is the constraint $\tau_{k-1} \leq \tau_k$. Thus, consider a relaxation to (9) with the constraints $\tau_{k-1} \leq \tau_k$ deleted; namely,

$$C_R(I^*,\infty) = \min_{\{i[k]\}\in S_{I^*}} \sum_{k=1}^{m} \min_{0 \leq \tau_k \leq T(I_{k-1})} \left\{ \sum_{t=\tau_0}^{\tau_k-1} [P(I_{k-1},t)-P(I_k,t)]e^{-rt} + c_{i[k]}e^{-r\tau_k} \right\}$$

$$+ \sum_{t=\tau_0}^{\tau_{m+1}-1} P(I^*,t)e^{-rt}. \tag{12}$$

For (12), the return function of DP is

$$g(I,l) = \min_{0 \leq \tau \leq \tau(I-i)} \{g(I,l,\tau)\} \tag{13}$$

$$= \min_{0 \leq \tau \leq T(I-i)} \sum_{t=\tau_0}^{\tau-1} [P(I-1,t) - P(I,t)]e^{-rt} + c_i e^{-r\tau}.$$

Notice that this return function (13) satisfies the monotonicity function and, hence, we can use DP to solve (12). Note that when using DP we need efficient mechanisms to evaluate the return function (13) since it will be evaluated frequently in the course of the algorithm.

Since we are only solving a relaxation to (9), we need a procedure to translate the solution of the relaxed problem into an optimal solution to the original problem. We do this by imposing some reasonable conditions on P which guarantee that the optimal solution to the relaxed problem is feasible and, hence, optimal to the original problem. To accomplish this, we need the following definition. If there are many τ''s that all minimize (13), then we define $\tau^* = \max \tau'$; i.e., τ^* is the *latest minimum damage* (*l*md) establishment time for (13). Combining (12) and (13), the relaxed formulation becomes

$$C_R(I^*,\infty) = \min_{\{i[k]\} \in S_I^*} \sum_{k=1}^{m} g(I_k,1[k]) + \sum_{t=\tau_0}^{\tau_{m+1}-1} P(I^*,t)e^{-rt}. \qquad (14)$$

Denoting $n = |I|$; i.e., the number of elements in (or *cardinality* of) the subset of project indices I, we define

$$f_n(I) = \min_{\{i[k]\} \in S_I} \sum_{k=1}^{n} g(I_k,1[k]). \qquad (15)$$

Finding $f_m(I^*)$ is equivalent to finding $C_R(I^*,\infty)$ since

$$C_R(I^*,\infty) = f_m(I^*) + \sum_{t=\tau_0}^{\tau_{m+1}-1} P(I^*,t)e^{-rt}. \qquad (16)$$

It is easily seen that the sequencing of the projects in I must result in the minimum net damage as defined in (15). Possibly several sequences of projects in some subset I may provide equal minimum net damage values. If this occurs, we adopt the tie-breaking rule of selecting the minimum net damage sequence that has the *latest l*md establishment time $\tau^*(I,1)$ for the last project 1 added to complete the set I. We shall call a sequence obtained by this rule a *latest minimum damage* solution for I and shall use *l*min to denote the lexicographic solution process of comparing solutions first on costs and second on the *l*md establishment time for the last project, 1. Notice that this definition of the latest minimum damage time for I is consistent with the definition for the *l*md establishment time for each 1.

We may determine a latest minimum damage solution for the relaxed formulation (14) using the following *functional equations* of DP

$$f_n(I) = l\min_{i \in I} \{g(I,1) + f_{n-1}(I-1)\}. \qquad (17)$$

In order to guarantee that the optimal solution to the relaxed formulation is feasible for (9) and, hence, optimal to (6), P must satisfy certain conditions. One such condition is

$$P(I,t) - P(I \cup 1,t) \geq P(I \cup 1',t) - P(I \cup 1' \cup 1,t). \qquad (18)$$

That is, the damage reductions obtained by adding project 1 to the set I are nonincreasing as additional projects are added to I. Another way of viewing this is

$$P(I,t) - P(I \cup 1 \cup 1',t) \leq [P(I,t) - P(I \cup 1,t)] + [P(I,t) - P(I \cup 1',t)]. \qquad (19)$$

That is, the projects have a subadditive relationship in terms of damage reduction: there is no *synergism*. For example, if we are considering constructing a levee and performing channel improvements, then the sum of the damage reductions associated with just constructing a levee and the damage reductions associated with just performing channel improvements, will be *greater* than the damage savings associated the implementation of *both* these structural measures.

The use of this condition (18) in guaranteeing the optimality of the solution given by (17) to the original problem (6) is easily established.

Proposition 1. *If (18) holds; i.e., the net damage reductions are subadditive in the solution to the relaxed formulation (12) defined via (17), then the timings for successive expansions satisfy the constraints $\tau_k \leq \tau_{k+1}$ and, hence, the solution also solves the original formulation (6).*

Proof. Follows from the proof of Proposition 2 of [36] *mutatis mutandis* and, hence, is omitted.‖

Thus, when the minimum net damage function P satisfies (18), the solution procedure defined by (17), is valid and $C(I^*,\infty) = C_R(I^*,\infty)$. Erlenkotter and Rogers [36] observe that this forward DP functional equation (17) generalizes the forward DP formulation for the simple expansion sequencing problem of [33, 34].

We next discuss procedures for evaluating the cost function, $g(I_k,1[k])$, as given in (13). Notice that this evaluation involves 1) determining the *lmd* establishment time τ_k^*, and 2) calculating the value $g(I_k,1[k]) = g(I_k,1[k],\tau_k^*)$. Note that it is possible that $g(I_k,1[k],\tau_k^*)$ is obtained in the process of determining τ_k^*, though this need not necessarily be the case. We have the following result on the latest minimum damage establishment time τ_k^*.

Proposition 2. *If $P(I,t) - P(I \cup l,t)$ is nondecreasing in t for each $I \in \mathscr{J}$ and $l \in I$, then the 'lmd' establishment timing τ_k^* for (13) is given by*

$$\tau^*(I,l) = \begin{cases} 0, \text{ if } T(I-1)=0 \text{ or } J(I,l,0) > 0, \\ \tau : J(I,l,\tau) \leq 0 \text{ and } J(I,l,\tau+1) > 0, \text{ if } J(I,l,T(I-1)) > 0, \text{ and} \\ T(I-1), \text{ otherwise,} \end{cases} \quad (20)$$

in which

$$J(I_k, 1[k], t) = P(I_{k-1}, t) - P(I_k, t) + c_{i[k]}(1-e^{rt}). \quad (21)$$

Proof. Refer to [83].

We have developed efficient procedures for determining $g(I,l,\tau^*(I,l))$ given $\tau^*(I,l)$ [83]. Details of a DP approach for the solution of model (3) of the Optimal Mix of Adjustments (OMA) Problem can also be found in reference [83]. We next present details of an application to a real-world problem that clearly demonstrate the superiority of our approach to conventional DP.

Application to a Real-World Problem

We will compare our DP approach to a conventional DP approach with time periods as stages used by Cortes-Rivera [19] on an Optimal Mix of Adjustments (OMA) Problem involving the Embarras River basin located in South-Central Illinois. We first describe the structural and non-structural measures considered, damage specification, and the constraints involved.

The structural components that were considered for construction are:

1. (a) A flood detention reservoir that can be one of seven different sizes.
   (b) The land required to be purchased for constructing the reservoir with different amounts for different sizes.
2. (a) A levee (denoted Levee-3) that can be one of nine different heights.
   (b) The purchasing of land that goes along with each of these.
3. (a) Another levee (denoted Levee-4), at a different location from the former, that can be one of 12 different heights.
   (b) The purchasing of land that goes along with each of these.

The nonstructural measures are agriculture land-use distribution in terms of *allocation of land* for (1) corn, (2) soybeans and (3) pasture.

After incorporating various hydrological considerations and performing linear programming optimizations over the agricultural land-use measures, Cortes-Rivera provides the function P(I,t) where P is the expected annual net income in year t for a given combination I of the structural measures. Other considerations involve (1) a rate of land-value differential inflation different from the normal rate of discount, (2) budget constraints in the form of a given budget supply schedule, and (3) institutional constraints in the form of a minimum time interval required between the construction of the two levees. The problem also provides for changes in levee height once a levee has been built. The objective is to determine which of the structural measures are to be built and, if they are built, when and at what rates should they be built. Furthermore, if there are to be changes in levee heights with time, then both when and how much change are also to be determined. The optimal land-use allocation in each year has already been determined in the process of determining the P function. All the above sequencing and timings of the structural measures are subject to the constraints referred to earlier.

In Cortes-Rivera's conventional DP approach the planning horizon (50 years) was split into 10 subperiods, five years each and, thus, construction decisions had to be made only on a subperiod basis. A conventional DP approach is taken with each stage corresponding to a subperiod. Each state in any stage is a 6-dimensional vector with the following attributes:

1. Storage capacity of the Detention Reservoir,

2. Elevation of Levee-3,

3. Elevation of Levee-4,

4. Land required for the construction of the Detention Reservoir,

5. Land required for the construction of Levee-3, and

6. Land required for the construction of Levee-4.

The decision variables for each state are represented by the magnitude of the enlargements or expansions of each of the structural components and by the additional land to be acquired for the construction of each structural components. Thus, there are six decision variables that are in one-to-one correspondence with the six state variables.

Since the state space turns out to be large (see Table 1), Cortes-Rivera [19] had to resort to differential discrete DP. Our DP approach is described next.

For now, we shall add an additional condition that once a levee has been built to a particular height, its height cannot be changed. This assumption provides for a much simpler solution procedure. In our DP approach, a stage corresponds to the *construction* of a structural measure. Thus, we only have three stages with each state in stage 1 being some structural measure at some capacity, states in stage 2 being all possible (unordered) pairs of structural measures and, finally, the states in stage 3 being all possible (unordered) triplets of the structural measures. The decisions associated with each state in stages 1 and 2 are what additional structural measure should be constructed.

Notice that the timing and the amount of land purchased for any structural measure can be incorporated into the cost of the measure in the form of

$$c_i(I,\tau) = c_i + c_{land} \min_{B(I,i) \leq t \leq \tau} \{(1+d)^t \cdot e^{-rt}\} e^{r\tau}, \qquad (22)$$

TABLE I. Comparison of the number of stages and states for two DP algorithms on Cortes-Rivera's problem.

Our DP Algorithm		Cortes-Rivera's (Conventional) DP Algorithm	
Stage	No. of states (worst case)	Stage	No. of states
1	28	1	7,125†
2	255*	2	7,125
3	756	.	.
		.	.
Total	1,039	.	.
		10	7,125
		Total	71,250

* $255 = 7 \times 9 + 7 \times 12 + 9 \times 12$

† $7125 = (7 + 1 + 7) \times (9 + 1 + 9) \times (12 + 1 + 12)$

where I is the resulting set of structural measures, l is the additional measure considered, d is the value of land-value differential inflation, c_{land} is the cost of the land required for measure l, c_i is the cost of the measure, r is the discount rate, and B(I,l) is the minimum possible time for purchasing as determined by the budget supply schedule and the current budget consumption. Notice that the optimization is trivial, since if $r < \log_e(1+d)$, then $t^* = B(I,l)$ and if $r \geq \log_e(1+d)$, then $t^* = \tau$.

The cost function associated with the addition of l to form I is

$$g(I,l) = \min_{\tau} \sum_{t=0}^{\tau-1} [P(I,t) - P(I-l,t)]e^{-rt} + c_i(I,\tau)e^{-r\tau}. \tag{23}$$

Since P has already been determined, it is fairly easy to obtain the return function as defined above.

The computational superiority of our approach to that of Cortes-Rivera is demonstrated in Table 1 where the number of states and stages required by both DP approaches are compared. Indeed, hand-calculations were all that was required by our DP approach on this application, as opposed to the extensive computer computations that were required in the Cortes-Rivera DP approach. We also note that our approach would be even more advantageous as the number of projects and periods grew. Finally, note that the *worst case* number of states for our DP approach is given in Table 1. The actual number of states evaluated was much less than this as a result of computational enhancements.

Our DP approach could even be further improved with the aid of computational refinements such as reaching [28, 29, 78] state elimination by bounds [73, 81] and *a posteriori* optimality analysis [82, 86] -- see [83] for details.

5. DISCUSSION

We have presented a very general model for the Optimal Mix of Adjustments Problem and have shown that this model fits naturally into a dynamic programming framework. This model is general enough to include numerous different types of damage functions. Furthermore, we proved that if the damage functions satisfy some reasonable and fairly mild conditions, then dynamic programming provides an exact solution algorithm. We discussed different ways of implementing the algorithm to account for and exploit properties of the different methods of estimating damage functions. Finally, we addressed issues of computational efficiency and pointed out various improvements which could be incorporated into the solution algorithm.

We next consider applications. A specific application of our solution approach was discussed in §4. In particular we indicated how our approach can be easily implemented even using hand-calculations. That is, the computational requirements were modest compared to the alternate DP solution approach originally used by Cortes-Rivera [19] to solve this problem. We also note that the effects of interdependence between the structural and nonstructural measures has been addressed qualitatively [115, 116], as depicted in Figures 1 and 2, and that data exist for specific applications -- see [19, 112] for example. However, to date there appear to be no readily available *general* quantitative methods (except for the approach we described in Appendix B of [83]) for estimating the damage functions that comprehensively evaluate interactions among the structural and nonstructural measures. Thus, one might critically ask if our model imposes excessive information requirements on the user when it requires such damage functions to be constructed. The answer, of course, is that if detailed quantitative decision making support is required, then it is essential that these damage functions be specified. In such cases we recommend our approach for constructing comprehensive damage functions described in [83]. However, if less detail is required, then it may be more reasonable to aggregate the effects of the nonstructural measures either in the constraints or in the damage function and explicitly consider only the structural measures as decision variables. One could then use DP [34, 77] to determine the sequencing of the structural measures. Once the sequencing has been determined, it may then be possible to assign different levels of the nonstructural measures using the decision maker's *qualitative* understanding of the effects of the nonstructural measures. However, not only would such an approach be heuristic, but it also would completely neglect interdependencies and interactions among the measures. We would, therefore, recommend our general model and DP algorithm for all except the most preliminary (or purely qualitative) phases of planning.

REFERENCES

1. V. Akileswaran, G.B. Hazen, and T.L. Morin, "Complexity of the Project Sequencing Problem", *Operations Res.* **31**, 772-778 (1983).
2. V. Akileswaran, T.L. Morin, and W.L. Meier, Jr., "Heuristic Decision Rules for Water Resource Planning", *Technical Report No.* **119** (Water Resources Research Center, Purdue University, West Lafayette, IN, 1979).
3. M.L. Albertson, M. Poreh, and G.A. Hurst, "Big Thompson Flood Damage Was Severe, But Some Could Have Been Prevented", *Civil Engr.* **48**, 74-77 (1978).
4. D. Anderson, "Models for Determining Least-Cost Investments in Electric Supply", *Bell J. Econ. & Mgmt. Sci.* **3**, 267-299 (1972).
5. R.D. Armstrong and C.E. Willis, "Simultaneous Investment and Allocation Decisions Applied to Water Planning", *Mgmt. Sci.* **23**, 1080-1088 (1977).
6. R.D. Armstrong and C.E. Willis, "Investment Sequencing and Allocation Decisions in Water Resource System Design: A Note on an Alternative Algorithm", *Water Resources Res.* **15**, 1279-1274 (1979).
7. N.V. Arvanitidis, R. Rosing, D.P. Petropoulos, D.G. Luenberger, and R.C. Lind, "A Computer Simulation Model for Flood Plain Development, Part I: Land Use Planning and Benefit Evaluation", *IWR Report* **72-1** (U.S. Army Engineer Institute for Water Resources, Fort Belvoir, VA 1972).
8. N.V. Arvanitidis, J. Rosing, D.P. Petropoulos, D.G. Luenberger, and R.C. Lind, "A Computer Simulation Model for Flood Plain Development, Part II: Model Description and Applications", *IWR Report* **73-1** (U.S. Army Engineer Institute for Water Resources, Fort Belvoir, VA 1972).
9. M.O. Ball, W.F. Bialas, and D.P. Loucks, "Planning the Location and Capacity of Flood Control Structures", *Technical Report No.* **101** (Water Resources and Marine Sciences Center, Cornell University, Ithaca, NY 1976).

10. M.O. Ball, W.F. Bialas and D.P. Loucks, "Structural Flood Control Planning", *Water Resources Res.* **14**, 62-66 (1978).
11. N. Balu and M. Caramanis, "EGEAS: Electric Generation Expansion Analysis System", in *Proceedings of the Conference on Generation Planning: Modeling and Decision Making* (University of Tennessee, Chattanooga, TN 1982) pp. 2-16.
12. W.F. Bialas, "Prescriptive Economic Model for Nonstructural Flood Control", *Technical Report No.* **97** (Water Resources and Marine Sciences Center, Cornell University, Ithaca, NY 1975).
13. W.F. Bialas and D.P. Loucks, "Nonstructural Floodplain Planning", *Water Resources Res.* **14** 67-74 (1978).
14. R.R. Booth, "Optimum Generation Planning Considering Uncertainty", *IEEE Trans. Power App. & Sys.* **PAS-91**, 70-77 (1972).
15. R.F. Boxley, et al., "The Relationship Between Land Values and Flood Risk in the Wabash Basin", *IWR Report* **69-4** (U.S. Army Engineer Institute for Water Resources, Fort Belvoir, VA 1969).
16. R.J. Brown, "*Flood Control, August 1977-August 1980 (Citations from the NTIS Date Base)*" (NTIS, Springfield, VA 1980).
17. W.D. Carson, *Estimating Costs and Benefits for Non-Structural Flood Control Measures* (U.S. Army Corps of Engineers, Hydrologic Engineering Center, Davis, CA 1975).
18. Colorado State University, "Analysis and Synthesis of Flood Control Measures", *Hydrology Papers* **76** (Department of Civil Engineering, University of Colorado, Fort Collins, CO 1975).
19. G. Cortes-Rivera, "Flood Control Project Planning by Mathematical Programming", *Ph.D. Dissertation* (University of Illinois at Urbana-Champaign, Urbana, IL 1973).
20. K.M. Dale, "Dynamic Programming Approach to the Selection and Timing of Generating Plant Additions", *Proc. Inst. Elec. Engr.* **113**, 803-811 (1966).
21. D.W. Davis, "Optimal Sizing of Urban Flood-Control Systems", *J. Hydraulics Div., ASCE* **101**, 1077-1092 (1975).
22. D.W. Davis, "Optimal Sizing of Urban Flood-Control Systems", *Technical Paper No.* **42** (Hydrologic Engineering Center, Davis, CA 1974).
23. J.C. Day, "An Activity Analysis of Non-Structural Plain Management Alternatives", in *Report,* Water Resources Center (University of Wisconsin, Madison, WI 1969) pp. 54-111.
24. J.C. Day, "A Recursive Programming Model for Non-structural Flood Damage Control" *Water Resources Res.* **6**, 1262-1271 (1970).
25. J.C. Day, "A Linear Programming Approach to Floodplain Land Use Planning in Urban Areas", *Am. J. Agr. Econ.* **55**, 165-174 (1973).
26. J.C. Day and R.N. Weisz, "A Linear Programming Model for Use in Guiding Urban Floodplain Management", *Water Resources Res.* **12**, 349-359 (1976).
27. T.N. Debo and G.N. Day, "Economic Model for Urban Watersheds", *J. Hydraulics Div., ASCE* **106**, 475-467 (1980).
28. E.V. Denardo, *Dynamic Programming: Models and Applications* (Prentice-Hall, Englewood Cliffs, NJ 1982).
29. E.V. Denardo and B.L. Fox, "Shortest-Route Methods: 1. Reaching, Pruning and Buckets", *Operations Res.* **27**, 161-186 (1979).
30. B. Eichert and D.W. Davis, "Sizing Flood Control Reservoir Systems by Systems Analysis", *Technical Paper* **44** (Hydrologic Engineering Center, Davis, CA 1976).
31. V.R. Eidman and R.D. Lacewell, "A Model for Estimating Agricultural Flood Damages", *Technical Bulletin* **T-136** (Agricultural Experiment Station, Oklahoma State University, Stillwater, OK 1974).
32. A.H. El-Ablad, T.L. Morin and Z.A. Yamayee, "A Hybrid Dynamic Programming/Branch-and-Bound Approach to Generation Planning", *Modeling and Simulation* **9**, 111-117 (1978).
33. D. Erlenkotter, "Sequencing of Interdependent Hydroelectric Projects", *Water Resources Res.* **9**, 21-27 (1973).
34. D. Erlenkotter, "Sequencing Expansion Projects", *Operations Res.* **21**, 542-553 (1973).

35. D. Erlenkotter and R.R. Trippi, "Optimal Investment Scheduling with Price-Sensitive Dynamic Demand", *Mgmt. Sci.* **23**, 1-11 (1976).
36. D. Erlenkotter and J.S. Rogers, "Sequencing Competitive Expansion Projects", *Operations Res.* **25**, 937-951 (1977).
37. G.W. Evans and T.L. Morin, "Hybrid Dynamic Programming/Branch-and-Bound Strategies for Electric Power Generation Planning", *IIE Trans.* **18**, 138-147 (1986).
38. G.W. Evans, T.L. Morin and H. Moskowitz, "Multiobjective Energy Generation Planning Under Uncertainty", *IIE Trans.* **14**, 183-192 (1982).
39. D.E. Evenson and J.C. Moseley, "Simulation/Optimization Techniques for Multi-Basin Water Resources Planning", *Water Resources Bulletin* **6**, 725-736 (1970).
40. C-J. Feng and T.L. Morin, "Optimal Operation of Flood Control Systems", *Research Report* **122** (Water Resources Center, University of Illinois at Urbana-Champaign, Urbana, IL 1977).
41. G.A. Forsythe, "Flood Mitigation Strategies: The Role of Insurance and Structural Measures", in *Proceedings Hydrology Symposium* (The Institute of Engineers, Australia 1976) pp. 10-14.
42. General Electric, "Descriptive Handbook: Optimized Generation Planning Program" (Electric Utility Systems Engineering Department, Schenectady, NY 1977).
43. A.M. Geoffrion, "Generalized Benders Decomposition", *J. Opt. Theory & Appl.* **10**, 237-260 (1972).
44. A.M. Geoffrion, "How Can Specialized Discrete and Convex Optimization Methods be Married?" *Annals Dis. Math.* **1**, 205-220 (1977).
45. A.M. Geoffrion and G.W. Graves, "Multicommodity Distribution System Design by Benders Decomposition", *Mgmt. Sci.* **20**, 822-844 (1976).
46. Great Lakes Basin Commission, *Great Lakes Basin Framework Study* (Ann Arbor, MI 1975).
47. N.S. Grigg and O.J. Helweg, "State-Of-The-Art of Estimating Flood Damage in Urban Areas", *Water Resources Bulletin* **11**, 379-390 (1975).
48. J.S. Haugh, "Floodplain Management in Project Planning", *Paper* **79-2560** (Winter Meeting of ASAE, New Orleans, LO 1979).
49. R.G. Healy, *Land Use and the States* (John Hopkins University Press, Baltimore, MD 1976).
50. G.E. Hollis, "The Effect of Urbanization on Floods of Different Recurrence Interval", *Water Resources Res.* **11**, 431-435 (1975).
51. L.D. Hopkins, E.D. Brill, Jr., J.C. Liebman and H.G. Wenzel, Jr., "Flood Plain Management Through Allocation of Land Uses - A Dynamic Programming Model", *Research Report No.* **117** (Illinois Water Resources Center, Urbana, IL 1976).
52. L.D. Hopkins, E.D. Brill, Jr., J.C. Liebman and H.G. Wenzel, Jr., "Land Use Allocation Model for Flood Control", *Working Paper* (University of Illinois, Urbana, IL 1980).
53. L.D. Hopkins, L.C. Goulter, K.B. Kurtz, H.G. Wenzel, Jr. and E.D. Brill, Jr., "A Model for Flood Plain Management in Urbanizing Areas", *Research Report No.* **146** (Water Resources Center, University of Illinois, Urbana, IL 1980).
54. Hydrocomp Co., "Flood Damage Mitigation" *Simulation Network News.* **8**, 1-18 (1976).
55. H.D. Jacoby and D.P. Loucks, "Combined Use of Optimization and Simulation Models in River Basin Planning", *Water Resources Res.* **8**, 1401-1414 (1972).
56. L.D. James, "A Time-Dependent Planning Process for Combining Structural Measures, Land Use and Flood Proofing to Minimize the Economic Cost of Floods", *Report* **EEP-12** (Institute of Engineering-Economic Systems, Stanford University, Palo Alto, CA 1964).
57. L.D. James, "Nonstructural Measures for Flood Control", *Water Resources Res.* **1**, 9-24 (1965).
58. L.D. James, "Economic Analysis of Alternative Flood Control Measures", *Water Resources Res.* **3**, 333-343 (1967).
59. L.D. James, "Economic Analysis of Alternative Flood Control Measures", *Research Report No.* **16** (Kentucky Water Research Institute, Lexington, KY 1968).

60. R.B. Jansen, *Dams and Public Safty,* Water and Power Resources Service (U.S. Department of the Interior, Washington, DC 1980).
61. R.T. Jenkins, "TARANTULA: A Generation Technology Evaluation Code to Analyze Electric Utility Alternatives and Consumer Options in the Eighties", in *Proceedings of the Conference on Electric System Expansion Analysis* (Ohio State University, Columbus, OH 1981) pp. 75-84.
62. R.T. Jenkins and D.S. Joy, "WIEN Automatic System Planning Package (WASP) - An Electric Utility Optimal Generation Expansion Planning Computer Code", **ORNL-4945** (Oak Ridge National Laboratories, Oak Ridge, TN 1974).
63. W.K. Johnson and D.W. Davis, "Analysis of Structural and Nonstructural Flood Control Measures Using Computer Program HEC-5C", *Training Document No.* **7** (Hydrologic Engineering Center, U.S. Corps of Engineers, Davis, CA 1975).
64. J.L. Kaul and C.E. Willis, "Applications of Integer and Quadratic Programming to Flood Plain Land Use Management" (ORSA/TIMS Meeting, San Juan, PR 1974).
65. V.A. Khavitch, "Mathematical Simulation and Optimization of Flood Control Systems", in *The Use of Computer Techniques and Automation for Water Resources Systems Vol. II* (Washington, DC 1974) pp. 126-129.
66. J.L. Knetch and P.C. Jennings, "Estimating the Influence of Large Reservoirs on Land Values", *Appraisal J.* **XXXII** (1964).
67. J. Kulper, *Water Resources Development* (Butterworths, London 1965).
68. J. Kulper and L. Ortolano, "A Dynamic Programming Simulation Strategy for the Capacity Expansion of Hydroelectric Power Systems", *Water Resources Res.* **9**, 1497-1510 (1973).
69. R.D. Lacewell and V.R. Eidman, "A General Model for Evaluating Agricultural Flood Plains", *Am. J. Agr. Econ.* **54**, 92-101 (1972).
70. S.T. Lee, "OPTGEN - Optimal Generation Expansion by Dynamic Programming", in *Regional Power System Planning; A State of the Art Assessment,* **DOE/RA/29144-01** (Washington, DC 1980) pp. 265-270.
71. D.L. Lott, "Optimization Model for the Design of Urban Flood-Control Systems", *Technical Report* (Center for Research in Water Resources, University of Texas, Austin, TX 1976).
72. J.C. Loughlin, "A Flood Insurance Model for Sharing the Costs of Flood Protection", *Water Resources Res.* **7**, 236-244 (1971).
73. R.E. Marsten and T.L. Morin, "A Hybrid Approach to Discrete Mathematical Programming", *Math. Prog.* **14**, 21-40 (1978).
74. J.A. McCrory, L.D. James and D.E. Jones, Jr., "Dealing with Variable Flood Hazard", *J. Water Resources Plan. & Mgmt. Div. ASCE* **102**, 193-208 (1976).
75. J.G. McNealy and R.D. Lacewell, *Flood Plain Management* (The Texas Agricultural Experiment Station, Texas A&M University, College Station, TX 1976).
76. W.L. Miller and S.P. Erickson, "The Impact of High Interest Rates on Optimum Multiple Objective Design of Surface Runoff Urban Drainage Schemes", *Water Resources Bulletin* **11**, 49-60 (1975).
77. T.L. Morin, "Optimal Sequencing of Capacity Expansion Projects", *J. Hydraulics Div. ASCE* **99**, 1605-1622 (1973).
78. T.L. Morin, "Computational Advances in Dynamic Programming", in *Dynamic Programming and its Applications,* M.L. Puterman, ed. (Academic Press, NY 1979) pp. 53-90.
79. T.L. Morin, "Monotonicity and the Principle of Optimality", *J. Math. Anal. & Appl.* **88**, 665-674 (1982).
80. T.L. Morin and R.T. Jenkins, "OPTIMIZER: An Enhanced Dynamic Program for Generation Planning", in *Proceedings of the Conference on Electric Generation System Expansion Analysis*" (Ohio State University, Columbus, OH 1981) pp. 238-252.
81. T.L. Morin and R.E. Marsten, "Branch-and-Bound Strategies for Dynamic Programming", *Operations Res.* **24**, 611-627 (1976).

82. T.L. Morin, R.E. Marsten and K.S. Nagaraj, "*A Posteriori* Optimality Analysis in Branch-and-Bound", *Working Paper* (School of Industrial Engineering, Purdue University, West Lafayette, IN 1986).
83. T.L. Morin, W.L. Meier, Jr. and K.S. Nagaraj, "Optimal Mix of Adjustments to Floods", *Technical Report* **139** (Water Resources Research Center, Purdue University, West Lafayette, IN 1981).
84. T.L. Morin and Y.S. Shin, "Optimal Expansion of Flood Control Systems", *Research Report No.* **121** (Water Resources Center, University of Illinois, Urbana, IL 1977).
85. E. Mosonyi, W. Buck, and W. Kiefer, "Methodology for Optimization of Flood Protection Measures Illustrated by an Example in the German Federal Republic", in *Flood Investigation: Vol. II, Proceedings, Symposium on River Mechanics* (International Association for Hydraulic Research, Bangkok, Thailand 1973) pp. 291-302.
86. K.S. Nagaraj, "The *A Posteriori* Approach for Discrete Optimization", *Ph.D. Dissertation* (School of Industrial Engineering, Purdue University, West Lafayette, IN 1983).
87. S. Nakamura, *et al.*, "CERES - Capacity Expansion and Reliability Evaluation System", in *Proceedings of the Conference on Electric Generating System Expansion Analysis* (Ohio State University, Columbus, OH 1981) pp. 356-369.
88. National Science Foundation, *Building Losses from National Hazards: Yesterday, Today and Tomorrow* (prepared for NSF by J.H. Wiggins Company, 1980).
89. National Science Foundation, *A Report on Flood Hazard Mitigation*, "Executive Summary" (Washington, DC 1980) pp. 1-8.
90. E.N. Oatman and H.J. Hamant, "A Dynamic Approach to Generation Expansion Planning", *IEEE Trans. Power App. & Sys.* **92**, 1888-1897 (1973).
91. Office of Water Research and Technology, *A Process for Community Flood Plain Management* (prepared for OWRT by Leman Powell Associates, Inc., Washington, DC 1979).
92. E.C. Penning-Roswell and J.B. Chatterton, "Assessing the Benefits of Flood Alleviation and Land Drainage Schemes, in *Proc. Inst. Civil Engr.*, London, 295-315 (1980).
93. E.R. Petersen, "A Dynamic Model for the Expansion of Electric Power Systems", *Mgmt. Sci.* **20**, 656-664 (1973).
94. D. Piper, A.R. Strohbehn and B.R. Boxely, "Analysis of Alternative Procedures for Evaluation of Agricultural Flood Control Benefits, Volume II", *IWR Report* **71-4** (U.S. Army Corps of Engineers, Institute for Water Resources, Fort Belvoir, VA 1971).
95. T.M. Rachford, "Economic Analysis of Alternative Flood Control Measures by Digital Computer", *Research Report* (Water Resources Institute, University of Kentucky, Lexington, KY 1961).
96. E.F. Renshaw, "The Relationship Between Flood Losses and Flood Control Benefits", *Research Paper No.* **65**, in *Papers on Flood Problems,* G.F. White, ed. (Department of Geography, University of Chicago, Chicago, IL 1960).
97. J.S. Rogers, "Optimal Generation Expansion: A Dynamic Model", *IEEE Paper No.* **C74-144-2** (Power Engineering Society Winter Meeting, New York 1974).
98. W.R.D. Sewell and H.D. Foster, "Flood Loss Management in Developing Countries: A Model for Identifying Appropriate Strategies", *Working Paper No.* **39** (UNDP-UN Inter-regional Seminar on Riverbasin and Inter-basin Development, Budapest, Hungary, 1975).
99. L.A. Shabman and D.I. Damianos, "Flood Hazard Effects on Residential Property Values", *J. Water Resources Plan. & Mgmt. Div. ASCE* **102**, 151-162 (1976).
100. K.P. Singh, "Agricultural Benefit Assessment from Flood Control", in *Environmental Aspects of Irrigation and Drainage: Proceedings of a Specialty Conference* (University of Ottawa, Ontario, Canada 1976) pp. 264-277.
101. J.F. Smiarowski, C. Willis and J.H. Foster, "Flood Plain Land Use Management: An Application of Operations Research Methodology", *Publication* **45** (Massachusetts Water Resources Research Center, Amherst, MA 1975).

102. State of Indiana, *The Indiana Water Resources: Availability, Uses, and Needs* (Governor's Water Resource Study Commission, G.D. Clark, ed. 1980).
103. U.S. Army Corps of Engineers, "A Methodology for Flood Plain Development and Management", *IWR Report* **69-3** (Institute for Water Resources, Fort Belvoir, VA 1969).
104. U.S. Army Corps of Engineers, *Wabash River Basin Comprehensive Study: Vol. V, Flood Problems and Solutions* (Louisville District, KY 1971).
105. U.S. Army Corps of Engineers, "Agricultural Flood Control Benefits and Land Values", *IWR Report* **71-3** (Institute for Water Resources, Alexandria, VA 1971).
106. U.S. Army Corps of Engineers, "Analysis of Theories and Methods for Estimating Benefits of Protecting Urban Flood Plains", *IWR Report* **74-4** (Institute for Water Resources, Fort Belvoir, VA 1974).
107. U.S. Army Corps of Engineers, "Analysis of Alternative Procedures for the Evaluation of Agricultural Flood Control Benefits", *IWR Report* **74-14** (Institute for Water Resources, Fort Belvoir, VA 1974).
108. U.S. Army Corps of Engineers, *Flood Damage Report Huntington - Ashland - Portsmouth Metro Area Study, Vol II* (prepared by Stanley Consultants, Huntington District 1976).
109. U.S. Army Engineer Division, *Flood Control in the Ohio River Basin, Appendix M of Ohio River Basin Comprehensive Survey* (U.S. Army Corps of Engineers, Cincinnati, OH 1967).
110. U.S. Water Resources Council, *The Nation's Water Resources: The First National Assessment of the Water Resources Council* (U.S. Government Printing Office, Washington, DC 1968).
111. R.N. Weisz, "A Methodology for Planning Land Use and Engineering Alternatives for Flood Plain Management", *Ph.D. Dissertation* (University of Michigan, Ann Arbor, MI 1973).
112. R.N. Weisz and J.C. Day, "A Methodology for Planning Land Use and Engineering Alternatives for Flood Plain Management: The Flood Plain Management System Model", *IWR Paper* **74-P2** (U.S. Army Corps of Engineers, Institute for Water Resources, Fort Belvoir, VA 1974).
113. R.N. Weisz and J.C. Day, "A Methodology for Planning Land Use and Engineering Alternatives for Flood Plain Management: The Value of Land in Alternative Urban Uses", *IWR Report* **74-P5** (U.S. Army Corps of Engineers, Institute for Water Resources, Fort Belvoir, VA 1974).
114. R.N. Weisz and J.C. Day, "A Regional Planning Approach to the Flood Plain Management Problem", *Ann. Reg. Sci.* **IX**, 80-92 (1975).
115. G.F. White, "Flood Hazard in the United States: A Research Assessment", *Monograph* (Institute of Behavioral Science, University of Colorado, Boulder, CO 1975).
116. G.F. White and J.E. Haas, *Assessment of Research on Natural Disasters* (MIT Press, Cambridge, MA 1975).
117. J.S. Windsor, "A Programming Model for the Design of Multireservoir Flood Control Systems", *Water Resources Res.* **11**, 30-36 (1975).
118. J.S. Windsor, "A Mathematical Model for the Analysis of Multi-Reservoir Flood Control Systems", in *Proceedings of the International Conference on Mathematical Models for Environmental Problems* C.A. Brebbia, ed. (John Wiley & Sons, New York, 1976) pp. 87-98.
119. K.R. Wright, P.D. Binney and M.M. Stokes, "Adequacy of Data Resources for the Needs of the Consulting Engineer to Local Government", in *Natural Hazards Data Resources, Uses and Needs; Program on Technology Environment and Man, Monograph* **27** (Institute of Behavioral Sciences, University of Boulder, CO 1979) pp. 44-53.
120. S. Yakowitz, "Dynamic Programming Applications in Water Resources", *Water Resources Res.* **18**, 673-696 (1982).

PART FIVE

REAL TIME AND STOCHASTIC RESERVOIR OPERATION MODELS

A MULTI-RESERVOIR MODEL WITH A MYOPIC OPTIMUM

MATTHEW J. SOBEL
SUNY at Stony Brook, Stony Brook, NY 11794-3775

ABSTRACT

A dynamic model with four reservoirs, autocorrelated inflows, and significant flow times between reservoirs is shown to possess a myopic optimum. If the model is solved as a dynamic program, it has eight state variables. The myopic approach consists of solving four static (one-period) scalar optimization problems which yield the solution of the dynamic program.

INTRODUCTION

The origins of dynamic programming are rooted partly in the development of dynamic reservoir optimization models [5]. The generic problem is to decide how much water to discharge, as time passes, in order to optimize an objective. The "curse of dimensionality" [1] has bedeviled the prospect of utilizing dynamic control models to solve real problems. That is, the "state variable" becomes a vector whose dimension grows with the number of reservoirs in the river basin, the number of layers modeled in the reservoir, and other attributes.

The curse has been exorcised in some cases. Highly structured multi-echelon inventory models [2] comprise one class of examples. Although inventory and reservoir models are closely related [6], simply solved multi-echelon inventory models correspond to bizarre reservoir models. For references to other approaches which accelerate and simplify the solution of large dynamic programming problems, see [4, pp. 302-305].

The curse of dimensionality can be exorcised from some dynamic programs by replacing them with static programs. In other cases, the curse is exorcised by recognizing that the same action is optimal in every state. The former approach was introduced first in inventory theory [9] and then developed for more general applicability [7]. The mathematical foundations of the latter approach were explored more recently [3]. The methods utilized in this paper are found in [8] which draws on both approaches.

Section 2 presents a model of a river basin which has four reservoirs, autocorrelated inflows, and delays for upstream discharges to reach down stream reservoirs. The model is manipulated in Section 3 to ellicit its affine structure. Section 4 summarizes the "myopic affine" approach in [8], and utilizes that approach in Section 5. As a consequence, a dynamic program with eight state variables and four decision variables can be solved by optimizing four scalar one period problems. Section 6 briefly suggests some generalizations. The assumptions in the model, its applicability, and this paper's results are summarized in Section 7.

2. A FOUR-RESERVOIR MODEL WITH AUTOCORRELATED INFLOWS

Figure 1 portrays a river basin with four reservoirs. Two tributaries merge upstream from reservoir #1. Reservoir #2 is located on one tributary, and reservoirs #3 and #4 are located on the other tributary with #4 upstream from #3. The model is discrete in time and has a one-period delay for a discharge to flow downstream to the next reservoir. For expository

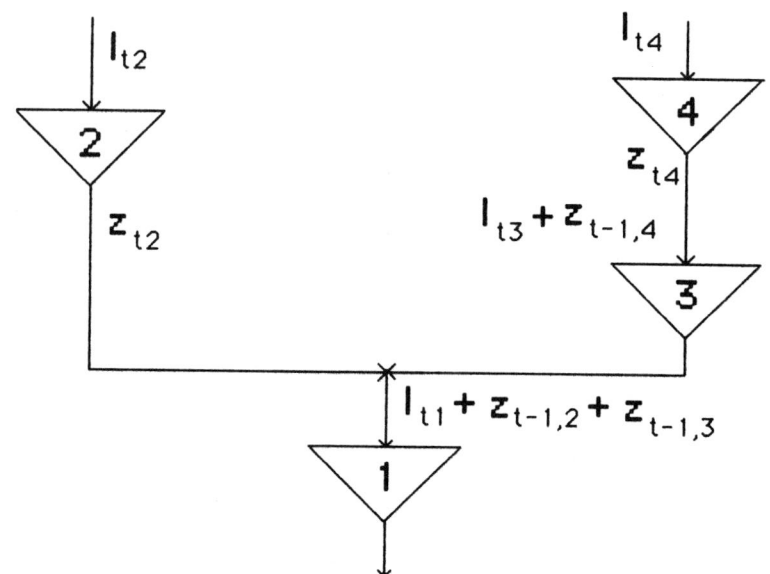

FIG 1. A River System with Four Reservoirs

convenience, inflows are concentrated at the reservoirs and evaporation and leakage are ignored.

Let x_{tj} be the amount of water in reservoir j at the beginning of period t, let z_{tj} be the quantity discharged from reservoir j during period t, and let I_{tj} be the inflow to reservoir j during period t (not including flow received from previous upstream discharges, if any). It is convenient to let $y_{tj} = x_{tj} - z_{tj}$ for each t and j.

The topology in Figure 1 and the definitions yield the following balance equations:

$$\begin{aligned} x_{t+1,1} &= x_{t1} - z_{t1} + I_{t1} + z_{t2} + z_{t3} \\ &= y_{t1} + I_{t1} + x_{t2} - y_{t2} + x_{t3} - y_{t3} \end{aligned} \qquad (1a)$$

$$x_{t+1,2} = x_{t2} - z_{t2} + I_{t2} = y_{t2} + I_{t2} \qquad (1b)$$

$$\begin{aligned} x_{t+1,3} &= x_{t3} - z_{t3} + I_{t3} + z_{t4} \\ &= y_{t3} + I_{t3} + x_{t4} - y_{t4} \end{aligned} \qquad (1c)$$

$$x_{t+1,4} = x_{t4} - z_{t4} + I_{t4} = y_{t4} + I_{t4} \qquad (1d)$$

Equations (1a-d) do not reflect the effects of finite reservoir capacity and possible overflow. For example, if K_2 is the capacity of reservoir #2, then one might wish to replace (1b) with $x_{t+1,2} = \min\{K_2, y_{t2}+I_{t2}\}$. However, the methods of this paper would not be appropriate. The approach in this paper is not applicable to flood management models.

The inflows are modelled in the following way:

$$I_{t+1,j} = \xi_{tj} + \rho_j(\mu_j - I_{tj}). \tag{2}$$

The scalars ρ_j and μ_j are assumed known (for each j) and $(\xi_{11},\xi_{12},\xi_{13},\xi_{14})$, $(\xi_{21},\xi_{22},\xi_{23},\xi_{24}),\ldots$ are assumed to be independent and identically distributed random vectors. It can be shown that $|\rho_j| < 1$ and $\rho_j\mu_j \neq 1$ imply that the asymptotic (as $t \to \infty$) autocorrelation of I_{tj} has the opposite sign of ρ_j, and the asymptotic mean and variance of I_{tj} are $[E(\xi_{1j}) + \rho_j\mu_j]/(1-\rho_j\mu_j)$ and $\text{Var}(\xi_{1j})/(1-\rho_j)^2$, respectively. The notation $E(X)$ and $\text{Var}(X)$ denotes the expected value and variance of a random variable X. If X is a random vector, $E(X)$ denotes the vector of expected values of components.

Net benefits in the model are assumed to depend on storage levels and discharges. Let B_t be the net benefit in period t. Then B_t is modelled as

$$B_t = \sum_{j=1}^{4}[g_j(y_{tj}) + c_j z_{tj}] \tag{3}$$

where c_j is the unit benefit of a discharge at reservoir j, and $g_j(y_{tj})$ is the single-period expected net benefit of having a storage level (after discharge) of y_{tj} at reservoir j.

The most important properties of (3) are the linearity of the term $c_j z_{tj}$ and the additive separation of the terms involving $\{z_{tj}\}$ and $\{y_{tj}\}$. The unit "benefit" c_j can be positive or negative and the separate storage level benefit functions can be replaced by $g(y_{t1},y_{t2},y_{t3},y_{t4})$.

The optimization problem is to maximize the expected value of the discounted benefits, namely,

$$\text{maximize} \quad E(\sum_{t=1}^{\infty} \beta^{t-1} B_t) \tag{4}$$

where $0 < \beta < 1$ is the discount factor.

3. MODEL MANIPULATION

Let s_t and a_t be the column vectors with transposes $s_t' = (x_{t1},x_{t2},x_{t3},x_{t4},I_{t-1,1},I_{t-1,2},I_{t-1,3},I_{t-1,4})$ and $a_t' = (y_{t1},y_{t2},y_{t3},y_{t4})$, respectively. These vectors are the state (s_t) and action (a_t) in period t. It follows from (1a-d) and (2) that

$$s_{t+1} = Qs_t + V_t \tag{5}$$

where the transpose of V_t is $(y_{t1}-y_{t2}-y_{t3}+\xi_{t-1,1}+\rho_1\mu_1,\ y_{t2}+\xi_{t-1,2}+\rho_2\mu_2,\ y_{t3}-y_{t4}+\xi_{t-1,3}+\rho_3\mu_3,\ y_{t4}+\xi_{t-1,4}+\rho_4\mu_4,\ \xi_{t-1,1}+\rho_1\mu_1,\ \xi_{t-1,2}+\rho_2\mu_2,\ \xi_{t-1,3}+\rho_3\mu_3,\ \xi_{t-1,4}+\rho_4\mu_4)$ and

$$Q = \begin{bmatrix} 0 & 1 & 1 & 0 & -\rho_1 & 0 & 0 & 0 \\ 0 & 0 & 0 & 0 & 0 & -\rho_2 & 0 & 0 \\ 0 & 0 & 0 & 1 & 0 & 0 & -\rho_3 & 0 \\ 0 & 0 & 0 & 0 & 0 & 0 & 0 & -\rho_4 \\ \hline 0 & 0 & 0 & 0 & -\rho_1 & 0 & 0 & 0 \\ 0 & 0 & 0 & 0 & 0 & -\rho_2 & 0 & 0 \\ 0 & 0 & 0 & 0 & 0 & 0 & -\rho_3 & 0 \\ 0 & 0 & 0 & 0 & 0 & 0 & 0 & -\rho_4 \end{bmatrix} \qquad (6)$$

Let $\Theta_j = E(\xi_{1j}) + \rho_j \mu_j$. From the definition of V_t, the transpose of $E(V_t)$ is $v(a_t)' = E(V_t)' = (y_{t1}-y_{t2}-y_{t3}+\Theta_1, y_{t2}+\Theta_2, y_{y3}-y_{t4}+\Theta_3, y_{t4}+\Theta_4, \Theta_1, \Theta_2, \Theta_3, \Theta_4)$ in which the notation $v(a_t)$ emphasizes the functional dependence on a_t. Therefore, for a given eight-vector s and four-vector $\underline{a}$, it follows from (5) that

$$E(s_{t+1} | s_t = s, a_t = a) = v(a) + Qs \qquad (7)$$

whose left side is a conditional expectation. The important feature of (7) is that it is an affine function of s whose linear coefficient Q does not vary with $\underline{a}$.

An expression similar to (7) can be obtained for the expected single-period net benefit, $E(B_t)$. From (3) and the identity $z_{tj} = x_{tj} - y_{tj}$,

$$B_t = \sum_{j=1}^{4} [g_j(y_{tj}) + c_j(x_{tj}-y_{tj})]$$

$$= w(a_t) + fs_t$$

where f is the row vector $(c_1, c_2, c_3, c_4, 0, 0, 0, 0)$ and

$$w(a_t) = \sum_{j=1}^{4} [g_j(y_{tj}) - c_j y_{tj}]. \qquad (8)$$

Therefore,

$$E(B_t | s_t = s, a_t = a) = w(a) + fs \qquad (9)$$

whose affine structure corresponds to that of (7).

4. AFFINE MYOPIC OPTIMA

The results in the next section exploit the following conclusions in [8]. Let B_t be the single-period net benefit in a stochastic discrete time control process in which s_t is the state vector and a_t is the action (vector, if necessary). Suppose that the elements of s_t are real numbers, the dynamics of the model satisfy an expression such as (7), the rewards satisfy an expression such as (9), and $I - \beta Q$ is nonsingular (where I is the identity matrix). Then

$$E(\Sigma_{t=1}^{\infty} \beta^{t-1} B_t) = E[\Sigma_{t=1}^{\infty} \beta^{t-1} G(a_t)] + J \qquad (10)$$

where

$$G(a) = w(a) + \beta f(I-\beta Q)^{-1} v(a) \qquad (11)$$

$$J = f(I-\beta Q)^{-1} s_1. \qquad (12)$$

Expression (10) is valid for all policies and initial states. It is apparent from (12) that the initial state affects the constant J. That is, J is constant with respect to the sequence of actions $a_1, a_2, \ldots$

Suppose that a* globally maximizes $G(\cdot)$, that is

$$G(a^*) \geq G(a) \qquad \text{for all a.} \qquad (13)$$

Then for all selections of the sequence of actions and initial states, (10) and (13) imply

$$E(\Sigma_{t=1}^{\infty} \beta^{t-1} B_t) \leq J + E[\Sigma_{t=1}^{\infty} \beta^{t-1} G(a_t^*)].$$

So a_t = a* for all t is optimal if it is feasible. A sufficient condition for feasibility is that a_1 = a* be feasible and, for every t, if a_t = a*, than a_{t+1} = a* is feasible (with probability one). It is observed in the next section that feasibility is easily resolved in the reservoir model.

5. THE RESERVOIR MODEL HAS A MYOPIC OPTIMUM

In order to apply the result just described, it is necessary to specify the function $G(\cdot)$. From (11), the steps are (i) compute $(I-\beta Q)^{-1}$, and (ii) use (11) to specify $G(a)$. After $G(\cdot)$ is specified, a global maximum, a*, is found. Then the feasibility of a_t = a* for all t is evaluated.

The last step, namely feasibility, is easily taken. Feasibility in this model corresponds to $z_{tj} \geq 0$ for all t and j. But $z_{tj} = x_{tj} - y_{tj}$; so a_t = a* is feasible if $y_j^* \leq x_{tj}$, j = 1,...,4 (where y_j^* is the j-th component of a*). If t = 1, this inequality stipulates that the initial reservoir levels be sufficiently high. If t > 1, notice in (1a-d) that $x_{t+1,j} \geq y_{tj} + I_{tj}$; so $y_{tj} = y_j^*$ implies $x_{t+1,j} \geq a_j^*$ which yields feasibility of $y_{t+1,j} = a_j^*$. In summary, if $y_j^* \leq x_{1j}$ for j = 1,...,4, then $y_{tj} = y_j^*$ for all t is feasible and optimal. This rule corresponds to discharges $z_{1j} = x_{1j} - y_j^*$ and

$$z_{t1} = I_{t-1,1} + I_{t-2,2} + I_{t-2,3} + I_{t-3,4} \qquad (t \geq 4)$$

$$z_{t2} = I_{t-1,2} \qquad (t \geq 2)$$

$$z_{t3} = I_{t-1,3} + I_{t-2,4} \qquad (t \geq 3)$$

$$z_{t4} = I_{t-1,2} \qquad (t \geq 2)$$

In order to compute $(I-\beta Q)^{-1}$, observe that Q in (6) can be partitioned

$$Q = \begin{bmatrix} U & -D \\ 0 & -D \end{bmatrix} \text{ so } (I-\beta Q)^{-1} = \begin{bmatrix} (I-\beta U)^{-1} & -(I-\beta U)^{-1}\beta D(I+\beta D)^{-1} \\ 0 & (I+\beta D)^{-1} \end{bmatrix}$$

in which U is upper triangular and D is diagonal. Recall that $f = (c,0,0,0,0)$ where $c = (c_1,c_2,c_3,c_4)$. Therefore, $f(I-\beta Q)^{-1} = c(I-\beta U)^{-1}[I,-\beta D(I+\beta D)^{-1}]$. It follows from the specific forms of U and -D (the northwest and northeast submatrices in (6), respectively) that

$$c(I-\beta U)^{-1} = (c_1, \beta c_1 + c_2, \beta c_1 + c_3, \beta^2 c_1 + \beta c_3 + c_4).$$

Finally, (8), (11), the expression for $v(a)$ ($E(V_t)$ between (6) and (7)), and rearrangement of terms yield

$$G(a) = \Sigma_{j=1}^{4}[g_j(y_j) - \alpha_{1j} - \alpha_{2j}y_j] \tag{14}$$

where $a = (y_1,y_2,y_3,y_4)$, $\alpha_{21} = (1-\beta)c_1$, $\alpha_{22} = (1-\beta)(c_2+\beta c_1)$, $\alpha_{23} = (1-\beta)(c_3+\beta c_1)$, $\alpha_{24} = (1-\beta)(c_4+\beta c_3+\beta^2 c_1)$, $\alpha_{11} = \beta\Theta_1 c_1/((1+\beta\rho_1))$, $\alpha_{12} = \beta\Theta_2(c_2+\beta c_1)/(1+\beta\rho_2)$, $\alpha_{13} = \beta\Theta_3(c_3+\beta c_1)/(1+\beta\rho_3)$, and $\alpha_{14} = \beta\Theta_4(c_4+\beta c_3+\beta^2 c_1)/(1+\beta\rho_4)$.

A global maximum of $G(\cdot)$, say $a^* = (y_1^*,y_2^*,y_3^*,y_4^*)$, is found by independently maximizing $g_j(y_j) - \alpha_{2j}y_j - \alpha_{1j}$ on the domain $0 \leq y_j$, for each j. If $g_j(\cdot)$ is specified analytically (rather than numerically), y_j^* may be derived analytically instead of being computed with a nonlinear programming algorithm). For example, if $g_j(\cdot)$ is a concave quadratic function, say

$$g_j(y_j) = \lambda_{1j} + \lambda_{2j} - \lambda_{3j}y_j^2$$

with $\lambda_{3j} > 0$, then $y_j^* = \max\{0, (\lambda_{2j} - \alpha_{2j})\}/(2\lambda_{3j})$

6. COMMENTS

Contrary to (3), suppose that the storage level benefits are not additively separable. For example, some of the recreational uses of reservoirs in the same region may permit a degree of substitutability or complementarity. Then (3) is replaced by

$$B_t = g(y_{t1},y_{t2},y_{t3},y_{t4}) + \Sigma_{j=1}^{4}c_j z_{tj}.$$

As a result, (14) is replaced by

$$G(a) = g(y_1,y_2,y_3,y_4) - \Sigma_{j=1}^{4}(\alpha_{1j}+\alpha_{2j}y_j) \tag{15}$$

where $\{\alpha_{1j},\alpha_{2j}\}$ is the same as in (14). The optimization of (15), unlike (14), does not uncouple to four separate maximizations, each with respect to a single scalar variable. Nevertheless, it is dramatically easier to optimize (15) than the corresponding dynamic program.

The coefficients $\{\alpha_{1j},\alpha_{2j}\}$ (specified after (14)) have a pattern that can be related to the topography in Figure 1. This pattern will be discussed elsewhere and is more general than the particular four reservoir model in this paper.

If inter-reservoir flow times had been two or more periods, instead of one period, then (7) would have been false. Nevertheless, results similar to (10) and (14) would have been valid. This generalization will be reported elsewhere.

7. SUMMARY

The model in this paper describes a river basin with tributaries, possibly several reservoirs on a tributary, and autocorrelated inflows. The decisions are the reservoir discharges; the benefits and costs derive from reservoir levels and discharges. The essential assumptions (and restrictions) are (i) the benefit and cost of a discharge are proportional to the quantity discharged, and (ii) the overflow effects of finite reservoir capacity are ignored. It follows from (ii) that the model should not be used for an application where flood management is a consideration.

Under the model assumptions, the usual dynamic programming formulation would have a "state variable" which is a vector with many components. For a wide range of initial conditions, it is shown that the dynamic program can be solved by optimizing a one-period model. This one-period model decomposes to scalar one-period models if the net **benefit of reservoir** storage is the sum of the benefits attributable to each reservoir's storage level.

REFERENCES

1. R. Bellman, Dynamic Programming, Princeton University Press, Princeton NJ (1957).
2. A.J. Clark, "An Informal Survey of Multi-Echelon Inventory Theory," Nav. Res. Logistic Quart., 19, 621-650 (1972).
3. E.V. Denardo and U.G. Rothblum, "Affine Structure and Invariant Policies for Dynamic Programs," Math. Oper. Res., 8, 342-365 (1983).
4. D.P. Heyman and M.J. Sobel, Stochastic Models in Operations Research, Vol. II (McGraw Hill, New York, 1984).
5. P. Massé, Les Réserves et la Régulation de l'Avenir dans la Vie Économique, 2 vols. (Hermann, Paris, 1946).
6. M.J. Sobel, "Reservoir Management Models," Water Resources Res., $\underline{11}$, 767-776 (1975).
7. _____, "Myopic Solutions of Markov Decision Processes and Stochastic Games," Oper. Res., $\underline{29}$, 995-1009 (1981).
8. _____, "Myopic Solutions of Affine Dynamic Models," unpublished manuscript, Georgia Institute of Technology (1985).
9. A.F. Veinott, Jr., "Optimal Policy for a Multi-Product Nonstationary Inventory Problem," Manage. Sci., $\underline{12}$, 206-222 (1965).

AN ADAPTIVE CONTROL MODEL FOR SINGLE RESERVOIR OPERATION

RAMESH SHARDA* AND MOHAMED ALI EL-TAYEB**
*Department of Management, Oklahoma State University, Stillwater, Oklahoma 74078; **University of Sudan, Khartoum, Sudan

ABSTRACT

This paper presents an extension of the adaptive control approach to single reservoir operation planning problems characterized by random inflows and discrete state and control variables. The model allows a decision maker to modify the release decisions not only on the basis of realized values of inflows but also on the basis of the past performance of the decision maker in attaining the target. Bayesian densities are used to capture continuously updated information. A dynamic programming formulation is presented which involves a large state space. The implementation of the model is made possible by decomposing the problem into smaller problems and discretizing some state variables. A post-optimality analysis is performed to ensure the optimality of the solution. The practical feasibility of the approach is demonstrated by taking data for a single reservoir out of a system and comparing the results of this model with two other models.

INTRODUCTION

Many techniques have been developed and applied to determine the optimum operation and regulation rules for different water resource systems. A great deal of literature focuses on deterministic linear, nonlinear and dynamic programming. The methods introduced by Chu and Yeh [3], and Roefs and Bodin [14] are based on the assumption of certain future inflows. However, a long-standing problem in reservoir operation and regulation is the risk involved due to the stochastic nature of the inflows and the value of information needed to reduce this risk. This issue has been considered via several approaches. An early approach to the stochastic nature of reservoir inflows was represented with explicit stochastic optimization methods such as chance-constrained programming [Eisel (7)], stochastic programming [Prêkopa (13)], and stochastic dynamic programming [Askew (1)]. These methods basically use the expected value in the objective function and the probability distribution of inflows in the constraints. Uncertainty associated with finite sample size has been dealt with in Bayesian decision approaches [Davis et al. (5)], value of information approaches [Close and Beard (4)], and various analytical probability models [Lloyd (11)].

This this paper we extend the two-state adaptive control approach in Bellman [2] and Kushner [9] to a more general multiple-states system, in particular a reservoir operation problem. This approach takes into account the probability distribution of inflows and continuously updated Bayesian densities which are derived from the system's past behavior. Incorporation of these two densities is likely to enhance the reliability of the final solution. Pekelman and Rausser [12] provide a survey of the methods and applications of adaptive control. To the best of our knowledge this is one of the first attempts to apply adaptive control to reservoir operation.

In the next section an adaptive control formulation of a single reservoir operation problem is presented. We then describe an approach to solving the problem. For the purpose of illustration we take a reservoir operated by the Tennessee Valley Authority.

MODEL FORMULATION

A typical single reservoir operation problem can be represented as follows:

$$\text{Maximize } J = E_{\xi_n} \left\{ \sum_{n=1}^{N} B_n(x_n, q_n, \xi_n) \right\} \quad (2.1)$$

subject to:

$$x_n = x_{n-1} - q_n + \xi_n \quad (2.2)$$

$$0 < a_n \leq x_n \leq b_n < c \quad n \in N \quad (2.3)$$

$$q_n \geq d_n \quad n \in N \quad (2.4)$$

$$B_n(x_n, q_n, \xi_n) = 0 \quad \text{for } n=0 \quad (2.5)$$

$$x_0 \text{ is given} \quad (2.6)$$

where N is the length of the horizon; x_n, q_n, and ξ_n are the water content of the reservoir at the end of period n, the release and the inflow during period n respectively; E_{ξ_n} is the expectation symbol with respect to the random variable ξ_n; a_n and b_n are predetermined lower and upper bounds on the water level during period n; C is the capacity of the reservoir; d_n is a predetermined minimum release requirement for period n and x_0 is the initial water content of the reservoir. B_n is a benefit function that measures the performance of the reservoir at different levels of x_n, q_n, and ξ_n. We want to determine the integer-valued control sequence: $q_n, \ldots, q_N$ that maximizes J in (2.1) subject to constraints (2.2) through (2.4) and initial conditions (2.5) and (2.6).

The model presented in (2.1) through (2.6) has been solved using many techniques under various assumptions. If it can be assumed that the underlying stochastic process $\{\xi_n, n=1,2,\ldots,N\}$ is a sequence of independent random variables with known probability distributions, the problem may be treated as a single state variable, dynamic programming problem. We wish to extend the model to include the adaptive learning on the part of the decision maker. This is accomplished through Bayesian densities. In essence, a decision maker can learn from his/her past decisions by keeping track of what decisions were made earlier under certain assumptions, and how many of those actions resulted in achieving the targets. This information, which is continuously updated, allows the decision maker to adapt to the system behavior. This adaptive control is incorporated into the model by defining the Bayesian densities of certain "counters," as described below.

The Posterior Densities

Let us start with the following definitions:

$U_n(q,x)$ = the decision to release q units during period n and plan to end up with x units of water level in the reservoir at the end of period n,

$K_n(q,x)$ = the number of times that target x has been chosen along the optimal trajectory during the first n lags of the planning period, i.e., the number of times decision $U_n(q,x)$ has been made,

$L_n(q,x)$ = the number of times that x has been realized given that it has been chosen K_n times during the first n lags of the planning period, i.e., the number of times decision $U_n(q,x)$ was chosen and realized.

$\Pi(q,x)$ = the true (unknown) probability of making decision $U_n(q,x)$, and

$\pi(q,x)$ = the prior estimate of $\Pi(q,x)$.

In a deterministic dynamic programming framework the optimal decisions for all future periods are made right in the beginning. In a stochastic environment the release decision is contingent upon the expected value of the random variable ξ_n. We propose to modify that decision further on the basis of our past performance as measured by K_n and L_n. When the model is solved at the beginning of the planning horizon, we do not know what decisions will actually be made in each period. This is why $U_n(q,x)$ is a random variable with a probability distribution $\pi(q,x)$. This probability distribution itself is unknown at the beginning, so we attempt to estimate this through the following.

We introduce an indicator variable $Y_n(q,x)$ such that,

$$Y_n(q,x) = \begin{cases} 1 & \text{if decision } U_n(q,x) \text{ is made,} \\ 0 & \text{otherwise.} \end{cases} \quad (3.1)$$

As the process evolves, information about U_n, K_n and L_n becomes available. Conditioning on K_n and L_n, the $\Pr[\Pi(q,x)|K_n,L_n]$ can be determined by using Bayes' theorem,

$$\Pr[\Pi(q,x)|K_n,L_n] = c \; \Pr[K_n,L_n|\Pi(q,x)] \cdot \Pr[\Pi(q,x)] \quad (3.2)$$

where c is a normalizing constant.

If we define $Y_n(q,x) = 1$ as a success with probability $\Pi(q,x)$ (i.e., the decision $U_n(q,x)$ is made), and $Y_n(q,x) = 0$ as a failure with probability $1-\Pi(q,x)$, then $[K_n,L_n|\Pi(q,x)]$ represents the event of obtaining L_n successes in K_n trials given that the last trial is a success. This implies that $\Pr[K_n,L_n|\Pi(q,x)]$ is a negative binomial density function. Let us denote this function by $F(q,x)$. Then,

$$F(q,x) = \binom{K_n-1}{L_n-1} [1 - \Pi(q,x)]^{K_n-L_n} \cdot [\Pi(q,x)]^{L_n} \quad (3.3)$$

Substituting (3.3) in (3.2) and replacing $\Pr[\Pi(q,x)]$ by $\pi(q,x)$, its prior estimate yields,

$$\Pr[\Pi(q,x)|K_n,L_n] = c\pi(q,x) \binom{K_n-1}{L_n-1} [1 - \Pi(q,x)]^{K_n-L_n} [\Pi(q,x)]^{L_n} \quad (3.4)$$

Upon integrating both sides of (3.4) and equating the integral to one we obtain,

$$c = \frac{K_n(K_n+1)}{L_n} \cdot \frac{1}{\pi(q,x)}$$

and (3.4) becomes,

$$\Pr[\Pi(q,x)|K_n,L_n] = \binom{K_n-1}{L_n-1} \left[\frac{K_n(K_n+1)}{L_n}\right] [1 - \Pi(q,x)]^{K_n-L_n} [\Pi(q,x)]^{L_n}. \quad (3.5)$$

We will denote the left hand side of this equation as $P(q,x)$ for simplicity. Now $\bar{\Pi}(q,x)$, the expected value of $\Pi(q,x)$ given K_n and L_n is,

$$\bar{\Pi}(q,x) = \int_0^1 \Pi(q,x) \, P(q,x) \, d\Pi(q,x) = \frac{L_n+1}{K_n+2}. \quad (3.6)$$

Since $\bar{\Pi}(q,x)$ is the expected value of the density $\Pi(q,x)$, it is necessary that,

$$\frac{L_n+1}{K_n+2} \leq 1 \text{ or } L_n \leq K_n+1$$

This condition is always satisfied since we decide on the releases and the targets at the beginning of the period and realize the targets at the end of that period. Finally it should be noted that $\bar{\Pi}$ does not depend on q and x, but it is a function of K_n and L_n only. Therefore $\bar{\Pi}(q,x)$ will be replaced by $\bar{\Pi}(K_n,L_n)$ in what follows.

The Recursive Relationship

First we determine the admissible sets Ξ_n, X_n and Q_n for ξ_n, x_n and q_n respectively. The set Ξ_n is given by

$$\Xi_n = \{\xi_{n,min}, \xi_{n,max}\}. \quad (4.1)$$

where $\xi_{n,min}$ and $\xi_{n,max}$ are the minimum and maximum observed inflows for period n respectively. The set X_n is given by (2.3) as,

$$X_n = \{a_n, b_n\}. \quad (4.2)$$

To obtain the admissible set Q_n we rewrite equation (2.2) as: $q_n = x_{n-1} - x_n + \xi_n$. The bounds on q_n are then given by the inequalities,

$$x_{n-1,min} - x_n + \xi_{n,min} \leq q_n \leq x_{n-1,max} - x_n + \xi_{n,maz}, \quad x_n \epsilon X_n. \quad (4.3)$$

Substituting $x_{n-1,min} = a_{n-1}$ and $x_{n-1,max} = b_{n-1}$ in (4.3) we get,

$$a_{n-1} - x_n + \xi_{n,min} \leq q_n \leq b_{n-1} - x_n + \xi_{n,max}, \quad x_n \epsilon X_n. \quad (4.4)$$

Combining (4.4) with inequality (2.4) we get,

$$Q_n = \{\max(d_n, a_{n-1} - x_n + \xi_{n,min}), b_{n-1} - x_n + \xi_{n,max}\}, \quad x_n \epsilon X_n. \quad (4.5)$$

When $n \geq 1$, the transition of the system from one state to another is governed by the decision to release q_n units resulting in the reservoir holding x_n units at the end of period n. Even though x_n is really a continuous variable, let us take it in terms of ranges. That is, we might redefine x_n values as between 0-500 acre-feet, 501-1000 acre-feet, and so on. This would allow us to discretize a state variable, i.e. this state variable will have a fixed number of states at any stage. This approach has been explored by many researchers. For example, see Klemes [8]. For instance, let the states of x_n be $x_n^{(i)}$ $i=1,2,3$. The actual content of the reservoir, Z_n, at the end of period n will be one of these states. Then, for fixed q_n, the realization of any one of these states depends upon the value of the random variable ξ_n as well as Z_{n-1}. However, Z_{n-1} can be any one of the states $x_{n-1}^{(i)}$ $i=1,2,3$. Consequently the state $x_n^{(1)}$ is accessible from $x_{n-1}^{(1)}$, $x_{n-1}^{(2)}$ and $x_{n-1}^{(3)}$; i.e., there are 3 possible paths connecting state $x_n^{(1)}$ with the states in period n-1. Figure 1 shows two possible outcomes for each one of these paths. If the system was at $x_{n-1}^{(1)}$ at the end of period n-1, i.e., $Z_{n-1} = x_{n-1}^{(1)}$, then a release of q_n units during period n may or may not result in the reservoir holding $x_n^{(1)}$ units at the end of period n. Our approach calls for keeping track of how many decisions to release q_n units to reach a target $x_n^{(i)}$ were successful. The decisions are counted in K_n and the successes are counted in L_n. In a way, this measures a decision-maker's predictive performance. Intuitively, one would like to modify the decision policy if the past performance (L_n/K_n) is poor. As discussed earlier, since the decision to release q_n to attain a target $x_n^{(i)}$ is itself a random variable with a probability distribution Π, the expected value of the benefit function is calculated as follows. If $Z_n = x_n^{(1)}$, a benefit of $B_n(q_n,x_n,\xi_n)$ is earned with a probability of $\Pi(q_n,x_n^{(1)})$. This case is represented by arc (1,3) in Figure 1. Arc (2,3) represents the case where $Z_n \neq x_n^{(1)}$ and no benefit is earned with a probability of 1 - $\Pi(q_n,x_n^{(1)})$.

FIGURE 1
THE STATE OF THE SYSTEM AND THE BENEFIT
EARNED DURING A TRANSITION FOR GIVEN K_n AND L_n

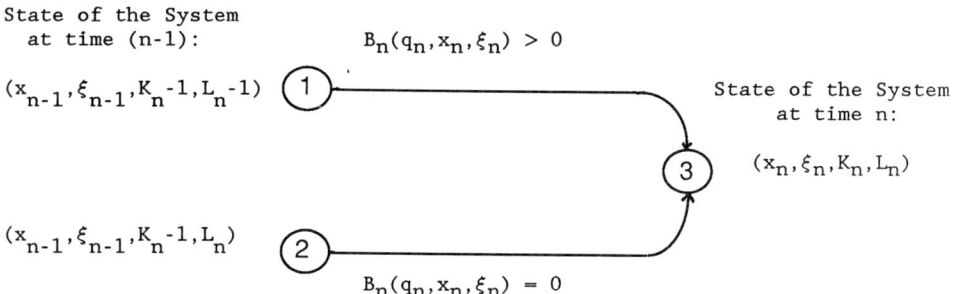

We start the derivation of the recursive relationship by determining the paths that connect x_{n-1} nodes with x_n using equation (2.2). Then, as shown in Figure 1, each path can be represented by two arcs. If $Z_n = x_n$, (i.e., a decision to release q_n and attain a target x_n was made and successful) then the expected benefit of the system is,

$$\int_0^1 \Pi(q,x) [B_n(q,x,\xi) + S_{n-1}(x_{n-1},\xi_{n-1},K_n-1,L_n-1)] P(q,x) d\Pi(q,x) \quad (4.6)$$

where $S_n(x_n,\xi_n,K_n,L_n)$ is the expected value of the benefit for the system at the end of period n when an optimal adaptive policy is used and when K_n choices of decision $U_n(q,x)$ along the optimal trajectory have led to L_n realizations of the target x_n in the first n periods. Notice that the arguments of S_{n-1} above include K_n-1 and L_n-1, both of which are incremented by 1 during this transition. Equation (2.2) gives $x_{n-1} = x_n - q_n - \xi_n$ and $P(q,x)$ is given by (3.5).

Summing up (4.6) for all paths connecting x_n with x_{n-1} and recognizing that ξ_n is a random variable, we obtain

$$\sum_{\xi_{n-1}} \sum_{\xi_n} Pr(\xi_n) \int_0^1 [\Pi(q,x) [B_n,(q,x,\xi) + S_{n-1}(x_{n-1},\xi_{n-1},K_n-1,L_n-1))] P(q,x) d\Pi(q,x) \quad (4.7)$$

where $n \in N$

$x_n \in X_n$

$q_n \in Q_n$

$\xi_n \in \Xi_n$

$x_{n-1} = x_n + q_n - \xi_n$

$K_n \in IK_n = \{1,2,\ldots,n-1\}$

$L_n \in IL_n = \{0,1,\ldots,K_n-1\}$

and $S_n(x_n,\xi_n,K_n,L_n) = 0$ for $n = 0$.

The double summation over ξ_n and ξ_{n-1} is due to the fact that ξ_n is conditioned on ξ_{n-1}.

An expression similar to (4.7) can be obtained for the case when $Z_n \neq x_n$, i.e., when the actual water content is different from the target x_n. The expected benefits then are

$$\sum_{\xi_{n-1}} \sum_{\xi_n} Pr(\xi_n) \int_0^1 [(1 - \Pi(q,x)) S_{n-1}(x_{n-1},\xi_{n-1}, K_n-1, L_n)] P(q,x) d\Pi(q,x) \quad (4.8)$$

Combining (4.7) and (4.8) and taking the maximum over q_n we obtain,

$$S_n(x_n,\xi_n,K_n,L_n) = \max_{q_n} \{ \sum_{\xi_{n-1}} \sum_{\xi_n} Pr(\xi_n) \int_0^1 [\Pi(q,x)B_n(q,x,\xi) + S_{n-1}(x_{n-1},\xi_{n-1},K_n-1,L_n-1) + (1-\Pi(q,x)) S_{n-1}(x_{n-1},\xi_{n-1},K_n-1,L_n)] P(q,x) d\Pi(q,x) \}. \quad (4.9)$$

Recursive relationship (4.9) can further be simplified by using $\overline{\Pi}(q,x)$ given in (3.6),

$$S_n(x_n,\xi_n,K_n,L_n) = \max_{q_n} \{ \sum_{\xi_{n-1}} \sum_{\xi_n} Pr(\xi_n) [\overline{\Pi}(K_n,L_n)B_n(q,x,\xi) + S_{n-1}(x_{n-1},\xi_{n-1},K_n-1,L_n-1) + (1 - \overline{\Pi}(K_n,L_n)) S_{n-1}(x_{n-1},\xi_{n-1},K_n-1,L_n)] \} \quad (4.10)$$

APPLICATION OF THE MODEL

The recursive formula in 4.10 suggests that there are four dimensions to the state vector. The K_n and L_n dimensions, as previously stated, let the decision maker adapt the policy on the basis of past performance. The problem now is a large stochastic dynamic programming problem. In practice, this approach would be applied as follows. First, all feasible values in the sets N, Ξ_n, x_n, Q_n, IK_n, IL_n are determined. These ar based on analysis of the historical data. IK_n and IL_n are usually determined on an arbitrary basis to begin with. Of course, the number of values in IK_n or IL_n would not exceed the number of possible targets. The optimal release plan is determined using the algorithm described next. This optimal trajectory is based on certain expected inflows. As the release plans are implemented, information on actual inflows is available. Also, we record how many successes we have had so far in reaching targets. Using this new information as it becomes available, a new trajectory is determined from the same output. Another decision is implemented, and the process continues. This requires that the basic model be solved only once.

The following six steps provide a general algorithm for the solution of the adaptive control problem. The sets N, Ξ_n, X_n, Q_n, IK_n, and IL_n have to be determined prior to the implementation of the algorithm.

Step (1):   Initialize the variables n, x_n, K_n and L_n by setting:

   (i)   n = 1
   (ii) $x_n = a_n$
   (iii) $K_n = 1$
   (iv) $L_n = 0$

Step (2):   First compute $\overline{\Pi}(K_n, L_n)$ in (3.6) and then $S_n\ (x_n, \xi_n, K_n, L_n)$ using equation (4.10) and the fact that $S_{n-1}\ (x_{n-1}, \xi_{n-1}, K_n, L_n) = 0$ for n = 0 (initial condition).

Step (3):   If the set IL_n is empty, go to step (4). Otherwise increment L_n and go to step (2).

Step (4):   If the set IK_n is empty, go to step (5). Otherwise increment K_n and go to step 1 (iv).

Step (5):   If the set X_n is empty, go to step (6). Otherwise increment x_n and go to step 1 (iii).

Step (6):   If the set N is empty, terminate. Otherwise set n = n+1 and go to 1 (ii).

The computational effort required to apply the above solution procedure is dependent upon the size of the sets N, Ξ_n, X_n, Q_n, IK_n, and IL_n. The planning horizon for an intermediate term problem is 52 weeks, i.e. N = 52. The release, storage and inflows are continuous variables, so the state space in (4.10) is astronomically large. One approach to reduce the problem to a manageable size is to discretize these variables by considering the ranges, rather than numerical values. These ranges are determined on the basis of historical values of variables for that week. Admittedly, this will introduce some error in the optimal policy, but a sensitivity analysis can be performed to ascertain the validity of the optimal trajectory. The planning horizon of 52 weeks itself may be divided into four 13-week subhorizons and thus smaller, solvable problems can be developed out of this large problem. One could argue that using these subhorizons could produce local optima less than what would be obtained otherwise. However, the 52-week horizon itself is an artificial division of the almost infinite horizon of operation of a reservoir. The use of subhorizons is only to reduce the state-space to a reasonable size. We anticipate no significant problems on this account.

COMPUTATIONAL EXPERIENCE

Sharda [16] applied a nonlinear stochastic programming model to Norris Reservoir, one of a large system of interrelated reservoirs operated by the TVA. Release plans were determined for a weekly time step over a 52-week horizon. His solution was based on the assumption that the operation policy for Norris Reservoir can be determined independent of the policies of the other reservoirs. The objective was to maximize a hydropower generation function. Appendix A describes this function. A similar function for TVA is also described in Rosenthal [15]. The function computes peak and off peak powers based on certain assumptions, turbine capacities, releases and storage levels in the reservoirs. We applied the adaptive control model to this same reservoir. Here the weekly inflows are assumed to be conditional random variables which are normally distributed. Although this assumption was verified by a Shapiro-Wilks tests of normality, we point out that it is not crucial to the application of this model.

The 52-week horizon was divided into four 13-week subproblems and the optimum solution for the first 10 weeks of each subproblem was accepted in the final solution of the problem. Moreover, 7 discrete levels of storage, release and inflow were used to determine the optimum weekly levels of release and storage. Doran [6] and Klemes [8] agree that a good discretization of storage variable can be accomplished by using 5-10 states. In addition, 3 levels of K and 2 levels of L were specified. This implied that the solution of the model required the investigation of $2 \times 3 \times 13 \times 7^3$ possible paths.

Since the original problem is a continuous one and in this formulation the variables had been discretized, a sensitivity analysis was performed to check if changing the discrete level of storage would alter the optimal solution. This was accomplished as follows:

(1) We formulated a new problem by fixing 3 storage levels on either side of the previously obtained optimal storage trajectory. Including the previous optimal storage levels, this new problem now had 7 storage levels. The problem was solved again using these discrete levels. The solution of this problem provided new optimal storage and release levels. This procedure was repeated until the value of the objective function did not improve substantially, i.e. did not change by more than a small prescribed quantity $\epsilon > 0$.

(2) We fixed the optimal storage levels obtained in (1). Then a search similar to the one described above was performed with respect to the optimal release trajectory. This procedure was stopped when no further improvement in the objective function was possible.

We point out that this steps described above cannot worsen the optimal solution of the original problem since the set of possible solutions of every new problem contains the optimal solution of the previous one. In the worst case, the new solution will be the same as the original solution.

The optimum solution applied to Norris Reservoir, was obtained in 5 iterations, on an IBM 370/168 computer system. The average run time was 95.12 seconds. Table (1) shows the optimum release and storage levels, the corresponding values of K_n, L_n and $\overline{\Pi}(K_n, L_n)$ of the solution of the adaptive

TABLE 1
ADAPTIVE CONTROL RELEASE POLICY (NORRIS RESERVOIR)

N Week	Release	Inflow (100 ft³/s-wk)	Storage	K(N)	L(N)	$\overline{\Pi}(K,L)$
0	-	-	400	-	-	--
1	4	11	412	2	1	0.5000
2	5	163	570	3	1	0.4000
3	5	88	650	1	0	0.3333
7	123	182	709	2	1	0.5000
5	93	114	730	3	2	0.6000
6	45	101	786	3	0	0.2000
7	112	140	814	3	2	0.6000
8	166	208	856	3	2	0.6000
9	181	202	877	2	1	0.5000
10	111	139	905	3	1	0.4000
11	91	168	982	3	1	0.4000
12	121	177	1038	2	0	0.2500
13	131	194	1101	1	0	0.3333
14	131	173	1143	1	0	0.3333
15	112	210	1241	2	1	0.5000
16	58	114	1297	1	0	0.3333
17	107	156	1346	1	0	0.3333
18	67	170	1449	1	0	0.3333
19	80	180	1549	3	2	0.6000
20	103	103	1549	2	0	0.2500
21	64	64	1549	2	1	0.5000
22	147	147	1549	2	0	0.2500
23	50	50	1549	2	1	0.5000
24	49	49	1544	3	0	0.2000
25	47	47	1544	3	2	0.6000
26	35	35	1549	2	1	0.5000
27	39	39	1549	3	2	0.6000
28	32	32	1549	1	0	0.3333
29	33	33	1549	1	0	0.3333
30	23	23	1549	1	0	0.3333
31	54	54	1549	1	0	0.3333
32	17	17	1549	1	0	0.3333
33	34	34	1549	1	0	0.3333
34	36	36	1549	1	0	0.3333
35	16	16	1549	1	0	0.3333
36	17	17	1549	1	0	0.3333
37	22	22	1549	1	0	0.3333
38	21	21	1549	3	2	0.6000
39	39	39	1549	1	0	0.3333
40	128	44	1465	1	0	0.3333
41	97	41	1409	1	0	0.3333
42	80	3	1332	1	0	0.3333
43	283	51	1100	3	1	0.4000
44	70	4	1034	3	2	0.6000
45	91	37	980	2	1	0.5000
46	91	25	914	3	1	0.4000
47	65	5	854	1	0	0.3333
48	182	254	926	2	1	0.5000
49	160	97	863	1	0	0.3333
50	318	269	814	2	1	0.5000
51	289	149	674	1	0	0.3333
52	265	217	626	2	0	0.2500

control model. For instance, given a storage level of 57,000 ft.^{3}/s* at the end of the second week, a release of 500 ft.^{3}/s (14.16 m^3/s) during the third week results in the reservoir holding 65,000 ft.^{3}/s at the end of the third week, since 8800 ft.^{3}/s are expected to flow into the reservoir during the week. Since the realization of 65,000 ft.^{3}/s at the end of the third week is uncertain due to the uncertainty of the inflows, the cumulative benefits obtained through the third week are multiplied by the inflows probabilities. Moreover, these benefits are also dependent on the number of times we set up a target of 73,000 ft.^{3}/s, K(3) and the number of times this target was realized during the 3-week period, L(3). This information is captured in $\bar{\Pi}(1,0) = 0.3333$.

While the feasibility of this approach is evident from this run, the comparison of the model's performance is somewhat more difficult. The main reason is that, in reality, Norris Reservoir is part of a multireservoir system. It is not operated in isolation. Notwithstanding the above fact, a simple comparison may be made. TVA [17] attempted to solve a number of release decision models for weekly time steps using various methods. The models were built for a six-reservoir subsystem and solved assuming a known inflow sequence for the 52 weeks. The optimum releases from Norris Reservoir based on their best model would have produced 1647.48 MW in power, all peaking. Sharda [16] solved a sequential decision model for Norris Reservoir. Releases based on this model for the same 52 week inflow sequence would have generated 3498.86 MW of hydropower (2246.31 peak, 1252.55 off-peak). The releases based on the adaptive control model presented here would have resulted in 2736.78 MW peaking and 747.77 MW off-peak power, for a total of 3484.55 MW. On the surface, the latter two models appear to do much better than the first one. But it must be remembered that the first model was solved as a multiple reservoir model, so that releases suggested or the objective function value are not really comparable to these models. These numbers do indicate that the sequential decision model in [16] and our model performed nearly as well. Further, the peak power that would have been generated was about 500 MW more than the adaptive control model. This test does not necessarily show that the AC model would always perform best, but the closeness of the objective values suggests the feasibility of the approach.

The advantages of this approach over the other two mentioned are as follows. Unlike the deterministic approach described in [17], no assumptions about the inflows need be made. If the realized inflows turned out to be different from the ones assumed in the run, the model would not have to be run again. The stochastic approach described in [16] is based on sequential decision making by running the model every week. In contrast, the AC approach would require for the model to be run just once. If the inflows are different, a new optimal trajectory can be obtained from the same run. The new trajectory is determined not only on the basis of changes in inflows, but also on the basis of frequency of the success of earlier decisions. This would permit a decision maker to adopt a different policy if the targets are being missed consistently. An added benefit also is to consider alternate release policies. The same run can also be used to investigate other release policies.

A salient feature of the optimum solution in Table (1) is the similar behavior of the optimum release, Q(N), and the inflow sequences, ξ(N). High levels of release occur during weeks 4 to 20 and 48 to 52, since high levels

---

*Storage level and releases are expressed as the average flow rate that can be provided by the volume of water over the selected time period. Storage of 1 ft^3/s-wk is equal to 13.884 acre-feet or 17128.6 m^3.

of inflow are expected during these weeks. Lower levels of release, on the other hand, occur during weeks 23 to 39 since the expected levels of inflows are relatively lower during these weeks. However, generally speaking, the solution is characterized by high release and storage levels. This can be attributed to the fact that the set Ξ_n of the adaptive control model contains high inflow levels. Since large amounts of water are expected to flow into the reservoir, a release of high levels of water is essential for the reservoir to meet its upper bound constraints. However, this is not a limitation on use of the model. The optimum solution trajectory can be obtained easily from the output from the same model if the realized inflow is not equal to the inflow assumed to be realized in the original run. This permits one to solve the intermediate term release planning problem just once and determine the optimal trajectory on the basis of the realized value of the state variable.

The model presented in this paper is a simplified one. It does not consider other benefits of reservoir operation explicitly. It also is limited to a single reservoir. The model also ignores evaporation losses. However, the focus here is on a method to solve a problem with an extremely large state space. This approach appears to have worked well. If the model is extended to multiple reservoirs, the state space will be considerably larger. The idea of decomposing the state space may still make the applications of AC to a multi-reservoir system feasible.

CONCLUSIONS

The model presented above was able to expand the application of the adaptive control approach to a class of reservoir operation problems. The model allows a decision maker to modify his decisions on the basis of not only the updated information on random inflows, but also on the basis of the past performance in terms of attaining the targets. The formulation accommodates a large state space. The practical feasibility of this approach was demonstrated by applying a discretization and decomposition process to a release decision problem for a single reservoir. The ex post results show that this approach would have done at least as well or better than a deterministic approach or another stochastic approach. The model does not require any assumptions about the probability distribution of inflows, in contrast to some other models. The ability to modify decisions on the basis of past performance should make the model attractive for actual applications. An extension of the model to a multiple reservoir system is possible, but further improvements in the computation algorithm would have to be made to accommodate the astronomically large state space. The decomposition and discretization steps used here do make it practical.

REFERENCES

(1) Askew, Arthur J., "Chance Constrained Dynamic Programming and the Optimization of Water Resources Systems," Water Resources Research, Vol. 10, No. 6, pp. 1099-1106, 1974.

(2) Bellman, R., Adaptive Control Processes, Princeton Univ. Press, Princeton, New Jersey, 1961.

(3) Chu, W. S., and Yeh, W. W-G, "A Non-Linear Programming Algorithm for Real Time Hourly Reservoir Operations," Water Resources Bulletin, 14, pp. 1048-1063, 1978.

(4) Close, E. R., Beard, L. R. and Dawdy, D. R., "Objective Determination of Safety Factor in Reservoir Design," J. Hydraulic Division, Amer. Soc. Civil Eng., Vol. 96, No. HY5, pp. 1167-1177, 1970.

(5) Davis, D. R., Kisiel, C. C., and Duckstein, L., "Bayesian Decision Theory Applied to Design in Hydrology," Water Resources Research, Vol. 8, No. 1, pp. 33-41, 1972.

(6) Doran, D. G., "An Efficient Transition Definition for Discrete State Reservoir Analysis: The Divided Interval Technique," Water Resources Research, Vol. 11, No. 6, pp. 867-873, 1975.

(7) Eisel, L. M., "Chance-Constrained Reservoir Models," Water Resources Research, Vol. 8, No. 2, pp. 339-347, 1972.

(8) Klemeš, V., "Discrete Representation of Storage for Stochastic Reservoir Optimization," Water Resources Research, Vol. 13, No. 1, pp. 149-158, 1977.

(9) Kushner, H., Introduction to Stochastic Control, Holt, Rinehard and Winston, Inc., New York, 1971.

(10) Larson, R. E. and Keckler, W. G., "Application of Dynamic Programming to the Control of Water Resource Systems," IFAC, Haifa Symposium on Automatic Control of Natural Resources, 1968.

(11) Lloyd, E. H., "A Probability Theory of Reservoirs with Serially Correlated Inputs," Journal of Hydrology, Vol. 1, pp. 99-128, 1963.

(12) Pekelman, D. and Rausser, G. C., "Adaptive Control: Surveys of Methods and Applications," TIMS Studies in the Management Science, 9, pp. 89-120, 1978.

(13) Prêkopa, A., "Optimal Control of Storage Level Using Stochastic Programming," Problems of Control and Information Theory, Vol. 3, No. 4, pp. 193, 1975.

(14) Roefs, T. G., and Bodin, L. D., "Multi-Reservoir Operation Studies," Water Resources Research, Vol. 6, No. 2, pp. 410-420, 1970.

(15) Rosenthal, R. E., "A Nonlinear Network Flow Algorithm for Maximization of Benefits in a Hydroelectric Power System," Operations Research, Vol. 28, No. 4, pp. 763-786, 1981.

(16) Sharda, R., "Stochastic Programming Applied to Reservoir Operations," Unpublished Ph.D. thesis, Univ. of Wisconsin, Madison, 1981.

(17) Tennessee Valley Authority Water Resources Management Methods Staff, "The Testing of Different Optimization Methods on a Reservoir Subsystem," Report B-16, TVA, Knoxville, 1977.

APPENDIX A - THE HYDROPOWER FUNCTION

The hydropower function employed in this study was developed by TVA [17]. The hydropower characteristics are defined as:

$$P = \min (q \cdot PC, PCAP)$$

where P is the weekly plant generation level, in megawatts,

q is the weekly plant release in 100 ft^3/sec,
PC is the plant power factor, defined as power output per unit release, in MW/(100cfs), and
PCAP is the plant capacity in MW.

Both PC and PCAP are functions of the reservoir head which is the difference between headwater and tailwater elevations. For computational ease, the PC and PCAP are represented as functions of storage, X. PC is a sixth-order polynomial and PCAP is a third-order polynomial in X. The coefficients of the polynomials are given in Tables A2 and A3 of TVA [17]. In each 168-hour weekly period, 64 hours are considered to make the peak period. Thus the peaking capacity for each plant is

$$PC = \min(168P/64, PCAP)$$

where PE is the weekly peak generation level. The off-peak generation is determined by

$$OPE = \max((168P-64PE)/164, 0),$$

where OPE is the weekly plant off-peak generation. Hydropower generation is calculated with some inaccuracy using these functions because they use beginning-of-the-week storage instead of average storage. But all the algorithms compared in this paper used the same hydropower function, so the comparisons are valid.

EXTENDED LINEAR QUADRATIC GAUSSIAN CONTROL FOR THE REAL TIME OPERATION OF RESERVOIR SYSTEMS

ARISTIDIS P. GEORGAKAKOS
School of Civil Engineering, Georgia Institute of Technology, Atlanta, GA 30332

ABSTRACT

The Extended Linear Quadratic Gaussian (ELQG) control method [9] is discussed and tested in the control of a single reservoir. The results indicate that the method is reliable, computationally efficient, and compares favorably with traditional reservoir operation schemes. ELQG is well suited for real-time reservoir control as well as for developing policy-making guidelines.

INTRODUCTION

The operation of a single reservoir has been a topic of extensive and fruitful research. Comparatively, the Markov-chain Stochastic Dynamic Programming has been the most comprehensive formulation [7,23,24,2,1]. Successful extensions to account for reliability constraints or forecasted information in real time have also been reported in [3,4,19,20,21,6,22]. However, the use of this formulation for multireservoir systems is seriously limited due to "dimensionality" problems.

In an effort to overcome this limitation, recent developments have moved away from this traditional approach and toward analytical reservoir operation schemes [26,16,10].

This paper reports computational experience with the Extended Linear Quadratic Gaussian (ELQG) control method [10] and compares this method's performance with that of some traditional formulations. The case study concerns the control of the High Aswan Dam in Egypt for which some interesting policy-making conclusions are also drawn.

SYSTEM DESCRIPTION

The High Aswan Dam (Figure 1) is by far the most effective control project in the Nile basin. It has a storage capacity of 168 billion cubic meters, 32.72 of which are allocated for silt deposits (dead storage). Water can be released at a controllable rate through a diversion channel which eventually divides into 24 branches. Twelve of these are feeding the power plant turbines while the rest are designed to bypass the turbines and, if necessary, discharge up to 30 billion cubic meters of water per month. However, the downstream river channel and water distribution network cannot transport releases higher than 7.59 billion cubic meters per month without severe damages due to bank erosion. To alleviate the possibility of excessive releases, the Toshka spillway was constructed on the western bank of the reservoir's lake (Lake Nasser). The spillway is a free-flow channel operating when water elevation exceeds 178 meters (above sea level), or, equivalently, when storage exceeds the volume of 137.72 billion cubic meters. It is designed to dispose of 7.5 billion cubic meters per month when reservoir elevation reaches 182.6 meters. The water is diverted into the Toshka depression where it evaporates. An emergency spillway is also situated on the reservoir's western bank and it begins to operate at 182.6 meters water level.

The reservoir's area-storage and elevation-storage curves are approximated sufficiently well by the following continuous functions [1]:

$$A(s(t)) = -3164.28 + 25.4914\ s(t) + 1092.92\ \ln(s(t)) \tag{1}$$

$$H(s(t)) = 79.9734 + 0.03698\ s(t) + 18.8705\ \ln(s(t)) \tag{2}$$

where area, $A(\cdot)$, and elevation, $H(\cdot)$, are obtained respectively in square kilometers and meters when storage, $s(t)$, is expressed in billion cubic meters.

Table 1 gives some Nile flow statistics at Wadi Halfa (entrance of Lake Nasser) based on observed discharges from 1912 through 1965. The apparent seasonal variability is a consequence of the different hydrologic responses characterizing the major Nile tributaries, the White and Blue Nile. The wide monthly flow range indicates also pronounced overyear variability.

Due to the semi-arid climate, HAD receives almost no precipitation and suffers heavy evaporation losses. The distribution of the average monthly evaporation depths presented in Figure 2 [6] results in a mean annual value of 2.7 meters and a loss of 10 to 15 billion cubic meters m^3 per year.

HAD's prime operational objective is to satisfy the Egyptian agricultural, municipal, and industrial water needs. On monthly basis, these downstream water supply requirements amount to the levels shown on Figure 3. They total 55.5 billion cubic meters per year.

Second but equally as important is HAD's function as energy supplier with a present output of almost half of Egypt's power demand. The GWH of energy produced when turbine release is u billion cubic meters per month and average reservoir elevation is H meters can be obtained from [18]

$$g(s(t),u(t)) = 0.517\ (u(t))^{1.194}\ [H(s(t)) - 108]^{1.268} \tag{3}$$

with a maximum of 1280 GWH per month.

The Nile is also of vital importance for the country of Sudan. By the Nile Water Agreement, Sudan is entitled to use up to 18.5 billion cubic meters per year. The estimated monthly Sudan abstractions in percentages of the yearly total [6] are shown on Figure 4.

THE CONTROL PROBLEM

In summary, the HAD operation problem can be posed as follows:
Find the control function sequence $\{\mu(\cdot,\cdot):u(t) = \mu(s(t),t), t\epsilon[t_0,t_T]\}$ which minimizes

$$J = E\left\{ \int_{t_0}^{t_T} (1280 - g(s(t),\mu(s(t),t))^2 dt + \ell_T(s(t_T)) \right\} \tag{4}$$

($E\{\cdot\}$ denotes expectation with respect to the joint probability density of the storage variables, $[t_0,t_T]$ is the control horizon, and $\ell_T(\cdot)$ is added to model storage preferences at the end of the control horizon)
subject to
(a) the reservoir dynamics,

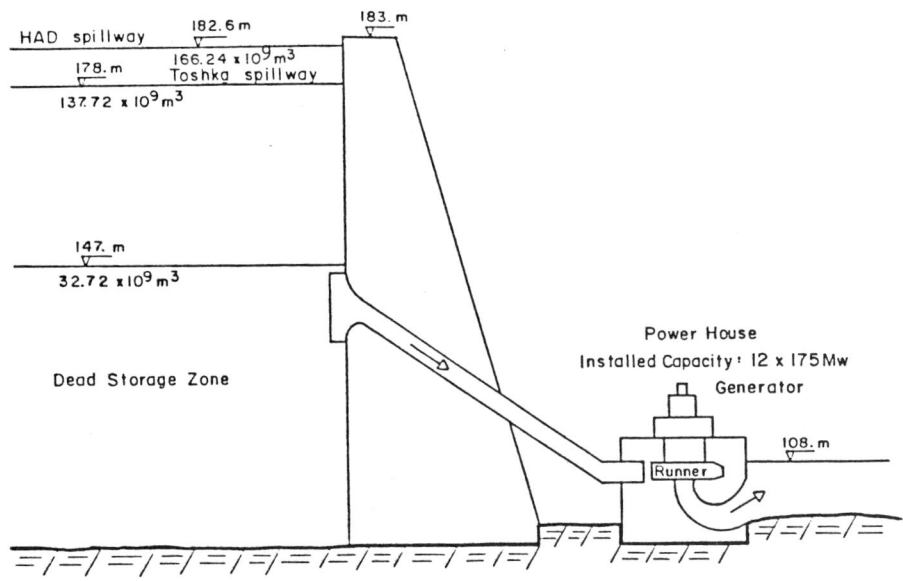

FIG. 1: The High Aswan Dam

TABLE 1: Monthly Nile Flow Statistics

Month	Expected Value ($\times 10^9$ m^3)	Standard Deviation ($\times 10^9$ m^3)	Correlation Coefficient	Minimum ($\times 10^9$ m^3)	Maximum ($\times 10^9$ m^3)
Jan.	3.507	0.734	0.9080	2.040	5.750
Feb.	2.452	0.681	0.7503	1.420	5.080
Mar.	2.275	0.661	0.8649	1.260	4.810
Apr.	2.042	0.685	0.8714	1.050	4.540
May	1.924	0.750	0.7953	0.880	4.340
June	2.073	0.772	0.4955	1.000	4.520
July	5.170	1.516	0.6377	2.230	10.000
Aug.	19.448	3.821	0.7115	7.680	27.100
Sept.	21.991	3.745	0.7851	13.400	31.700
Oct.	14.605	3.125	0.8070	7.860	24.200
Nov.	7.166	1.752	0.8948	4.140	12.200
Dec.	4.538	0.857	0.9206	2.990	7.060

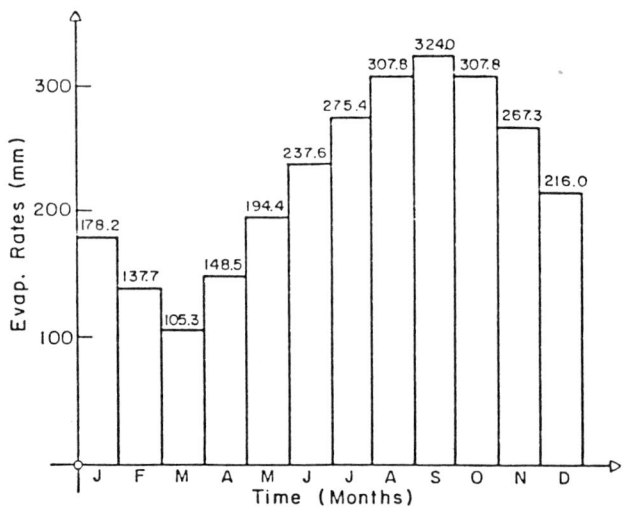

FIG. 2: Monthly Average Evaporation Rates

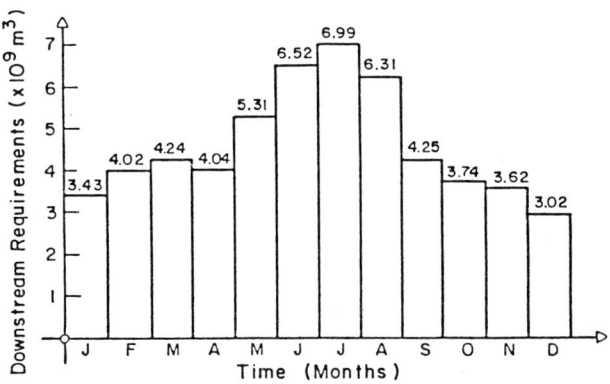

FIG. 3: Monthly Water Supply Requirements

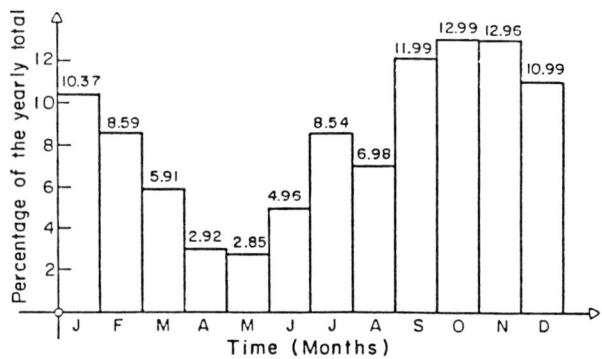

FIG. 4: Monthly Sudan Shares

$$\frac{ds(t)}{dt} = -e(t) A(s(t)) - \mu(s(t),t) + w(t), \quad t\epsilon[t_0, t_T], \qquad (5)$$

($e(t)$ represents the evaporation rate, and $w(t)$ the uncertain inflow process)
(b) the lower and upper constraints on the releases,

$$u^{min}(t) \leq \mu(s(t),t) \leq u^{max}(t), \quad t\epsilon[t_0,t_T], \qquad (6)$$

($u^{min}(t)$ represents the water supply requirements, and $u^{max}(t)$ is the highest discharge not causing bank erosion),
and (c) the lower and upper storage reliability constraints,

$$\int_{-\infty}^{s^{min}(t)} p(s(t),t) \, ds(t) \leq \gamma^{min}(t), \quad t\epsilon[t_0,t_T] \qquad (7)$$

$$\int_{s^{max}(t)}^{+\infty} p(s(t),t) \, ds(t) \leq \gamma^{max}(t), \quad t\epsilon[t_0,t_T] \qquad (8)$$

($s^{min}(t)$ corresponds to the dead storage, and $s^{max}(t)$ to Toshka's operating threshold; $\gamma^{min}(t)$ and $\gamma^{max}(t)$ are probabilistic tolerance levels determined by the reservoir manager based on his risk attitude).

Although this problem was motivated by the operation of the High Aswan Dam, it is very typical in the operation of reservoir systems. Its basic idiosyncracies are that it is stochastic (due to the uncertainty of inflows), nonlinear ($-e(t)A(s(t))) = F(s(t),t)$ is a nonlinear function of $s(t)$), constrained with respect to both the release and storage variables, and seeking to optimize a general-type functional (the running cost $(1280 - g(s(t),u(t))^2 = \ell(s(t),u(t))$ is a general nonlinear function of $s(t)$ and $u(t)$). In fact, it can be shown [10] that the operation problem of a general multipurpose reservoir system can also be cast in the above format. And, furthermore, the algorithm to be discussed for this problem's solution is generally applicable as well.

EXTENDED LINEAR QUADRATIC GAUSSIAN (ELQG) CONTROL

Although stochastic control problems are solvable in principle [5], their solution process can in practice be a formidable task due to numerical difficulties. In fact, for general problems such as the one formulated previously, it may be safely stated that the optimal solution, consisting of the optimal feedback functions $\{\mu(\cdot,\cdot):u(t) = \mu(s(t),t), t\epsilon[t_0,t_T]\}$, is impossible to obtain. Fortunately, in such cases, one may resort to suboptimal approaches and obtain adequate approximations of the optimal solution.

ELQG is based on the most successful suboptimal control methodology: the Open Loop Feedback control approach. Rather than generating the entire feedback solution sequence and then applying the controls $u(t) = \mu(s(t),t)$, this method obtains adequate approximations of $u(t_k)$ at each decision time t_k by performing the following operations:
(i) Determine the control trajectory $\{u(t), t\epsilon[t_k,t_T]\}$ minimizing

$$J(t_k) = E\left\{ \int_{t_k}^{t_T} \ell(s(t), u(t))dt + \ell_T(s(t_T)) \right\} \quad (9)$$

subject to

$$\frac{ds(t)}{dt} = F(s(t), t) - u(t) + w(t), \quad (10)$$

$$u^{min}(t) \leq u(t) \leq u^{max}(t), \quad (11)$$

$$\int_{-\infty}^{s^{min}(t)} p(s(t), t) ds(t) \leq \gamma^{min}(t) \quad (12a)$$

$$\int_{s^{max}(t)}^{+\infty} p(s(t), t) ds(t) \leq \gamma^{max}(t), \quad t \epsilon [t_k, t_T], \quad (12b)$$

(ii) apply the first portion of the OLF-optimal control trajectory, $\{u(t), t\epsilon[t_k, t_{k+1}]\}$, and repeat steps (i) and (ii) at the next decision time t_{k+1}.

The advantage of the OLF procedure is that it identifies a control sequence $\{u(t), t\epsilon[t_k, t_T]\}$ as opposed to a series of feedback functions $\{\mu(s(t), t), t\epsilon[t_0, t_T]\}$, the disadvantage being that this has to be repeated at each decision time of the control horizon. Its suboptimality is due to the fact that each such optimization does not recognize that only a part of the identified control trajectory will actually be applied and that the process will be repeated thereafter. However, the OLF procedure makes stochastic problems amenable to an efficient deterministic optimization philosophy.

The common basis of the most efficient deterministic control methods (see, for instance, [15,12,11,13,17,25]) is that they start from some initial control sequence and the corresponding state trajectory and successively generate new sequences improving on the performance index value. The approach breaks down when stochastic systems are considered, for there is no unique state trajectory that the system is guaranteed to follow when a certain control sequence is applied.

However, each Open Loop Feedback Control problem can be reformulated with a probabilistic state; the probability density of the physical state variable [5]. The dynamics of this new state are deterministic and generate a unique trajectory as a response to a certain control sequence; thus, the iterative approach of the deterministic optimization methods becomes theoretically applicable. Unfortunately, the new system dynamics is in the form of a nonlinear partial differential equation (the Fokker-Planck equation, [14]) whose solution is usually a numerical challenge.

To facilitate the computations involved under the above reformulation, ELQG performs local linearizations of the nonlinear problem and considers the first two statistical moments of the state's probability density. The new system equations describe the evolution of these two statistical moments and pose no numerical difficulty. The linearizations are around nominal control and state trajectories which are updated at each iteration. The method iterates according to the Newton direction when there are no control or state constraints' violations, uses the Projected Newton direction if control bounds are exceeded, and employs a penalty function or a multiplier method in case state constraints are binding. The iterations are based on analytically computed optimization directions, a feature which makes the method computationally efficient. ELQG is applicable to control problems of any time scale (e.g., monthly, weekly, daily, or shorter) because in the solution process the continuous time formulation (of the previous section) is discretized over time intervals arbitrarily specifiable by the user.

An algorithmic step-by-step description of this method as implemented for the HAD operation problem is provided in Appendix A. The interested reader is also referred to [9,10] for more comprehensive ELQG discussions. The following section describes some control experiments with the High Aswan Dam system.

ELQG CONTROL

A typical ELQG iteration cycle is shown on Tables 2 to 7. This application seeks to determine the optimal HAD release sequence over a 36-month period. The Nile inflows to Lake Nasser are assumed to follow their historical monthly distributions. (The monthly means are reduced by the monthly Sudan shares.) The monthly evaporation rates and downstream water supply requirements are those reported earlier. Reservoir storage at time 0 (beginning of the control horizon) is set at 95×10^9 m^3, and, at any time, it is not to exceed 137.72×10^9 m^3 (Toshka's operating threshold) or fall below 32.72×10^9 m^3 (dead storage zone ceiling) with probability higher than 0.025. Maximum allowable downstream release is set at 7.59×10^9 m^3 per month. The objective is to maximize expected energy generation given the previous constraints and the requirement to have the expected end-of-the-control-horizon reservoir storage at 95×10^9 m^3. The initial nominal control trajectory is taken equal to the downstream water supply requirements. Tables 2 to 7 report the following used or computed iteration data. "HYDROPOWER" gives the energy in GWH which will be produced if the nominal control and mean state trajectories are realized. The associated value of the penalty function minimized is given under "PENALTY." Apart from penalizing energy production lower than 1280 GWH per month (HAD's installed capacity), this function also penalizes terminal mean storages away from 95×10^9 m^3 according to $\ell(s(t_k)) = (s(t_k) - 95 \times 10^9)^2 \times 10^9$. "WI" is quantity w_i of ELQG's termination criterion, "M" is the integer m_i in the Armijo rule which determines the stepsize α_i via $\alpha_i = \beta^{m_i}$, and "NCCV" gives the number of binding control constraints. Parameters β, σ, and ϵ of the algorithm were set equal to 0.5, 10^{-4}, and 9×10^{-4} respectively. The columns from left to right display the months, mean storage values, nominal releases, first derivatives of the performance index with respect to the controls at the nominal sequences, second derivatives of the performance index with respect to the controls at the nominal sequences, the Projected Newton Direction, the upper and lower standard deviations and the binding constraint set. A lower binding control variable is indicated by "-1", an upper binding control variable by "1", and a nonbinding one by "0". The positive second derivatives ("DUUJ") are an indication of the problem's convexity while the negative first derivatives ("DUJ") show that the cost function can be decreased by increasing the nominal releases. The high "WI" value implies that the initial nominal control trajectory is far from being optimal. For comparison with subsequent iterations, the standard deviations are computed in an Open Loop fashion. (See comment following ELQG algorithm.) The first iteration produces the trajectories shown in Table 3. After cutting in half the original stepsize, this iteration reduced the penalty function value, increased energy generation, and set the expected terminal storage (from 107.7×10^9 m^3) to 93.6×10^9 m^3. The standard deviations are now computed in the constrained-feedback manner

TABLE 2: Initial Trajectories

```
                HYDROPOWER ( GWH ) = .20791970E+05
                PENALTY            = .80932077E+12

                Z    =  1.960
                LT   =  36
                EI   =  .9000E-03
                WI   =  .1935E+02
                M    =  1
                NCCV =  0
```

L	E[S]	U	DUJ	DUUJ	DU	STDEU	STDEL	U+
1	95.000	3.430	-.9400E+11	.5459E+10	.3182E-02	.00	.00	0
2	92.622	4.018	-.9455E+11	.5524E+10	-.5162E+00	.74	.74	0
3	89.075	4.240	-.9499E+11	.5574E+10	-.6856E+00	1.01	1.01	0
4	85.724	4.040	-.9545E+11	.5629E+10	-.4877E+00	1.21	1.21	0
5	82.674	5.312	-.9609E+11	.5704E+10	-.1502E+01	1.39	1.39	0
6	78.096	6.521	-.9691E+11	.5803E+10	-.2437E+01	1.58	1.58	0
7	72.003	6.999	-.9793E+11	.5925E+10	-.2863E+01	1.75	1.75	0
8	67.867	6.311	-.9911E+11	.6068E+10	-.2353E+01	2.31	2.31	0
9	78.808	4.251	-.1003E+12	.6221E+10	-.1368E+00	4.51	4.51	0
10	93.306	3.741	-.1015E+12	.6370E+10	.6549E+00	5.85	5.85	0
11	100.706	3.619	-.1026E+12	.6505E+10	.8725E+00	6.60	6.60	0
12	100.930	3.019	-.1035E+12	.6620E+10	.1936E+01	6.77	6.77	0
13	99.681	3.430	-.1043E+12	.6716E+10	.1239E+01	6.78	6.78	0
14	97.273	4.018	-.1049E+12	.6794E+10	.4976E+00	6.77	6.77	0
15	93.702	4.240	-.1053E+12	.6856E+10	.3030E+00	6.77	6.77	0
16	90.332	4.040	-.1058E+12	.6921E+10	.6127E+00	6.78	6.78	0
17	87.256	5.312	-.1065E+12	.7012E+10	-.6680E+00	6.78	6.78	0
18	82.644	6.521	-.1074E+12	.7131E+10	-.1712E+01	6.77	6.77	0
19	76.508	6.999	-.1085E+12	.7278E+10	-.2111E+01	6.75	6.75	0
20	72.322	6.311	-.1098E+12	.7450E+10	-.1473E+01	6.85	6.85	0
21	83.209	4.251	-.1112E+12	.7634E+10	.8711E+00	7.81	7.81	0
22	97.653	3.741	-.1125E+12	.7814E+10	.1583E+01	8.60	8.60	0
23	105.004	3.619	-.1136E+12	.7977E+10	.1755E+01	9.07	9.07	0
24	105.187	3.019	-.1146E+12	.8117E+10	.3025E+01	9.16	9.16	0
25	103.905	3.430	-.1154E+12	.8233E+10	.2202E+01	9.13	9.13	0
26	101.469	4.018	-.1161E+12	.8328E+10	.1368E+01	9.10	9.10	0
27	97.877	4.240	-.1166E+12	.8402E+10	.1181E+01	9.09	9.09	0
28	94.491	4.040	-.1172E+12	.8482E+10	.1559E+01	9.07	9.07	0
29	91.392	5.312	-.1179E+12	.8592E+10	.2108E+00	9.05	9.05	0
30	86.750	6.521	-.1189E+12	.8734E+10	-.7904E+00	9.02	9.02	0
31	80.576	6.999	-.1201E+12	.8911E+10	-.1093E+01	8.97	8.97	0
32	76.345	6.311	-.1215E+12	.9118E+10	-.3948E+00	9.00	9.00	0
33	87.184	4.251	-.1229E+12	.9339E+10	.1860E+01	9.71	9.71	0
34	101.580	3.741	-.1244E+12	.9557E+10	.2512E+01	10.32	10.32	0
35	108.888	3.619	-.1257E+12	.9755E+10	.2664E+01	10.69	10.69	0
36	109.034	3.019	-.1267E+12	.9923E+10	.4074E+01	10.73	10.73	0
37	107.722					10.69	10.69	

TABLE 3: Iteration #1

HYDROPOWER (GWH) = .22653652E+05
PENALTY   = .10261323E+11

Z   = 1.960
LT   = 36
EI   = .9000E-03
WI   = .3821E+01
M   = 0
NCCV = 15

L	E[S]	U	DUJ	DUUJ	DU	STDEU	STDEL	U+
1	95.000	3.431	.1052E+11	.5405E+10	-.4599E+00	.00	.00	0
2	92.621	4.018	.1059E+11	.5469E+10	-.1936E+01	.74	.74	-1
3	89.074	4.240	.1063E+11	.5520E+10	-.1927E+01	.99	.99	-1
4	85.722	4.040	.1069E+11	.5573E+10	-.1917E+01	1.15	1.15	-1
5	82.672	5.312	.1076E+11	.5648E+10	-.1905E+01	1.28	1.28	-1
6	78.095	6.521	.1085E+11	.5745E+10	-.1889E+01	1.45	1.45	-1
7	72.001	6.999	.1096E+11	.5867E+10	-.1869E+01	1.62	1.62	-1
8	67.865	6.311	.1110E+11	.6009E+10	-.1847E+01	2.21	2.21	-1
9	78.807	4.251	.1123E+11	.6160E+10	-.1824E+01	4.44	4.44	-1
10	93.305	4.068	.1137E+11	.6307E+10	-.3478E+00	5.63	5.63	0
11	100.379	4.055	.1149E+11	.6441E+10	-.2080E+00	6.15	6.15	0
12	100.172	3.987	.1159E+11	.6555E+10	-.1035E+00	6.04	6.04	0
13	97.965	4.049	.1167E+11	.6651E+10	-.1351E+00	5.59	5.59	0
14	94.951	4.267	.1174E+11	.6729E+10	-.3109E+00	5.23	5.23	0
15	91.144	4.392	.1180E+11	.6790E+10	-.3902E+00	4.97	4.97	0
16	87.633	4.347	.1185E+11	.6856E+10	-.3029E+00	4.72	4.72	0
17	84.266	5.312	.1193E+11	.6947E+10	-.1717E+01	4.43	4.43	-1
18	79.677	6.521	.1203E+11	.7066E+10	-.1703E+01	4.29	4.29	-1
19	73.568	6.999	.1216E+11	.7214E+10	-.1685E+01	4.22	4.22	-1
20	69.415	6.311	.1230E+11	.7387E+10	-.1665E+01	4.37	4.37	-1
21	80.337	4.687	.1246E+11	.7572E+10	-.3446E+00	5.70	5.70	0
22	94.383	4.532	.1260E+11	.7752E+10	-.3671E-01	6.46	6.46	0
23	100.984	4.496	.1274E+11	.7916E+10	.5681E-01	6.70	6.70	0
24	100.332	4.532	.1285E+11	.8057E+10	.6164E-01	6.38	6.38	0
25	97.581	4.531	.1294E+11	.8174E+10	.1128E+00	5.66	5.66	0
26	94.089	4.702	.1302E+11	.8270E+10	-.3038E-02	5.08	5.08	0
27	89.853	4.831	.1308E+11	.8346E+10	-.6065E-01	4.61	4.61	0
28	85.909	4.820	.1314E+11	.8428E+10	.3234E+00	4.17	4.17	0
29	82.080	5.417	.1323E+11	.8541E+10	-.4804E+00	3.67	3.67	0
30	77.403	6.521	.1334E+11	.8689E+10	-.1536E+01	3.35	3.35	-1
31	71.315	6.999	.1348E+11	.8872E+10	-.1520E+01	3.10	3.10	-1
32	67.187	6.311	.1365E+11	.9088E+10	-.1502E+01	3.16	3.16	-1
33	78.137	5.181	.1382E+11	.9317E+10	.6936E-01	4.71	4.71	0
34	91.719	4.997	.1398E+11	.9541E+10	.2807E+00	5.20	5.20	0
35	97.888	4.951	.1413E+11	.9745E+10	.3441E+00	4.76	4.76	0
36	96.814	5.056	.1426E+11	.9920E+10	.2883E+00	3.20	3.20	0
37	93.569					.87	.87	

TABLE 4: Iteration #2

HYDROPOWER (GWH) = .22671973E+05
PENALTY   = .44969357E+10

Z   = 1.960
LT   = 36
EI   = .9000E-03
WI   = .3133E+01
M   = 0
NCCV = 20

L	E[S]	U	DUJ	DUUJ	DU	STDEU	STDEL	U$^+$
1	95.000	3.430	.6968E+10	.5418E+10	-.1286E+01	.00	.00	-1
2	92.622	4.018	.7009E+10	.5482E+10	-.1279E+01	.74	.74	-1
3	89.075	4.240	.7042E+10	.5533E+10	-.1273E+01	.97	1.01	-1
4	85.724	4.040	.7076E+10	.5587E+10	-.1267E+01	1.13	1.21	-1
5	82.674	5.312	.7123E+10	.5661E+10	-.1258E+01	1.24	1.39	-1
6	78.096	6.521	.7184E+10	.5759E+10	-.1247E+01	1.40	1.58	-1
7	72.003	6.999	.7260E+10	.5880E+10	-.1235E+01	1.56	1.75	-1
8	67.867	6.311	.7347E+10	.6023E+10	-.1220E+01	2.16	2.31	-1
9	78.808	4.251	.7439E+10	.6174E+10	-.1205E+01	4.39	4.51	-1
10	93.306	3.741	.7527E+10	.6322E+10	-.1191E+01	5.54	5.85	-1
11	100.706	3.847	.7606E+10	.6456E+10	-.7603E-01	6.04	6.29	0
12	100.703	3.883	.7674E+10	.6570E+10	-.7158E-01	5.90	6.12	0
13	98.595	3.914	.7729E+10	.6666E+10	-.7342E-01	5.55	5.75	0
14	95.711	4.018	.7774E+10	.6744E+10	-.1153E+01	5.18	5.37	-1
15	92.148	4.240	.7810E+10	.6805E+10	-.1148E+01	4.84	5.01	-1
16	88.785	4.044	.7847E+10	.6871E+10	-.8576E-01	4.51	4.66	0
17	85.713	5.312	.7899E+10	.6962E+10	-.1135E+01	4.14	4.28	-1
18	81.113	6.521	.7966E+10	.7080E+10	-.1125E+01	3.91	4.32	-1
19	74.991	6.999	.8048E+10	.7227E+10	-.1114E+01	3.77	4.35	-1
20	70.822	6.311	.8143E+10	.7400E+10	-.1101E+01	3.89	4.57	-1
21	81.727	4.342	.8244E+10	.7583E+10	-.8077E-01	5.31	5.95	0
22	96.099	4.495	.8341E+10	.7763E+10	-.5922E-01	6.09	6.55	0
23	102.717	4.553	.8428E+10	.7927E+10	-.5624E-01	6.36	6.73	0
24	101.992	4.593	.8502E-10	.8067E+10	-.5630E-01	6.03	6.34	0
25	99.166	4.644	.8564E+10	.8183E+10	-.5974E-01	5.49	5.77	0
26	95.552	4.699	.8614E+10	.8279E+10	-.5739E-01	4.92	5.17	0
27	91.311	4.770	.8653E+10	.8355E+10	-.5850E-01	4.36	4.57	0
28	87.422	4.852	.8695E+10	.8436E+10	-.6310E-01	3.79	3.97	0
29	83.552	5.312	.8753E-10	.8549E+10	-.1024E+01	3.18	3.33	-1
30	78.968	6.521	.8828E+10	.8696E+10	-.1015E+01	2.66	2.77	-1
31	72.866	6.999	.8920E+10	.8878E+10	-.1005E+01	2.32	2.85	-1
32	68.721	6.311	.9027E+10	.9092E+10	-.9928E+00	2.44	3.21	-1
33	79.652	5.251	.9139E+10	.9320E+10	-.6626E-01	4.32	5.02	0
34	93.146	5.277	.9248E+10	.9543E+10	-.6621E-01	4.88	5.21	0
35	99.020	5.295	.9346E+10	.9746E+10	-.6608E-01	4.48	4.63	0
36	97.592	5.344	.9429E+10	.9921E+10	-.6206E-01	2.85	2.91	0
37	94.053					.87	.87	

TABLE 5: Iteration #3

```
                    HYDROPOWER ( GWH ) =  .22561963E+05
                    PENALTY            =  .39186596E+08

                    Z    =   1.960
                    LT   =   36
                    EI   =   .9000E-03
                    WI   =   .2836E+00
                    M    =   0
                    NCCV =   21
```

L	E[S]	U	DUJ	DUUJ	DU	STDEU	STDEL	U$^+$
1	95.000	3.430	.5015E+09	.5422E+10	-.9250E-01	.00	.00	-1
2	92.622	4.018	.5045E+09	.5486E+10	-.9196E-01	.74	.74	-1
3	89.075	4.240	.5068E+09	.5537E+10	-.9154E-01	.97	1.01	-1
4	85.724	4.040	.5093E+09	.5591E+10	-.9109E-01	1.12	1.21	-1
5	82.674	5.312	.5127E+09	.5666E+10	-.9050E-01	1.23	1.39	-1
6	78.096	6.521	.5172E+09	.5763E+10	-.8973E-01	1.38	1.58	-1
7	72.003	6.999	.5226E+09	.5885E+10	-.8880E-01	1.53	1.75	-1
8	67.867	6.311	.5289E+09	.6027E+10	-.8774E-01	2.13	2.31	-1
9	78.808	4.251	.5354E+09	.6179E+10	-.8665E-01	4.37	4.51	-1
10	93.306	3.741	.5417E+09	.6327E+10	-.8563E-01	5.49	5.85	-1
11	100.706	3.771	.5474E+09	.6461E+10	-.7730E-02	5.91	6.60	0
12	100.779	3.812	.5523E+09	.6575E+10	-.7550E-02	5.71	6.31	0
13	98.741	3.841	.5563E+09	.6671E+10	-.7705E-02	5.30	5.84	0
14	95.930	4.018	.5595E+09	.6749E+10	-.8291E-01	4.87	5.36	-1
15	92.366	4.240	.5621E+09	.6810E+10	-.8253E-01	4.49	5.38	-1
16	89.001	4.040	.5648E+09	.6876E+10	-.8214E-01	4.15	5.40	-1
17	85.932	5.312	.5685E+09	.6967E+10	-.8160E-01	3.77	4.87	-1
18	81.331	6.521	.5734E+09	.7085E+10	-.8093E-01	3.56	4.89	-1
19	75.206	6.999	.5793E+09	.7232E+10	-.8010E-01	3.44	4.91	-1
20	71.035	6.311	.5861E+09	.7404E+10	-.7916E-01	3.59	5.09	-1
21	81.938	4.261	.5933E+09	.7588E+10	-.7580E-02	5.11	6.35	0
22	96.387	4.436	.6003E+09	.7768E+10	-.5349E-02	5.88	6.75	0
23	103.061	4.497	.6066E+09	.7931E+10	-.4718E-02	6.15	6.83	0
24	102.388	4.537	.6119E+09	.8071E+10	-.4626E-02	5.80	6.38	0
25	99.616	4.584	.6163E+09	.8188E+10	-.4661E-02	5.23	5.74	0
26	96.058	4.642	.6199E+09	.8284E+10	-.4695E-02	4.64	5.08	0
27	91.871	4.711	.6227E+09	.8359E+10	-.4778E-02	4.03	4.40	0
28	88.038	4.789	.6257E+09	.8440E+10	-.4838E-02	3.40	3.70	0
29	84.228	5.312	.6299E+09	.8553E+10	-.7365E-01	2.75	2.98	-1
30	79.639	6.521	.6353E+09	.8699E+10	-.7303E-01	2.30	3.06	-1
31	73.531	6.999	.6419E+09	.8881E+10	-.7228E-01	2.03	3.13	-1
32	69.378	6.311	.6495E+09	.9094E+10	-.7142E-01	2.25	3.45	-1
33	80.301	5.184	.6576E+09	.9321E+10	-.3967E-02	4.26	5.17	0
34	93.853	5.211	.6654E+09	.9544E+10	-.3169E-02	4.84	5.26	0
35	99.785	5.229	.6724E+09	.9747E+10	-.2857E-02	4.46	4.66	0
36	98.415	5.282	.6784E+09	.9921E+10	-.2765E-02	2.86	2.94	0
37	94.932					.87	.87	

TABLE 6: Iteration #4

```
             HYDROPOWER ( GWH ) = .22553820E+05
             PENALTY            = .15992801E+08

                 Z    =   1.960
                 LT   =   36
                 EI   =   .3485E-05
                 WI   =   .3485E-05
                 M    =   1
                 NCCV =   21
```

L	E[S]	U	DUJ	DUUJ	DU	STDEU	STDEL	U+
1	95.000	3.430	.9585E+04	.5422E+10	-.1768E-05	.00	.00	-1
2	92.622	4.018	.2635E+05	.5487E+10	-.4802E-05	.74	.74	-1
3	89.075	4.240	.3204E+05	.5537E+10	-.5786E-05	.97	1.01	-1
4	85.724	4.040	.2419E+05	.5591E+10	-.4326E-05	1.11	1.21	-1
5	82.674	5.312	.6472E+05	.5666E+10	-.1142E-04	1.22	1.39	-1
6	78.096	6.521	.1079E+06	.5764E+10	-.1872E-04	1.37	1.58	-1
7	72.003	6.999	.1199E+06	.5885E+10	-.2038E-04	1.52	1.75	-1
8	67.867	6.311	.8912E+05	.6028E+10	-.1478E-04	2.12	2.31	-1
9	78.808	4.251	.2093E+05	.6179E+10	-.3387E-05	4.36	4.51	-1
10	93.306	3.741	-.2490E+04	.6327E+10	.3936E-06	5.47	5.85	-1
11	100.706	3.763	-.6351E+04	.6461E+10	-.7903E-04	5.88	6.60	0
12	100.786	3.804	-.6407E+04	.6576E+10	-.7646E-04	5.65	6.27	0
13	98.756	3.833	-.6454E+04	.6671E+10	-.7574E-04	5.19	5.75	0
14	95.952	4.018	-.1459E+04	.6749E+10	.2161E-06	4.73	5.22	-1
15	92.388	4.240	.4454E+04	.6811E+10	-.6539E-06	4.32	5.24	-1
16	89.024	4.040	-.3784E+04	.6876E+10	.5502E-06	3.97	5.26	-1
17	85.955	5.312	.3862E+05	.6967E+10	-.5543E-05	3.62	5.28	-1
18	81.353	6.521	.8379E+05	.7086E+10	-.1182E-04	3.43	5.30	-1
19	75.228	6.999	.9654E+05	.7233E+10	-.1335E-04	3.32	5.31	-1
20	71.057	6.311	.6421E+05	.7405E+10	-.8672E-05	3.49	5.47	-1
21	81.959	4.254	-.6884E+04	.7588E+10	-.4392E-04	5.05	6.65	0
22	96.416	4.431	-.6966E+04	.7768E+10	-.2825E-04	5.84	6.96	0
23	103.094	4.492	-.7039E+04	.7932E+10	-.2105E-04	6.11	6.99	0
24	102.426	4.532	-.7101E+04	.8071E+10	-.1630E-04	5.77	6.52	0
25	99.658	4.579	-.7152E+04	.8188E+10	-.1198E-04	5.20	5.86	0
26	96.104	4.637	-.7194E+04	.8284E+10	-.7775E-05	4.61	5.18	0
27	91.922	4.707	-.7227E+04	.8360E+10	-.3762E-05	4.00	4.49	0
28	88.094	4.784	-.7262E+04	.8441E+10	.1625E-05	3.38	3.77	0
29	84.289	5.312	.9155E+04	.8553E+10	-.1070E-05	2.74	3.03	-1
30	79.699	6.521	.5294E+05	.8699E+10	-.6086E-05	2.30	3.11	-1
31	73.590	6.999	.6490E+05	.8881E+10	-.7307E-05	2.03	3.18	-1
32	69.437	6.311	.3281E+05	.9094E+10	-.3608E-05	2.25	3.50	-1
33	80.359	5.180	-.7635E+04	.9322E+10	.7498E-04	4.26	5.20	0
34	93.914	5.208	-.7726E+04	.9544E+10	.7295E-04	4.84	5.27	0
35	99.848	5.226	-.7808E+04	.9747E+10	.7686E-04	4.46	4.66	0
36	98.481	5.280	-.7877E+04	.9921E+10	.8540E-04	2.86	2.94	0
37	95.000					.87	.87	

TABLE 7: Iteration #5

```
              HYDROPOWER ( GWH ) =  .22553828E+05
              PENALTY            =  .15992801E+08

              Z     =   1.960
              LT    =   36
              EI    =   .4900E-06
              WI    =   .4900E-06
              M     =   0
              NCCV  =   21
```

L	E[S]	U	DUJ	DUUJ	DU	STDEU	STDEL	U+
1	95.000	3.430	.1598E+05	.5422E+10	-.2947E-05	.00	.00	-1
2	92.622	4.018	.3278E+05	.5487E+10	-.5974E-05	.74	.74	-1
3	89.075	4.240	.3850E+05	.5537E+10	-.6953E-05	.97	1.01	-1
4	85.724	4.040	.3068E+05	.5591E+10	-.5487E-05	1.11	1.21	-1
5	82.674	5.312	.7126E+05	.5666E+10	-.1258E-04	1.22	1.39	-1
6	78.096	6.521	.1145E+06	.5764E+10	-.1986E-04	1.37	1.58	-1
7	72.003	6.999	.1266E+06	.5885E+10	-.2151E-04	1.52	1.75	-1
8	67.867	6.311	.9586E+05	.6028E+10	-.1590E-04	2.12	2.31	-1
9	78.808	4.251	.2776E+05	.6179E+10	-.4492E-05	4.36	4.51	-1
10	93.306	3.741	.4419E+04	.6327E+10	-.6985E-06	5.47	5.85	-1
11	100.706	3.763	.6301E+03	.6461E+10	-.3985E-04	5.88	6.60	0
12	100.786	3.804	.6356E+03	.6576E+10	-.3856E-04	5.65	6.27	0
13	98.756	3.833	.6401E+03	.6671E+10	-.3820E-04	5.19	5.75	0
14	95.953	4.018	.5678E+04	.6749E+10	-.8412E-06	4.73	5.22	-1
15	92.388	4.240	.1162E+05	.6811E+10	-.1706E-05	4.32	5.24	-1
16	89.024	4.040	.3420E+04	.6876E+10	-.4973E-06	3.97	5.26	-1
17	85.955	5.312	.4587E+05	.6967E+10	-.6584E-05	3.62	5.28	-1
18	81.353	6.521	.9110E+05	.7086E+10	-.1286E-04	3.43	5.29	-1
19	75.229	6.999	.1039E+06	.7233E+10	-.1437E-04	3.32	5.30	-1
20	71.057	6.311	.7169E+05	.7405E+10	-.9681E-05	3.49	5.47	-1
21	81.959	4.254	.6820E+03	.7588E+10	-.2212E-04	5.06	6.65	0
22	96.416	4.431	.6898E+03	.7768E+10	-.1423E-04	5.84	6.96	0
23	103.095	4.492	.6969E+03	.7932E+10	-.1060E-04	6.11	6.99	0
24	102.426	4.532	.7029E+03	.8071E+10	-.8207E-05	5.77	6.52	0
25	99.658	4.579	.7079E+03	.8188E+10	-.6032E-05	5.20	5.86	0
26	96.105	4.637	.7119E+03	.8284E+10	-.3913E-05	4.61	5.18	0
27	91.923	4.707	.7151E+03	.8360E+10	-.1894E-05	4.00	4.48	0
28	88.094	4.784	.7184E+03	.8441E+10	.8144E-06	3.38	3.77	0
29	84.289	5.312	.1719E+05	.8553E+10	-.2010E-05	2.73	3.03	-1
30	79.699	6.521	.6104E+05	.8699E+10	-.7017E-05	2.30	3.11	-1
31	73.591	6.999	.7308E+05	.8881E+10	-.8229E-05	2.03	3.17	-1
32	69.437	6.311	.4109E+05	.9094E+10	-.4519E-05	2.25	3.50	-1
33	80.359	5.180	.7536E+03	.9322E+10	.3769E-04	4.26	5.20	0
34	93.914	5.208	.7623E+03	.9544E+10	.3666E-04	4.84	5.27	0
35	99.848	5.226	.7702E+03	.9747E+10	.3861E-04	4.46	4.66	0
36	98.481	5.280	.7768E+03	.9921E+10	.4289E-04	2.86	2.94	0
37	95.000					.87	.87	

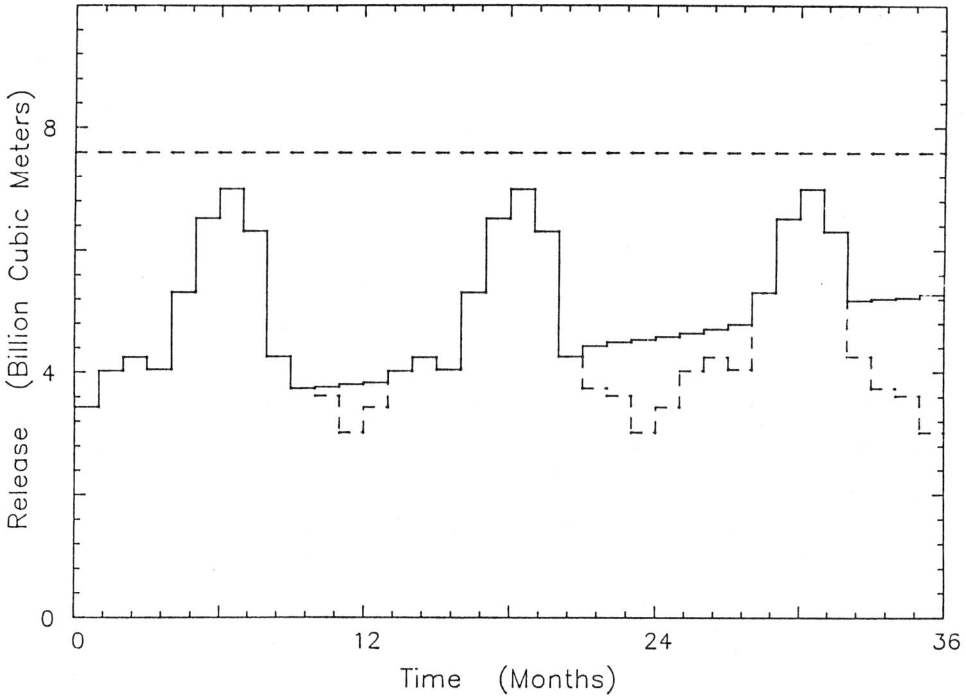

FIG. 5: Optimal Release Trajectory

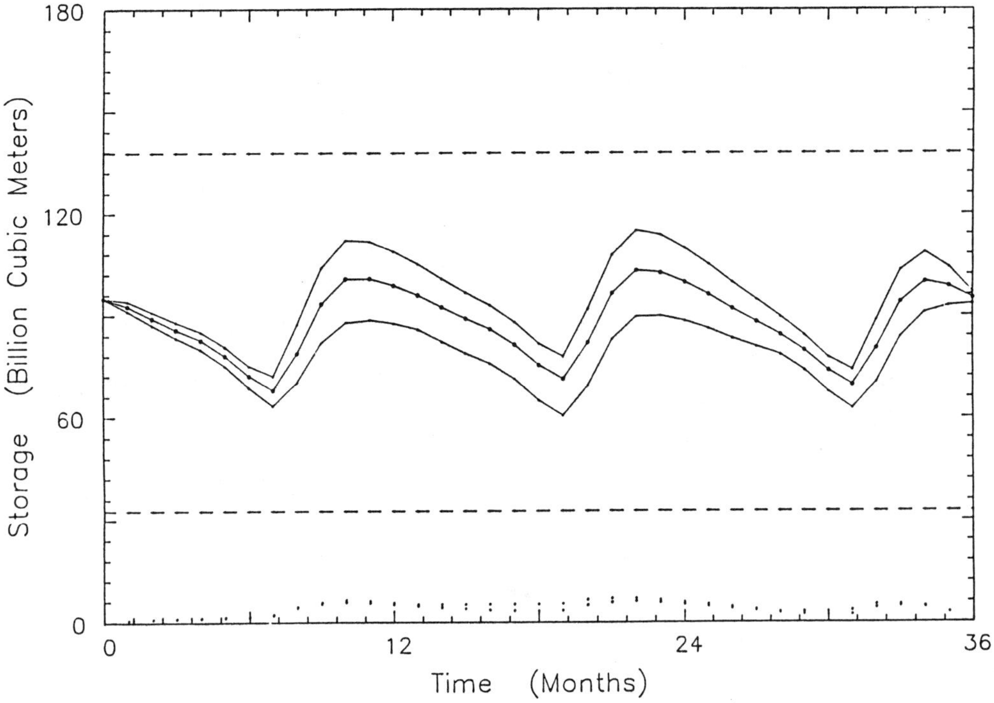

FIG. 6: Optimal Storage Trajectory

presented in Appendix A, and are significantly less than the Open-Loop values. The new binding constraint set includes 15 binding constraints and the magnitude of "WI" has decreased but is still significant. Iteration 2, shown on Table 4, places the expected terminal storage at 94.05×10^9 m^3, reduces the penalty function value, and increases energy generation further. The binding constraint includes now 5 additional constraints. Iteration 3, Table 5, builds up higher expected terminal storage and identifies one more binding constraint. Iteration 4, Table 6, shares the same binding constraint set with Iteration 3 and, as theoretically predicted, materializes an impressive convergence rate essentially terminating the search. This is indicated by the negligible "WI" and "DU" values and is readily concluded by comparing its trajectories with those of Iteration 5, Table 7. Optimal control and state trajectories are shown on Figures 5 and 6. In these figures, dashed lines indicate upper and lower boundaries, while solid lines represent the optimal trajectories. Apart from the mean state trajectory, the upper and lower probabilistic bounds have also been plotted. The dotted points are the corresponding standard deviations. No binding state constraints were found in this experiment. The above iterations require 25 seconds of CPU time on a Honeywell 68/DPS digital computer or 2.5 seconds on a CDC 180/855 digital computer. It is notable that only in the first and the last iterations it became necessary to cut in half the original stepsize avoiding multiple objective function evaluations and adding to the overall computational efficiency. Lastly, a very good approximation of the optimal solution has already been obtained at the 3rd iteration.

To test the method's performance in larger control horizons, the previous problem was considered with a 360-month control horizon. The Toshka spillway is not assumed operational and no probabilistic constraints are imposed. ELQG converges to the trajectories shown in Figures 7 and 8 at the 6th iteration. Some characteristic quantities from the 6 iterations are reported on Table 8. As in the previous experiment the algorithm converges to the optimal control trajectory in one iteration after the binding constraint set (298 constraints) is identified. Notice the significant difference between upper and lower standard deviations and the use of unity stepsize for most of the iterations. This experiment requires 4 minutes and 35 seconds CPU time on a Honeywell 68/DPS system or 20 seconds on a CDC 180/855 system. The optimal trajectories show that, from a long-run energy perspective, it is best to build up reservoir storage by releasing according to the downstream requirements and operate under high hydraulic head rather than at higher release rate. The high releases toward the end of the control horizon are necessary to draw the storage down to 95×10^9 m^3 target.

Following is an experimental run where the problem includes probabilistic state constraints. The additional requirement is to keep reservoir storage lower than 137.72×10^9 m^3 and higher than 32.72×10^9 m^3 with 97.5% reliability. The new optimal trajectories are shown on Figures 9 and 10. There are 20 state and 264 binding control constraints and the energy production has now dropped (with respect to the state unconstrained optimum) to $0.2371 4417 \times 10^6$ GWH. The penalty parameter C was taken equal to 10. This experiment requires 2 minutes CPU time on a CDC 180/855 computer. Although overall reliable, the handling of state constraints via the penalty function method is, in general, less efficient than the treatment of control constraints via the Projected Newton method. The optimal release trajectory follows again the minimum requirements until reservoir storage rises to its upper bound. Subsequently, the releases are adjusted to maintain highest feasible storage, and, as before, the last part draws the reservoir down to the 95×10^9 m^3 expected storage target. For this run, ELQG was implemented to solve the control constrained problem completely for every value of the penalty parameter. See comment following ELQG algorithm in [10].

TABLE 8: HAD Control with 360-Month Horizon

	HYDROPOWER (GWH)	PENALTY	WI ($10^9 m^3$)	M	NCCV	ES(LT+1) ($10^9 m^3$)
Initial Trajectories	235427.9	0.11434E+20	61.2	1	0	142.820
Iteration #1	234391.4	0.52137E+18	10.2	0	156	84.789
Iteration #2	240153.7	0.30724E+18	10.4	0	248	87.161
Iteration #3	241408.1	0.32364E+17	9.6	0	286	92.456
Iteration #4	241589.1	0.90557E+12	0.17	0	298	94.987
Iteration #5	241591.1	0.14559E+09	0.6E-04	0	298	95.000

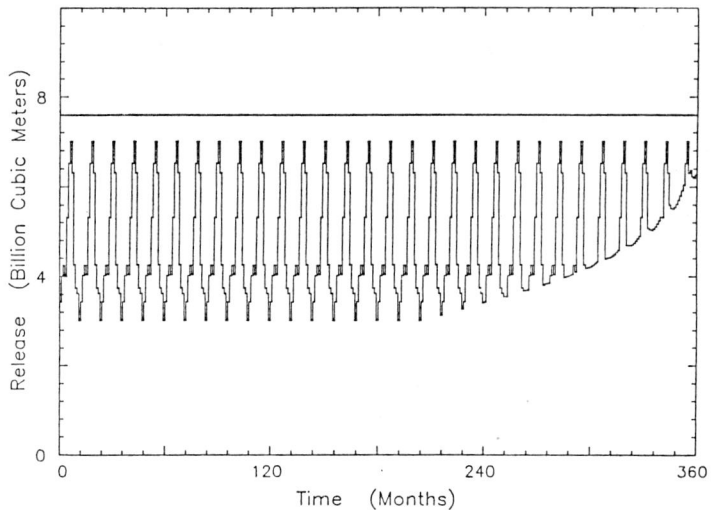

FIG. 7: Optimal Release Trajectory

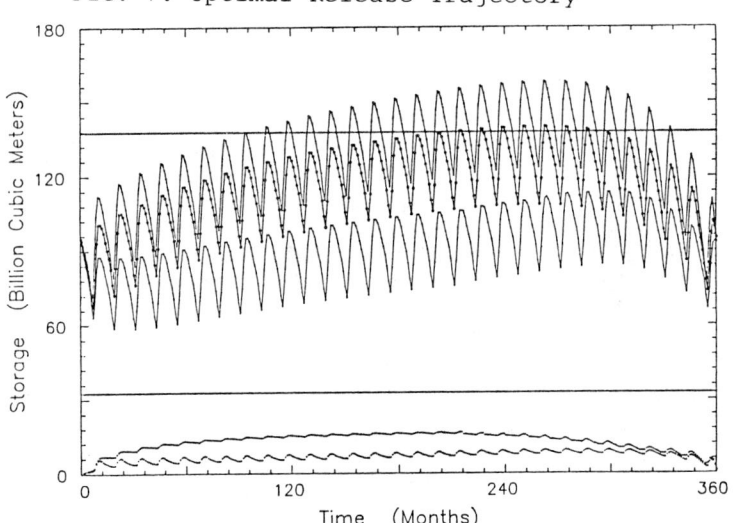

FIG. 8: Optimal Storage Trajectory

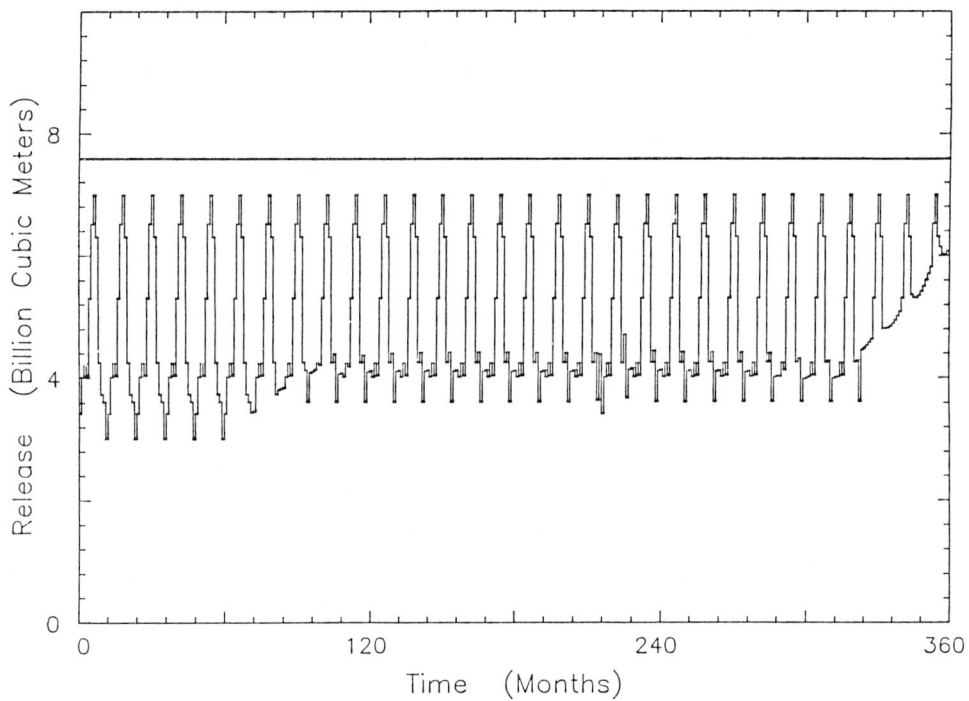

FIG. 9: Optimal Release Trajectory

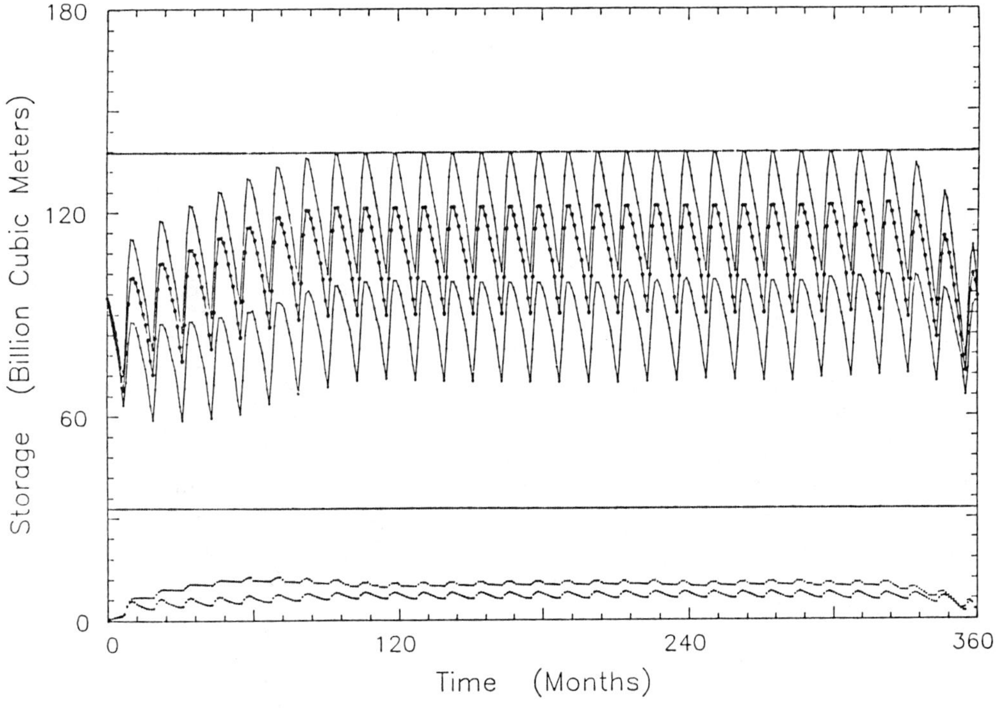

FIG. 10: Optimal Storage Trajectory

By contrast, if the unmodified Open-Loop variance is used, the storage trajectory on Fig. 10 looks like the one on Figure 11. If the probabilistic constraints were to be satisfied with this variance, the expected storage trajectory would have to be considerably lower which would reduce energy generation even further. The fact that the Open-Loop variance attains stable levels is due to the negative evaporation term in the system dynamics. If it were not for this term, the variance would grow unbounded which brings up the necessity of the modification scheme.

The previous computational runs were performed to demonstrate ELQG's efficiency. In a real time application, a much shorter control horizon is required. As mentioned in the preceding paper, the question becomes one of finding an appropriate terminal state penalty term. (The influence of this term on the optimal trajectories is apparent in all previous experiments.) To this end, it was noted that in the long run it pays (in energy generation) to build up high reservoir elevations by releasing the minimum downstream requirements. This release pattern is modified, and higher releases are recommended, either due to the influence of the terminal penalty term or when the upper storage bound is binding. It turns out that this behavior is independent of the initial storage and, therefore, leads to the following operation policy: release the minimum requirements until the upper storage bound becomes binding; thereafter, optimize the releases to maintain the highest feasible reservoir storage. Georgakakos and Marks, [9], verified the validity of this conclusion by performing control experiments for a variety of hydrologic and operational conditions. When the Toshka Spillway is active, the upper storage bound should be placed at its operating threshold to avoid excessive spills by passing higher (yet feasible) discharges through the power plan turbines. Thus, in a real time HAD control application, the terminal penalty term should seek to guide the reservoir at the highest feasible elevation. Note that it is necessary to look at the long term system behavior to establish the upper storage bound and the terminal penalty term. These specifications cannot be done with short-run investigations because over a short control horizon it is energy-optimal to empty the reservoir.

SIMULATION EXPERIMENTS

To further test ELQG in real time applications and to compare its performance with other control procedures, we run simulation experiments. The data base for these experiments was the historical streamflow record at Wadi Halfa (at the entrance of the HAD reservoir) from January 1912 to December 1965. The recorded levels were adjusted by the estimated Sudan abstractions having the monthly distribution of Figure 4 and 16.5×10^9 m^3 yearly total. The downstream water supply requirements, channel degradation threshold, evaporation rates, etc., were as in the previous sections. ELQG was implemented with a 12 month control horizon, a value of 10 for the penalty parameter, and a terminal cost term penalizing any terminal storage deviation from the upper bound. We allowed for two forecasting possibilities:

The first is based on the a priori monthly statistics (A-S) and simply forecasts the monthly means with the corresponding variance. The second is the Thomas-Fiering lag-1 univariate seasonal autoregressive model (T-F). Results from using the Steady State Markov Dynamic Programming (SSMDP) and Adaptive Markov Dynamic Programming (AMDP) methods in the same simulation experiments are also reported here from [6], for purposes of comparison. SSMDP [1] uses a Markov-Chain inflow model, while AMDP is a sequential modification of SSMDP with a multivariate seasonal autoregressive inflow predictor.

The storage reliability constraints for the ELQG control models were set to either 2.5% or 50%, the second case implying deterministic (expected value) optimization.

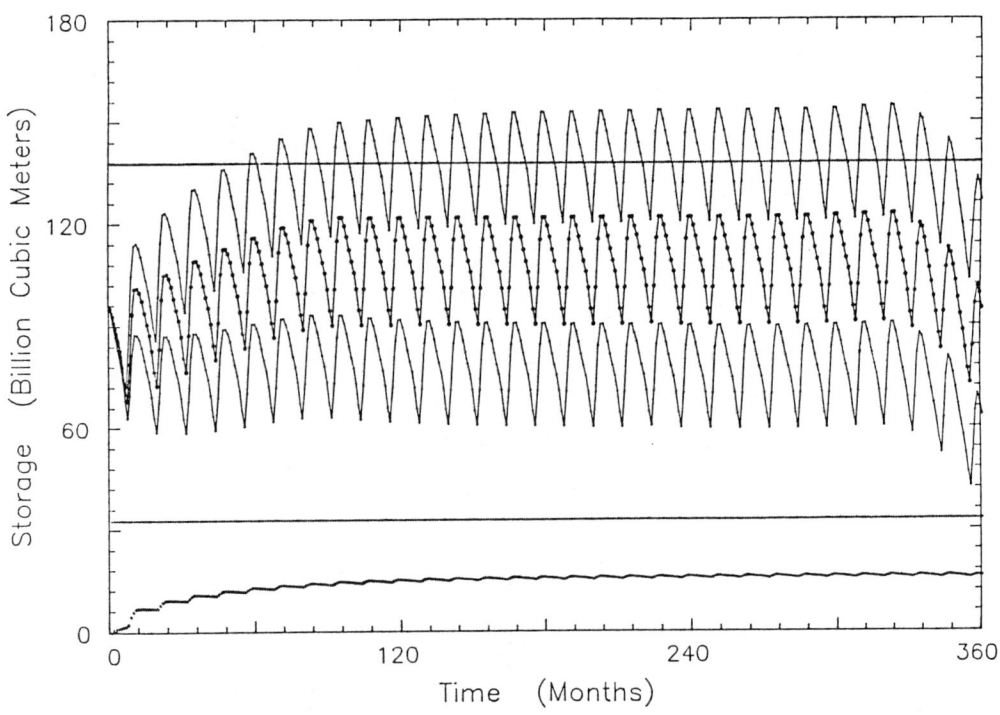

FIG. 11: Open Loop Variance

TABLE 9: Simulation Experiments

	WATER SUPPLY DEFICITS (10^9 m^3)	EXCESSIVE RELEASES (10^9 m^3)	TOSHKA SPILLS (54-YRS) (10^9 m^3)	MEAN ANNUAL ENERGY (GWH)	TERMINAL STORAGE (10^9 m^3)	ADJUSTED MEAN ANNUAL ENERGY (GWH)	MEAN ANNUAL EVAPORATION (10^9 m^3)
SSMDP	0.000	0.000	100.700	7,787.000	--	--	13.600
AMDP	0.000	0.000	100.700	7,787.000	--	--	13.600
Y=50% A-S	0.000	0.000	2.199	8,081.433	126.572	8,104.096	13.017
Y=50% T-F	0.000	0.000	1.086	8,086.420	122.065	8,097.194	12.796
Y=2.5% A-S	0.000	0.000	1.338	8,061.083	122.946	8,074.169	12.725
Y=2.5% T-F	0.000	0.000	0.560	8,053.812	117.930	8,053.812	12.498
PERFECT FORECAST	0.000	0.000	0.265	8,046.319	135.819	8,093.818	12.937

Table 9 presents results from 7 simulation runs (each column representing a 648-month (54-year) simulation experiment. The quantities reported are the total volume of water supply deficits (i.e., the amount by which the models failed to meet the downstream water supply requirements), the total volume of water released in excess of 7.59×10^9 m^3 per month, the total volume of spills to Toshka, the mean annual energy production (GWH), the reservoir storage at the end of the simulation period, the adjusted mean annual energy generation based on terminal storage differences and the mean annual evaporation losses. All runs were started at 180 meters reservoir elevation or 149.55×10^9 m^3 reservoir storage. The correction of the mean annual energy generation was necessary since the model which maintained higher terminal storage could, alternatively, generate more energy. The adjustment was as follows: The lowest terminal storage was subtracted from the rest, and the resulting amount was assumed to generate energy at maximum release for as long as it lasted. The evaporation rate was that of January. The energy produced was divided by 54 and added to the mean annual energy production. Terminal storage information was not available for the SSMDP and AMDP methods, and, therefore, no energy correction was attempted.

It is seen that all models meet the downstream water supply requirements and prevent channel degradation at all times. The ELQG models cause substantially less Toshka spills and, on the average, produce about 300 GWH more energy per year over SSMDP and AMDP. The basis of this result is ELQG's ability to explicitly account for control and state constraints. Each ELQG simulation experiment required approximately 30 minutes CPU time on a Honeywell 68/DPS or 4.7 minutes on a CDC 180/855 digital computer. AMDP requires 50 Honeywell CPU hrs., while SSMDP, which is not a sequential scheme, takes 20 sec. Key attribute of ELQG's computational efficiency is its analytical structure.

With regard to the ELQG models, it is noted that these with 2.5% reliability constraints incur less Toshka spills and evaporation losses but also produce less energy over those with 50%. These are observed because tighter probabilistic constraints force reservoir operation at lower elevations. The same results are noted when comparing the performance of the ELQG (T-F) model combinations with that of the ELQG (A-S) ones. Namely, the models with better foresight are able to manage the inflow process better.

The last simulation experiment on Table 9 was run to provide an upper bound of the sequential models' performance. It had perfect knowledge of the 12 upcoming inflows and, consequently, incurred minimal Toshka spills. (The 0.265×10^9 m^3 of Toshka spillage was the result of starting the simulation from a level exceeding Toshka's operating threshold.) However, notice that the performance of the other ELQG models compares well with these results despite their imprecise information on future inflows. The conclusion from this observation is that, under the current operating and hydrologic conditions, forecasted information is not of crucial significance to the HAD operation. The reasons are that (1) most of the time the reservoir has enough available storage to accommodate the incoming flows and (2) the presence of the Toshka spillway has eliminated the threat of severe flooding. However, the worth of forecasting is expected to rise at a future time when water supply requirements will grow and the goal will be to minimize the effects of water shortages.

Figures 12 and 13 present the storage and release trajectories which resulted from the (γ = 2.5%, T-F) simulation experiment. As the simulation experiments have shown, HAD suffers heavy evaporation losses (approximately 13 billion cubic meters are lost per year). On the other hand, its storage capacity is more than adequate for the current water supply purposes. If it were possible to maintain lower reservoir elevations while always meeting the water supply requirements, there would be water gains from evaporation at the expense of energy losses due to the lower hydraulic head. This is the tradeoff we wish to determine in the

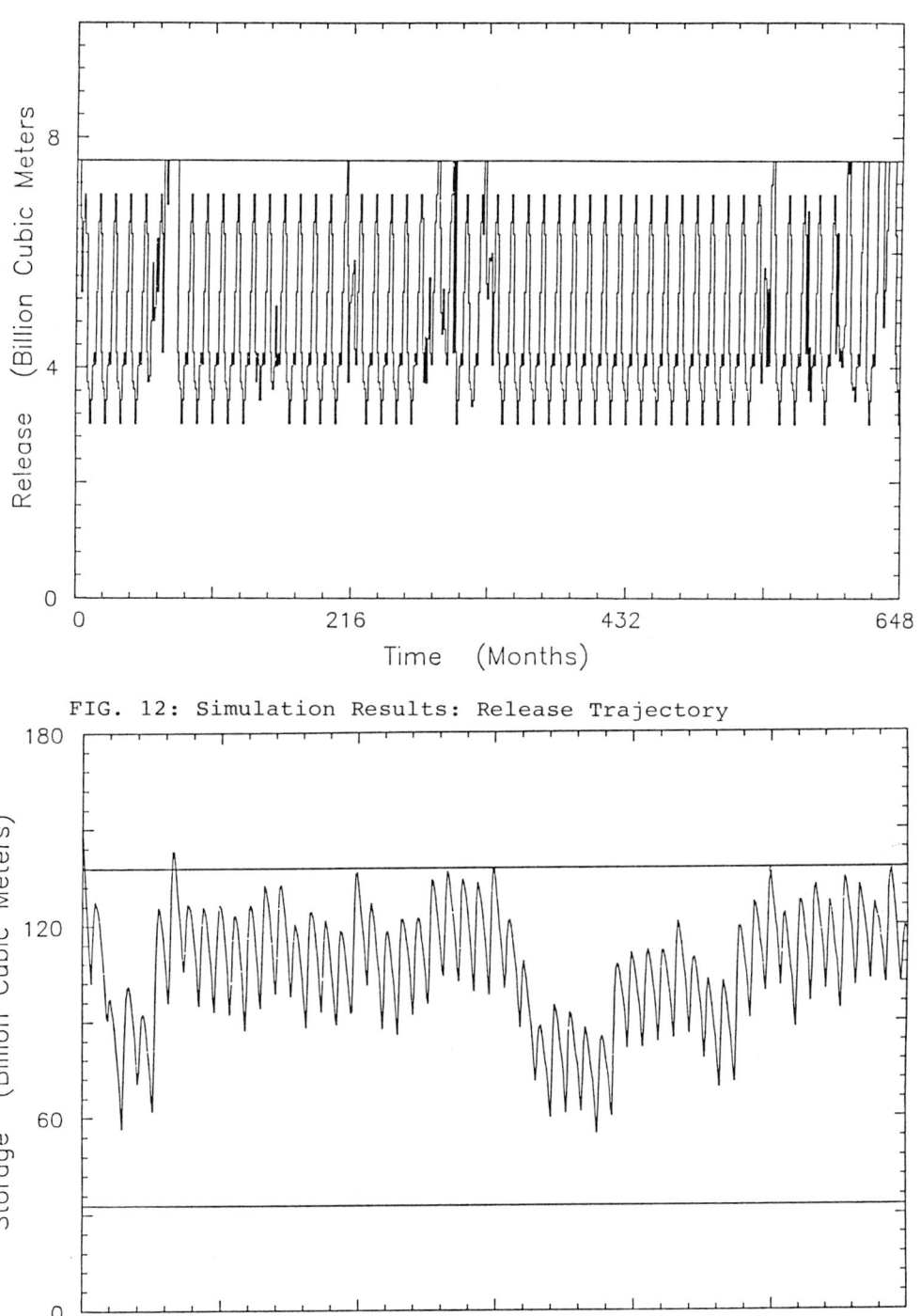

FIG. 12: Simulation Results: Release Trajectory

FIG. 13: Simulation Results: Storage Trajectory

TABLE 10: Energy vs Evaporation Tradeoff

	WATER SUPPLY DEFICITS (10^9 m^3)	EXCESSIVE RELEASES (10^9 m^3)	MEAN ANNUAL ENERGY (GWH)	MEAN ANNUAL EVAPORATION (10^9 m^3)
137.72 (178.008)	0.000	0.000	8,047.676	12.769
132.000 (177.000)	0.000	0.000	7,978.403	12.327
126.50 (175.989)	0.000	0.000	7,899.718	11.926
121.25 (175.000)	0.000	0.000	7,814.218	11.531
116.25 (174.016)	0.000	0.000	7,723.255	11.132
111.50 (173.053)	0.000	0.000	7,627.997	10.737
106.50 (172.002)	0.000	0.000	7,513.618	10.290
102.00 (171.020)	3.101	0.000	7,406.982	9.882

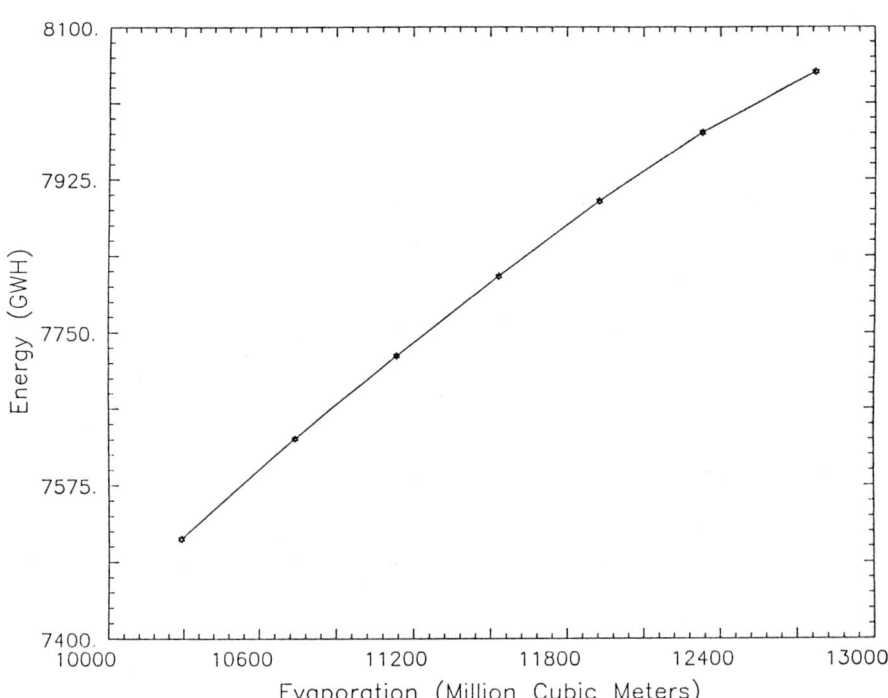

FIG. 14: Energy vs Evaporation Tradeoff

second part of this section. It has a practical interest because in the years to come water rather than hydropower availability will be the limiting factor in the development of the Egyptian economy. Furthermore, reservoir operation at low levels may become mandatory if the seismic activity affecting HAD continues to intensify. The procedure adopted here is to place the upper state bound at increasingly lower levels and perform simulation experiments. ELQG was used with ($Y = 50\%$, T-F). The initial storage was assumed equal to the corresponding upper bound. Table 10 reports the results of these experiments. The maximum allowable elevation was decreased from 178 m (137.72×10^9 m^3 storage) to the point where water supply deficits could not be avoided (at approximately 171 meters). The results show that, on the average, about 2.5 milliard m^3 can be saved from evaporation at the expense of 500 GWH per year (storage bound at 172 m). With Egypt and Sudan already involved in costly expenditures on water conservation projects (Egyptian Ministry of Irrigation, Main Report and TR 5, 1981), for fear of shortages, these results become particularly significant. Note, further, that compared to the other control methods as well as the currently employed heuristic operation policy [6] ELQG control results in over 2 milliard m^3 water benefits at 7,800 GWH annual energy production (compare also with Table 9 although in those runs the initial reservoir storage was equal to 149.55×10^9 m^3). The derived tradeoff curve is shown on Figure 14. It should also be noted that a more detailed analysis would require simulation experiments with synthetically generated inflow series.

CONCLUSIONS

The ELQG control method was tested in several computational experiments for the control of the High Aswan Dam. The problem was to maximize expected energy generation subject to release and storage reliability constraints imposed from operational requirements (water supply and flood control objectives). The method displayed reliability and computational efficiency even for very long control horizons. Constrol constraints were accounted for within a few iterations (5 or 6), while the handling of storage reliability constraints, although satisfactory overall, was less efficient. ELQG was also compared favorably with models of the Markov-Chain Stochastic D.P. philosophy in simulation experiments. Some interesting policy making issues for the HAD operation were discussed, and a potentially attractive tradeoff between energy losses and evaporation gains was identified.

REFERENCES

1. Alarcon, L. and Marks, D., "A Stochastic Dynamic Programming Model for the Operation of the High Aswan Dam," Ralph M. Parsons Lab. for Water Resources and Hydrodynamics, Dept. of Civil Engineering, M.I.T., TR No. 246, 1979.
2. Arunkumar, S. and Yeh, W.W-G., "Probabilistic Models in the Design and Operation of a Multi-Purpose Reservoir System," Contribution No. 144, California Water Researches Center, University of California, Davis, December 1973.
3. Askew, A., "Chance-Constrained Dynamic Programming and the Optimization of Water Resource Systems," Water Resources Research, Vol. 10, No. 6, December 1974a, pp. 1099-1106.
4. Askew, A., "Optimum Reservoir Operating Policies and the Imposition of a Reliability Constraint," Water Resources Research, Vol. 10, No. 1, February 1974b, pp. 51-56.
5. Bertsekas, D., "Dynamic Programming and Stochastic Control," Academic Press, New York, 1976.

6. Bras, R., Buchanan, R., and Curry, K., "Real Time Adaptive Closed Loop Control of Reservoirs with the High Aswan Dam as a Case Study," Water Resources Research, Vol. 19, No. 1, 1983, pp. 35-52.
7. Butcher, W., "Stochastic Dynamic Programming for Optimum Reservoir Operation," Water Resources Bulletin, Vol. 7, No. 1, February 1971, pp. 115-123.
8. Curry, K.D., and Bras, R.L., "Multivariate Seasonal Time Series Forecast with Application to Adaptive Control," Technical Report #253, Ralph M. Parsons Laboratory for Hydrology and Water Resources, Dept. of Civil Engineering, M.I.T., 1980.
9. Georgakakos, A.P., and Marks, D.H., "Real Time Control of Reservoir Systems," Ralph M. Parsons Laboratory for Hydrology and Water Resources, Department of Civil Engineering, M.I.T., Technical Report No. 301, May 1985, 313 p.
10. Georgakakos, A.P., and Marks, D.H., "A New Method for the Operation of Reservoir Systems. Part I: Theory," Water Resources Research, 1986.
11. Heidari, M., Chow, V.T., Kotovic, P.V., and Meredith D.D., "Discrete Differential Dynamic Programming Approach to Water Resources System Optimization," Water Resources Research, Vol. 7, No. 2, April 1971, pp. 273-283.
12. Jacobson, D. and Mayne, D., "Differential Dynamic Programming," American Elsevier, New york, 1970.
13. Jamshidi, M. and Heidari, M., "Application of Dynamic Programming to Control Khuzestan Water Resources System," Automatica, Vol. 13, pp. 287-293, 1977.
14. Jazwinski, A., "Stochastic Processes and Filtering Theory," Academic Press, New York, 1970.
15. Larson, R. and Keckler, W., "Applications of Dynamic Programming to the Control of Water Resource Systems," Automatica, Vol. 5, pp. 15-26, 1969.
16. Loaiciga, H.A., and Marino, M.A., "An Approach to Parameter Estimation and Stochastic Control in Water Resources with an Application to Reservoir Operation," Water Resources Research, Vol. 21, No. 11, 1985, pp. 1575-1584.
17. Murray, D. and Yakowitz, "Constrained Differential Dynamic Programming and its Applications to Multireservoir Control," Water Resources Research, Vol. 15, No. 5, pp. 1017-1027, 1979.
18. Oven-Thompson, K., Alarcon, L., and Marks, D., "Agricultural versus Hydropower Trade-offs in the Operation of the High Aswan Dam," Water Resources Research, Vol. 18, No. 6, 1982, pp. 1605-1614.
19. Sniedovich, M., "A Variance-Constrained Reservoir Control Problem," Water Resources Research, Vol. 16, No. 2, 1980a, pp. 271-274.
20. Sniedovich, M., "Analysis of a Chance-Constrained Reservoir Control Model," Water Resources Research, Vol. 16, No. 5, 1980b, pp. 849-853.
21. Sniedovich, M., "Reliability-Constrained Reservoir Control Problems: 1. Methodological Issues," Water Resources Research, Vol. 15, No. 6, 1979, pp. 1574-1582.
22. Stedinger, J., Sule, B., Loucks, D., "Stochastic Dynamic Programming Models for Reservoir Operation Optimization," Water Resources Research, Vol. 20, No. 11,1984, pp. 1499-1505.
23. Su, S. and Deininger, R., "Generalization of White's Method of Successive Approximations to Periodic Markovian Decision Processes," Oper. Res., 20(2), 1972, pp. 318-326.
24. Su, S. and Deininger, R., "Modeling the Regulation of Lake Superior under Uncertainty of Future Water Supplies," Water Resources Research, Vol. 10, No. 1, 1974, pp. 11-25.
25. Turgeon, A., "Optional Short-Term Hydro Scheduling from the Principle of Progressive Optimality," Water Resources Research, Vol. 17, No. 3, 1981, pp. 481-486.

26. Wasimi, S. and Kitanidis, "Real-Time Forecasting and Daily Operation of a Multireservoir System During Floods by Linear Quadratic Gaussian Control," Water Resources Research, Vol. 19, No. 6, 1983, pp. 1511-1522.

APPENDIX A

ELQG CONTROL OF A SINGLE RESERVOIR

a. Forecast the continuous time inflow statistics $\bar{w}(t)$, $Q_w(t)$ (mean and spectral density) over the control horizon $t\epsilon[t_k, t_T]$ via an available streamflow model.

Assume that as a result of i previous iterations nominal trajectory $u^i = u_k^i, \ldots, u_{T-1}^i$ has been obtained. Then the operations that follow are performed during iteration i+1.

b. Compute that state's mean and variance trajectories ($\bar{s}^i(\tau)$, $P_s^i(\tau)$, $\tau\epsilon[t_k, t_T]$), as well as the coefficients of the discrete time perturbation model by integrating the following system of differential equations over the intervals $[t_\ell, t_{\ell+1}]$, $\ell=k,\ldots,T-1$:

$$\frac{d\bar{s}^i(\tau)}{d\tau} = F(\bar{s}^i(\tau),\tau) - u^i(\tau) + \bar{w}(\tau) \tag{A.1}$$

$$\frac{dP_s^i(\tau)}{d\tau} = 2F(\bar{s}^i(\tau),\tau) P_s^i(\tau) + Q_w(\tau) \tag{A.2}$$

$$\frac{d\phi^i(\tau,t_\ell)}{d\tau} = f(\bar{s}^i(\tau),\tau) \phi(\tau,t_\ell) \tag{A.3}$$

$$\frac{dB^i(\tau,t_\ell)}{d\tau} = \frac{1}{\phi^i(\tau,t_\ell)} \tag{A.4}$$

where $F(\cdot,\cdot)$ is the state transition function and

$$f(\bar{s}^i(\tau),\tau) = \left.\frac{\partial F(s(\tau),\tau)}{s(\tau)}\right|_{s(\tau)=\bar{s}^i(\tau)} \tag{A.5}$$

The initial conditions for the mean state and variance equations are the results of the previous integration step: $\bar{s}^i(t_\ell)$, $P_s(t_\ell)$, while for (A.3) and (A.4) there holds

$$\phi^i(t_\ell, t_\ell) = 1 \text{ and } B^i(t_\ell, t_\ell) = 0 \tag{A.6}$$

for any ℓ, i. At time t_k, $P_s(t_k) = 0$ (perfect state information) and $s^i(t_k)$ equals the currently observed $s(t_k)$.

Then the discrete time quantities are denoted and given by

$$\bar{s}^i_{\ell+1} = \bar{s}^i(t_{\ell+1}), \tag{A.7}$$

$$P^i_{s(\ell+1)} = P^i_s(t_{\ell+1}), \quad (A.8)$$

$$\phi^i_\ell = \phi^i(t_{\ell+1}, t_\ell), \quad (A.9)$$

$$B^i_\ell = \phi^i_\ell B(t_{\ell+1}, t_\ell), \quad (A.10)$$

$$\ell = k, \ldots, T-1.$$

and the associated perturbation system model is as follows:

$$\delta \bar{s}^i_{\ell+1} = \bar{s}_{\ell+1} - \bar{s}^i_{\ell+1} = \phi^i_\ell \delta \bar{s}^i_\ell + B^i_\ell \delta u^i_\ell, \quad (A.11)$$

$$\ell = k, \ldots, T-1$$

where

$$\delta u^i_\ell = u_\ell - u^i_\ell. \quad (A.12)$$

c. Descritize and perform a quadratic approximation of the performance index.
Descritization:

$$\int_{t_\ell}^{t_{\ell+1}} E\{\ell(s(t), u(t))\} dt \cong \hat{\ell}(\bar{s}^i_\ell, P^i_{s\ell}, u^i_\ell)(t_{\ell+1} - t_\ell) \quad (A.13)$$

(For the evaluation of $\hat{\ell}(\cdot,\cdot,\cdot)$, see [9]. Taylor's series quadratic approximation around $\bar{s}^i$, u^i, $\ell = 0, \ldots, T-1$:

$$\tilde{\ell}(\delta \bar{s}^i_\ell, \delta u^i_\ell) = N^i_{s\ell} \delta \bar{s}^i_\ell + N^i_{u\ell} \delta u^i_\ell + \frac{1}{2} N^i_{ss\ell}(\delta \bar{s}^i_\ell)^2 + \frac{1}{2} N^i_{uu\ell}(\delta u^i_\ell)^2$$
$$+ N^i_{us\ell}(\delta \bar{s}^i_\ell)(\delta u^i_\ell) + \text{constants} \quad (A.14)$$

for $\ell = k, \ldots, T-1$ where

$$N^i_{s\ell} = \frac{\partial \hat{\ell}(\bar{s}^i_\ell, u^i_\ell)}{\partial s} \Delta t \quad (A.15a)$$

$$N^i_{u\ell} = \frac{\partial \hat{\ell}(\bar{s}^i_\ell, u^i_\ell)}{\partial u} \Delta t \quad (A.15b)$$

$$N^i_{ss\ell} = \frac{\partial^2 \hat{\ell}(\bar{s}^i_\ell, u^i_\ell)}{\partial s^2} \Delta t \quad (A.15c)$$

$$N_{uu\ell}^i = \frac{\partial^2 \hat{\ell}(\bar{s}_\ell^i, u_\ell^i)}{\partial u^2} \Delta t \qquad \text{(A.15d)}$$

$$N_{us\ell}^i = \frac{\partial^2 \hat{\ell}(\bar{s}_\ell^i, u_\ell^i)}{\partial u \partial s} \Delta t \qquad \text{(A.15e)}$$

$$\Delta t = t_{\ell+1} - t_\ell$$

and similarly for the terminal cost term

$$\tilde{\ell}(\delta \bar{s}_T^i) = N_s^i \delta \bar{s}_T^i + \frac{1}{2} N_{ssT}^i (\delta \bar{s}_T^i)^2 . \qquad \text{(A.16)}$$

d. Check the validity of the probabilistic state constraints:

$$\bar{s}_\ell^i \geq s_\ell^{min} - z_\ell^{min} \sqrt{P_{s\ell}^i} , \qquad \text{(A.17a)}$$

$$\bar{s}_\ell^i \leq s_\ell^{max} - z_\ell^{max} \sqrt{P_{s\ell}^i} , \qquad \text{(A.17b)}$$

$$\ell = k+1, \ldots, T.$$

where $\{z^{min}, z^{max}\}_{\ell=k+1}^T$ are specified from the Standard Normal Distribution based on the probabilistic allowances $\{\gamma^{min}, \gamma^{max}\}_{\ell=k+1}^T$. If some constraint at time m is violated, modify the corresponding quadratic costs in Step c by

$$N_{ssm}^{i'} = N_{ssm}^i + C_i N_{ssm}^i \qquad \text{(A.18)}$$

$$N_{sm}^{i'} = N_{sm}^i + C_i N_{ssm}^i \Delta s_m \qquad \text{(A.19)}$$

where

$$\Delta s_m = \begin{cases} \bar{s}_m^i - (s_m^{min} - z_m^{min} \sqrt{P_{sm}^i}), & \text{for lower constr. violation} \\ \\ \bar{s}_m^i - (s_m^{max} - z_m^{max} \sqrt{P_{sm}^i}), & \text{for upper constr. violation} \end{cases} \qquad \text{(A.20)}$$

and $C_i = C^i$ (here is the exponent, $C \in [4,10]$). (A.21)

e. Compute the Gradient vector of the performance index

$$\tilde{J} = \sum_{\ell=k}^{T-1} \tilde{\ell}(\delta \bar{s}_\ell^i, \delta u_\ell^i) + \tilde{\ell}(\delta \bar{s}_T^i) \qquad \text{(A.22)}$$

at the current trajectories, $\{\delta \bar{s}_\ell^i = 0, \delta u_\ell^i = 0\}$, $\delta \bar{s}_T^i = 0$, by

$$\frac{\partial \tilde{J}}{\partial u_\ell}\bigg|_i = N^i_{u\ell} + B^i_\ell p^i_{\ell+1}, \qquad (A.23)$$

$$\ell = k, \ldots, T-1$$

where
$$p^i_T = N^i_{sT}$$

$$p^i_j = N^i_{sj} + \phi^i_j p^i_{j+1}, \qquad (A.24)$$

$$j = T-1, \ldots, k+1$$

 f. Compute the diagonal elements of the Hessian matrix at the ith nominal trajectories from

$$\frac{\partial^2 \tilde{J}}{\partial u_\ell^2}\bigg|_i = N^i_{uu\ell} + (B^i_\ell)^2 G^i_{\ell+1}$$

$$\ell = k, \ldots, T-1 \qquad (A.25)$$

where
$$G^i_T = N^i_{ssT}$$

$$G^i_j = N^i_{ssj} + (\phi^i_j)^2 \qquad (A.26)$$

$$j = T-1, \ldots, k+1.$$

 g. Test the optimality of the ith nominal trajectory by the following criterion:
Compute

$$w_i = \left[\sum_{\ell=k}^{T-1} \left\{ u^i_\ell - \left[u^i_\ell - \left[\frac{\partial \tilde{J}}{\partial u_\ell} \bigg/ \frac{\partial^2 \tilde{J}}{\partial u_\ell^2} \right]_i \right]^{++} \right\}^2 \right]^{1/2} \qquad (A.27)$$

where

$$\left[u_\ell^i - \left[\frac{\partial \tilde{J}}{\partial u_\ell}\bigg/\frac{\partial^2 \tilde{J}}{\partial u_\ell^2}\right]\right]_i = \begin{cases} u_\ell^{min}, & \text{if } u_\ell^i - \frac{\partial \tilde{J}}{\partial u_\ell}\bigg/\frac{\partial^2 \tilde{J}}{\partial u_\ell^2}\bigg|_i < u_\ell^{min} \\[2mm] u_\ell^{max}, & \text{if } u_\ell^i - \frac{\partial \tilde{J}}{\partial u_\ell}\bigg/\frac{\partial^2 \tilde{J}}{\partial u_\ell^2}\bigg|_i > u_\ell^{max} \\[2mm] u_\ell^i - \frac{\partial \tilde{J}}{\partial u_\ell}\bigg/\frac{\partial^2 \tilde{J}}{\partial u_\ell^2}\bigg|_i, & \text{otherwise} \end{cases}$$

$\ell = k, \ldots, T-1$

(A.28)

Assuming that the problem is convex, if w_i is negligibly small, then the ith nominal trajectory is a global minimum. This holds because $w_i \cong 0$ implies either that $\partial \tilde{J}/\partial u_\ell = 0$ or that u equals one of its bounds with the descent direction pointing toward infeasible regions. Given that $w_i \cong 0$, if C_i is sufficiently high, the ith nominal trajectory is the optimal solution to the control and state constrained problem. If C_i is not as high, some state constraints may be violated. In that case set

$$u_\ell^{i+1} = u_\ell^i$$

$\ell = k, \ldots, T-1$

(A.29)

and repeat Steps d, e, f, g.

If the problem is not convex, then $\partial^2 \tilde{J}/\partial u_\ell^2$ should be replaced by a positive number (unity could be a choice). In this case, $w_i \cong 0$ implies a point satisfying the first order necessary conditions for optimality.

If w_i is not negligible, continue to Step h.

h. Determine the binding constraint set $A^{++}(\underline{u}^i)$ from

$$A^{++}(\underline{u}^i) = \{j/u_j^{min} \leq u_j^i \leq u_j^{min} + \varepsilon_i \text{ and } \left[\frac{\partial \tilde{J}}{\partial u_j}\right]_i > 0$$

$$\text{or } u_j^{max} - \varepsilon_i \leq u_j^i \leq u_j^{max} \text{ and } \left[\frac{\partial \tilde{J}}{\partial u_j}\right]_i < 0 \quad \text{(A.30)}$$

$$j = k, \ldots, T-1\}$$

where $\varepsilon_i = \min\{\varepsilon, w_i\}$, with ε being a small positive number.

i. Compute the two parts (nonbinding and binding) of the Newton's direction as follows:

For the nonbinding control variables (i.e. $\ell \notin A^{++}(\underline{u}^i)$)

$$d_\ell^i = \frac{(N_{us\ell}^i + B_\ell^i K_{\ell+1}^i \phi_\ell^i)\delta\bar{s}_\ell^i + N_{u\ell}^i + B_\ell^i \kappa_{\ell+1}^i}{N_{uu\ell}^i + (B_\ell^i)^2 K_{\ell+1}^i} \quad \text{(A.31)}$$

where

$$K_T^i = N_{ssT}^i \tag{A.32}$$

$$\kappa_T^i = N_{sT}^i \tag{A.33}$$

and for $\ell = T-1, \ldots, k+1$, if $\ell \notin A^{++}(\underline{u}^i)$

$$K_\ell^i = N_{ss\ell}^i + (\phi_\ell^i)^2 K_{\ell+1}^i - \frac{[N_{us}^i + B_\ell^i K_{\ell+1}^i \phi_\ell^i]^2}{N_{uu\ell}^i + (B_\ell^i)^2 K_{\ell+1}^i} \tag{A.34}$$

$$\kappa_\ell^i = N_\ell^i + \phi_\ell^i \kappa_{\ell+1}^i - \frac{(N_{us\ell} + B_\ell^i K_{\ell+1}^i \phi_\ell^i)(N_{u\ell}^i + B_\ell^i \kappa_{\ell+1}^i)}{N_{uu\ell}^i + (B_\ell^i)^2 K_{\ell+1}^i} \tag{A.35}$$

while if $\ell \in A^{++}(\underline{u}^i)$

$$K_\ell^i = N_{ss\ell}^i + (\phi_\ell^i)^2 K_{\ell+1}^i \tag{A.36}$$

$$\kappa_\ell^i = N_{s\ell}^i + \phi_\ell^i \kappa_{\ell+1}^i . \tag{A.37}$$

The state deviations $\delta \bar{s}_\ell^i$ in (A.31) are obtained from

$$\delta \bar{s}_{\ell+1}^i = \phi_\ell^i \delta \bar{s}_\ell^i + B_\ell^i d_\ell^i \tag{A.38}$$

$$\ell = k, \ldots, T-1$$

with $\delta \bar{s}_k^i = 0$ and d_ℓ^i as given by (A.31) if $\ell \notin A^{++}(\underline{u}^i)$ and zero otherwise. The binding Newton's direction (i.e. when $\ell \in A^{++}(\underline{u}^i)$) is given by

$$d_\ell^i = - \left[\frac{\partial \tilde{J}}{\partial u_\ell} \right] \left[\frac{\partial^2 \tilde{J}}{\partial u_\ell^2} \right]_i^{-1} \tag{A.39}$$

(In nonconvex problems $\partial^2 \tilde{J}/\partial u_\ell^2$ should be replaced by some positive number.)

j. Select a stepsize α_i from the following Armijo stepsize selection rule:

$$\alpha_i = \beta^{m_i} \tag{A.40}$$

where m_i is the first nonnegative integer m for which

$$\tilde{J}^i - \tilde{J}([\beta^m d_\ell^i]^{++}) \geq$$

$$-\sigma \left[\beta^m \sum_{\ell \not\in A^{++}(\underline{u}^i)} \left[\frac{\partial \tilde{J}}{\partial u_\ell} \right]_i d_\ell^i + \sum_{\ell \in A^{++}(\underline{u}^i)} \left[\frac{\partial \tilde{J}}{\partial u_\ell} \right]_i [\beta^m d_\ell^i]^{++} \right] \quad (A.41)$$

with $\beta \epsilon (0,1)$ $\sigma \epsilon (1, 1/2)$, and

$$[\beta^m d_\ell^i]^{++} = \begin{cases} u_\ell^{min} - u_\ell^i, & \text{if } u_\ell^i + \beta^m d_\ell^i < u_\ell^{min}, \\ u_\ell^{max} - u_\ell^i, & \text{if } u_\ell^i + \beta^m d_\ell^i > u_\ell^{max}, \\ \beta^m d_\ell^i, & \text{otherwise} \end{cases} \quad (A.42)$$

$\tilde{J}^i$ is the performance index value at the nominal trajectories, and $\tilde{J}([\beta^m d^i]^{++})$ is its value for $u_\ell = u_\ell^i + [\beta^m d_\ell^i]^{++}$.

k. Perform the iteration

$$u_\ell^{i+1} = u_\ell^i + \alpha_i d_\ell^i$$

$$\ell = k, \ldots, T-1$$

and repeat Steps b to k until an optimal trajectory $\{u_\ell^*\}_{\ell=k}^{T-1}$ (according to the criterion in Step g) is identified. (Care should be taken so that the specified probabilistic levels $\{\gamma_\ell^{max}, \gamma_\ell^{min}\}_{\ell=k}^{T-1}$ allow for a nonempty feasible solution set.)

l. Apply the optimal action u_k^* and repeat the previous operations at time k+1.

Some comments regarding this algorithm are now noted:

1. In the version outlined the method at each iteration simultaneously accounts for both control and state constraints. This is expected to be effective for convex problems with linear or mildly nonlinear dynamics and convex cost functions (as, for example, the HAD operation case). The theoretically correct approach is to completely solve the control constrained problem for each new increase of the penalty parameter. Although this will add computational load, it will also be a safer route to follow.

2. In the event of "ill-conditioning" where C_i has been raised to numerically problematic levels and there are still more iterations needed for the algorithm to converge, we have found it reliable to restart the C_i increment cycles until convergence is induced. The idea is that increasingly better nominal trajectories are being identified and used as initial choices. However, this may be an issue in problems of very long control horizonts (>1000 time steps).

3. In accordance with the OLF control approach, the trajectory of the state's covariance is computed based on Open Loop control sequences, yet the actually applied controls are of Feedback nature. Had the true feedback laws been taken into account in the computation of the covariance matrix, its diagonal elements would grow at a rate lower than the Open

Loop, and they would stabilize at some finite level. By contrast, the Open Loop covariance propagation may, for certain types of systems, result in diagonal elements growing unbounded over time (unstable systems). Evidently, using Equation (A.2) results in suboptimal (overly conservative) control strategies, for the applied controls are forced to meet tighter probabilistic constraints than those actually required. A modification to remedy this suboptimality hinges on obtaining a local approximation of the actual feedback laws and using it in propagating the covariance matrix. Essentially, at each iteration, ELQG solves a local approximation of the control problem and obtains the optimization direction in feedback form, Equation (A.31). At the optimal trajectories, this equation constitutes a linear approximation of the true feedback laws. Using it in the covariance computations results in the following discrete time covariance propagation equation.

$$P_s(t_{\ell+1}) = (\phi_\ell - B_\ell D_\ell L_\ell)^2 P_s(t_\ell) + Q_\ell \quad (A.44)$$

$$\ell = 0,\ldots,T-1,$$

where ϕ_ℓ and B_ℓ are as in Equations (A.9) and (A.10), and D_ℓ and L_ℓ are given by

$$D_\ell = 1/(N^i_{uu\ell} + (B^i_\ell)^2 K^i_{\ell+1}) \quad (A.45)$$

$$L_\ell = N^i_{us\ell} + B^i_\ell K^i_{\ell+1} \phi^i_\ell \quad (A.46)$$

$$Q_\ell = \int_{t_\ell}^{t_{\ell+1}} {}^2(\phi(t_{\ell+1},\tau)) \; Q_w(\tau) d\tau \quad (A.47)$$

When the control elements are not binding, Equation (A.44) propagates the state's covariance in Closed Loop fashion, that is, by considering the feedback dependence of the controls on the states; however, with respect to controls that are binding, Equation (A.44) unfolds as in the Open Loop procedure, the rational being that the binding controls are fixed as in the Open Loop sequences.

However, there is still a discrepancy with the above scheme. When a control variable is binding, it will be so with respect to one of the associated constraints (Equation (11)), implying that it is free to move toward the nonbinding bound. Therefore, if the variance from Equation (A.44) is used in Step d of the ELQG control algorithm, it will underestimate the feasible solution region. To correct for this suboptimality, we propose to generate two more variance trajectories. Both are computed from Equation (A.44), but each keeps track of the corresponding upper or lower binding constraints. Each variance is then used accordingly in ELQG's Step d.

A DUAL APPROACH TO STOCHASTIC DYNAMIC PROGRAMMING FOR RESERVOIR RELEASE SCHEDULING

E.G.READ
University of Canterbury, Christchurch, New Zealand

ABSTRACT

In a mixed hydro thermal power system, optimal reservoir operating rules may be summarised by a set of "guidelines" which can be developed by a backwards recursion equivalent to conventional SDP. Because it uses a dual-based discretisation which directly reflects the structure of decision-making, this method is more accurate and yields significant insights. It will also be faster for problems with a sufficiently compact set of guidelines and has been used to develop a module which produces operating rules for twin reservoir problems with over a thousand stages, as part of a long term simulation package.

INTRODUCTION

Read and Boshier [8] describe an iterative technique based on marginalistic Stochastic Dynamic Programming (SDP) which has been used for some years now to optimise operating rules for New Zealand's hydro reservoirs over an annual time horizon. More recently Baker and Daellenbach [1] investigated long term coal stockpiling strategies, using a two reservoir SDP model to develop quarterly hydro reservoir operating rules over a 15 year time horizon, then simulating system operation using these.

A similar philosophy has been adopted for the much more detailed expansion planning model of [9], which simulates optimal management of up to 50 historical inflow sequences using a weekly time step over a 30 year horizon, assuming reservoir operating rules optimised by the model developed here. Since this simulation must reflect the way in which the system is really operated, strict accuracy may not be necessary and there is no point pursuing accuracy beyond that of current scheduling models. But it is important that the comparisons between plans are not distorted by the reservoir operating rules being better for one plan than an other. Thus heuristic rules were discarded in favour of an optimisation which would automatically adjust to changes in the system. To be realistic the stochastic nature of the inflows must be modelled (see [8]), and the decision period should be no longer than one month. The power system is briefly described in [8], but an important feature not modelled there is that the transmission link between the pure hydro South Island system and the mixed hydro/thermal North Island system constrains operations significantly. Since the expansion of inter-island transmission capacity, and the balance of supply in the two islands, are major long term planning issues, it is crucial that the North and South Island reservoirs be modelled separately.

Since the model must be run as part of a larger package which is in constant use, computational efficiency was a major concern. Although a two reservoir version of the model in [8] is used for weekly scheduling over an annual time horizon, it would be computationally out of the question to run this model regularly over a 30 year time horizon. The model of [1] could have provided approximate (quarterly) rules in about 60 seconds cpu on the IBM 3033 used for this project. But a significantly faster and more accurate program was developed using a combination of strategies. These include the efficient separation of the uncertainty and decision phases and the use of precomputed system schedules, as explained in the next Section. But the major conceptual change is the modification of SDP so that, instead of searching for an optimal release pair corresponding to each storage pair, the optimal decision rules for the beginning of a period are constructed directly from those for the end of that period. This approach is developed first for a single reservoir deterministic model, then extended to handle two reservoirs under uncertainty.

DP is ideally suited to single reservoir problems, but the "curse of dimensionality" has made it difficult to apply to realistic multi-reservoir systems, particularly under uncertainty. Although ignored here, the dimensionality of many SDP models for reservoir optimisation is further increased by the need to model serial correlation using Markov chains as in [12]. Yeh [14] and Yakowitz [13] review many developments designed to reduce these computational problems, both concluding that methods which do not require discretisation of the state space show considerable promise. Labadie and Fontane [4] show that multi-dimensional deterministic problems can be solved by discretising the objective function space instead, as long as a more-or-less unique decision can be associated with each objective function value at each stage. Our problem does not satisfy this condition, and instead we exploit the piece-wise quadratic nature of the DP value function to construct an optimal release policy for the whole state space. Our method is also related to that of Gal [3] who applied "parameter estimation" to develop approximations to the optimal value functions for a stochastic multi-reservoir problem, and to "Constrained Differential DP" [5], which works with a locally valid quadratic approximation to the value function to find an optimal storage trajectory for a deterministic problem.

ORGANISING THE MODEL EFFICIENTLY

Let $t=1,...T$ index time periods, $b=1,...B$ thermal stations in order of increasing fuel cost, $MC(b)$, and $h=1,...H$ a representative set of inflow sequences. For convenience we will take these to be the set of all historical inflows, and thus treat them as being equiprobable. We model the hydro system to the same degree of accuracy used in actual release scheduling. There is only one major storage reservoir in the North Island, but South Island storage capacity and inflows are expressed in terms of their potential generation and represented by an equivalent aggregate reservoir, as in [8]. Head effects are insignificant because virtually all generation plant is downstream from the storage reservoirs. For this time scale, it is reasonable to ignore the detailed chracteristics of individual machines and assume output to be linear in release, although some adjustments are made during the precomputations described below. Flows are assumed to be independent for successive periods, but we preserve any cross correlations by always considering each historical set of North and South Island reservoir and tributary flows $(NIF, NIT, SIF, SIT)_h$ together. For convenience let EF^t denote the expected inflow vector, and F_h^t the random components only. Long term demand uncertainty is accounted for by running the entire simulation package for a number of demand scenarios. Short term uncertainty is dealt with by the precomputations described below.

A decision is made at the beginning of each period to aim at an end-of-period storage target, ES^{t+1}, assuming expected inflows. At the end of the period the actual inflows become apparent, producing a range of possible storage pairs around the target. Thus the calculations at each stage can be divided into two phases; an "uncertainty adjustment", in which the expected value, EV, of storing water is calculated as a function of the target storage; and a "decision" phase in which a release decision is selected for each grid point, so as to achieve an optimal trade off between the benefits of releasing water and those of storing it for later use, as measured by EV at the target storage point. If the same grid is used throughout, each uncertainty adjustment requires H interpolations to determine EV for a grid point, plus an average of, say, A interpolations, one for each release pair considered, to determine future storage values in the "decision phase". Since H is 20, and tests done by the author show A to be about 15 for the model of [1], this represents a significant saving over their approach which requires HA interpolations, one for each inflow/release pair considered.

In the New Zealand power system, release priorities and thermal base-loading are normally reviewed on a weekly basis. So this decision-first formulation is quite accurate if weekly periods are used, and provided that tributary flows are accounted for by adjusting reservoir flows appropriately during the uncertainty adjustment phase. It is also a reasonable approximation for monthly periods, although for longer decision periods it is unduly pessimistic, because it does not allow for the fact that release strategies will be modified as inflows become apparent during the period.

Since our model has a continuous state space, the optimal release (pair) in each period can be identified by setting the marginal value of releasing water (in each island) equal (or as close as the release constraints will allow) to the marginal value of storing it for future use, EMV, as in [8]. For a deterministic problem this marginal value would simply be the dual variable or Lagrange multiplier on the water conservation constraint, which could be determined from the solution of a mathematical programming problem. For this stochastic problem EMV may also be thought of as a dual variable in the sense of Rockafellar and Wets [10] and [11]. (The appendix demonstrates this for a simple model.) EMV could be determined by differentiating the EV function produced by conventional SDP, but, for a single reservoir model, it can be just as easily updated directly by backwards recursion, using marginalistic SDP as in [8].

For a two reservoir problem, marginalistic SDP requires twice as many interpolations as the conventional approach, because two marginal water value surfaces must be updated. The expected marginal value of storing water in the North, ENV, is a function of South and North Island storage (SIS and NIS), while ESV(SIS,NIS) defines the value of water in the South. On the other hand the marginalistic approach reduces the order of each interpolation. Thus our model which interpolates linearly on the marginal value surfaces, effectively provides a piece-wise quadratic approximation of the underlying value function. In general, if $MV^{t+1}(S)$ represents the (vector of) marginal water value(s) at the beginning of period t+1, then $EMV^t(S)$, the (vector of) expected marginal water value(s) at the end of t can be formed using the following "uncertainty adjustment":

A. <u>Uncertainty Adjustment Phase:</u>

For each point, S, in some regular or irregular grid:

$$EMV^t(S) = \sum_h (MV^{t+1}(S + F_h^t)) / H \qquad (1)$$

The distribution of loads experienced within a period is commonly summarised by a "Load Duration Curve" (LDC), and a good approximation to the optimal generation schedule can be achieved by "filling the LDC in merit order", that is using cheap capacity to meet base loads as far as possible, while reserving more expensive capacity to meet peaks. In a single reservoir model, the marginal value of water for any period determines the position of hydro release in the merit order. All stations for which MC(b) is less than EMV should be used in preference to hydro release and may be regarded as "base-loaded". Those with higher fuel costs should only be used when hydro is being employed to the maximum. It is easy to compute the optimal hydro release, R, for a period as a function of EMV by filling the LDC in this way, and this optimal release can only take on B discrete values, one for each (group of) station(s) with a distinct marginal cost. Thus all possible situations can be covered by pre-computing R(b,t) for all b=1,...B and t=1,...T. This only requires each LDC to be filled twice, once with and once without hydro in the merit order. The release requirements can then be determined by differencing. (Short term demand uncertainty and random unit outages are accounted for by adjusting the LDC during this process as in [2]). Then, once the expected marginal value of water in a period has been determined, the optimal release can be obtained directly from this table.

If there are a number of hydro systems, each should appear in the merit order at the position determined by its water value. But, in our system, there is no thermal plant in the South, so the value of releasing water in the South depends on the reduction this allows in North Island costs, either by reducing North-South transfer or by increasing South-North transfer. This means that South Island hydro effectively appears in the North Island merit order in a position which depends on the direction of transfer on the inter-island link. Thus all the possible situations can be covered by precomputations parameterised on t, b, and i = 0,1,2 ; corresponding to maximum South-North maximum South-North transfer, no transfer, and maximum North-South transfer. Thus we fill 120 North Island LDC's six times each, taking 2 seconds cpu. (Even though the optimisation uses shorter time steps, data is only provided on a quarterly basis, with changes being assumed to occur abruptly at the end of each quarter, or linearly over the quarter. The quarterly data is then scaled to produce average monthly, or weekly data, interpolating between end-of-quarter values where appropriate.)

Precomputation is obviously valuable when each "situation" is encountered several times in the course of a model run, as it is during the simulation process, when 20-50 years of inflow data are used on the same system model, or in traditional SDP models which consider several release pairs before determining the optimum. But the model developed here actually only considers each release pair once at each stage, and does it in a structured way which effectively mirrors the precomputation process. Thus precomputation is only useful here because our model is able to use the schedules precomputed for the simulation.

DIRECT CONSTRUCTION OF OPERATING RULES FOR ONE RESERVOIR

Figure 1(a) shows a typical end-of-period expected marginal water value curve, $EMV^t(s^t)$ such as could be inferred fom the output of any DP model of a single reservoir system. Since the optimal decision rule is to base-load all thermal stations whose fuel cost is less than EMV, the reservoir operating rules can be expressed using "guidelines", L_b, ranging from b = 0 representing spill (reservoir full) to b= B+1 representing shortage (reservoir empty), as shown in the Figure. If these guideline levels are plotted over the planning horizon, as in [8], station b should be base-loaded if, and only if, storage falls below L_b^t in period t.

Ideally, release decisions should be revised as the expected marginal value of water changes during each decision period. The optimal solution under this regime would always involve hydro at a specific position in the merit order, with all cheaper thermal stations fully base-loaded. But, if decisions are only revised weekly, they should be based on the marginal value of water at the <u>end</u> of the week. Since this depends on the end-of-period storage level, which depends on the release decision itself, some iteration may be required to determine the optimal release. In practice the system is actually operated using L^t (rather than L^{t+1}) to determine which

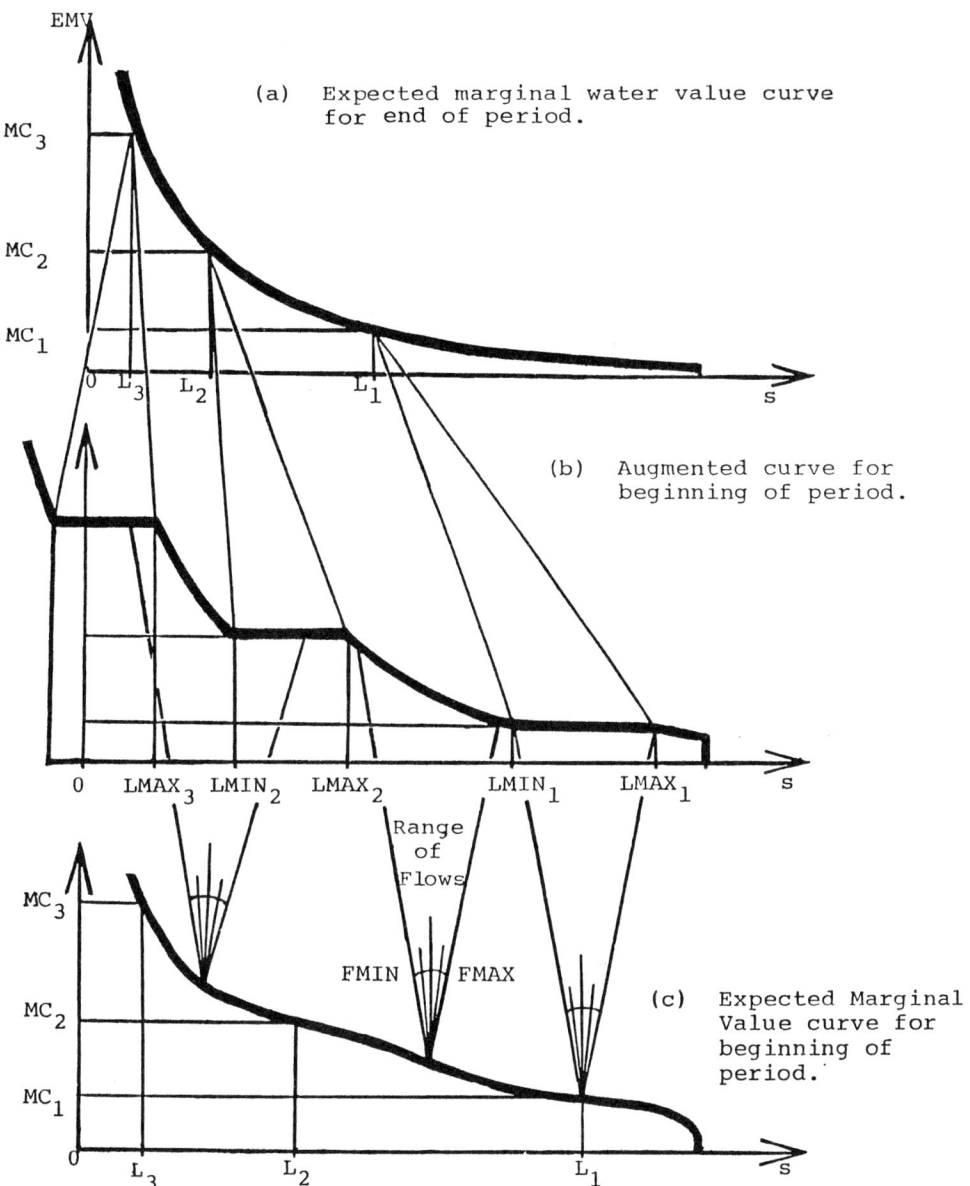

FIG. 1: Forming a New EMV Curve

stations should be fully base-loaded on the basis of s^t. This is reasonable for weekly decision periods, but inadequate for longer periods, during which water values can be expected to change significantly.

Figure 2 demonstrates the kind of situation which causes problems. If all stations up to and including b are fully base-loaded in t, storage will fall below L_{b+1}, so that EMV is higher than MC(b+1), by the end of the period, suggesting that b+1 should be base-loaded. But if it is fully base-loaded, storage will rise above L_{b+1}, indicating that some, at least, of the water being conserved is worth less than the cost of conserving it, MC(b+1). In fact, ignoring discounting, the optimal solution is to partially base-load b+1, at a cost of MC(b+1) per unit so that end-of-period storage exactly equals L_{b+1}, and thus the marginal value of the water stored also equals MC(b+1). Rather than iterate on the release levels, this decision can be made on the basis of the initial storage level alone, using "augmented guidelines", LMIN and LMAX, constructed directly using:

$$LMAX_b^t = L_b^{t+1} + EF^t - R(b-1,t) \qquad (2)$$

$$LMIN_b^t = L_b^{t+1} + EF^t - R(b,t) \qquad (3)$$

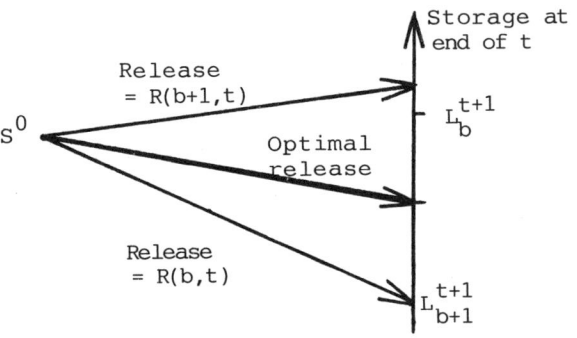

FIG. 2. Partial Base-loading

Now, if s^t equals $LMAX_b^t$, it is optimal to aim for storage target L_b^{t+1} at the end of the period with expected inflows and only stations up to b-1 base-loaded. If s^t lies below $LMAX_b^t$ but above $LMIN_b^t$, b should also be partially base-loaded to achieve that target. The precise loading level can be found by linear interpolation, so that if, say, s^t is half way between LMIN and LMAX, half loading would be appropriate, or more precisely, the station should appear above hydro in the merit order for half of the period and below it for the other half. For initial storage levels at or below $LMIN_b^t$, b should be fully base-loaded, while for those above $LMAX_b^t$ it should not be used at all.

The augmentation process also generates the marginal value curve for the end of t. Clearly the marginal value of extra water will be zero if the reservoir is already full, and equal the shortage cost, or at least the cost of meeting the load without hydro, if it is empty. In between those limits the (augmented) marginal water value curve for the beginning of a period, MV^t, will have the form shown in Figure 1(b), with significant "flats" bounded by the augmented guidelines, over which the marginal value is constant at the marginal cost of the corresponding thermal station. If the EMV curve were discontinuous, two flats might be adjacent, but otherwise the curve between two flats is identical to the corresponding section of the end-of-period curve, appropriately translated. That is:

$$MV^t(s^t) = EMV^{t+1}(s^t + EF^t - R(b,t)) \qquad (4)$$
$$\text{for } LMAX_{b+1}{}^t < s^t < LMIN_b{}^t$$

This discussion assumes that there will be a single storage level, $L_b{}^{t+1}$, at which the end-of-period water value equals MC(b). If the value of water in storage at the end of the planning horizon is specified by a concave increasing function of s^T, the marginal value function at the end of each period will clearly be monotone (implying a concave value function) but may not be strictly so. In fact, in the absence of uncertainty, the marginal value curve for the end of t-1, being just the curve for the beginning of t as in Figure 1(b), will have a flat over which the marginal water value exactly equals MC(b). Thus $L_b{}^t$ will not be unique, and $LMAX_b{}^{t-1}$ and $LMIN_b{}^{t-1}$ should be generated using (2) and (3), but substituting $LMAX_b{}^t$ and $LMIN_b{}^t$, respectively for $L_b{}^t$. Thus repeated application of (2) and (3) generates the solution to the deterministic problem (without discounting), by backwards recursion without any explicit optimisation, or even any precise knowledge of the value function between the guidelines.

But for our stochastic model the EMV curve for the end of period t-1 is derived from that for the beginning of t via the uncertainty adjustment described above, and may also be discounted, if desired. The general effect of this will be to remove the flats, producing a much smoother curve as in Figure 1(c). Then we must identify the new end-of-period guideline levels before applying (2)-(4). Thus the "decision phase" in the traditional (marginalistic) SDP algorithm can be entirely replaced by:

B. **Augmentation Phase** (i.e. Optimisation Phase):

a) Derive guidelines from the expected (discounted marginal) value surface for the end of period t, as in Figure 1(a).
b) Augment guidelines to form a complete set of decision rules for the beginning of period t, using (2) and (3).
c) Form an augmented (marginal) value surface for the beginning of period t/end of period t-1, using (4), as in Figure 1(b).

Thus, given a value function for water in storage at the end of the horizon, our problem can be solved by backwards recursion, using this augmentation process in conjunction with the uncertainty ajustment. In principle, this algorithm should produce exactly the same solution as would be found by conventional SDP using a sufficiently fine grid. However it has advantages in both accuracy and efficiency. These result from the fact that, rather than approximating continuous value functions using a regular grid of points in the primal state space, it works with the marginal cost of the thermal stations, which are particularly significant points in the "dual space" of expected marginal water values. These critical values directly determine the optimal release strategy, which may then be summarised by the corresponding points in the primal state space, i.e. the guideline levels.

Firstly, for deterministic problems, this method produces a solution which is as accurate as the assumed value function of the final period will allow. Traditional methods approximate this solution by choosing finer and finer grids and placing them as close to the expected guideline positions as possible. For stochastic problems, both approaches lose some accuracy by performing the uncertainty adjustment for an abitrary grid in the primal space, even though our model effectively uses a piece-wise quadratic approximation to the value function at each stage. (The accuracy of this phase can be increased arbitrarily, by including "guidelines" for intermediate marginal costs.) The "optimisation phase" of our method is always more accurate, defining the optimal policy exactly, rather than approximating it by interpolation between solutions at grid points.

Secondly, depending on the system, the number of points needed to describe the decision rules may be lower than the number of regular grid points required to give a reasonably accurate description of the value curves, and this reduces the number of decisions which need to be made, and the number of expected values which need to be computed. For a single reservoir model of the New Zealand system 6-10 guideline levels are required to define each expected marginal value curve.

Thirdly, once the guideline levels have been determined the "optimisation" can be achieved exactly without any search through the possible release levels. For a single reservoir deterministic model, in addition to the arithmetic involved in (2) and (3), the method only requires that the releases be determined by filling LDC twice for each period. It seems unlikely that any more efficient method can be found, using the same number of grid points. The stochastic case requires an uncertainty adjustment, as for the traditional method. But the computational requirements of this phase may be higher for an irregular grid than for a regular one, and an extra interpolation is needed to identify each guideline. Thus the relative effort required by the two methods will depend on the particular application.

A TWO RESERVOIR MODEL

In principle the same approach can be applied to any multi-reservoir problem, but this will only be practicable if the optimal operating rules can be expressed in a convenient way. We have already noted that, for a two reservoir model, both the South and North Island marginal water values need to be considered and both depend on storage levels in both islands. However, consideration of the marginal conditions defining the optimum reveals that the (end-of-period) decision rules for a twin reservoir model of the New Zealand system can be summarised as in Figure 3. Here the thermal guidelines are interpreted as for L above, except that now they depend on the South Island storage level as well as the North. Thus if South Island storage is high, the operators should be prepared to let North Island storage fall lower before base-loading b. Of course, if water in the two islands were interchangeable there would be no need to treat them separately in a two reservoir model. But in fact the inter-island link imposes a significant restriction, particularly on South-North transfer, so that when South Island storage is high, there is very little benefit from additional water in the South Island, making the thermal guidelines nearly horizontal.

Guidelines also tend to the horizontal when South Island storage is so low that additional water in the South makes no difference to the North because maximum North-South transfer is inevitable. Basically, inter-island transmission is scheduled so as to keep storage in the two islands as near to balance as possible, that is to keep their marginal water values, ESV and ENV, as nearly equal as possible. Thus if storage is disproportionately high in the South Island (North Island), i.e. storage is to the right (left) of the South-North (North-South) transfer guideline, then maximum South-North (North-South) transmission should be employed to balance up the storage levels. In between these two transfer guidelines the losses on the link exceed the discrepancy in water values, making further transfer unprofitable. For long term planning purposes, marginal losses on a link of this type can reasonably be assumed to be constant at ML. (Non-linear losses can be handled using extra guidelines if they are important). Thus the marginal cost of South Island hydro power delivered in the North Island is simply ESV/(1-ML) per unit delivered. Similarly the marginal cost of <u>not</u> sending North Island power South (i.e. the marginal value of any power sent)

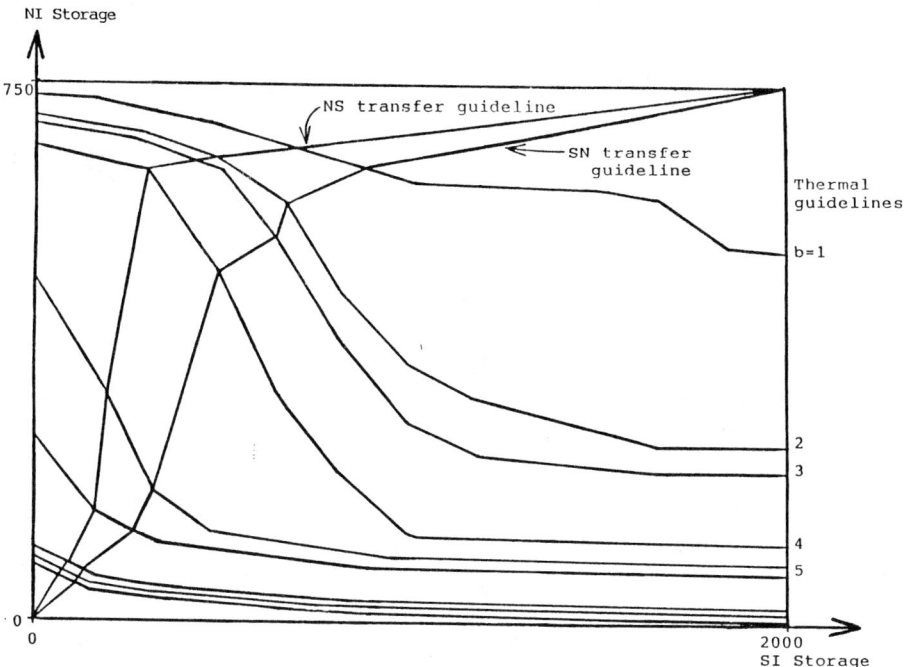

FIG. 3: Guideline Diagram for end of a Week

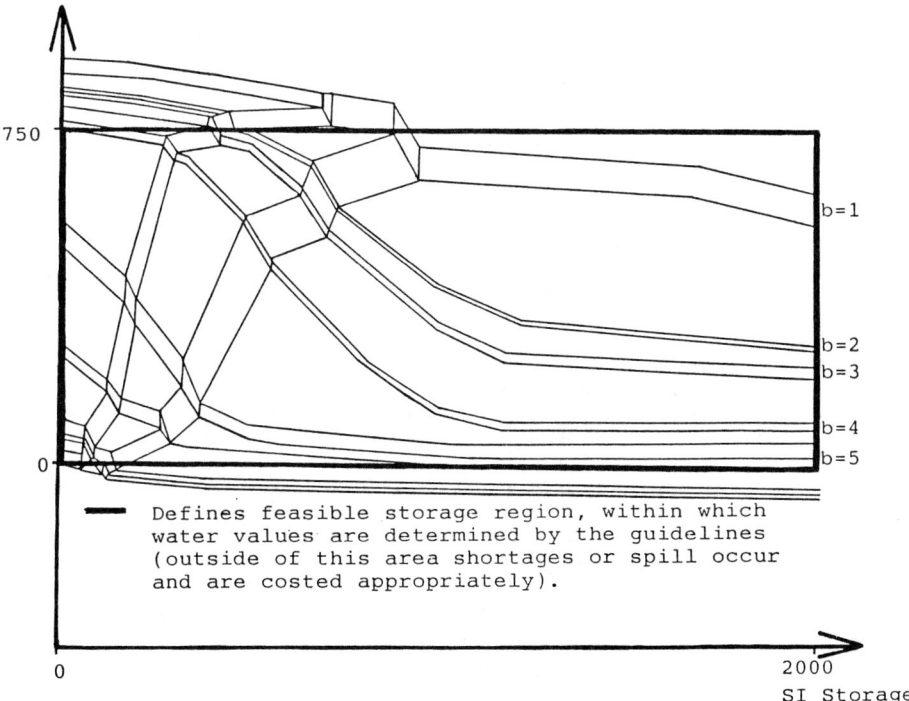

FIG. 4: Augmented Guideline Diagram for Beginning of same Week

is ESV*(1-ML). Under this regime no transfer should occur while:

$$(1-ML) < ESV/ENV < 1/(1-ML) \qquad (5)$$

Thus these critical ratios define the North-South and South-North transfer guidelines. But the situation in the North Island is not as simple as that in the South, and energy transferred to the North will often end up displacing thermal power, which may be more or less valuable than North Island hydro. This can be handled by placing transfer in the merit order in the position indicated by its marginal cost. Thus North-South transfer may be employed at off-peak times to build up South Island storage using cheap North Island thermal power, but this will cease as soon as the marginal cost of North Island production rises to exceed ESV*(1-ML). At peak times in the same period when the North Island marginal cost exceeds ESV/(1-ML), South-North transfer will be employed to reduce North Island costs. Thus the transfer guidelines really indicate when the relative positions of North and South Island hydro swap in the merit order, which in the model determines which set of precomputations (i=0, 1 or 2) is used. For instance, with storage in between the guidelines, i=1, and North Island hydro release appears between cutting back on North-South transfer and initiating South-North transfer in the merit order of ways to supply more power to the North.

Naturally, partial loading levels are sometimes optimal for the link too, and by similar reasoning to the above it may be seen that the diagram in Figure 3 can be augmented to form a complete set of decision rules for the beginning of the period as in Figure 4. For instance, the "plateau" corresponding to L_{ib}^t, the intersection of transfer guideline i with thermal guideline b for the end of period t, is a quadrilateral defined by:

$$LMAXMAX_{ib}^t = L_{ib}^t - EF^t + R(b-1, i-1, t) \qquad (6)$$

$$LMAXMIN_{ib}^t = L_{ib}^t - EF^t + R(b, i-1, t) \qquad (7)$$

$$LMINMAX_{ib}^t = L_{ib}^t - EF^t + R(b, i-1, t) \qquad (8)$$

$$LMINMIN_{ib}^t = L_{ib}^t - EF^t + R(b, i, t) \qquad (9)$$

Here, L, EF, R and LMAXMAX etc. are 2-dimensional vectors of South and North Island values. If the initial storage pair lies in the quadrilateral defined by (6)-(9) then partial thermal and link loadings should be determined, by linear interpolation as before, to ensure that the expected end-of-period storage pair will be exactly L_{ib}. Rather than using the L_{ib} values directly to represent the expected marginal value surfaces, each thermal guideline is represented by a set of North Island storage levels, NIL(k,b) defined over a regular grid of South Island storage levels, SIS_k, k = 1,...,K, as in Figure 3. ESV(k,b) records the marginal value of South Island storage for the same point, so that, for K=12, 12B grid points are used, and 24B values stored. This representation provides a comparable level of detail to that of [1], although, as argued above, the overall accuracy is greater. For any value of SIS the guideline levels provide an accurate representation of the marginal value of North Island storage just as in Figure 1(a). The ESV values provide a rather more arbitrary representation of the marginal value of South Island storage, but this inaccuracy is not too important because these values do not depend strongly on NIS. Value surfaces for the final period are supplied as input, but they can be refined by running one or two iterations of the model on the final year's data before commencing the main optimisation.

The augmentation procedure for this model involves tracing out these guidelines, translated appropriately, using formulae analogous to (2) and (3) to model South and North Island release. One dimensional linear interpolation between these guideline levels, holding South Island storage fixed at SIS_k, for $K=1,...,K$, is then used to define South and North Island marginal water value surfaces over a regular grid of storage pairs, and this representation is used during the uncertainty adjustment. Although this involves a loss of accuracy, it allows the uncertainty adjustment to be performed efficiently, using a single set of interpolation weights for each inflow pair to update ENV and ESV for all gridpoints in a single pass.

But this particular procedure is in no way integral to the method developed here, and may not be the best. All that we require is some way to define an augmented marginal value surface, some way to perform the uncertainty adjustment on it, and some way to form a new augmented surface from the adjusted values. Thus the value function itself could be updated in the uncertainty adjustment, as in traditional, non-marginalistic, SDP, the guidelines identified by differentiation, the augmentation performed using (6)-(9), and a new value surface formed from the augmented guidelines. Or the 2-dimensional equivalent of (4) could be applied directly to a regular, or irregular, grid of points, without forming guidelines at all. The uncertainty adjustment could then be performed on the resultant irregular grid, or it could be converted into a regular grid first.

But the whole "optimisation" phase, including forming the guidelines from a regular grid representation, augmenting them and reducing the implied marginal value surface back to a regular grid 120 times to solve a quarterly model over 30 years, only takes 0.3 seconds cpu. Thus, even if using 1560 weekly time steps increases this to about 4 seconds, there is little incentive to pursue further efficiency in this phase. Since the data, and hence the release requirements, only change quarterly, a deterministic weekly version of our 30 year problem could be solved in less than 6 seconds, including 2 seconds for the precomputations. Over an annual planning horizon, with data for each week, a total of 1 second would be required.

But the overheads of setting up the data etc. take 4.2 seconds and the uncertainty adjustment takes 2.5, using 20 inflow pairs at each stage of a quarterly model, giving a total cpu requirement of about 7 seconds, or 9 seconds including the precomputations. Because the overheads remain the same, a monthly model can be run in 15 seconds and a weekly model in 60 seconds. The model is now used on a regular basis, usually with a monthly time step, as part of the simulation package. Hoped for developments include the addition of a third reservoir representing a coal stockpile, and the use of the model, over an annual time horizon, for actual release scheduling. In this mode the stochastic problem could be solved in about 2 seconds, as compared to several minutes for the two reservoir model of [8].

CONCLUSIONS

Apart from the insights it yields into the nature of reservoir operation, the technique developed here should be more accurate than traditional SDP using the same number of grid points. Moreover it is more efficient for the deterministic case, and also for the stochastic case unless the overhead involved in performing the required "uncertainty adjustment" outweighs the savings made in the "optimisation" phase. Although we did not model current inflows as an extra dimension, our method should be able to handle this particular kind of multi-dimensionality with relative ease, because the number of decision alternatives does not increase.

In general the method shows promise for problems where the primal state space is large, but the decision rules can be expressed relatively compactly, reflecting the fact that decisions are determined by relatively few critical values and relationships between the dual variables. (For example an inventory problem with a linear production/consumption technology at each stage.) For such problems the decision rules can be expressed by "guidelines" analogous to those used here and the augmentation performs the optimisation required at each stage efficiently and with complete accuracy.

REFERENCES

1. W.R. Baker and H.G. Daellenbach, European J. OR 18, 304-314 (1984).
2. R.A. Bloom and L Charny, IEEE Trans. Power Apparatus and Systems PAS-102, 2861-2869 (1983)
3. S. Gal, Water Resources Research 15, 737-749 (1979)
4. J.W. Labadie and D.G. Fontane (this volume)
5. D.M. Murray and S.J. Yakowitz, Water Resources Research 15, 1017-1027 (1979)
6. E.G. Read, PhD thesis (University of Canterbury 1979)
7. E.G. Read, CBA working paper 152 (University of Tennessee 1982)
8. E.G. Read and J.F. Boshier (this volume)
9. E.G. Read, J.G. Culy, T.S. Halliburton, and N.L. Winter, Proc. Intl. Federation of OR Societies, Beunos Aries (to appear, 1987)
10. R.T. Rockafellar and R.J-B. Wets, in Mathematical Programming Study 6, R.J-B. Wets ed. (North Holland 1976) pp. 170-187
11. R.T. Rockafellar and R.J-B. Wets, SIAM J. Control and Optimization 16, 16-36 (1978)
12. J.F. Stedinger, B.F. Sule and D.P. Loucks, Water Resources Research 20, 1499-1505 (1984)
13. S. Yakowitz, Water Resources Research 18, 673-696 (1982)
14. W.W-G. Yeh, Water Resources Research 21, 1797-1818 (1985)

APPENDIX

The relationship between expected marginal water values may be seen from the following simple problem, faced by a manager who has s^t units of water at the beginning of t, thermal costs given by $C^t(R^t)$, and a concave differentiable end-of-period water value function, EV^{t+1}, assessed via SDP:

$$\text{MAX}_R \sum_h EV^{t+1}(ES^{t+1} + F_h^t)/H - C^t(R^t) \qquad (10)$$

Such that:
$$ES^{t+1} \leq s^t + EF^t - R^t \qquad (11)$$

Form a Lagrangian

$$\ell(R, ES, \lambda) = \sum_h EV^{t+1}(ES^{t+1} + F_h^t)/H - C^t(R^t)$$
$$- \lambda * (ES^{t+1} - s^t - EF^t + R^t) \qquad (12)$$

Now, in order for (R,ES) to be optimal for the original problem, the derivative of ℓ with respect to both must be zero. This yields:

$$-\partial C^t(R^t)/\partial R^t = \lambda = \sum_h \partial EV(ES^{t+1} + F_h^t)/\partial ES^{t+1} = EMV_{t+1}(ES^{t+1}) \qquad (13)$$

A more detailed discussion in terms of the duality theory of [10] and [11] may be found in [6] or [7].

Acknowledgement

This research was carried out at the New Zealand Ministry of Energy, but all opinions expressed are those of the author. Particular thanks are due to M.W.Davies who programmed the model.

ACCURACY OF THE FIRST-ORDER APPROXIMATION TO THE STOCHASTIC OPTIMAL CONTROL OF RESERVOIRS

PETER K. KITANIDIS* AND ROKO ANDRICEVIC**
*Associate Professor, St. Anthony Falls Hydraulic Laboratory, Department of Civil and Mineral Engineering, Mississippi River at 3rd Avenue SE, Minneapolis, Minnesota 55414-2196; **Research Assistant, St. Anthony Falls Hydraulic Laboratory, Mississippi River at 3rd Avenue S.E., Minneapolis, Minnesota 55414-2196

ABSTRACT

Optimization of the operation of a multireservoir system may be formulated as a stochastic optimal control problem which can often be solved through stochastic dynamic programming. This paper describes the application of a new approximate method which can be used for the solution of problems with many state and control variables. The optimal solution is given by the solution of the deterministic feedback control plus a caution (or hedging) term. The caution term is analytically approximated by the leading term of an asymptotic expansion obtained by assuming that the variances of the random inputs are small. The developed approximation makes use of the first two statistical moments of the random inputs and of the first three derivatives of cost functions. Its computational requirements do not exhibit the exponential growth exhibited by discrete DP. It can be used as an approximate solution to problems for which it is not feasible to use classical discrete stochastic dynamic programming. The paper presents an evaluation of the method through Monte Carlo simulations. A comparison with the exact solution and with deterministic feedback control is very encouraging, showing that the new method gives near optimal results even when the "small-perturbation" assumption is only approximately met and that it is superior to deterministic feedback control.

INTRODUCTION

In a previous paper [3] a small-perturbation approximation was proposed for the solution of a class of explicit stochastic optimization (or stochastic optimal control) problems. The developed methodology was named First-Order Approximation (FOA). The approach decomposes the problem into two parts: in the first one the deterministic feedback solution is obtained and in the second one stochasticity is accounted for using analytical small-perturbation techniques.

The purpose of this paper is to illustrate the applicability and potential advantages of FOA in the stochastic optimal control of reservoir systems. FOA is compared through Monte Carlo simulations with the two most commonly used methods: deterministic feedback control (DFC), and discrete stochastic dynamic programming (DSDP). In the following section the problem is mathematically formulated and the recently proposed FOA method of calculating optimal controls is reviewed. Numerical examples and conclusions are given in the last three sections.

MATHEMATICAL DEFINITION OF THE PROBLEM AND OVERVIEW OF THE FIRST-ORDER APPROXIMATION

The condition of the riverine system at time t may be represented by the amount of water stored in each reservoir, in each river reach, and in the soil. All such variables may be arranged in a vector know as the state vector $x(t)$. Mathematical description of the system thus becomes equivalent to description of the evolution of the state vector as it is affected by input and is regulated, to the extent possible, through the control variables. In the discrete-time case the planning horizon is divided into time intervals. The problem is to determine the optimal control variables (releases) at each interval taking into account the condition of the system at the beginning of that interval.

In mathematical terms, the evolution of the reservoir system is given by the state transition equation. For many applications it is sufficient to assume that the transition equation is of the form:

$$x(k+1) = \Phi(k) \, x(k) + \Psi(k) \, u(k+1) + \mu(k+1) + w(k+1) \tag{1}$$

where: $\Phi(k)$ is given nxn matrix (known as the transition matrix); $\Psi(k)$ is a given nxm matrix; $x(k)$ is the nx1 vector of the state variables of the reservoir system at the beginning of the interval (or stage) k; $u(k+1)$ is the nx1 vector of control variables; $\mu(k+1)$ is an nx1 vector of known inputs; $w(k+1)$ is an nx1 vector of random variables. In this representation, $w(k)$, $k = 1,\ldots,N$ is a process of zero-mean uncorrelated random vectors, i.e.

$$E[w(k)] = 0, \qquad k = 1,\ldots,N \tag{2}$$

$$E[w(k)w^T(\ell)] = \begin{cases} Q(k), & \text{if } k = \ell \\ 0, & \text{if } k \neq \ell \end{cases} \tag{3}$$

Examples of representing reservoir systems in the form of Equations (1) - (3) can be found in Maidment and Chow [5], Wasimi and Kitanidis [8], Loaiciga and Marino [4], Georgakakos and Marks [2], as well as this paper.

At each period, k, there is a cost associated with the state of the system, $x(k)$, at the beginning of this period and with the control vector, $u(k+1)$, during that period. This is called the stage-wise cost function and denoted by $c_k(x(k),u(k+1))$. The cost associated with the state of the system at the end of the operating horzon, $x(N)$, is called the terminal cost and is represented by a given function $f_N(x(N))$. All cost functions are assumed to have continuous third derivatives.

Assume that we are at the beginning of period k and that $x(k)$ is given. The problem is to determine the controls $u(k+1),\ldots,u(N)$ which minimizes the cost of operation in the remaining periods. However, the cost associated with a sequence of controls, being a function of the random inputs $w(k+1),\ldots,w(N)$, is a random variable itself. In the prototypical optimal control problem we adopt the criterion of minimizing the expected value of the cost during the remaining periods

$$f_k(x(k)) = \min_{u(k+1),\ldots,u(N)} \left\{ \sum_{i=k+1}^{N} c_{i-1}(x(i-1),u(i)) + E[f_N(x(N))] \right\} \tag{4}$$

We refer to $f_k(x(k))$ as the cost-to-go function.

The control variables must also satisfy some inequality constraints. For example, releases must be nonnegative and cannot exceed some values determined by the spillway discharge-elevation curves. Thus

$$u^{min}(x(i)) \leq u(i+1) \leq u^{max}(x(i)) \tag{5}$$

where $u^{min}(x(i))$ and $u^{max}(x(i))$ are specified values.

Similarly, storages $x(k+1)$ must satisfy nonnegativity and capacity constraints which, in turn, limit the range of the values the releases $u(k+1)$ may take. However, in discrete-time stochastic systems described by Equation (1), where $u(k+1)$ is considered the decision variable, $x(k+1)$ is a random variable given the state at time k. Constraints on some storage $x_j(k+1)$ may thus be introduced in a probabilistic sense, such as by introducing the requirement that

$$\Pr[x_j(k+1) \geq X_j^{max} | x(k)] \leq \alpha_j \tag{6a}$$

and

$$\Pr[x_j(k+1) \leq X_j^{min} | x(k)] \leq \beta_j \tag{6b}$$

where α_j and β_j, $j=1,\ldots,n$ are given probabilities and X_j^{min} and X_j^{max} are specified limits. The deterministic equivalents of such chance constraints in terms of the control variables must be satisfied by the control variables.

Since both the cost functions and the constraints are separable in periods, dynamic programming is the choice method. The problem is solved in stages, one stage for each time period. Once the solution is obtained, the first-period control is applied.

In the case of uncertain future inflows, the future controls $u(k+2),\ldots,u(N)$ cannot be determined at k since they depend on the yet unknown future state of the system. That is, $u(k+2)$ depends on $x(k+1)$, $u(k+3)$ depends on $x(k+2)$, and so on. What can be determined is the optimal policy, i.e. the control $u(i+1)$ as a function of the state $x(i)$, $i=k+1,\ldots,N$. Since the cost-to-go valuedepends on the unknown functions it is a functional.

Discrete dynamic programming is the most common numerical method of solution of such functional optimization problems. In practice, the exponential growth of computing requirements with n limits the applicability of this methodology to systems with few state variables ($n=1$ or 2 for most reported applications). Approximate methods of solution have been proposed to overcome such difficulties. Of particular interest is a widely used method known as deterministic feedback control. Given all information at the beginning of period k, which in our case means the state $x(k)$, the best estimates of future inputs are used as perfectly known quantities in a deterministic optimization problem. Only the first-period control, $u(k+1)$, is applied. At the next period forecasts are updated on the basis of new information and another deterministic optimization problem is solved. The main advantage of this approach is that the resulting deterministic optimization problems can be solved without discretization using successive approximation methods such as differential dynamic programming or the principle of progressive optimality. The disadvantage of this approach is that it disregards "hedging" or "caution" and consequently is generally suboptimal.

In the first-order approximation the control is expanded in powers of σ

$$u(i+1) = u_0(i+1) + u_1(i+1)\sigma + u_2(i+1)\sigma^2 + \ldots, \quad i=k, \ldots, N \tag{7}$$

where σ is a scaling factor of the forecasting-error covariance matrix (for example, $\sigma^2 = \frac{1}{n} \text{Tr}[Q]$). $u_0(i+1)$, $u_1(i+1)$, and $u_2(i+1)$ are control policies, i.e. functions of the state vector $x(i)$. The ellipsis dots of the expansion represent the neglected terms of order σ^3 or higher. In a likewise fashion the cost-to-go function is expanded into

$$f_i(x(i)) = D_i(x(i)) + T_i(x(i))\sigma + S_i(x(i))\sigma^2, \quad i=k,\ldots,N-1 \tag{8}$$

Following a standard approach [7], Kitanidis [3] has shown that: $u_0(i)$ and D_i correspond to the deterministic feedback control obtained for the given initial state and neglecting the forecasting uncertainty represented by the w terms; $u_1(i)$ and T_i are zero; and $u_2(i)$ and S_i can be analytically calculated.

The caution term u_2 as calculated by the FOA method depends on the first two moments of the random terms and on the first three derivatives of the cost functions, calculated along the state trajectory specified by the deterministic feedback control with the given initial state. The methodology is applied in a feedback mode. That is, calculations are repeated at the beginning of each period to account for the observed state of the system.

A SINGLE RESERVOIR CASE

The simplest and probably most studied case is of a single reservoir with uncorrelated inflows. This is a scalar case (n=m=1) for which accurate numerical solutions may be obtained at a reasonable computational cost using discrete stochastic dynamic programming. Neglecting evaporation and seepage losses and assuming that the mean inflow is equal to 3 units of storage while the standard deviation of the forecasting error is σ, the state transition equation is given by the continuity equation

$$x(k+1) = x(k) - u(k+1) + 3 + w(k+1) \tag{9}$$

where $x(k)$ represents reservoir storage at the beginning of period k; $u(k+1)$ is the volume released during period k; $w(k)$ is the deviation from the mean value of the inflow. Note that (9) is a special case of Equation (1), with n=m=1, $\Phi(k) = 1$, $\Psi(k) = -1$, $\mu = 3$, and $Q(k+1) = \sigma^2$.

The performance criterion at the beginning of period k(k = 0,...,4) is minimization of the expected value of the cost-to-go functional

$$f_k(x(k)) = \min_{u(k+1),\ldots,u(N)} \{ \sum_{i=k+1}^{4} \frac{1}{2}(u(i)-1)^4 + E[(x(4)-3)^6] \}$$

$$k = 0,\ldots,4 \qquad (10)$$

The usable capacity of the reservoir is taken equal to 15 units. Deterministic equivalents of the nonnegativity and capacity constraints are enforced, with $\alpha = \beta = 0.05$ in Equation (6). The variance σ^2 is assumed to be in the range between 0.25 and 1.5.

The performance of an approximate method of optimizing the operation of a reservoir system with uncertain inputs can best be evaluated through Monte Carlo simulations, as pointed out by Gal [1]. For each value of variance, σ^2, one thousand inflow sequences were generated. Each inflow sequence is a realization of the stochastic process which represents the inflows, i.e., independent and normally distributed with mean 3 and variance σ^2. For each inflow sequence, the control was calculated using first-order analysis, discrete stochastic dynamic programming (DSDP), and deterministic feedback control (DFC). The actual value of the cost function was then calculated. Finally, the results from all realizations were pooled together and the statistics of the actual cost function were estimated. In this study, we will report the results for x(0) given equal to 11.

The mean actual cost of operation is depicted on Figure 1 as a function of the variance, σ^2. As expected, DSDP gave the lowest average cost of operation for any value of σ^2. The DSDP solution closely approximates the true optimum because, in this scalar case, it was feasible to use a fine discretization grid ($\Delta x = 0.1$). The average cost of operation is slightly higher when FOA control is applied and the difference increases with σ^2. Note that for small values of σ^2, FOA performs almost as well as the optimal solution. DFC has a much higher cost of operation than the other two methods and this difference increases drastically with σ^2. For $\sigma^2=1$, the average cost of operation with the suboptimal DF is 34% higher than for the optimum, as opposed to only 4% for the FOA.

Of course, the performance varies from one realization to another. Although the objective, as set forth in Equations (4) and (10), is to minimize the average cost of operation, it is of great practical interest to evaluate how the performance varies from realization to realization. The standard deviations of cost for each operation are shown on Figure 2 as functions of σ^2. The lowest values correspond to FOA which thus is the most consistent of the three methods. DSDP control, although optimal in terms of the expected-cost criterion, exhibits higher variability than FOA. The standard deviation of cost when DFC is applied is much higher than the other two methods. The histogram and duration curve of the actual cost of operation for $\sigma^2 = 0.5$ for each of the tested procedure are depicted on Figures 3 and 4, respectively. It is obvious that the chance of the actual cost of operation exceeding a high value is smaller for FOA than for any of the other two procedures. The actual cost when operating according to the FOA solution is between 350 and 450 units in about eighty percent of the cases.

Actual cost statistics are also shown in Table 1. It is obvious that FOA has the smallest difference between maximum and minimum cost of the three methods while DF has the largest one. In most water resources applications, the uniformity of performance of the FOA control is a definite advantage.

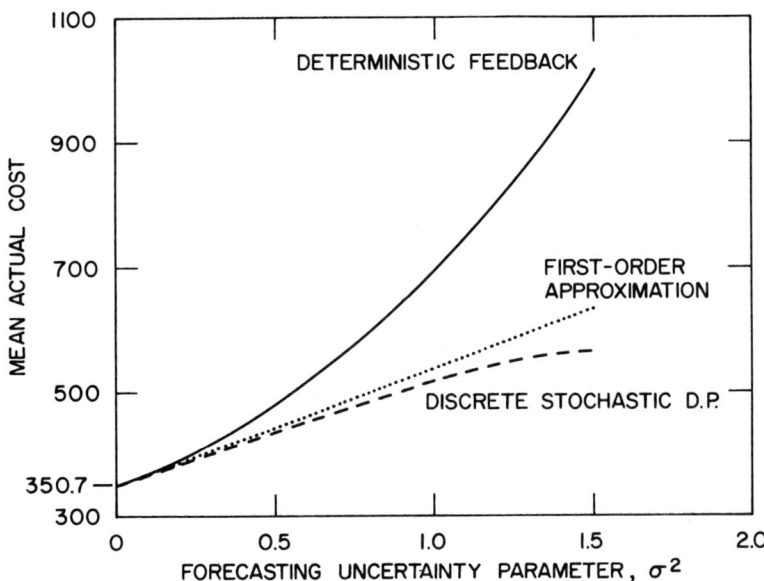

FIG. 1. Mean actual cost vs. input variance, σ^2, determined through Monte Carlo simulations with one thousand replicates applying DFC, FOA, and DSDP.

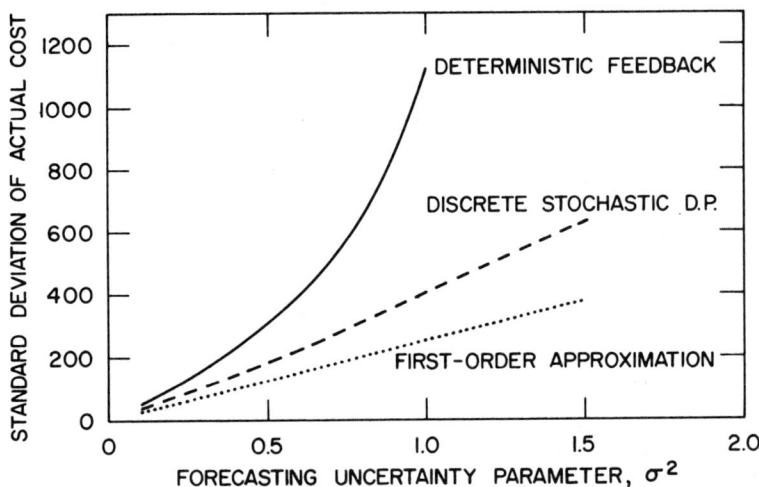

FIG. 2. Standard deviation of the actual cost vs. input variances, σ^2, for DFC, FOA, and DSDP methods.

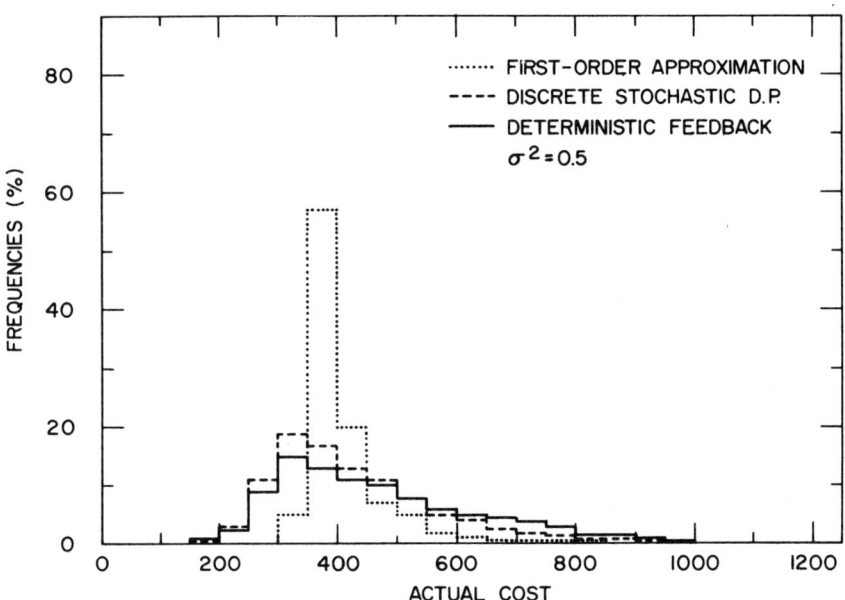

FIG. 3. Actual-cost histogram for DFC, FOA, and DSDP with input variance equal to 0.5.

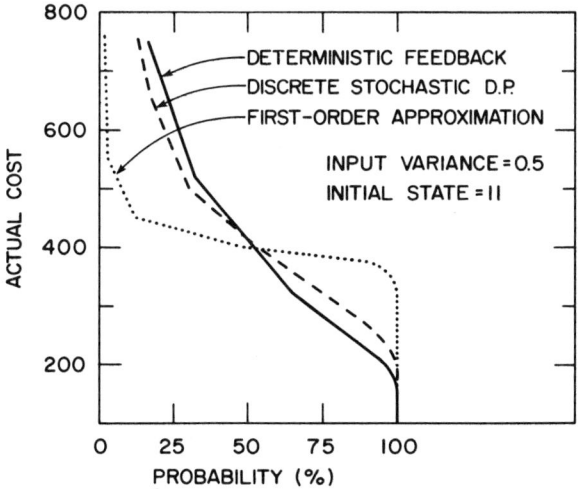

FIG. 4. Duration curve for the three applied methods for input variance equal to 0.5.

TABLE I. Statistics of actual cost for scalar case, obtained from one thousand realizations.

	SAMPLE STATISTICS FOR ACTUAL COST [N = 1000]				
INPUT VARIANCE	SELECT METHOD	MAXIMUM COST	MINIMUM COST	ST. DEV.	MEAN
0.1	FOA	695.2	338.5	30.6	370.1
	DFC	851.8	226.4	69.9	370.4
	DSDP	801.5	359.1	69.1	398.7
0.25	FOA	1296.1	369.4	69.1	398.7
	DFC	1944.0	187.9	155.7	405.7
	DSDP	1286.0	210.3	111.1	396.1
0.50	FOA	2721.1	398.0	128.3	446.2
	DFC	5045.3	159.1	310.3	481.3
	DSDP	2629.2	186.4	171.7	439.6
0.75	FOA	3347.6	439.0	157.5	492.7
	DFC	9973.7	140.5	687.7	587.7
	DSDP	3128.6	178.3	221.4	479.8
1.00	FOA	8207.3	483.1	255.6	541.0
	DFC	16966.6	127.8	1127.6	699.3
	DSDP	7565.8	165.2	413.1	520.1
1.50	FOA	9961.7	581.3	380.6	635.2
	DFC	38041.8	112.1	2411.7	1016.3
	DSDP	8342.5	143.2	624.7	569.7

FOA: First-Order-Analysis
DFC: Deterministic Feedback Control
DSDP: Discrete Stochastic D.P.

A MULTI-STATE CASE

Consider the reservoir system shown on Figure 5, consisting of four reservoirs indicated by triangles. The system is similar to the one considered in Murray and Yakowitz [6]. The input consists of the inflows into the two most upstream reservoirs. The inflows are modeled as a realization of bivariate autoregressive process of order one, AR(1).

$$x_1(k+1) - 3 = 0.7(x_1(k) - 3) - 0.296(x_2(k) - 4) + 0.839\ \epsilon_1(k+1) \tag{11a}$$

$$x_2(k+1) - 4 = -0.1(x_1(k) - 3) + 0.563\ (x_2(k) - 4) + 0.834\ \epsilon_1(k+1) + .299\ \epsilon_2(k+1) \tag{11b}$$

and

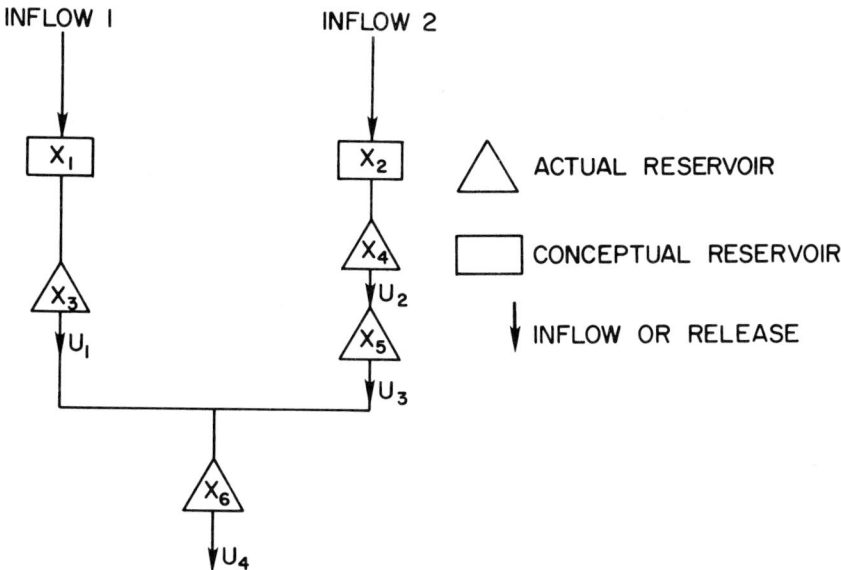

FIG. 5. Schematic representation of the reservoir system considered in the multi-state case.

$$E[\epsilon_1(k+1)] = 0 \qquad (12a)$$
$$E[\epsilon_2(k+1)] = 0 \qquad (12b)$$
$$E[\epsilon_1^2(k+1)] = \sigma^2 \qquad (12c)$$
$$E[\epsilon_2^2(k+1)] = \sigma^2 \qquad (12d)$$
$$E[\epsilon_1(k+1)\epsilon_2(k+1)] = 0 \qquad (12e)$$

Using the continuity principle, the reservoir storage volumes are given by

$$x_3(k+1) = x_3(k) - u_1(k+1) + x_1(k+1) \qquad (13a)$$

$$x_4(k+1) = x_4(k) - u_2(k+1) + x_2(k+1) \qquad (13b)$$

$$x_5(k+1) = x_5(k) - u_3(k+1) + u_2(k+1) \qquad (13c)$$

$$x_6(k+1) = x_6(k) - u_4(k+1) + u_1(k+1) + u_3(k+1) \qquad (13d)$$

For the sake of simplicity, we have neglected evaporation and seepage losses, local inflow, and the time delay caused by flow routing. These assumptions do not involve loss of generality [8].

Combining, the system transition equation is represented in the form of Equation (1) with n = 6, m = 4, and

$$\Phi = \begin{bmatrix} 0.70 & -0.296 & 0 & 0 & 0 & 0 \\ -0.10 & 0.563 & 0 & 0 & 0 & 0 \\ 0.70 & -0.296 & 1 & 0 & 0 & 0 \\ -0.10 & 0.563 & 0 & 1 & 0 & 0 \\ 0 & 0 & 0 & 0 & 1 & 0 \\ 0 & 0 & 0 & 0 & 0 & 1 \end{bmatrix} \quad (14)$$

$$\Psi = \begin{bmatrix} 0 & 0 & 0 & 0 \\ 0 & 0 & 0 & 0 \\ -1 & 0 & 0 & 0 \\ 0 & -1 & 0 & 0 \\ 0 & 1 & -1 & 0 \\ 1 & 0 & 1 & -1 \end{bmatrix} \quad (15)$$

$$Q = \begin{bmatrix} 0.704 & 0.700 & 0.704 & 0.700 & 0 & 0 \\ 0.700 & 0.785 & 0.700 & 0.785 & 0 & 0 \\ 0.704 & 0.700 & 0.704 & 0.700 & 0 & 0 \\ 0.700 & 0.785 & 0.700 & 0.785 & 0 & 0 \\ 0 & 0 & 0 & 0 & 0 & 0 \\ 0 & 0 & 0 & 0 & 0 & 0 \end{bmatrix} \sigma^2 \quad (16)$$

$$\mu = \begin{bmatrix} 2.08 \\ 2.04 \\ 2.08 \\ 2.04 \\ 0 \\ 0 \end{bmatrix} \quad (17)$$

The problem is to operate the reservoirs so that the objective function

$$f_k(\underline{x}(k)) = \sum_{i=k+1}^{4} \frac{1}{2}(u_1(i) - 1)^4 + \sum_{i=k+1}^{4} \frac{1}{2}(u_2(i) - .5)^4$$

$$+ \sum_{i=k+1}^{4} (u_3(i) - 1)^4 + \sum_{i=k+1}^{4} 1.5(u_4(i) - 1)^4$$

$$+ E\{(x_1(4) - 3)^6 + (x_2(4) - 3)^6 + (x_3(4) - 2)^6$$

$$+ (x_6(4) - 5)^6\} \quad (18)$$

is minimized subject to transition equation (1) with coefficients as specified in this section and the control and state constraints, as given with Equations (5) and (6), respectively. For this example, it was assumed that the initial vector x(0) is given and perfectly known (no measurement errors)

$$x(0) = \begin{bmatrix} 3 \\ 4 \\ 11 \\ 10 \\ 10 \\ 4 \end{bmatrix}$$

Furthermore, capacity of the reservoirs is taken equal to 15 units for x_e, e = 3,4,5, and to 20 units for x_6 (Figure 5). Deterministic

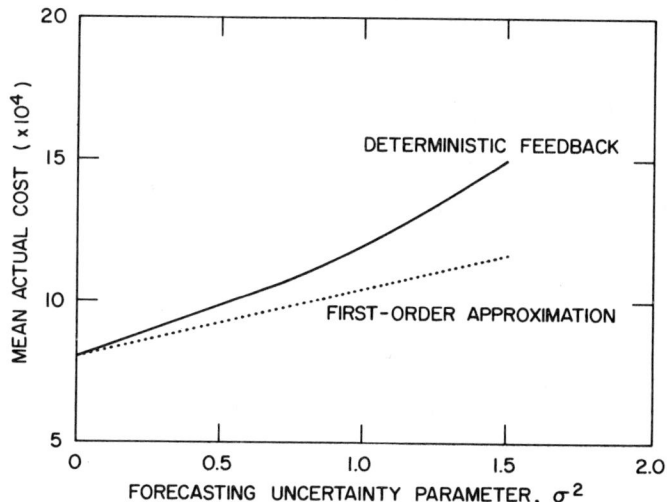

FIG. 6. Mean actual cost vs. input variance, σ^2, determined through Monte Carlo simulations with three hundred replicates, applying FOA and DFC.

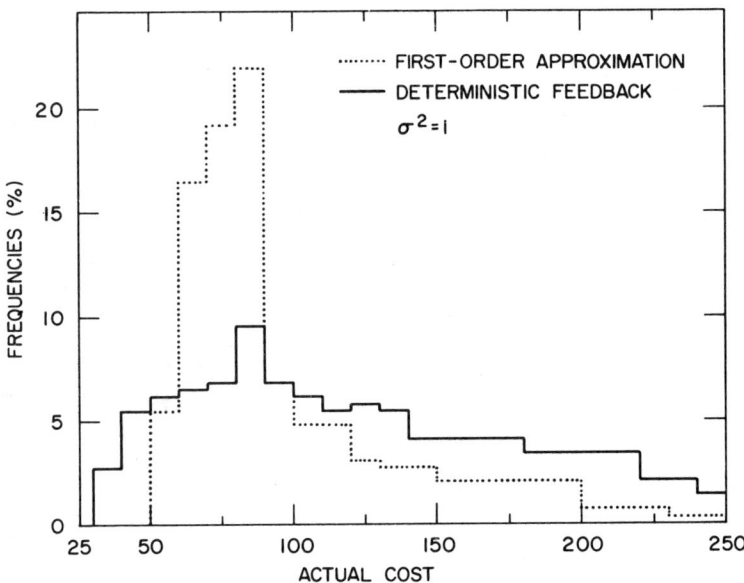

FIG. 7. Actual-cost histogram for DFC and FOA with input variance equal to 1.

equivalents of the state constraints were obtained for $\alpha_i = \beta_i = 0.05$. Following the proposed approach, differential dynamic programming (DDP) [6] was first used to calculate the optimal deterministic feedback control. Then, the caution term ($\mathbf{u}_2$) was computed analytically. The first-period control was applied and the calculations were repeated at the next period, k. The methodology was applied with three hundred different realizations. For purposes of comparison, the DFC solution was separately applied to the same realizations. Comparison with DSDP was not attempted because of the high computational cost associated with the application of this method to problems with many state variables. In fact, the authors are not aware of any application of DSDP to systems represented by six state variables.

The results are given on Figures 6 and 7. As in the single case, the FOA method performs significantly better than DFC. Not only is the average cost of operation smaller when using the FOA solution, but also the variance (from one realization to another) of the cost is significantly reduced.

DISCUSSION AND CONCLUSIONS

The problem of optimizing the real-time operation of complex reservoir systems with uncertain inflows is very difficult and no practical general method of solution is currently available. The applicability of discrete stochastic dynamic programming is limited by the dimensionality curse to problems simpler than many cases of practical interest. As a result, methods which separate the stochastic optimization problem into a stochastic estimation and a deterministic optimization part are recently commanding much attention. Firstly, the estimation problem is solved and then some of its results, such as the best predictions, are used as known inputs in deterministic optimization. There are, of course, systems for which estimation and control may be solved separately without loss of optimality. The best known representative of such systems, which are called certainty equivalent, is the class of LQG systems (Linear transition equation, Quadratic cost function, and Gaussian inputs).

In some of the most important problems of reservoir operation, however, it is not possible to neglect the interaction between estimation and optimization. These are cases for which there is significant need for "caution" or "hedging". First-order approximation is a new approximate method developed for the solution of some stochastic optimization problems for which caution is important. An important advantage of this method is that its analytical nature allows us to develop a better grasp of the essential features of caution-affected stochastic optimal control problem. Thus, it illustrates that the caution effect is conditional on the presence of forecasting uncertainty but also depends on system characteristics, such as the shape of cost functions and constraints.

This paper has presented the results of Monte Carlo simulations in which the performance of the first-order analysis was compared with the performance of deterministic feedback control (which assumes certainty equivalence) and, in one of the two cases, discrete stochastic dynamic programming. Attention was limited to two caution-affected stochastic optimization problems, one scalar and one multistate. The results illustrate the superiority of FOA, which accounts for caution albeit approximately, over the deterministic feedback solution, which implicitly assumes certainty equivalence. This superiority is both in terms of average cost of operation and improved consistency in performance from one possible realization of inflows to another.

Comparison with the exact solution, closely approximated through stochastic dynamic programming with a very fine discretization grid, indicates that the analytical solution is nearly optimal for a wide range of the values of the variance of forecasting uncertainty. However, an advantage of the analytical solution is that its applicability is not limited to problems with few state variables. This has been illustrated by presenting a case with six state variables, a case beyond the range of applicability of conventional discrete stochastic dynamic programming. Thus, the analytical approach can be applied to problems of practical significance as an alternative to deterministic feedback control.

Although limited to two particular cases, the extensive Monte Carlo simulations presented in this paper confirm the usefulness of FOA analysis in obtaining near-optimal solutions to caution-affected stochastic optimal control problems. Thus, they illustrate the robustness of the small-perturbation approximation made in the derivation of the analytical solution. In fact, the most interesting results of this study may be that as the forecasting uncertainty increases, so does the superiority in performance of the first-order analysis solution over the deterministic feedback control.

ACKNOWLEDGEMENTS

The material is based upon work supported by the National Science Foundation under Grants CEE-8212066 and CEE-8420664.

REFERENCES

1. Gal, S., Optimal management of a multireservoir water supply system, Water Resour. Res., 15(4), 737-749, 1979.

2. Georgakakos, A. P., and D. H. Marks, Real-time control of reservoir systems, Technical Report No. 301, R. M. Parsons Laboratory for Hydrology and Water Resources, MIT, May 1985.

3. Kitanidis, P. K., A First-order approximation to closed-loop stochastic optimal control of reservoirs, St. Anthony Falls Hydraulic Laboratory Technical Report, University of Minnesota, Minneapolis, Minnesota, 1985.

4. Loaiciga, H. A., and M. A. Marino, An approach to parameter estimation and stochastic control in water resources with application to reservoir operation, Water Resour. Res., 21(11), 1575-1584, 1985.

5. Maidment, D. R., and V. T. Chow, Stochastic state variable dynamic programming for reservoir systems analysis, Water Resour. Res., 17(6), 1578-1584, 1981.

6. Murray, D., and S. Yakowitz, Constrained differential dynamic programming and its applications to multireservoir control, Water Resour. Res., 15(5), 1017-1027, 1979.

7. Nayfeh, A. H., Perturbation Methods, J. Wiley-Interscience Series on Pure and Applied Mathematics, 425 pp., 1973.

8. Wasimi, S. A. and P. K. Kitanidis, Real-time forecasting and daily operation of a multi-reservoir system during floods by linear quadratic Gaussian control, Water Resour. Res., 19(6), 1511-1522, 1983.

BIASES IN STOCHASTIC RESERVOIR SCHEDULING MODELS

E.G.READ[*] AND J.F.BOSHIER[**]
*University of Canterbury, PB Christchurch, New Zealand
**Ministry of Energy, PB Wellington, New Zealand

ABSTRACT

The New Zealand power system is currently operated using guidelines produced by a model which combines marginalistic Stochastic Dynamic Programming with forward simulation. This approach was selected after a comparative test confirmed theoretical predictions that other approaches to modelling uncertainty would bias the solution towards unduly high or low releases. The results of this study are presented, the reasons for the effect are discussed and the implications for reservoir scheduling drawn out.

INTRODUCTION

The aim of reservoir management in a power system is to minimise expected costs while maintaining an adequate security of supply. In a purely hydro-electric system, there is no fuel cost and security of supply is the only concern. But in mixed hydro/ thermal systems, balancing short term costs with long term economic and security considerations is a complex and important problem which has attracted considerable attention over the years. Yeh [30] provides a comprehensive, state-of-the-art review of alternative approaches to this problem, including a variety of linear and non-linear programming models, simulation, and DP, concluding that there is no universally applicable method.

The model developed here, based on Stochastic Dynamic Programming (SDP), is used to optimise an hydro release policy for the New Zealand power system, on a weekly basis, so as to achieve economy and security throughout the year. Short term schedulers then dispatch generation on an hourly basis throughout the week so as to minimise costs, given the overall operating policy determined by the long term model. Before this study the system was scheduled using a heuristic, but the first oil crisis stimulated research into improved methods. Daellenbach and Read [7] reviewed representative examples of the practical implementation of Linear Programming (Pacific Gas and Electric), non-linear decomposition using a "trajectory method" (Electricité de France), and SDP (Swedish State Power Board [12], [24]).

Two deterministic multi-reservoir models were developed, one based on Linear Programming [4] and the other using non-linear decomposition [18], but the random nature of the inflows is a major feature of the New Zealand power system, and it was not clear whether such models were appropriate. Thus a simple, one reservoir, power system model was used as a test-bed for several approaches. The program was designed for flexibility, and testing focussed on the quality of the solutions produced, rather than computational efficiency. It was concluded that, although deterministic models have been successfully applied elsewhere (see [23] and [10] for example), they are not adequate for our system. An "iterative method", based on SDP, was the best method tested, producing significant savings. Boshier et al [5] describe a more efficient implementation which is now used in practice. Stochastic variations on the trajectory method were also tested, because they can be generalised to handle many reservoirs relatively easily, thus offering a potential way of bypassing the "curse of dimensionality".

These tests confirmed that the biases predicted by [15] and [17] are significant for the New Zealand system. That is models, which are too optimistic about the ability of management to cope with random inflow variations tend to store insufficient water, while those which are too pessimistic store too much. This is in line with the experimental results of Loucks and Dorfman [13] with respect to linear decision rules, and also of Wunderlich and Giles [28] and Halliburton and Sirisena [9]. The "iterative method" used here was based on that of Stage and Larsson [24] and Lindqvist [12], following early work on "Marginal Expectations" by Masse [14]. Lanna [11] has recently provided a more formal statement of this theory, which we refer to as "Marginalistic SDP". Yakowitz [29] surveys a variety of DP models which have been applied to water resources problems, while recent applications of DP to practical reservoir scheduling in power systems include [8] and [26].

A SIMPLE MODEL OF THE NEW ZEALAND POWER SYSTEM

There are seven reservoirs in New Zealand with significant capacity to store water for use later in the year, but they are represented here as one equivalent reservoir by summing all inflows and capacities, expressed in terms of potential electric generation (GWH) as in [2]. Clearly this is less than ideal, and a two reservoir representation is used in practice (see [5]), but it provides a convenient environment in which to test algorithms without introducing other complicating factors. Over the long term, hydro contributes about 20,000 GWH annually, or 80 percent of the electricity generated in New Zealand. Of this about half may be considered to flow into the aggregate storage reservoir, which has a capacity of about 3000 GWH. Load levels peak in Winter, while inflows peak in Spring, so that reservoirs tend to be full at the beginning of Winter and empty at the end. Thus the "long term" scheduling horizon is only one year, and this may be divided into weekly periods indexed by $t = 1,...T$. If F^t denotes flows into the aggregate storage reservoir in t, q^t denotes aggregate release in t, and s^t aggregate storage at the end of t, then:

$$s^t = s^{t-1} + F^t - q^t \qquad t=1,...T \qquad (1)$$

Inflows vary significantly, with a 5% probability that aggregate annual flows will be less than 85% of the mean, while weekly deviations can be greater than 50% for individual flows. In continental climates, weather patterns or ice thickness can give a good guide to inflows over coming months, but this is not true in New Zealand's insular climate. Weather forecasts are not available more than a day or two ahead, and snow melt is unpredictable. Although some correlations undoubtedly exist, management considered that none had been identified with sufficient reliability to affect scheduling. Thus our model, unlike many in the literature (eg [26] and those surveyed in [25]) does not use a Markov model of the inflow process and the state of the system at any time is described solely by the reservoir level. Instead correlations are implicitly accounted for by simulating historical inflow sequences in the iterative method.

Usable storage is physically bounded above and below, and "buffer zones" are added to these bounds to reflect the fact that reservoirs can not all be exactly full or empty simultaneously. Spill or shortage is assumed to begin when storage enters the top or bottom buffer zone. Although these bounds will be modified later, they can be expressed by:

$$\underline{s}^t \leq s^t \leq \overline{s}^t \qquad t=1,...T \qquad (3)$$

There is a significant component of geothermal plant which is always base-loaded, but apart from this, the thermal system, which includes coal, gas, fuel oil and distillate fired plant, is primarily used to provide back up to the hydro system at peak times and in dry years. Thus the thermal load varies considerably from year to year, from nearly zero in a wet year, to about 35% of total requirements in a dry one. Consequently system marginal costs also fluctuate markedly, giving considerable incentive to optimise release scheduling, and to model the stochastic aspects in particular.

The range of loads expected in any week can conveniently be summarised by a Load Duration Curve (LDC), and the optimal generation schedule for any week can be approximated by "filling the LDC in merit order". That is, the cheapest sources are used as much as possible, leaving the peaks to be met by more expensive plant. Reservoir release is treated as a thermal station with a position in the merit order, for any particular week, determined by an "expected marginal water value", EMV, estimated by the model described here. Thus all thermal stations with marginal costs below EMV are "base-loaded" for as long as possible during the week. The remaining load is met, as far as possible, by reservoir releases, but this is constrained by limits on river flows and machine capacity, so that more expensive thermal plant is also required to meet peak loads. Allowance is also made for generation from the uncontrolled flows and for the special role of hydro in providing "spinning reserve" to meet short term variations in load (see [5]). Thus as EMV rises, more thermal stations are base-loaded so that less water is released. By filling the LDC for a range of EMV values we obtain the desirable reservoir release, q^*, and the total system cost, C, both as a function of EMV and U, the uncontrolled inflows. In order to formulate a standard SDP, we express this relationship as a cost function, $C^t(q,U)$.

THE GUIDELINE APPROACH

In the early 1970's security of supply was the major requirement of the New Zealand power system, and a "guideline" or "rule curve" approach was used to determine when thermal generation was required. As in [6], the "Basic Rule Curve" is the locus of storage levels at which maximum thermal generation (GMAX) needs to be scheduled in order to avoid an energy deficiency under the worst inflow conditions considered likely to occur, FMIN and UMIN. In our case this sequence is the "one-in-twenty Design Dry Year", so that there is a 5% chance that shortage will occur even if all thermal stations are base-loaded when storage falls to the Basic Rule Curve, (or "bottom guideline" or "forbidden zone"). Substituting these values into (1) and (2), rearranging and letting $SMIN^T$ be the lowest acceptable end-of-year storage, this can be constructed by working back on a weekly basis, via:

$$SMIN^t = MAX\{ \underline{s}^t, SMIN^{t+1} - UMIN^t - FMIN^t - GMAX^t + D^t \} \quad t=T-1,\ldots 0 \quad (4)$$

System controllers are expected to make every effort to conserve storage if it falls below this guideline. During past crises, this has involved restricting supplies as well as base-loading all thermal plant. This load reduction is modelled as a fictitious station, costed at the shortage cost. At certain times of the year storage must be held below the maximum for flood control reasons and a "Limited Spill Curve", $SMAX^t$, can be defined similarly. These bounds then replace the physical bounds in (2). Theoretically, the economic cost of electricity shortages should be balanced directly against the cost of avoiding it. But, when this approach was tested, the solution became very sensitive to the assumed shortage cost, which is notoriously difficult to estimate. Thus the forbidden zone has been used in all of the models described here, an approach which also assures management that traditional security levels have not been compromised.

As long as all thermal plant in New Zealand had similar fuel costs, only one guideline was required, but, when thermal costs became more diverse following the first oil crisis, there was considerable motivation to find a method whereby the early use of cheap thermal plant would save the use of expensive thermal plant later. This was done using several guidelines which acted as switching curves for each plant type. If storage dropped below the "coal guideline", for example, then coal fired plant would be base-loaded. This implies that, for any point on this guideline, the expected value of storing an extra unit of water was considered to exactly equal the cost of generating an extra unit at a coal fired station. Thus these guidelines implicitly defined the expected marginal water value, EMV, as a function of storage and the time of the year.

Heuristic reasoning was used to derive the guidelines shown in Figure 1, which were in use prior to this study. System operation using these guidelines was simulated using forty 40-year sequences of weekly data, constructed from historical records, and assuming the system to be in equilibrium (so that the 1963 simulation, say, starts from the final storage attained in the 1962 simulation). This gave an expected fuel cost of $24.8 million per year, with shortages being no more frequent than the one-in-twenty criterion would imply.

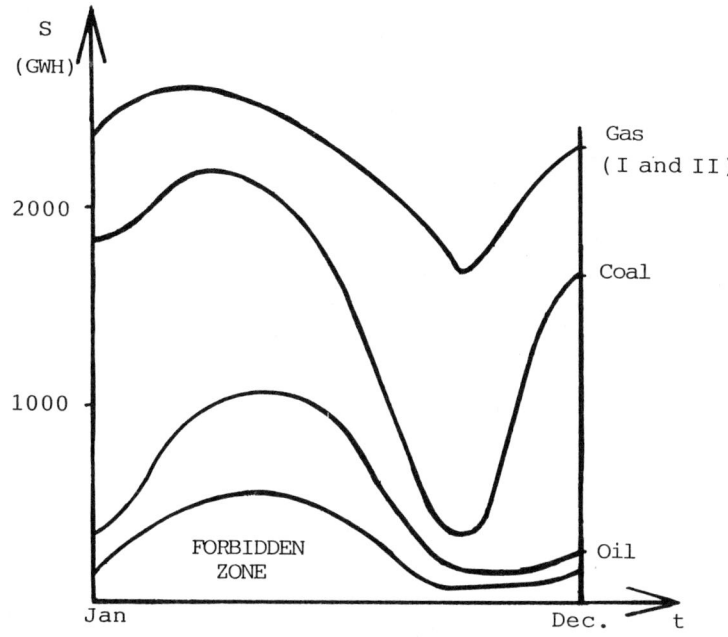

FIG. 1 Traditional Guidelines

SDP AND THE ITERATIVE METHOD

Let $EV^t(s^t)$ specify the expected value of water stored at the end of t, with EV^T, the value of storing water at the end of the planning horizon for use in the next year, given. Let h=1,...H index the historical inflows and k=1,...K a grid of storage levels. Then it is easy to state an SDP algorithm for the single reservoir model, ignoring inflow correlations and assuming that release decisions are made and carried out on the basis of expected flows, EF^t and EU^t, before the inflows are observed. For grid point s_k^{t-1}, the optimal release in t, and the value of stored water, are determined by:

$$V^{t-1}(s_k^{t-1}) = \underset{q_k^t}{\text{MAX}} [EV^t(s_k^{t-1} + EF^t - q_k^t) - C^t(q_k^t, EU^t)] \qquad (5)$$

This formulation assumes that the random components of the inflows, f_h^t and u_h^t, will only affect storage levels at the end of the period. This is reasonable for the New Zealand system, except that it does not allow for the way in which variations in uncontrolled inflows will affect thermal costs. This could be modelled by determining the expected thermal cost via simulation, or by assuming that inflows are observed before decisions are made. Neither approach is entirely satisfactory, but the issue is not important here, because the iterative model accounts for these effects via simulation. For the basic model the recursion is completed by defining the expected value of water in storage at the grid points, using:

$$EV^t(s_k^t) = [\sum_h V^t(s_k^t + f_h^t + u_h^t)] / H \qquad (6)$$

Naturally interpolation between these values is required to determine EV in (5). Now, assuming the C and EV functions to be sufficiently smooth and letting EMV^t denote the derivative of EV^t with respect to storage level, s^t, (5) may be solved to find the optimal release, q_k^{t*}, by setting:

$$d\,C^t / d\,q_k^{t*} = -d\,EV^t / d\,q_k^{t*} = EMV^t(s_k^{t-1} + EF_h^t - q_k^{t*}) = \lambda \qquad (7)$$

That is, the marginal value of releasing water must be equated to the expected marginal value of keeping it in storage. In fact the entire optimisation can be performed by backwards recursion using these marginal value functions, instead of the value functions themselves. Thus (5) is replaced by (7), which defines not only the optimal release in t, but also MV^{t-1}, the marginal value of storage at the beginning of t, since this obviously also equals λ. EMV^{t-1} is then estimated from MV^{t-1}, using an equation analagous to (6). Eventually this recursion will yield an EMV function for the first period in the planning horizon, implying an optimal release for the current period. But, if the system is in equilibrium with an annual planning cycle, the EMV function for the beginning of the year should be the same as that for the end of the year, and, if it is not, the end-of-year EMV function must have been mis-estimated. Thus the recursion is continued for another year, using the EMV function estimated for the beginning of the year as the new end-of-year EMV function. This process is continued until the beginning and end of year EMV functions are indistinguishable. This marginalistic approach to SDP is developed formally by [11]. It was used by [12] and [24], on which our model was based.

Now it is obvious from the way the LDC is filled that dC/dq must equal the marginal production cost of some thermal station, MC(b), b=1,...B. Thus the search for q^* is considerably simplified by assuming that it must correspond to fully base-loading all thermal stations up to some level. Although [19] shows that this assumption is not valid when the decision periods are of significant length, it is reasonable for weekly periods, and was adopted in all models considered here. Also, it is easy to see from [19] that the EMV curves will be monotone, so that each successive increment of water in storage is worth no more than its predecessors. Thus the output of the SDP model can be summarised using "guidelines" defined by:

$$L_b^t = \text{MAX} \{ s : EMV^t(s) \leq MC(b) \} \qquad (8)$$

Now station b should be base-loaded if storage falls below L_b^t in t, because this implies that EMV, the value of keeping water in storage, is greater than the marginal cost of b. The guidelines shown in Figure 2 were derived from the output of a marginalistic SDP model, constructed as above, but with one major change: rather than only looking forward one period at a time, as in conventional SDP, the iterative method introduced by [12] and [24] uses extensive forward simulation to capture the inflow correlations in the historical data. The algorithm starts with an estimated EMV curve for each month in the planning horizon. In practice these are defined on eight storage levels for each month, concentrated in the areas where guidelines are likely to lie. Working backwards, the EMV at each grid point is then re-estimated by simulating the (weekly) future management of each historical inflow sequence assuming those EMV curves (interpolating as appropriate) for, say, 20 weeks, unless SMAX or SMIN is reached earlier. The marginal value of water for each sequence is then estimated from the assumed EMV curves, or set to zero if the SMAX buffer was reached (ie water was spilled), or to the shortage cost if the forbidden zone was reached. The average of these values then becomes the new estimate for EMV at the original point. The re-estimation process continues until the set of EMV curves from two successive iterations are equal.

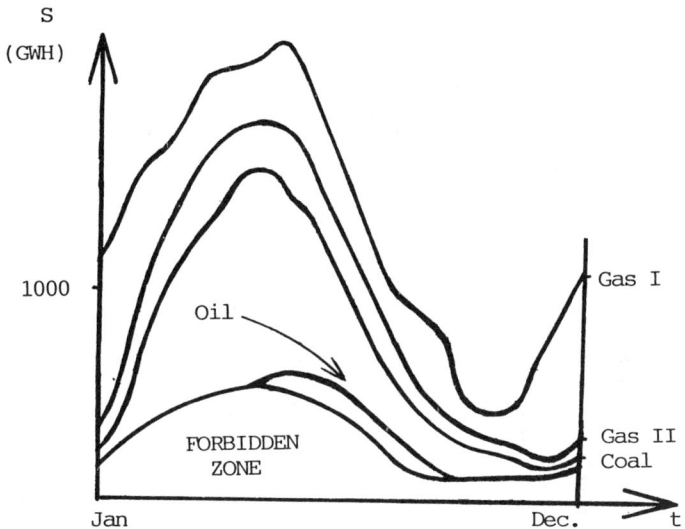

FIG. 2. Optimal SDP Guidelines

In order to dampen oscillations from iteration to iteration it proved necessary to average the results of successive iterations, with 10 to 15 iterations being required to achieve convergence from a reasonable initial solution. An alternative method, in which each simulation was stopped as soon as another guideline was encountered, was much more stable, but converged too slowly. However it can be used to refine the guidelines produced by the standard iterative process. Total cpu time is 2-3 minutes on an IBM 3033, but, although release decisions are made on a weekly basis, the guidelines need only be updated as the system changes, in practice routine updates are performed quarterly. (If the system is expected to change significantly over the planning horizon, the iterative process should really be used to find equilibrium values for the period after the change. Then operating rules for the period leading up to the change can be derived in a single iteration, working backwards from the change point and using those values to define end-point values for each sequence as appropriate.) Extensive simulation on the guidelines in Figure 2 indicated an expected annual fuel cost of $22.9 million, which was the lowest of any method tested, giving a saving of $1.9 million or about 8% over the traditional guidelines without any increase in shortage.

ALTERNATIVE METHODS

Water released from a reservoir through a power station has value because it displaces thermal generation. But the water could be stored instead, and the value which is assigned to this option depends on the assumptions made about the future, and in particular about the future inflow distributions and the ability of management to react to them. Rockafellar and Wets ([21] and [22]) have developed a very general theory of optimisation for problems of this kind, and their results are applied to our problem in [15] and [16] to show that optimal release decisions require an accurate determination of EMV, which requires the determination, explicitly or implicitly, of the optimal "non-anticipative" future utilisation of every possible inflow sequence. Here non-anticipativity means that each future simulated release decision is required to be made on the basis of information which would actually be available to the manager at that stage. That is a non-anticipative model reflects the realities of the decision making process. Since the only data available is the historical inflow sequences, it may be adequate to require optimal non-anticipative management of these. But many "stochastic" models do not involve determining a sequence of future decisions made in the light of circumstances as they occur, or, if they do, do not require them to be strictly optimal. Indeed any model which is not internally consistent, in the sense that future release decisions are assumed to be made using the model itself, can not be truly accurate. This study examines the impact of this inaccuracy.

For this purpose, stochastic reservoir models can usefully be divided into two categories; those which are over optimistic about the ability of management to cope with random inflow variations and/or underestimate that variability, and those which are over pessimistic. Clearly optimistic models will tend to underestimate future costs, while pessimistic models overestimate them. But we have already seen that release decisions are based on _marginal_ costs, so that, even if costs are mis-estimated, the releases recommended by a model will only be biased if the expected _marginal_ water value is mis-estimated. It is easy to demonstrate (see eg [18] or [20]) that optimal deterministic reservoir management involves equating marginal costs as far as possible across the planning horizon, and this intuition generalises to the stochastic case ([16]), via the duality theory of [22], as we have already seen from our consideration of marginalistic SDP. Thus we expect that, since reservoir managers will strive to keep marginal costs as nearly equal as possible across time, optimistic models will underestimate the variability of marginal costs, while pessimistic models overestimate it.

If the marginal cost curve were linear (ie a quadratic return function), this would not matter, but if it is convex, it is easy to see that greater dispersion will lead to higher expected values. Thus optimistic models will tend to underestimate expected future marginal costs, and hence EMV, leading to higher releases, while pessimistic models will do the opposite. Now, marginal cost curves can not be truly convex because they generally involve significant steps, as in Figure 3, but they will be approximately convex if cheap base-load power is available in large blocks, backed up by relatively smaller stations at higher cost. This is commonly the case and, while it is not clear that the curve in Figure 3 is sufficiently convex, this is implied by the fact that the EMV curves produced by our models are basically convex, and tests confirm the direction of the predicted biases. Thus optimistic assumptions about future management lead to underestimation of storage requirements, while pessimistic assumptions have the opposite effect, which seems intuitively reasonable.

SDP, being a special case of the general theory developed in [21], does provide an internally consistent methodology, producing optimal non-anticipative solutions for the inflow sequences it implicitly assumes. These may be quite a good approximation to reality, depending on the extent of correlations and the way they are handled. Here, for instance, the state of the system has been described by the storage level alone, so that all combinations of high and low inflows in successive weeks are implicitly assumed to be equally likely. If there really is a positive correlation between weekly inflows, this method implicitly overstates the proportion of moderate sequences, and so underestimates expected future costs, since the system cost function is always convex. If the marginal cost curve is also convex, it may also bias the solution toward unduly high releases. This tendency could be reduced by including current inflow levels in the description of the state space.

On the other hand, the iterative method described above uses the true historical inflow distribution, but makes the pessimistic assumption that, future releases will be decided on the basis of storage levels alone, ignoring any information about current inflows. Thus, given (approximately) convex marginal costs, it may be expected to err on the side of caution by holding more water than is strictly necessary. Management actually preferred this cautious bias, reinforcing the decision to adopt the method.

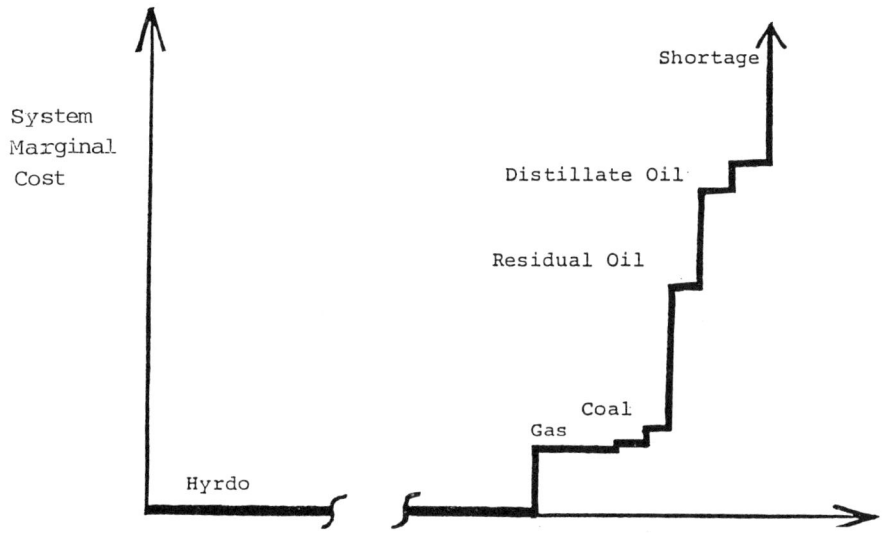

FIG. 3 System Marginal Cost Curve

Perhaps the most common alternative to SDP is deterministic optimisation, simply because it can provide solutions for large complex system models in reasonable computation times, often using standard software. (see eg [23] and [10]) However such models, by assuming perfect foresight, may make insufficient provision for adverse flows, particularly when storage is near its bounds. Thus, for instance, the model of [18] tends to hold storage near the bounds for considerably longer than would be considered wise in reality. In general, if marginal costs are approximately convex, the assumption of perfect foresight will tend to reinforce the optimistic bias caused by the fact that such models typically assume mean inflows. Thus, in view of the extensive use of such models, it is important to determine whether this bias is significant in practice.

The "trajectory method" developed by Electricité de France [1] is efficient for deterministic problems of this type. The application of the

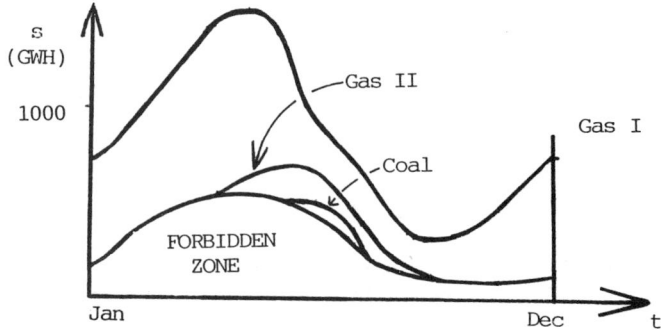

FIG. 4: "Optimal" Deterministic Guidelines

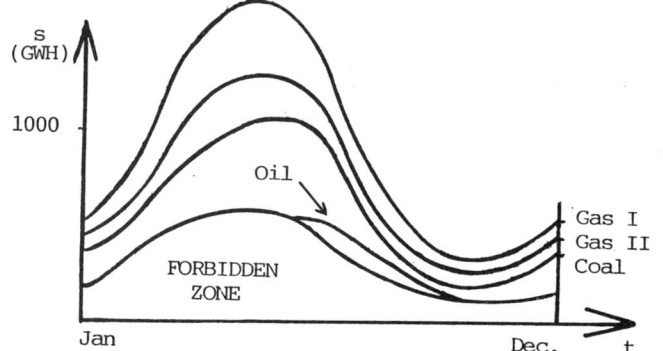

FIG. 5: "Optimal" Guidelines, Averaged Trajectory method.

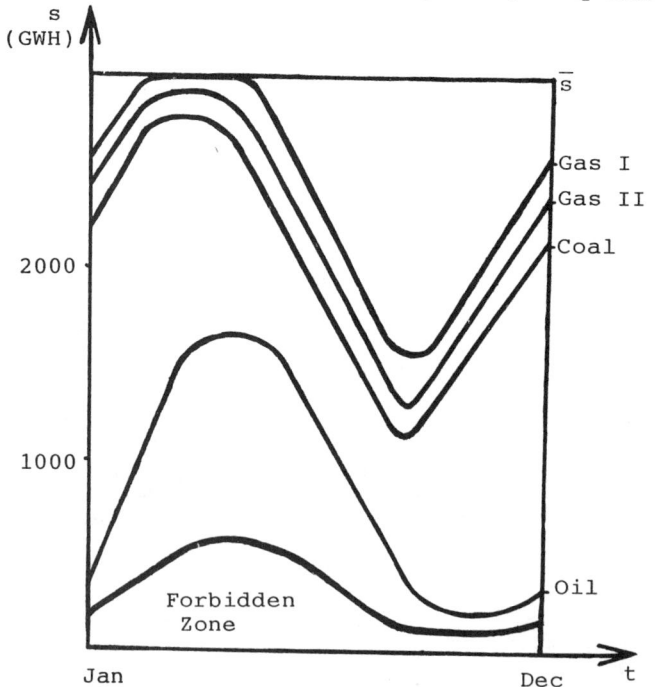

FIG. 6: Guidelines Produced by Simulating Sub-optimal Management

method to our system is described in [18], while [27] contains a recent statement of the method and computational comparison with DP. It involves a process of trial and error, iterating on the marginal water value (or "Lagrange multiplier on the storage conservation constraint") until a trajectory of storage levels is found to satisfy the optimality conditions of the problem. The guidelines shown in Figure 4 were produced by applying this technique to determine the marginal value of water at each of the storage grid points used by the SDP model. These guidelines are clearly less cautious than those in Figure 2, letting storage fall much lower before base-loading each thermal station. Extensive simulation confirmed this bias, with costs of $24.7 million, 8% higher than for the SDP method, and unacceptably high shortage probabilities. The extent of the bias is also revealed by the fact that, in order for the deterministic method to produce guidelines comparable to the SDP method, Winter inflows 25% below the mean had to be assumed. A similar bias was noted by Halliburton and Sirisena [9], who compared the results from deterministic and stochastic DP models for this same system. Thus the use of deterministic methods, assuming mean inflows, can not be recommended for reservoir scheduling in New Zealand, because they tend to store insufficient water.

Intuitively, much of this bias results from the unduly optimistic assumption that future inflows will be moderate rather than varying between extremes. This can be remedied by finding an "optimal" deterministic marginal water value for each of the recorded historical inflow sequences and averaging these to yield EMV. (This is not equivalent to averaging the first period release decisions.) The advantage of this "averaged trajectory" approach is that it can be applied to multi-reservoir systems by repeated applications of deterministic models which are relatively easy to solve. But it is still optimistic because, although it models the range of inflows, it still assumes that future management will be able to predict and avoid any potentially troublesome situations. Thus it will still underestimate the variation of marginal costs, and hence, if the marginal cost curve is sufficiently convex, EMV. Intuitively, unrealistically optimistic assumptions about future management lead to insufficiently cautious recommendations.

The guidelines shown in Figure 5 confirm this prediction, lying between those of Figure 2 and Figure 4. But, despite the fact that these operating rules differ significantly from those produced by the iterative model, the expected cost of this bias was negligible for this system, with simulated expected costs and shortage levels from the two models being indistinguishable. Tests with smoother marginal cost curves did reveal a discernible impact on expected costs, but it would appear that this approach could reasonably be applied to the New Zealand system without unduly distorting release decisions. The true optimum presumably lies somewhere between the pessimistic guidelines produced by the iterative method and the optimistic ones produced by the averaged trajectory method, but the optimum appears to be rather flat, with very little bias due to either method.

Finally, many "stochastic" models make unrealistically pessimistic assumptions about future decision-making, often assuming that future releases will be determined by a decision rule having some simple form which does not reflect the true marginal cost structure of the problem. Although such a model may find the best possible "target trajectory" or "linear decision rule", say, it can not find the true optimal decision rule, which, for this type of system, must be a step function defined by the guidelines. For such a model the EMV can again be estimated by averaging water values determined by simulating management of historical sequences using its recommendations. By assuming poor future management such models will tend to exaggerate the dispersion of future marginal costs, and, if they are sufficiently convex, their mean also. Thus it is not surprising that the linear decision rules tested by [13] were so conservative.

To test this hypothesis, a realistic, but pessimistic, model was proposed. Future management was assumed to use a deterministic optimisation to determine releases. Thus each historical sequence was simulated, starting from each grid point, using the deterministic guidelines in Figure 4. End-point marginal water values were then averaged to yield EMV. Figure 6 shows the guidelines implied by these EMV's, which were surprisingly conservative. Simulation confirmed the extent of this bias, yielding a cost of $27.7 million, higher than for the traditional guidelines. This method was clearly unsuitable for application to cur system.

Note that all of the alternative methods need either a storage target or a function giving the value of water in storage at the end of the year, to ensure reasonable operating conditions in the next year. In this study, the value function developed by the iterative method was used, or a target was set at the expected end of year storage level found by simulation on the guidelines produced by the iterative method. (In practice this proved to be quite insensitive to the starting storage level, confirming the appropriateness of an annual planning horizon for this system.) However the setting of appropriate targets or value functions is a significant problem if any of these methods, or standard SDP, are to be applied on their own.

CONCLUSIONS

The iterative method, based on SDP, is quite suitable for the single reservoir problem, and has been shown to produce significant savings over traditional methods. Although it may be slightly conservative, this is seen as a favourable feature by management, and it has been adopted in practice. But the computational requirements of this method increase dramatically when it is extended to the multi-reservoir case. On the other hand, alternative approaches to handling uncertainty may bias release recommendations.

Our results show that direct application of deterministic methods for scheduling purposes can not be recommended, at least for this system, because they consistently recommend holding too little water in storage. To the extent that this bias is due to assuming mean inflows, it can be overcome by adopting the "averaged trajectory method", which is relatively easy to apply, even in a multi-reservoir context, using repeated runs of a deterministic model. It is less conservative than the ideal, but the impact of this bias was negligible for this system. On the other hand the alternative of assuming that future decisions will be made using a deterministic model was far too conservative to be of any use.

Examples can doubtless be constructed for which one method or another works particularly well, and the biases discussed here may even be reversed with other marginal cost curves. But we have shown that care must be exercised, since the way in which uncertainty is handled can have a major impact on the quality of the results, even if the approximations employed seem reasonable. Although our conclusions were derived for a specific power system, they are reinforced by the fact that similar biases were observed for the, very different, TVA system by Wunderlich and Giles [28]. They studied the degree to which foresight should be assumed in their DP based model, and concluded that, using the same "long run guides" (and implied marginal water values), "short sighted" policies lead to more conservative releases than those which assume some foresight by simulating deterministic management for the first few weeks of the planning horizon.

Finally, since New Zealand consists of two islands, with limited capacity to transfer power between them, there is a significant incentive to model at least two aggregate reservoirs, representing North and South Island storage. A generalisation of the iterative approach to this case is described in [5]. In practice, either this model, or a single reservoir model of the North Island on its own, are used, depending on the balance of storage between the islands at the time. Aggregate release requirements are determined by these models and apportioned among the individual reservoirs so as to equalise the probability of spill as far as possible. More recently [3] and [19] have also used twin reservoir SDP models to derive operating rules for long term studies.

REFERENCES

1. M. Adnet et. al, paper 32-12 in CIGRE, Paris Vol II (1968)
2. N.V.Arvantidis and J.Rosing, IEEE Trans. Power Apparatus and Systems PAS-89, 319-326 (1970)
3. W.R.Baker and H.G.Daellenbach, European J. OR 18, 304-314 (1984)
4. J.F.Boshier and R.J.Lermit, New Zealand OR 5, 85-100 (1977)
5. J.F.Boshier, G.B.Manning and E.G.Read, Trans. Inst. Professional Engineers In New Zealand 10, 33-41 (1983)
6. R.M.Brudenell and J.H.Gilbreath, AIEE Trans. June, 136-156 (1959)
7. H.G.Daellenbach and E.G.Read, New Zealand OR Soc. Conf., Auckland (1976)
8. J.E. Giles and W.O. Wunderlich, J.Water Resource Planning and Management Division, AMSCE 107(WR2), 495-511 (1981)
9. T.S.Halliburton and H.R.Sirisena, Optimal Control Applications and Methods 6, 91-103 (1985)
10. Y. Ikura, G. Gross and G.S. Hall, Interfaces 16, 65-82 (1986)
11. A.E. Lanna, Engineering Optimisation 6, 51-57 (1982)
12. J. Lindqvist, AIEE Trans. PAS-81, 1-6 (1962)
13. D.P. Loucks and P.J.Dorfman, Water Resources Research 11, 777-782 (1975)
14. P. Masse, Les Reserves et la Regulation de l'Avenir dans la vie Economique (Hermannn,Paris 1946)
15. E.G. Read, PhD Thesis (University of Canterbury, 1979)
16. E.G. Read, CBA working paper 152 (University of Tennessee, 1982)
17. E.G. Read, CBA working paper 153 (University of Tennessee, 1982)
18. E.G. Read, New Zealand OR 11, 125-142 (1983)
19. E.G. Read, (this volume)
20. J.C. Riley and C.R.Sherer, Water Resources Research 15, 233-239 (1979)
21. R.T. Rockafellar and R.J-B. Wets, in Mathematical Programming Study 6, R.J-B. Wets (ed) (North Holland 1976), pp 170-187
22. R.T. Rockafellar and R.J-B. Wets, SIAM J. Control and Optimisation 16, 16-36 (1978)
23. R.E. Rosenthal, Operations Research 29, 763-786 (1981)
24. S.Stage and Y.Larsson, AIEE Trans. Part III, (PAS) 80, 301-365 (1961)
25. J.F. Stedinger, B.F. Sule and D.P. Loucks, Water Resources Research 20, 1499-1505 (1984)
26. L.A. Terry, M.V.F. Pereira, T.A. Ariripe Neto, L.C.F.A Silva and P.R.H. Sales, Interfaces 16, 16-38 (1986)
27. S.H. Wan, R.E. Larson and A.I. Cohen, IEEE Trans. Power Apparatus and Systems PAS-103, 1163-1169 (1984)
28. W.O.Wunderlich and J.E.Giles in: Experiences in Operation of Reservoirs, T.E.Unny and E.A.McBean eds. (Water Resources Publications, Littleton, Colorado, 1982) pp. 223-243
29. S. Yakowitz, Water Resources Research 18, 673-696 (1982)
30. W.W-G. Yeh, Water Resources Research 21, 1797-1818 (1985)

Acknowledgement

This research was carried at the New Zealand Ministry of Energy, but all opinions expressed are purely personal. Particular thanks are due to C.B.Lusk, who programmed the model.

NOTES ABOUT AUTHORS

NOTES ABOUT AUTHORS

Roko Andricevic ("Accuracy of the First-Order Approximation to the Stochastic Optimal Control of Reservoirs") received his B.S. in civil engineering in 1979, and obtained his Master's degree in 1985 from the University of Zagreb, Yugoslavia. His interests include stochastic control and optimization problems in water resources systems. He joined the Department of Civil Engineering at Stanford University, Stanford, California in 1987. He was previously at the St. Anthony Falls Hydraulic Laboratory of the University of Minnesota, Minneapolis.

Nina Bellman ("On the Origin of the Name Dynamic Programming") received the B.A. in political science from the University of California, Los Angeles in 1960. She worked at the Rand Corporation, Santa Monica until her marriage in 1964 to Richard Bellman.

J. F. Boshier ("Biases in Stochastic Reservoir Scheduling Models") is a professional engineer employed by the Ministry of Energy, Wellington, New Zealand. His expertise is in power systems design, analysis and optimization with special emphasis on hydropower systems.

Osman Coskunoglu ("Knowledge Based Dynamic Programming for Water Resources Management") received his B.S. (1970) and M.S. (1972) in civil engineering at the Middle East Technical University, Ankara, Turkey. He did his doctoral studies at the School of Industrial and Systems Engineering, Georgia Institute of Technology, and earned his Ph.D. (1979) in operations research. Since then he has been with the Department of General Engineering, University of Illinois at Urbana-Champaign. Currently, he is an associate professor of operations research. Dr. Coskunoglu's papers have been published in a number of engineering and operations research journals. The focus of his current research is on coupling symbolic and numeric computation for decision aiding and automation.

Bryan Coulbeck ("Dynamic Programming for Optimized Control of Water Supply and Distribution Systems" and "Dynamic Programming for Optimization of Pump Selection and Scheduling in Water Supply Systems") was born in Yorkshire, England and has received various awards in the field of electrical engineering, including an M.Sc. from Loughborough University of Technology and the Ph.D. from the University of Sheffield. In addition to his recent teaching and research appointments he has had eighteeen years industrial experience with various national and international companies in the UK, USA and Canada. He is currently professor of water control systems and director of the Water Control Unit at Leicester Polytechnic, Leicester, England. His special research interests include computer control of water distribution systems and hierarchical control of large-scale industrial processes.

Mohamed Ali El-Tayeb ("An Adaptive Control Model for Single Reservoir Operation") is at the University of Sudan, Khartoum. He received his doctorate degree from the University of Wisconsin-Madison. His work has appeared in *International Journal of Hydroelectric Energy*.

J. Hugh Ellis ("A Variable State-Space Dynamic Programming Model for Optimizing Industrial Waste Treatment Sequences") received his Ph.D. in civil engineering (water resources) from the University of Waterloo in 1984. Since then, he has been assistant professor in the Department of Geography and Environmental Engineering of the Johns Hopkins University. His research interests involve environmental systems analysis with special emphasis on stochastic mathematical programming. Dr. Ellis' specific research interests currently include stochastic water quality optimization, the development of stochastic, multiobjective acid rain abatement strategies and procedures to incorporate generalized measures of robustness in stochastic programming.

Augustine O. Esogbue (*editor and author of several articles in this book*) is professor of industrial and systems engineering at the Georgia Institute of Technology, Atlanta, Georgia. He earned his Ph.D. in engineering (operations research and systems engineering) from the University of Southern California, Los Angeles in 1968, his M.S. in industrial engineering and operations research from Columbia University, New York in 1965 and his B.S. in electrical engineering from the University of California, Los Angeles in 1964. Professor Esogbue was formerly development engineer at the Water Resources Center, UCLA (where he worked with Warren Hall and William Butcher), a research associate (engineering and medicine) at the University of Southern California (with Richard Bellman), an assistant professor of operations research at Case Western Reserve University, Cleveland, Ohio, and an associate faculty member of Columbia University's Seminar on Water Resources and Pollution. Dr. Esogbue is on the editorial board of several scientific journals including *The International Journal of Fuzzy Sets and Systems* and the *Journal of Mathematical Analysis and Applications*. He has served on various panels of the National Research Council, National Academy of Sciences, and the National Science Foundation. Dr. Esogbue's publications have appeared in numerous scientific journals including several on dynamic programming and water resources, fuzzy sets and optimization of nonserial and neural networks. He is a coauthor with Richard Bellman and I. Nabeshima of the book, *Mathematical Aspects of Scheduling and Applications* published in 1982 by Pergamon Press.

Darrell Fontane ("Objective-Space Dynamic Programming Approach to Multi-dimensional Problems in Water Resources") is assistant professor of civil engineering at Colorado State University, Fort Collins, Colorado. He received his B.S. in civil engineering from Louisiana State University, Baton Rouge, Louisiana in 1968, an M.S. in civil engineering from the Georgia Institute of Technology, Atlanta, Georgia in 1970 and the Ph.D. in civil engineering (water resources) from Colorado State University in 1982. Dr. Fontane's principal research interests are water resources planning and management, systems engineering and mathematical modelling. He has consulted in these areas internationally.

Aristidis P. Georgakakos ("Extended Linear Quadratic Gaussian Control for the Real Time Operation of Reservoir Systems") is an assistant professor of civil engineering at the Georgia Institute of Technology. He received his Ph.D. in civil engineering from the Massachusetts Institute of Technology in 1984. His doctoral dissertation involved the development of a new control methodology for the optimal operation of large reservoir systems. He obtained a Bachelor's degree from the National Technical University of Athens (Greece) in 1980, and a Master's degree from the Massachusetts Institute of Technology in 1983, both in civil engineering. Dr. Georgakakos' field of specialty is optimal forecasting and control of water resources systems. His research interests include watershed modeling and streamflow forecasting, optimal design and operation of reservoir systems, real-time control of wastewater treatment plants, efficient utilization of ground water aquifers, and scheduling of hydropower systems. Dr. Georgakakos' research work has been supported by the U.S. Agency for International Development, U.S. Geological Survey, U.S. Army Corps of Engineers, and the electrical utility industry. He is a member of the American Geophysical Union, American Society of Civil Engineers, American Water Resources Association, Institute of Electrical and Electronic Engineers, Sigma Xi, and Chi Epsilon (member and faculty advisor).

Warren A. Hall ("*Dynamic Programming and Practical Water Resources Systems Engineering*") is professor emeritus at the University of California, Los Angeles, as well as Colorado State University at Fort Collins, Colorado where until his retirement recently, he was Elwood Mead Professor of Civil Engineering. He received the B.S. in civil engineering from the California Institute of Technology and the Ph.D. from the University of California, Los Angeles. Prior to occupying the chair at CSU, he was director of the Office of Water Resources and Technology of the U.S. Department of Interior. He had also served as a Scientific Advisor on water resources to President Nixon. Dr. Hall was the director of the Dry Lands Research Institute and the Water Resources Center at UCLA. He is a coauthor of the book *Water Resources Systems Engineering* and numerous journal articles in the field. Dr. Hall is a pioneer in the application of optimization techniques, particularly dynamic programming to water resources. He has consulted extensively nationally and internationally on various aspects of water resources development and management.

Bernd Kahn ("*Water Problems and Issues of the State of Georgia: Program Goals and Priorities*") is professor of nuclear engineering and health physics at Georgia Institute of Technology. He is also director of the Environmental Resources Center which is responsible for developing and steering water resources research programs in various universities within the State of Georgia. He earned his Ph.D. in chemistry in 1960 from the Massachusetts Institute of Technology, the M.S. in physics from Vanderbilt University in 1952 and a B.S. in chemical engineering from Newark College of Engineering in 1950. His research interests include treatment of radioactive waste, movement of radionucleides in the environment and water resources management.

Lov Kumar Kher ("*Identification of Demand Models from Noisy Observations: An Application to Water Resources*") received the B.E. degree in civil engineering in 1975 and the M.E. degree in environmental engineering in 1977, both from the University of Roorkee, India. He completed the Ph.D. degree in systems engineering from Case Western Reserve University, Cleveland, Ohio, in 1985. From 1977 to 1979 he worked with the Environmental Science and Engineering Group at the Indian Institute of Technology, Bombay; and from 1979 to 1981 he worked with PHE Consultancy

Services, Bombay as part of a consortium for a World Bank project. He joined AT&T Bell Laboratories in 1985 and is currently working in the Business Analysis Systems Center at Murray Hill, New Jersey. His current research interests include noisy system identification, parameter estimation, optimization, communication systems and integrated services digital networks. Dr. Kher is a member of IEEE and the Operations Research Society of America.

Peter K. Kitanidis ("Accuracy of the First-Order Approximation to the Stochastic Optimal Control of Reservoirs") received the B.S. in civil engineering with high distinction from the National Technical University of Athens, Greece, and both his M.S. in civil engineering and the Ph.D. in water resources from the Massachusetts Institute of Technology in 1976 and 1978 respectively. His interests include the solution of sequential optimization problems under uncertainty, arising in monitoring and control. He is currently associate professor of civil engineering at Stanford University, Stanford, California. Prior to joining Stanford, he was associate professor of civil and mineral engineering at the University of Minnesota, Minneapolis, Minnesota.

John W. Labadie ("Objective-Space Dynamic Programming Approach to Multi-dimensional Problems in Water Resources") is professor of civil engineering at Colorado State University, Fort Collins, Colorado. He received the B.S. in 1966, the M.S. in 1968, both in engineering from the University of California at Los Angeles, and the Ph.D. in civil engineering in 1972 from the University of California, Berkeley. Professor Labadie's areas of special interest include water resource systems analysis; real-time forecasting and control of water systems; urban water management; conjunctive use of surface and groundwater; computer aided decision support systems; mathematical programming and dynamic optimization.

Chae Young Lee ("Optimal Design of Large Complex Water Resources Conveyance Systems Via Nonserial Dynamic Programming") is assistant professor of management science and information systems at the Korea Institute of Technology, Taejon, Korea. He received the B.S. in engineering from Seoul National University, S. Korea in 1979, the M.S. and Ph.D. both in operations research from the Georgia Institute of Technology, Atlanta, Georgia in 1981 and 1985 respectively. Dr. Lee's research interests include mathematical programming, computational complexity, software reliability, and management information systems.

Wilbur L. Meier, Jr. ("Dynamic Programming for Flood Control Planning: The Optimal Mix of Adjustments to Floods") is chancellor of the University of Houston System. He received a B.S.C.E., M.S.C.E., and the Ph.D. in operations research from the University of Texas at Austin. Dr. Meier previously served as assistant head of the Department of Industrial Engineering at Texas A&M University, Chairman of the Department of Industrial Engineering at Iowa State University, head of the School of Industrial Engineering at Purdue University, and dean of engineering at the Pennsylvania State University. He is a member of AIIE, NSPE, ASCE, and ORSA, and is a past president of AIIE.

Thomas L. Morin ("*Dynamic Programming for Flood Control Planning: The Optimal Mix of Adjustments to Floods*") is professor of operations research in the School of Industrial Engineering at Purdue University. He received a B.S. in civil engineering from Rutgers University and the Ph.D. degree in operations research from Case-Western Reserve University, and has served on the faculty of Northwestern University. Professor Morin, a Fullbright Scholar, is known for his work in dynamic programming and multiple-objective optimization. He was the first editor of the *Journal of the Water Resources Planning and Management Division* of the ASCE, is a member of ORSA, TIMS, ASCE, AIIE, MAA, and ACM, and currently is principal investigator for Purdue's five-year $4.25 million ONR-sponsored University Research Initiative in computational combinatorics.

Kolinjawadi S. Nagaraj ("*Dynamic Programming for Flood Control Planning: The Optimal Mix of Adjustments to Floods*") is a computer systems specialist in the Office of the City Administrator, Washington, D.C. He received a B.S. in civil engineering from the Indian Institute of Technology, Madras, an M.S. and the Ph.D. in operations research (industrial engineering) from Purdue University, and has served on the faculty of the University of Iowa. Dr. Nagaraj is a member of ORSA, TIMS, AIIE, and the Mathematical Programming Society.

Toshio Odanaka ("*On Optimal Pumping Policies for Groundwater*") is professor of applied mathematics and engineering at the Tokyo Metropolitan Institute of Technology, Tokyo, Japan, where he has served since 1955. He received his doctorate from the Tokyo Institute of Technology in 1967. He was formerly a visiting research associate at the University of Southern California. Dr. Odanaka has published several books and numerous articles primarily on inventory theory and optimal control.

Chun-Hou Orr ("*Dynamic Programming for Optimization of Pump Selection and Scheduling in Water Supply Systems*") received the B.Sc. degree in electronics from the Chinese University of Hong Kong, Hong Kong, in 1975, and both the M.Sc. and Ph.D. degrees in electrical and electronic engineering from Queen Mary College, University of London, London, in 1977 and 1980 respectively. Since 1979 he has been with the Water Control Unit, Leicester Polytechnic, Leicester, England. His current research interests include the application of computing technologies and control techniques in the optimization of large-scale water distribution systems.

José A. Ramos ("*Closed-Loop Control, Balancing, and Model Reduction of Large Scale Water Resource Systems*") received the BSCE degree from the University of Puerto Rico, Mayaguez, in 1978, and the MSCE and Ph.D. degrees in hydrosystems and systems engineering from the Georgia Institute of Technology, Atlanta, in 1979 and 1985, respectively. From 1979 to 1980 he was a research engineer for the Water Resources Research Institute at the University of Puerto Rico. From 1984 to 1985 he was a technical consultant for the Georgia Pacific Corporation, Atlanta, Georgia. Dr. Ramos has been an associate research engineer since 1985 at United Technologies Optical Systems, Inc., West Palm Beach, Florida, where he is involved in the application of modern estimation and signal processing techniques to space defense systems. His research interests are in the areas of estimation and control theory, robust modeling and approximation of large systems, and mathematical programming.

E. Grant Read ("A Dual Approach to Stochastic Dynamic Programming for Reservoir Release Scheduling" and "Biases in Stochastic Reservoir Scheduling Models"), born and educated in New Zealand, received the Ph.D. in operations research from the University of Canterbury in 1979. He worked at the New Zealand Ministry of Energy on a variety of topics including oil stockpiling and gas depletion models, but concentrating on optimization of planning, operations, pricing and organizational structure for the electricity industry. He has been a consultant to New Zealand Electricity and the Tennessee Valley Authority, and taught at universities in the United States. Currently, he is teaching at the University of Canterbury, and pursuing his interests in electricity economics and planning methods, as well as in dual approaches to dynamic programming.

Ramesh Sharda ("An Adaptive Control Model for Single Reservoir Operation") is associate professor of management science in the College of Business Administration at Oklahoma State University, Stillwater, Oklahoma. He received a B. Engg. from the University of Udaipur (India), an M.S. from the Ohio State University, an M.B.A. and the Ph.D. from the University of Wisconsin-Madison. His papers have appeared in *Management Science, Interfaces, Computers and Operations Research, Annals of Operations Research, Management Science and Policy Analysis, Environment and Planning A, Simulation and Games*, and other journals. His research interests are in stochastic programming, financial applications of management science models, DSS, integration of microcomputers in OR/MS, and policy analysis. He is a member of ORSA, TIMS, DSI, and ABSEL.

Moshe Sniedovich ("Dynamic Programming and Non-Separable Water Resources Problems") is a chief specialist researcher at the Center for Advanced Computing and Decision Support of the CSIR, Pretoria 0001, South Africa. He received a B.Sc. from the Technion, Haifa, Israel, and the Ph.D. from the University of Arizona, Tucson, Arizona. His publications have appeared in journals such as *Water Resources Research, Advances in Water Resources, Operations Research, Management Science, Operations Research Letters, Journal of the Operational Research Society, Engineering Optimization, Journal of Optimization Theory and Applications, Journal of Mathematical Analysis and Applications*, and *APL Quote Quad*. Dr. Sniedovich's current research focuses on sequential decision processes, computers in mathematical education, and interactive computing modelling and analysis.

Matthew J. Sobel ("A Multi-Reservoir Model With A Myopic Optimum") is leading professor at the State University of New York at Stony Brook where he is in the Institute for Decision Sciences, the Department of Applied Mathematics and Statistics, and the W. Averell Harriman School for Management and Policy. His research concerns stochastic models in operations research and their applications. The primary application areas are water resources (since 1962), logistics and manufacturing, and economics. Dr. Sobel's degrees are from Columbia University and Stanford University and his previous permanent faculty positions have been at Yale University and the Georgia Institute of Technology.

Soroosh Sorooshian ("*Identification of Demand Models from Noisy Observations: An Application to Water Resources*") received the B.S. in mechanical engineering from California State Polytechnic University, San Luis Obispo, in 1971, the M.S. in operations research in 1973 and the Ph.D. in systems engineering in 1978, both from the University of California at Los Angeles. He was on the faculty of Case Western Reserve University in Cleveland, Ohio until 1982 as assistant professor of systems engineering and civil engineering. He is currently professor of engineering at the University of Arizona in Tucson, Arizona. Dr. Sorooshian is an associate editor of the *Water Resources Research* journal (American Geophysical Union), program chairman of the Hydrology Section of AGU's Fall Meeting, and a member of the Working Group on Water Resources of the IFAC Technical Committee on Systems Engineering (SECOM). His research interests include hydrologic modeling, linear and nonlinear systems with noisy observations, nonlinear compartment models, flash-flood and acid rain modeling, and application of statistical models to ground-water problems related to the nuclear waste disposal sites.

Hiroshi Sugiyama ("*Improving Water Quality by Optimal Aeration Control via Dynamic Programming*") received the D.Sc. in mathematics from Kyushu University in 1960 and the Doctor of Medical Science from Osaka City Medical School in 1961. From 1962-1985 he was chair professor of industrial mathematics at National Osaka University and chairman of the Department of Applied Physics (1982-1984). He has held many visiting research appointments in the United States including the Rand Corporation in Santa Monica, the University of Southern California and recently Kansas State University, Manhattan, Kansas. Dr. Sugiyama's publications dealing mostly with statistics, dynamic programming, medical and water resources systems have appeared in Japanese and U.S. journals.

George Tauxe ("*Multi-Objective Dynamic Programming in Water Resources*") is associate professor of civil engineering at Oklahoma University, Norman, Oklahoma. He received his Ph.D. in 1973 in water resources systems engineering from the University of California, Los Angeles, California. He has served on various committees of the American Geophysical Union and the American Society of Civil Engineers. He is president of Tauxe and Associates and has consulted nationally and internationally. Dr. Tauxe's research publications which have primarily been in the areas of water resources and hydrologic systems have appeared in numerous journals.

AUTHOR INDEX

AUTHOR INDEX

A

Adams, B.J.	288,296
Agthe, D.E.	174
Akilesaran, V.	94,118
Andricevic, R.	63
Argaman, Y.	53,65
Aris, R.	252,266
Arrow, K.J.	185,192
Arvantidis, N.V.	412
Askew, R.J.	47,48,72,329,341
Athans, M.	154,162

B

Baker, W.R.	354
Bard, Y.	174
Barr, A.	132
Beard, L.R.	329,341
Beattie, B.R.	172
Becker, L.	51,95,154,244,247
Bellman, N.	3
Bellman, R.E.	3,8,19,20,29,36,37,47,80,93,185,192,283,327,329,341
Bediment, P.B.	53,74
Beightler, J.C.	47,48,72
Bennett, M.S.	53,66
Bertsekas, D.P.	36,156
Billings, R.B.	174
Biswas, A.K.	131
Bodin, L.	329,340
Bogan, R.H.	51,72
Boshier, J.F.	63,375,401,412
Box, G.P.	280,284
Bras, R.L	50,156
Brown	51
Buchanan, R.	154
Buras, N.	21,48,131,186,192
Burt, O.R.	54,185,192
Butcher, W.	21

C

Childs, S.W.   112
Chow, V.T.   81,388,399
Chu, W.S.   266,329
Clark, A.J.   327
Close, E.R.   329,341
Cohen, G.   93
Cohon, J.L.   76
Cooper, M.   93,95
Cooper, L.   93
Cole, A.   49
Collins, D.   37,41
Cortes-Rivera, G.   301,311,312,313,314
Coskunoglu, O.   121
Coulbeck, B.   193,223
Croley, T.E.   50
Curry, K.C.   154

D

Daellenbach, H.G.   354,401
Damlanos, D.J.   302,317
Dandy, G.C.   174
Davis, D.R.   329,341
Dawdy, D.R.   341
Day, J.C.   301,314
DeLucia   54
DeMare, L.   174
Denardo, F.V.   327
Dinkelbach, W.   138
Doran, D.G.   336,340
Dorfman, P.J.   402,412
Dreyfus, S.E.   93,173
Dudley   54
Duckstein, L.   341
Durling   4,37,39
Dynkin   4

E

Eisel, L.M.	329,341
El Tayeh, M.A.	63,329
Ellis, J.H.	287
Esogbue, A.O.	1,4,21,31,34,37,39,42,51,53,94,243,253,262,266
Erlenkotter, D.	303,305,315

F

Farber, M.A.	109
Feigenbaum, E.A.	131
Fontane, D.G.	93,376
Foster, H.S.	174
Freidenfelds, J.	173
Fults, D.M.	47

G

Gal, S.	50,376,391
Garber, S.	174
Georgakakos, A.P.	63,343,388,399
Giles, J.E.	402,412
Glass, R.	239
Goicoechea, A.	79
Goldstein, J.D.	154,160
Grimes, D.W.	112
Grygier, J.C.	94

H

Haith, D.	51
Haimes, Y.	9,21,79,80
Hall, W.	4,8,19,21,29,46,94,131,186,192,243,244,246
Halliburton, T.S.	402,410,412
Hancock, L.F.	47
Hanke, S.M.	174
Hanks, R.J.	112

Hansen, R.D.	174
Hayes-Roth, F.	131
Heidari, M.	94
Heyman, D.P.	327
Hillier, F.S.	81
Hinomoto	56
Hodges, S.D.	174
Hogen, R.M.	131
Hopkins, L.D.	300,314
Houck, M.	154,155
Howard, R.	36,41
Huber, G.P.	131
Hullett, W.	272,283,284

I

Inman, R.R.	79

J

James, D.L.	301,315
Jansen, R.B.	298,316
Jonckheere, E.A.	154,167,168

K

Kaiper, J.	55
Kalaba, R.E.	23,36
Kalman, R.E.	162,173,174,175,176
Keckler, W.G.	244,246
Keen, P.G.	131
Kher, L.K.	173,174,177,184
Kindler, J.	174
Kitanidis, P.K.	63,154,155,159,160,387,390
Klemes, V.	333,336,340
Klepper, S.	174
Kokotovic, P.V.	119
Krishnan, P.	288,296
Kushner, H.	329,340

L

Labadie, J.W.	93,94,376
Lanna, A.E.	402,412
Larson, R.E.	4,47,94,160,244,245,402,412
Latimore, W.E.	154,159
Law, A.M.	93
Leamer, E.E.	174
Lee, C.Y.	53,243
Liberman, G.J.	81
Lindqvist, J.	402,412
Little, J.D.C.	131,185,192
Lloyd, E.H.	329,341
Loucks, D.	154,402,412
Loaiciga, H.A.	388,399

M

Mades, D.M.	79
Maidment, D.R.	174,250,266,388,399
Martin, D.L.	111
Martin, Q.	248,249,266
Marino, M.R.	388,399
Marks, D.	388,399
Masse, P.	185,192,327
Mawer, P.	48,71
Mays, L.	53,250,260,266
Mehra, R.K.	174
Meier, W.L.	47,48,56,160,297
Merritt, L.M.	52,72
Minns,	239
Moncur, J.E.	56,72
Moore, B.	165
Moore, P.G.	174
Morin, T.L.	4,37,39,51,94,164,297
Murray, D.M.	47,94,394,399

N

Nagaraj, K.S.	297
Narayanan, R.	174
Nayfeh, A.H.	399
Nemhauser, G.L.	119,252
Newell, A.	132
Nopmongcol, P.	47,72

O

Odanaka, T.	185
O'Neill, S.	239
Orlovski, S.	80
Orr, C.H.	217,223
Ortolano, L.	55,70

P

Panagiotakopoulos, D.	288,296
Parikh, S.	21
Parzen, E.	174
Pekelman, D.	329,341
Pereira, M.V.	94
Perry, B.R.J.	239
Pinto, L.M.	94
Pleban, S.	93
Prekopa, A.	329,341
Puterman, M.	37,73

R

Ramos, J.A.	153,154,159,168,169
Rao, S.G.	153
Rausser, G.C.	329,341
Read, E.G.	63,375
Reid, R.W.	79,82
Rockafeller, R.T.	377,386,407,412

Roefs, T.	21,329
Rogers, J.S.	303,315
Rosenthal, R.	336,340
Rosing, J.	410
Rossman, L.	94,114
Rothblum, U.G.	327
Russell, C.S.	174

S

Schweig, Z.	48
Sen, S.	173
Shabman, L.A.	302,317
Sharda, R.	63,329,338,340
Shepard, R.	4,21,244,246
Shin, C.S.	56,288,296
Silverman, L.M.	154,165,166
Simon, H.A.	132
Sirisena, H.R.	402,410,412
Snell, G.H.	239
Sniedovich, M.	93,133,151
Sobel, M.	36,51,63,321,327
Sorooshian, S.	173
Stage, S.	402,412
Stedinger	36,50,94,154,386
Sule, B.F.	8,11,154
Sugiyama, H.	55,271

T

Tauxe, G.	21,79,82
Tether, A.J.	160
Thorn, D.	48
Trott, W.	46
Turgeon, A.	50

V

Veinott, A.F.	327
Vemuri, V.V.	79
Verriest, E.I.	154,165,167,168,169

W

Walsh, S.	51,75
Warsi, N.	41,253,262
Wasimi, S.	154,155,159
Wasimi, S.A.	388,399
White, G.F.	302,218
Wilson, C.	3
Williams, M.	239
Wiederhold, G.	132
Wilde, D.J.	252,266
Wunderlich, W.O.	94,402,412
Weisz, R.N.	301,314
Wets, R.J.	377,386,407,412

Y

Yakowitz, S.	5,8,31,47,50,63,93,133,151,376,394
Yeh, W.	46,95,153,154,167,244,329,376,401
Yen, B.C.	250,260,266
Young, G.K.	50,75
Yushkevich	4

SUBJECT INDEX

SUBJECT INDEX

A

Accuracy	387
Actual cost	394
Adaptive control model	329
Admissible control region	161
Aeration control	271,272,282,284
feasible	271,278,289
incidental	277,290,292,293
Aeration control rate	274,276,280,283
Affine myopic optima	324
Allegheny River	106
Aposteriori optimality analysis	312
Approximation methods	46
Aquatic and environmental pollution	14
Augmentation phase	381
Augmented guideline diagram	383
Autonomous system	164

B

Balanced realizations	164
closed loop	164
open loop	164
Binary isotonic operator	33
BOD	271,272
Box's Hill-climbing Method	280,281,282,283
Branch and bound strategies	307

C

C-Programming	134,135,139,150
California	20
Central Valley Project	8,246,247
Multi reservoir system	21,46
Sacramento-San Joaquin delta	87
Shasta reservoir	86
Capacity constraint	389
Capacity expansion	173

Caution term	387
Chance constrained programming	329
Chemical waste specific path	293
Cholesky factorization	177
Civil and environmental engineering	8
Classification Matrix	60
Closed loop, balancing	154,165
policy	160
stochastic optimal control problem	160
system	154
Composition function	135
Computational experience	336
Computational solution	31
Computer output	263
Computer technology	
flexible computer program	239
Fortran computer program	291
hardware	9,193
micro computer	9,114
programmable palm calculator	277
software	9,193
super computer	9
Conceptual reservoir	395
Continuity principle	395
Control, feedback	159,167
constraint	396
policies	167
variable	387,388,389
vector	167,388
Controller	154,163
Convex optimization problem	226
Cost functional	159
Cost per removal function	289
Cost to go function	387,388,389
Covariance matrix	173,174,175,179,182,390
Criterion functional	271,272,283
Curse of dimensionality	133,153

D

Dams	
High Aswan	343,345
Kinzua	106
Decision	
factors	57
makers	122,155
making	155,173
model	155
modeling	122
system	155,159,160
Decomposition	
conditions	96
Dantzig-Wolfe method	44
recomposition	24
scheme	135
Demand	153,173,174,175,182
Demand analysis and prediction	199
Deterministic equivalent	390,392
Deterministic feedback control	387,381,397
Dimensionality	278,283
Disaggregation-reaggregation	24
Distribution network	212
Disturbance	
model	158
system	155,158
DO	271,272
Downstream target temperature	107
Dual approach to stochastic dynamic	
programming	375
Dump energy	90
Duality	163
Duality theory	386
Duration curve	391
Dynamic programming	3,5,9,19,31,153,160,173
Algorithms	
approximation techniques	94

differential	8,390
discrete differential	9,94,311
fathoming	94
forward	223,239
imbedded state spaces	8,9
Lagrange multipliers	94,116
nearest neighbor techniques	94
new	121
nonserial	9,243,250,251,253
reaching procedures	94
state increment	8,119,194,216
successive approximations	144
variable state space	287
Applications to	
a real world problem	310
flood control planning	297
industrial waste treatment systems	287
pump selection	223
underground water inventory	115,185
water supply and distribution systems	193,223
Bellman's Principle of Optimality	3,23,33
computational complexity	262
computational efficiency	262,312
curse of dimensionality	8,93,321
guide to problem formulation	27
knowledge based	121
methodology	9
model overview	288
models of water resources systems	31
multi dimensional problems	93
multi objective aspects	79,80,82
nonseparable	133
origins and interconnections with water resources	5,31
objective space	93,95,105,110,115
overview	32
reduction of dimensionality	93

structures 32
    discrete deterministic nonserial
    discrete deterministic serial
    continuous deterministic nonserial
    continuous stochastic nonserial
    continuous deterministic serial

E

Efficient data management structures	94
Effluent contaminant standards	288
Egypt	344
Eigenvalues	165
Eigenvectors	165
ELQG Control	349
ELQG control of a single reservoir	367
Electricity cost model	203
Elementary regression vectors (ERV)	179,182
Energy vs. evaporation tradeoff	364
Environmental impact	26
Equation, algebraic Riccati	163,165
augmented state	158
continuity	157
Lyapunov	165
Riccati	154
Error	180
in-equation model	174
in-variable model	174
Estimation theory	173,174,175
Evaporation	390
Existence and uniqueness questions	33
Expert system	154
Expected value	388
Expected marginal value curve	379
Exponential growth	387
Extended linear quadratic guassian control	343,347,349
Experimental design	281

F

Feedback control	283
First order approximation	387,390,393
Finite element Galerkin matrices	14
Forecast	184
Forecasting	167
Forecasting error/covariance matrix	390
Fortran program	88,291
Fractional programming	137,138
Floods	
control	156,157
control planning	297,299
damage mitigation program	297
dynamic programming algorithm for mitigation	307
latest minimum damage solution	309
matrix of interaction of adjustments	301
optimal mix of adjustments to	297,302
structural and nonstructural adjustment measures	298,299
trends and limits to adjustments	298
Flow routing	395
Forecast	184
streamflow	154
weather	154
Full order, model	167

G

G-uniqueness	97
Gains	162,163
Gaussian noise	381
Generalized Benders Decomposition	303
Groundwater	
geophysical methods for location	17
optimal pumping poliies	185
pumping	17
treatment for contaminated	17
Guideline approach	403
Guideline diagram	383

H

Harza Engineering	21
Harzardous waste facility	13
Hedging	387
Heuristic reasoning	404
High Aswan Dam	50
Hilbert space	283
Histogram	394
Honduras	22
Human decision making process	122
Hydrology, critical period	
Hydropower	349,354
Hydropower function	340

I

Identification	173,175,177
Illinois-South Central	310
Incomplete state information	162
India	21,27
India's Central Water Commission	20
Inventory control	185
Inventory theory	321
Mahanadi	24,26
Inequality constraint	389
Infinite dimensional Riccatti	
equation	283
Influent	
contaminant concentrations	287
data	288
strength variability	292
waste regimes	287
Infow	389,390,394,395
Initial trajectories	350
Input variance	392,393,394
Inventory problem	185
Iraq	21
Irrigation scheduling	95,110
Iterative method	402

K

Kalman filter	162,163
Knapsack problem	146
Knowledge	124
Knowledge and dynamic programming	126
Knowledge and optimization	125
Knowledge in action	129
Knowledge versus data	125

L

Lagrange multipliers	94,189
Large scale systems	10
Latin America	20
Limited spill curve	403
Linear programming	3,6,305,311,401
Linear relationships	175,177,179
Linear systems theory	167
Load duration curve	403
LQG	398

M

Management science	121
Marginal cost	407
Marginal cost curve	408
Marginalistic dynamic programming	377
Maritime cargo research	20
Mathematical	
descriptions	289
models	19,22
programming	99,173,182
Matrices	
controllability	164,165
Gramian	164,165
observability	164,165
Maximum cost	394
Maximum output power	164
Mean	390,394
Measurement error	175,182

Measurement system	155
Middle East	20
Minimize	388
Minimum cost	394
Model	
efficient organization	376
manipulation	323
reduction	154,165,167
Monotonicity	96
Monte Carlo simulation	387,398
Multi converging branch systems	255,258,260
Multi-level hierarchical techniques	10
Multi objective water resources development	26
Multiple regression	174
Muskingum method	156
Myopic affine	321
Myopic optima	321

N

National Science Foundation	298
Network analysis and simulation	194
Network configuration	245
Network control	214
New Zealand Reservoir and Power System	375,401,402
Nile	
Basin	50
River	344
river flow statistics	345
Noise	173,174,175,178
Noise in variable model	173,175,181,182
Noisy realization theory	174,175,177
Noninferior solution set	79,84,85
Nonlinear	
boundary value problem	191
integer programming	303
optimization problem	139
programming	5,175,182,303

Nonseparability	144
Nonseparable water resources problems	133

O

Objective function (modified)	135
Open loop feedback control	348
Open loop variance	361
Operation horizon	388
Operations research	121
Optimal design	243
control	9
guidelines	409
law	161
LQG tracking	154,163
policy	389
problem	154,162
Optimality principle	274,283
Optimization	
combined simulation models	302
explicit stochastic	48
implicit stochastic	48
master model	20
multi objective via dynamic	
programming	27
procedure	230,232
stochastic model	289
Optimized	
control	193,204
pump scheduling	227
schedules	206,208

P

Pareto optimum	27
Parametric problem	144
Partial base-loading	380
Perfect state information	161

Perman method	112
Perturbation	387
Peru	22
Policy making guidelines	343
Pollutant discharge	172
Pontryagin Principle	283
Posterior densities	330
Practical considerations	105
Prediction scheme	177,178,182
Principle of progressive optimality	390
Probabilistic descriptions	287
Probability distribution	185,329
Problem formulation	21,24
Problem structure	122
Professional practice	22
PROLOG program	130,131
Public utilities	193
Pump analysis and simulation	194
Pump selection optimization	223,224
Pumping policies for groundwater	185

Q

Quadratic criterion function	283
Quadratic gaussian control	343
Quantity of water	26

R

Rayleigh-Ritz-Galerkin Methods	283
Rand Corporation, Santa Monica	3
Recursive relationship	332
Reduced order, controllers	154
LQG models	154
state estimators	154
state space model	154
streamflow models	167

Reservoir
- adaptive control model for operation 329
- detention 311
- deterministic models 401
- direct construction of operating rules 378
- ELQG control of 367
- four model with autocorrelated inflows 321
- hydro systems 375
- hydroelectric generation 106
- linear 155,156,162
- multi model 321
- multi purpose operation 8
- operations for water quality control 106
- optimal operation 17
- real time operation 343
- release scheduling 375
- several interconnected systems 202,211
- single systems 193,202,205
- stochastic scheduling models 401
- storage 154,155,390,396
- surface and underground 186

Response surface 281
Return function 271,273,283
Riccati equation 283
Riverine system 388

S

San Antonio-Guadalupe River System 249
Saturation 272
Separability 96
Separable convex C-programming 141
Separation principle 154,162
Sequential decision 185
- model 185
- problem 103
- process 23,185

Sequencing and scheduling problems	8
Sewer systems (branched) optimization	260,262
Shasta Reservoir	86
Simulating sub-optimal management	409
Simulation modeling	293
Simulation experiments	360
Simulation results	363
Small perturbation approximation	387,388
Social and political constraints	306
South Atlantic Gulf Region	14
Spillway discharge elevation curve	389
Stagewise cost function	388
Standard deviation	391
State constraints	396,397
State transition equation	388
Statistical moment	387
Steady state system	162
Steepest ascent (descent)	281,282
Stochastic approximation-Kiefer Wolfowitz method	283
Stochastic dynamic programming	154,160,387,398,401
Stochastic models	401,410
Stochastic realization problem	154,161
Stochastic systems	10
Storage, constant	156
levels	155
Sufficient conditions	95
Surrogate with trade off method	80,91
Switzerland	21
System control model	201
System data	235
System equations	201
Systems analysis and simulation	194
Systems engineering	
indispensable phases of	23
practical applications	23
techniques	22
Systems planning	173,174

T

Target levels	159
Taxonomy scheme	57
Technology transfer	22
Tennessee Valley Authority	330
Terminal cost	388
Third derivative	388
Time series	174
Transition equation	388,390,396
Transition matrix	388
Trajectory	
initial	350
optimal release	356,358,359
optimal storage	356,358,359
Transformation, balancing	166,167
inverse balancing	167
Turnpike horizon policies	185,188

U

Uncertainty adjustment phase	377
Unit hydrograph	155

V

Variance	391
Variance type objective function	137
Variance separation techniques	137

W

Waste water facility	
chemical waste	295
hazardous treatment systems	287
industrial and municipal	16
metal and tannery wastes	294
treatment and distribution	13
treatment sequence	287
Water conveyance systems	243

Water management
   basin water cycle understanding   16
   ground   13
   quality protection studies   16
   urban   5
Water quality   6,15,16,95,271,280,284
Water quality management   95
Water quantity   6,14,16,26
Water resources   173,174,178,182,183
   large scale system   167
   planning   243
   systems engineering   19,20
   urban   7
Water supply   6,13,209,233
   actual system   233
   and distribution   193,211,212
   capacity expansion   8
White river system   248

X Y Z

Zero-mean   388